William Fairbairn:
the experimental engineer

William Fairbairn: the experimental engineer

A study in mid 19th-century engineering

Richard Byrom

RAILWAY & CANAL HISTORICAL SOCIETY

First published 2017
by the Railway & Canal Historical Society

www.rchs.org.uk

The Railway & Canal Historical Society was founded in 1954
and incorporated in 1967.
It is a company (no.922300) limited by guarantee
and registered in England as a charity (no.256047)
Registered office: 34 Waterside Drive, Market Drayton TF9 1HU

ISBN 978 0 901461 64 3

Designed and typeset by
Malcolm Preskett
Printed and bound in Great Britain by
Short Run Press, Exeter

Frontispiece:

William Fairbairn, CE, FRS, by Benjamin Rawlinson Faulkner, c.1840.
Bequeathed to the Royal Society in Fairbairn's will, 1874.
Image used by permission of the Royal Society.

Contents

Abbreviations used in source references

The following are abbreviations used for frequently occurring sources in the endnotes.
Other sources are given in full at the first use in each chapter's endnotes.

Ahrons, 'Famous Firms'	E.L. Ahrons, 'Short Histories of Famous Firms, No.II, W. Fairbairn and Sons, Manchester', *The Engineer,* 129, Jan.–July 1920, 184–5.
BAAS1833	*Report of the Meeting of the British Association for the Advancement of Science* [for the year stated]
CE&AJ	*Civil Engineer and Architect's Journal.*
Fairbairn, *Application*	W. Fairbairn, *On the Application of Cast and Wrought Iron to Building Purposes* (2nd ed. 1857–8).
Fairbairn, *B&CTB*	W. Fairbairn, *An Account of the Construction of the Britannia and Conway Tubular Bridges with a complete history of their progress* ... (1849).
Fairbairn, *M&MWI*	W. Fairbairn, *Treatise on Mills and Millwork Part I, On the Principles of Mechanism and on Prime Movers (*2nd ed. 1864).
Fairbairn, *M&MWII*	W. Fairbairn, *Treatise on Mills and Millwork Part II, On Machinery of Transmission and the Construction and Arrangement of Mills* (2nd ed. 1865).
Fairbairn, *UIfE*	W. Fairbairn, *Useful Information for Engineers, being a series of lectures delivered to the working engineers of Yorkshire and Lancashire* ... (1856).
Fairbairn, *UIfE2*	W. Fairbairn, *Useful Information for Engineers, Second Series* (1860).
Fairbairn, *UIfE3*	W. Fairbairn, *Useful Information for Engineers, Third Series* (1866).
Fortunes Made in Business	[J. Burnley and J. Hogg, eds], *Fortunes Made in Business: A Series of Original Sketches Biographical and Anecdotic from the recent history of industry and commerce,* Vol. II (1884).
Hayward, 'Fairbairns'	R.A. Hayward, 'Fairbairns of Manchester' (UMIST MSc Thesis 1971).
IAR	*Industrial Archaeology Review.*
ILN	*Illustrated London News.*
Life	W. Pole (ed. and completor), *The Life of Sir William Fairbairn, Bart., partly written by himself* (1877).
MG	*Manchester Guardian.*
MM	*Mechanics' Magazine.*
MPICE	*Minutes of Proceedings of the Institution of Civil Engineers.*
MPIME	*Minutes of Proceedings of the Institution of Mechanical Engineers.*

Rosenberg and Vincenti, *Britannia*	N. Rosenberg and W.G. Vincenti, *The Britannia Bridge: The Generation and Diffusion of Technological Knowledge* (1978).
Smith, *Contribution*	A.I. Smith, *The Contribution of Sir William Fairbairn to the Advancement of Mechanical Engineering* (1974).
TNS	*Transactions of the Newcomen Society.*
Ure, *Philosophy*	A. Ure, *Philosophy of Manufactures or an Exposition of the Scientific, Moral, and Commercial Economy of the Factory System* (1835).
Walker, *Unwin*	E. G. Walker, *The Life and Work of William Cawthorne Unwin* (1947).
[Woodcroft] I–X	[B. Woodcroft], 'Fairbairn and his Times', Ten articles in *The Engineer,* 44, July–Nov. 1877. No. X is incorrectly numbered IX (i.e. there are two No. IXs) – a correction in this respect has been made.

I	13 July 1877, p.19.
II	20 July 1877, p.38.
III	27 July 1877, p.57.
IV	10 August 1877, p.95.
V	17 August 1877, p.111.
VI	7 September 1877, p.163.
VII	14 September 1877, p.181.
VIII	12 October 1877, pp.253–4.
IX	26 October 1877, p.291.
X	2 November 1877, pp.309–10.

Preface

I acknowledge my thanks to Derek Dearden, an excellent tutor at Manchester University School of Architecture who, during my fourth year as an undergraduate, introduced me to William Fairbairn, whose derelict premises remained within walking distance of the School. Also to the outstanding architectural historian, John Archer, at that School, who subsequently encouraged my ongoing interest by supervising my MPhil. Having decided to write this book it seemed prudent to tackle a PhD in preparation. By a circuitous route I ended up in the History Department at Huddersfield where I received valued guidance from Professors Martin Hewitt and Keith Laybourn – both historians of considerable standing in their own fields, and both very conscientious in the task they took on. The book is unquestionably the better for the academic input. That said, the responsibility for misjudgements, errors and omissions remains entirely mine. It is fair to record that John Archer advised against doing a PhD – academia had 'lost its way since the 1970s'. I have no regrets in not taking that advice, but I now see where he was coming from – the prescribed format, strict 'date for completion', limit on the number of words, monthly progress reports, monthly swiping-in on campus, and pressure to produce 'papers'. All this would surely have been treated with disdain by scholars in the Victorian era. But those scholars would be wholly envious of the electronic access to many academic journals provided by the University Library.

There are many people who have gone out of their way to give me help and encouragement in this research to whom I am extremely grateful. There are too many to name them all but they include Michael Bailey, Matthew Barry, Tony Bonson, John Broadley, Eamonn Cannon, Adrian Clement, John Davies, Julia Elton, Keith Fenwick, Celia Ferguson, Ron Fitzgerald, David George, John Glithero, Rhonda Grantham, Jan Hallbæck-Andersen, Richard Hartree, Ron Hayward, John Hodgson, Roger Holden, Giel van Hooff, Paula Howard, Andy King, Michael Lewis, Steve Little, Kenneth Major, Warren Marshall, Claire Mayoh, Ian Miller, Carol Morgan, Richard Morris, Sergio Nepoti, Anne-Mary Paterson, Norman Redhead, Peter Rowlands, Kanji Tamagawa, Robert Thorne, John Vignoles, Peter Walker, Mark Watson, Terry Wyke, and many others.

Three people are deserving of my especial thanks: my wife, Sue Byrom, who has driven me on when the whole project seemed overwhelming, and driven me around almost all the Fairbairn archaeology in the British Isles; Sir James Brooke Fairbairn, 6th Baronet of Ardwick, whose generous encouragement has been in the great tradition of the 1st Baronet; and David Joy, Chairman of the R&CHS

Publications Committee, whose lifetime of experience in journalism and publishing has been immensely helpful.

Researching and writing this book has been an exhilarating, if at times stressful, exercise. Exhilarating because there was so much, previously unknown, to discover about Fairbairn, and because of the exposure to so much wider 19th-century history which formed the cultural environment within which millwrights and engineers fulfilled their callings, and which their achievements helped to mould.

I hope that this book will spur further study of those who laid the foundations of today's prosperity and, in particular that it might encourage more detailed studies of the various sectors and phases of Fairbairn's exceptional career. I hope too that it will encourage many young engineers, scientists and business leaders in their careers.

Richard Byrom,
Manchester, 2016

Introduction

THIS book is about the millwright and engineer William Fairbairn (1789–1874) and the company he founded – their intertwined ascent to fame and influence, and the eventual decline and demise of the company. From an apprenticeship as a colliery millwright near Newcastle, Fairbairn became one of the best known and most influential engineers in the half-century 1820–1870. He was active in many areas of engineering – a leading builder of mills and millwork throughout his career, an iron shipbuilder during the formative decade of iron shipbuilding, 1835–1844, a builder of waterwheels, steam engines, boilers, locomotives, bridges and cranes, and an exceptional experimental engineer. His many achievements brought widespread recognition, although his career was not without its set-backs.

The book is written for academic historians – economic, social, business and technological – to bring to life the engineer to whom many make passing reference, and to provide a source of illustrative material. It is written for practising engineers – civil, mechanical, structural, hydraulic, mining, electrical, aeronautical, materials scientists, and craftsmen engineers – to provide inspiration from the past for the future. It is also written for the tens of thousands of members of historical and heritage engineering groups who are enthusiastic about 19th-century railways, maritime history, bridges, steam engines, industrial buildings, waterwheels and cranes; and for those interested in Manchester's history.

Chapter endnotes are included, quite profusely. This reflects the frustration experienced during years of research with books which give no references to sources. Abbreviations used in references are listed on pages 8–9. Illustrations are included, possibly reflecting the writer's initial training in architecture where much information is conveyed visually; and perhaps as a reaction to those areas of academia where tables and graphs seem to be considered scholarly, but illustrations less so. Fairbairn's own books were liberally embellished with illustrations. Sources of illustrations are listed at the end of each chapter. Some quotations are also included as they often give an authentic flavour, in a way which a précis might not. Extensive technical details and descriptions are omitted, as they would make the book too long, and are only likely to be of relevance to those with discrete specialist interests (who will be well able to follow up references in the end notes and appendices). Similarly it is only feasible to look at selected examples of Fairbairn's work – a mixture of the typical and the leading examples, sometimes dictated by a lack of information. Mathematical formulae are also omitted, not so much in deference to Fairbairn who was not comfortable with algebra and calculus, but because mathematics has its own language, like music or organic chemistry or

Japanese, which only a minority of readers will be conversant with. Secondary literature, post 1913, is omitted from the bibliography for economic reasons – there were over 700 papers and books. Most of these can be found in the chapter endnotes, and thus readers with specialist interests should be able to find relevant references to those interests from the notes that follow their areas of interest. Imperial units were used by Faibairn and this usage is retained. For those who do not have time or inclination to read the whole book, the gist of it can be gleaned from this Introduction, the introductions and conclusions to each chapter, and the Conclusion.

A respected historian describes Fairbairn as 'a worthy heir to John Smeaton',[1] who is widely referred to as 'the father of civil engineering'.[2] Smeaton was the first to achieve distinction as an engineering scientist. More than any other, he laid the foundations of the civil engineering profession in Britain.[3] Another distinguished historian has written that it can 'readily be defended that Fairbairn contributed more than any other single individual to the emergence of mechanical engineering in the first half of the 19th century'.[4] How remarkable it is therefore that, apart from the Britannia Bridge, so little has been written about Fairbairn and so little is generally known about what he built, the experiments he undertook and what he wrote, or about the famous firm he founded and its subsequent rapid demise. There has been no published comprehensive study of this exceptional man since a Victorian eulogy was written soon after his death 140 years ago.[5] Why is this so? R.A. Buchanan laments that 'engineers as a whole have received very little attention from historians'.[6] Christine MacLeod highlights 'the much neglected mechanical engineering industry'.[7] Is this attributable, as some say, to Wiener's 'decline of the industrial spirit' which 'threw earlier enthusiasms for technology into disrepute' – the world's workshop replaced by the English garden? [8] MacLeod attributes today's ignorance of the names and achievements of most 19th-century engineers to 'cultural programming' and quotes Stefan Collini's contention that 'our concept of culture has been shaped by historians', notably Arnold Toynbee, the Webbs, the Hammonds, and, much later, E.P. Thompson, with their 'catastrophist' view of the industrial revolution, an approach which notably began to change with T.S. Ashton's *The Industrial Revolution* (1948), but was influential until the end of the century.[9] To study British industrialisation was to study the evils of capitalism and the degradation of working people – hence the dearth of scholarly studies of engineers and industrialists. Thus, too, the 'dismissive or condescending' attitude to Fairbairn's first biographer (and fellow Scot), Samuel Smiles, who offered his readers 'the example of the great inventors, scientists and businessmen who had risen above humble beginnings and conquered adversity',[10] an example which, in the new century many academics considered 'no longer admirable'.[11] But in the 21st century a rehabilitation of Smiles is happening,[12] and there is a renewed interest in engineers.[13]

There has also been a tendency to dismiss much history of technology as 'Whiggish',[14] in that it tends to emphasise 'certain principles of progress'.[15] Mokyr responds that it is 'neither Whiggish nor naive' to suggest that the accelerating growth of useful knowledge since 1750 'has affected the world more than all other social and political changes taken together'.[16] Also, given the difference between the standard of living in the West today and three centuries ago, criticisms of Whiggishness are easy to ward off.[17] 'Economists' remind us that 'the rise of the western economies based on economic growth and technological progress is the central event of modern history', to which 'nothing else comes even close',[18] and historians note that in the middle decades of the 19th century 'poverty, disease, squalor and injustice' were beaten back 'more than at any previous time in the history of Britain'.[19] Mayr, noting that the pejorative label 'Whiggishness' has been used in 'an irresponsible and often unjustified manner', hopes 'that it will disappear altogether from the literature of scientific historiography'.[20] Similarly there is a tendency to refer dismissively to solid research as 'antiquarianism'.[21] Of course the historian must assess causality and significance, but this can only be done effectively if there is adequate evidential data, which, in part, this book and its appendices seek to provide. It is often only when detailed data is assembled that trends, patterns and links become apparent.[22]

Another factor is the division between the arts and sciences – C.P. Snow's *The Two Cultures* – with history seen as an 'arts' subject.[23] There are difficulties in writing engineering history without some technical background and few current academic historians have knowledge of engineering such as Donald Cardwell had.[24] Thus during much of the 20th century, whilst there were notable exceptions, there has been a tendency for historians to major on issues of labour, feminist and literary history, to the neglect of 19th-century scientists, engineers and industrial leaders. A comparison of the material available about Fairbairn with that about his Manchester contemporary, Elizabeth Gaskell, emphasises the point. From 1991 to 2005 there were over 100 American doctoral theses devoted to Mrs Gaskell.[25] During the same period of time there were no doctoral theses about the work of William Fairbairn. However the emphasis is changing as it is increasingly appreciated that it was the late-18th and 19th-century scientists, engineers and industrial leaders whose work has led to the prosperity enjoyed in developed nations today – water supply, drainage, electricity, ample food, modern medicine, fast transport, worldwide communications, heating and air-conditioning, widespread literacy, and state pensions. And tragically the weapons of war. This indeed was Fairbairn's view: in a lecture, delivered when he was in his seventies, perhaps with memories of his own experiences of poverty in his youth, he said:

> Now, we are better housed, better fed, and better clothed than at any former period; and what were luxuries to our forefathers, have to us become the necessaries of life. Nothing, therefore, but a state of war and anarchy would again drive us back to the cheerless homes, and rude enjoyments of a previous state. ... compare [the] improved health, increase of years, and a more intellectual and a higher state of existence ... but we have not in our present state arrived at that degree of perfection which awaits us on a more extended application of ... science.[26]

There was a flowering of technical and industrial history in Manchester in the 1960s and '70s – Musson, Robinson, Farnie, Chaloner, Henderson at the University, Cardwell and Hills at UMIST. Hopefully we can look forward to a renewal of such a flowering in various 21st-century universities. Engineering history will also be enriched if academic historians do not stand aloof from the engineering heritage societies; and if these societies cease to ignore the way in which professional historians look at industrial history.[27]

As technological transformation gathered momentum in the first half of the 19th-century, Britain's engineering leadership was unchallenged, increasingly shaping the Western world and beyond. 'Technology is the engine of social change' and without that technological creativity 'we would all still live ... short lives of toil, drudgery and discomfort'.[28] Based for sixty years in Manchester at the time when it became the centre of the industrial world, William Fairbairn was both an agent and eye-witness of that transformation. There was a progression in the scope of the work he undertook as external factors changed, notably as water, wind and horse gave way to steam power; and the advent of main-line railways, steamships and telegraphy transformed travel and communications. At the same time workshops were revolutionised by machine tools, and engineering products by the increasing availability of wrought iron. Fairbairn's work was diverse and complex in that it involved on-site construction of mills, waterwheels, steam-engines, bridges and cranes, but also involved a factory where the metal components for all these products were made, along with locomotives. In addition there was a shipyard, and his on-going experimental engineering. *Table 1* illustrates changes in the main areas of Fairbairn's activity as new opportunities arose and circumstances led to the decline or abandonment of others. To bring all together in one book presents a challenge, but it is an important challenge because the breadth and the interconnectedness of Fairbairn's activities are not generally understood. Inevitably this approach means that some areas are not treated in the depth they deserve and it is hoped that this book will stimulate in-depth studies in discrete areas.

History in the 20th century has tended to have been marked by perceived lines of demarcation between different branches, but there is evidence that these lines are becoming less distinct and that business and economic history can be 'mutually reinforcing'.[29] Increasingly technical, economic, political, social and other categories of research are seen to overlap, interwoven in a 'seamless web'.[30] Bijker and Pinch refer to the 'entangled character of technology, society and science'.[31] Others have stressed the combined need for narrative coherence, causal analysis and social contextualisation.[32] Indeed there is a renewed emphasis on narrative.[33] Donald Cardwell understood:

> that history of science must be concerned with practice as well as theory, that local studies are enormously useful in exploring the interplay of content and context, and that we do well not to divide the histories of science, technology and medicine from each other, or from economic and social history.[34]

Failures of both businesses and technological innovations have often been neglected by historians, but there is increasing appreciation of the value of the study of failure and 'things that turned out to be dead ends', as well as of success.[35] It is the aim of this book to provide an overall understanding of Fairbairn, and his firm, within a technical, social and economic context. In doing so, three main questions are addressed. First, how is the rise of Fairbairn and his company to a position of leadership and influence within the engineering industry accounted for? Secondly, in what respects were Fairbairn and his company important and influential, and how was that influence spread? Thirdly what caused one of the

	1817–1819	1820–1824	1825–1829	1830–1834	1835–1839	1840–1844	1845–1849	1850–1854	1855–1859	1860–1864	1865–1869	1870–1875
Mills and Millwork												
Research/investigations												
Waterwheels												
Steam Engines/boilers												
Cast-iron bridges												
Iron steamships at M/cr												
Iron s/ships at Millwall												
Locomotives												
Tubular and t/g bridges												
Tubular cranes												
Lattice/trussed bridges												
W/I Roof Structures												

Table 1: Timescales of the main Fairbairn spheres of activity.
The intensity of activity is represented by the density of the shading.

most successful engineering companies, with a global reputation, to cease to trade within a year of its founder's death? Answering these questions provides the opportunity to re-assess the range and significance of Fairbairn's contributions to 19th-century engineering.

The records of the Fairbairn company were destroyed in 1899.[36] Their destruction has dictated the approach to the book, making an economic business history, such as Floud's of Greenwood & Batley or Cantrell's of James Nasmyth, impossible.[37] What the material does facilitate is an understanding of Fairbairn and the company through a biographical approach and a methodology reconstructing a picture of the company through its products. The detail of the methodology was first to identify what Fairbairn built and when – the mills, waterwheels, steam engines, ships, bridges, locomotives, cranes – from whatever sources could be found. The same was done for Fairbairn's experiments, and his books, papers and reports. An attempt was then made to reconstruct Fairbairn's life and that of his firm in its various phases, identifying its resources – premises, money, machines and men – and the commissions undertaken. The picture is far from complete, but it does provide a much more reliable basis for an understanding and interpretation of Fairbairn than has been available hitherto. Three areas where disappointingly little material has been found, beyond the first years of the business, are the organisation of the drawing office, the integration of design and manufacture, and the organisation of the works and workforce.[38] The available material inevitably limits economic argument, but the approach has validity, not least because the narrative of the firm shows Fairbairn himself to have been critical to its growth and reputation. There are gaps, too, in the illustrations – as yet no images of Dorothy Fairbairn have emerged; nor of any of the Fairbairn homes in Manchester.

Fairbairn's own writings were extensive. They include ten books and parts of at least six others, and many reports. He read papers to groups ranging from the

Royal Society to Ancoats Lyceum. Records exist of more than 200 papers and reports. Whilst some contain duplicated material, this immense output, entirely atypical for a working engineer, is hugely important in providing details of Fairbairn's work and experiments and, on occasions, of his thinking on wider issues. A note of caution needs to be sounded. Fairbairn was far from unaware of the publicity value that books and papers could have. He presented copies of his books to many people of influence. This may have led to some selectivity and exaggeration of his contributions, and possibly a failure to give adequate attribution to others.[39] His best known book, *Mills and Millwork,* is in some ways a catalogue showcasing outstanding examples of his work, but its wealth of illustrations and technical descriptions of 19th-century mills and millwork, much of it unavailable elsewhere, are of great value. Fairbairn's standard textbook, *Iron: Its History, Properties, and Processes of Manufacture* is an objective standard work. Fairbairn's papers divide into three main types, those which describe works he had constructed, those that set out the findings of his research, and those to mechanics' institutions and the like which embrace history of engineering and self-improvement as well as technical subjects. Many of the papers recording his experiments contain extensive, detailed and precise measurements. In summary, Fairbairn's writings are a bountiful and, apart from a few oft-quoted texts, largely untapped resource. They have formed a significant part of the research for this book.

In 1851 Fairbairn wrote some autobiographical notes, mainly covering the period up to around 1840.[40] Alert to public relations, and possibly pressed by the spur of fame, it seems he passed some form of these notes to Smiles, as some of their content features in Smiles' *Industrial Biography* of 1863.[41] Around the same time *The Imperial Dictionary of Universal Biography* contained an entry for Fairbairn by Rankine.[42] The only full-length biography is that written after Fairbairn's death, 'at the request of the family', by the engineer and polymath William Pole. Fairbairn's notes were included by Pole, but in an edited form as 'guided by the judgment and wishes of the family', that is by Fairbairn's son Thomas, who sought advice from his father's friend Thomas Romney Robinson of Armagh. Robinson wrote that 'after Bodner the narrative is by no means so carefully drawn up and contains many expressions which I have no doubt he himself in revision would have materially modified ... some of the expressions about both Ashworth and Lillie are too strong for publication'.[43] Fairbairn's autobiographical notes comprise only around twenty per cent of Pole's book of over 500 pages.[44] They are differentiated by being written in the first person and by being in a smaller typeface than the remainder.[45] Other than the points referred to by Robinson and the circumstances of the dissolution of the partnership with Lillie, it is unclear what has been edited out – perhaps the disappearance of eldest son John in the early 1840s, and the sequence of visits to Turkey around the same time.[46] Nevertheless, accepting that the genre of autobiographical testimony is subject to the problems of self-interest, Fairbairn's notes are of great importance as a first-hand account of an early 19th-century apprentice and journeyman, and contain much about him which is unavailable elsewhere, illustrating the difficulties as well as the successes of his early life. James Brittain's review of the 1970 reprint notes that 'the book conveys unusually well the sense of euphoric self-satisfaction and the optimistic faith in the future of progress through technology which eminent engineers such as Fairbairn and Pole felt as they assessed the extent and implications of the dramatic changes they had done so much to bring about'.[47]

Part of the task of a historian is to recognise and correct imbalances in past accounts. This is assisted in this present case by reading Pole in conjunction with a series of ten anonymous and incisive articles in *The Engineer*, which formed an extended and critical review of the book. It was a scathing review.

> Disruptions of partnerships, like divorces in domestic life, are never pleasant subjects to dwell upon; but either those circumstances which we are informed had been fully described and discussed by Mr Fairbairn in his autobiography should have been given in his own words without garbling or curtailment, or they should not have been mentioned at all, which latter course would have been scarcely consistent with a work professing to be a life of Fairbairn. In this respect, and in many other passages throughout the volume, great regret is forced upon us that Mr Fairbairn's autobiography was not published in full, just as he wrote it. The large omissions which, as editor, Mr Pole has no doubt felt himself justified in making, partly in deference to persons still living who may have been referred to, and the scattered portions of narrative which have been editorially supplied to fill these lacunae and connect the broken fragments of the autobiography, seriously mar that which should have constituted the main feature of the entire work'.[48]

Throughout his long book, Pole allowed himself but one criticism of his subject:

> He had perhaps an excessive ambition for popularity and fame; but this foible had one redeeming feature, namely, that he aimed not so much at obtaining the applause of the million as at standing well with the good and the wise.[49]

The Engineer, publishing its review anonymously, and not beholden to the family, was more explicit:

> The love of fame is not the sole nor by any means the loftiest feeling which urges the efforts of the philosopher, nor perhaps amongst the highest order of minds is it ever thought of as the ultimate aim and reward.[50]

R. Angus Smith, who knew Fairbairn, wrote, 'He was a most lovable man, gentle and agreeable ... People instinctively believed in him ... If he was a little vain of his progress, it is only that which nearly all men who have risen so rapidly are blamed for showing'.[51] Fairbairn gives credence to these criticisms – 'I had no master, nor had I the privilege of Greek or Latin, but I had a craving appetite for distinction'.[52] Yet Unwin, who was his assistant for seven years, found him 'very ready to recognise conscientious work in others',[53] and the lengthy anonymous obituary in the Memoirs of the Institution of Civil Engineers described him thus:

> Affable and accessible, he was ever ready to communicate information and to give advice to all who sought it; buoyant and cheerful ... he was an inborn gentleman in mind, taste and manner, conscious of his own

> strength, and gratified by the approbation of others. He was, however, singularly modest and unassuming.[54]

Who wrote the articles in *The Engineer*? From them we learn that as a child their author visited Fairbairn's home with his mother; as a young man he saw iron ships being built at Fairbairn's works and spoke with Fairbairn about them; he visited the Canal Street works around 1840 and in 1845; and had a professional understanding of engineering. By a process of elimination the contention here is that the author was Bennet Woodcroft (1803–79). He was born in Heaton, South Manchester, but by 1824 was living at Ardwick Green, near to the Fairbairns. He studied chemistry under Dalton.[55] Fairbairn records that about 1830 Woodcroft showed him a plan of his invention of a screw propeller.[56] Woodcroft was one of the group which met at Fairbairn's home in the mid-1830s.[57] He joined the Manchester Literary & Philosophical Society in 1841 and organised a Mechanical Exhibition at the British Association meeting in Manchester in 1842 (when Fairbairn was a vice-president of the Mechanical Section).[58] From 1852, Woodcroft was Superintendent of Specifications at the Patent Office and organised the Patent Office Museum. He became a Fellow of the Royal Society and Professor of Machinery at University College London. He also wrote biographies. Thus he had ample authority to review Pole's book.[59] It is a piece of intense irony that twelve years before Pole's book, Fairbairn wrote to Unwin:

> Have you seen Mr Pole's 'History of the Tubular Bridges' in Jefferson and Pole's Life of Robert Stephenson? All the experimental researches are given to Stephenson and a more garbled statement of facts I have seldom read. Everything done by Stephenson although it is well known he never was present at any of the experiments but twice and that only for half an hour at a time. But Pole and [Clark?] were both of them employees of his.

Pole's biography of Fairbairn, notwithstanding valid criticisms that can be made of it, provides a great deal of information. Apart from the autobiographical section, information relating to networks is particularly valuable. *The Engineer*'s extended review is much more engineering-orientated. As such it amplifies Pole, as well as providing corrective emphases. In welcoming the reprint of Pole in 1970, Brittain adds, 'a new critical biography of Fairbairn would be even more welcome'.[61]

Most of the significant 20th-century Fairbairn studies date from the 1970s and '80s when the most important contributions were A.I. Smith on Fairbairn's experiments,[62] Rosenberg & Vincenti on the Britannia Bridge – written to 'shed some light on the historical learning process and thus to illuminate one critical dimension of the dynamics of industrial growth',[63] R.A. Hayward's research, particularly on ships and the limited liability company,[64] and Ron Fitzgerald's detailed work on cast-iron structures.[65] From fifty years earlier there is Ahrons' paper on Fairbairn locomotives.[66]

An additional purpose of the book is to correct some of the many factual errors and misleading statements about Fairbairn that appear in recognised works by well-known authors. It would be ungracious to identify these in source notes, but all the works were published 1950–2016. Fairbairn was not an apprentice wheelwright; nor did he serve an apprenticeship as a miner. In Manchester he did not 'obtain employment as a production engineer'. Nor, 'in the mid 1820s' was William Fairbairn's firm 'the largest engineering contractor for artefacts of both wrought and cast iron ranging from machine tools, water wheels, boilers and steam engines, to industrial machinery'. Fairbairn & Lillie were not based in Leeds. The first mills known to incorporate Hodgkinson beams were not built in the mid-1820s. He did not 'pioneer the belt-drive system in factories' nor was it he who made 'a breakthrough replacing cast-iron shafting with wrought iron. In the 1830s he did not work seven days a week and he most certainly did not have 'a knowledge of mathematics that rivalled the best university professors of the day'. Fairbairn never employed as many as 2,000 men at Millwall and railway locomotives were never built there. He did not patent a flooring system which used iron in conjunction with concrete, and he did not use such a system in 1845 in 'a sugar refinery in Manchester' or in 'an English refinery'. The Britannia Bridge experiments were not undertaken 'at his laboratory in Manchester'. He did not fabricate the beams for the Britannia Bridge. He did not 'retire from the field of bridgebuilding' after the disagreement with Stephenson. In fact he and his company constructed over 1,000 iron bridges (not 100). The Saltwater Creek Bridge near Melbourne was a Fairbairn tubular-girder; it was not 'designed by Brunel, similar to his Balmoral Bridge'. Fairbairn was not 'President' of the Aeronautical Society. There was no Mechanics' Institute in Ancoats. Fairbairn was not 'a devout Unitarian lay preacher'. He never had a London home, at Holland Park or any other address. Nor did the Fairbairns live at The Polygon in the 1830s, and their first child was Anne, not Andrew. There is confusion between William and his son Thomas, of whom it is said, 'He worshipped at the city's Unitarian Church Street Chapel [*sic*] and involved himself with the Literary and Philosophical Society'. Thomas, unlike his father, did neither. Nor was Thomas 'a third-generation Mancunian philanthropist' and he did not use his experience as a commissioner of the Great Exhibition to raise funds for the 1857 Art Treasures Exhibition – indeed he did not become a Commissioner until 1861.[67] J.C. Deane was neither wealthy nor a cotton manufacturer. The 1862 Exhibition building was not known as the 'Brompton Boilers'.[68] There is confusion too with another son William Andrew. Fascinating as it would have been to find Fairbairn marrying into the Kennedy family in 1864, it was his son William Andrew Fairbairn who did so![69] William's brother Peter was never Mayor of Manchester. The tubular crane at Dover is not 'of unidentified make' – it was built by Fairbairns, as its maker's plate states. William Fairbairn did not catch a chill in 1870 from which he never recovered (it was in 1873) and he was not buried at Prestwick, Northumberland. All that said, it is almost inevitable that a book with as wide a diversity of material as this present one will also contain factual errors, of which please advise the author.[70]

With the constraints imposed by the absence of company records, the laborious assembly, so far as was reasonably possible, of a record of Fairbairn's life and work was an essential prerequisite to understanding what he achieved and to any assessment of his contribution to the various wider current academic debates, six of which are referred to below.

Science and Technology

THERE are unresolved differences about how much technological progress during the Industrial Revolution '*really* depended on scientific expertise'.[71] Did technological progress begin with 'pure' scientific research, leading to 'applied' research, industrial development and diffusion,[72] or did inventors during the Industrial Revolution owe very little to contemporary science?[73] Developing Ashton's suggestion of the late 1940s that the Industrial Revolution may be seen as an intellectual movement, Muson and Robinson in their 'magisterial' *Science and Technology in the Industrial Revolution* (1969) argued for an applied-science model for understanding technological innovation in late 18th-century Britain.[74] Nine years later Rosenberg and Vincenti used the Britannia Bridge to illustrate their thesis that innovation in technology does not involve the application of knowledge derived from science,[75] and by 1994 Rosenberg believed that 'everyone knew that the linear model was dead'.[76] It was not, for in 1997 Margaret Jacob was writing of the elements of the natural world encoded in science as central to industrialisation and western hegemony,[77] inspiring much of Mokyr's thinking in his *The Gifts of Athena* (2002).[78] He uses 'Industrial Enlightenment' to describe 'that part of the Enlightenment which believed that material progress and economic growth could be achieved through increasing human knowledge of natural phenomena and making this knowledge accessible to those who could make use of it in production'. Both Mokyr and R.C. Allen see the process of industrialisation rooted in the concepts and culture of Newtonian science, where experimentation is applied to solve mankind's problems.[79] Alexander von Humboldt (1769–1859), whom Fairbairn admired and had met, advocated an approach similar to the Newtonian – 'the accurate, measured study of widespread but interconnected real phenomena in order to find a definite law and a dynamical cause'.[80] It was also the approach of the British Association, with which Fairbairn was so much involved, where 'proper science' was to be based on 'slowly cumulating inductive observations and hard won experimental results; only on this basis could true, mathematical generalisations be securely erected'.[81] There are also issues relating to chronological aspects of the debate, with 1850 seen as a watershed by S.F. Mason who argues that before 1850 technical advances were not greatly dependent on the science then known, but thereafter the application of science to technology became important in the advance of industry, such that most subsequent outstanding technical discoveries stemmed from scientific research.[82]

However, this whole debate may be finding its resolution in a much closer integration of science and technology. Barnes wrote of science and technology 'enmeshed in a symbiotic relationship'.[83] Jacob sees science and technology as twins, with 'different *personae*, different looks, but ... profoundly related'.[84] This was not a new analogy – in a lecture in 1852 Fairbairn referred to 'theory and practice' as 'twin sisters'.[85] Wengenroth points to a 'systemic interrelatedness' between science and technology; and Mokyr writes, 'Rather than posing the question of whether it was the theorists or practical people who brought about technological progress, we need to see the fundamental complementarity between them'.[86] The relationship between economic growth and technological change has been described as 'useful knowledge' – a term probably first used by Kuznets in

1965 to describe the source of modern economic growth – or 'useful and reliable knowledge', these being seen as flexible, inclusive terms, less burdened by association than 'science' or 'technology'.[87] 'Useful and Reliable Knowledge' according to Inkster, 'lay beyond formal science and embraced the knowledge that was brought to bear at points of significant technological advancement'.[88]

Invention, Innovation and Diffusion

ANOTHER area of debate relates to how technological invention, innovation and diffusion take place. One respected approach, that of Schumpeter, envisages a clear division between 'invention' and the main entrepreneurial function, 'innovation'. There may be a time lag between 'invention' and 'innovation'. The 'innovator' who sets up the new production function may not have been the 'inventor'.[89] It is in this sense that 'innovation' and 'innovator' are used in this book. However, some scholars consider this a pointless distinction and hold the view that major innovations are built up of numerous, almost imperceptible, incremental improvements, made over lengthy periods of time by a large number of anonymous people – or as Jones styles it, 'technological drift' – the cumulative importance of which is of decisive economic significance.[90] This has some affinity with Usher's 'cumulative synthesis approach', drawing on Gestalt psychology – 'the whole is greater than the sum of its parts' – with major inventions emerging from the synthesis of relatively simple inventions, each of which required an individual 'act of insight'.[91] Mokyr's questioning of incremental theories is apposite:

> As a reaction to heroic theories of invention in which all improvements are attributed to individual geniuses, the drift theory has been justly influential. But is it not possible to go too far in the other direction and give too little credit to major inventions made by a vital few?[92]

Buchanan goes further, endorsing 'the image which Smiles so successfully portrayed of a generation of heroic engineers by whose labours the face of the nation had been transformed'.[93]

Mokyr attributes the on-going inventiveness of the Industrial Revolution to the Scientific Revolution and the Enlightenment.[94] R.C. Allen summarises the four important aspects of Mokyr's Industrial Enlightenment. First, macro-inventions – inventions in which a radical new idea, without clear precedent – emerge more or less *ab nihilo*; and are distinguished from micro-inventions – small, more frequent, incremental steps that improve, adapt and streamline existing techniques already in use, accounting for most gains in productivity. Secondly, social networks – a set of bridges between intellectuals and producers. Thirdly, the application of the scientific method to the study of technology, through experimentation. Fourthly, that the Industrial Revolution did not come from artisans, 'The Industrial Enlightenment ... was confined to a fairly thin sliver of highly trained and literate men'.[95]

There is considerable scepticism about theories in general. Redlich stressed continuity: 'it is impossible to draw lines of demarcation between innovation and re-innovation, between primary and derivative innovation, between derivative innovation and copying, and between copying and routine'. He also noted that entrepreneurs and their work are unpredictable.[96] Half-a-century later Rosenberg called for more attention to be paid to the disorderly process of technological innovation.[97] Pinch and Bijker argue that simplistic models and generalisations are unhelpful, as technological innovation takes place in a wide range of circumstances.[98] Did Fairbairn's work fit a pattern in respect of invention and innovation, or was there an unpredictability and uncertainty about it?

Diffusion is the process in which knowledge of an invention or innovation is spread.[99] David Jeremy argues that diffusion is marked by unpredictability and uncertainty – a complex process which often defies economic prediction. The typical approach has been to trace diffusion through agents which differ from case to case – networks, apprenticeships, printed information, letters and drawings, visits, consultancy, skilled artisans changing employer or 'working away' or emigrating, copying of products, exhibitions, lectures, societies, licences, espionage and war.[100]

Scottish Connections – Networks and Networking

ONE of the questions about diffusion, significant in business history, is the role of networks. In the last twenty years there has been a surge of academic interest in networks and networking.[101] Networks – financial, informational or personal – based on common values and attitudes, have a profound influence in business activity.[102] Such networks may relate to places of origin, religious affiliation, educational, business or familial contacts. Science-based societies were used by technicians and manufacturers 'as a vehicle for upward social mobility'.[103] Mokyr, and Allen, see the Industrial Enlightenment as having created network bridges between intellectuals and producers. At the apex, information was exchanged at the Royal Society. More people were involved in provincial scientific societies and engineering institutions, 'but what counted especially were informal relationships and correspondence in which producers sought access to the best knowledge available at the time'.[104]

An astonishing number of inventors, engineers, chemists, mathematicians, and physicians were Scottish. Hobsbawm wrote:

> The Lowlands not merely adapted themselves to economic development; they welcomed it and led it ...
>
> How much of this extraordinary readiness of the Lowland Scots for industrial society, was due to their Calvinism, or perhaps more exactly to the democratic and almost universal educational system it created, is a complex question ... but it must be difficult to deny that the remarkable record of Scotsmen in the 19th century ... had some connection with the institutional system the country had acquired in the Revolution of 1559, which was made under the banner of Calvin and John Knox ... Few if any areas of the world have contributed proportionately more to industrialism than Scotland.[105]

Arthur Herman, linking the influence of the Scottish diaspora to the 'Scottish Enlightenment', goes further: 'They came from every part of Scotland and from every social condition ... They brought with them a generally high level of literacy, of job skills, and a strong sense of moral discipline and personal initiative. They would transform every society they touched.'[106]

Manchester was well endowed with Scots. Second-generation Manchester engineer, John Galloway, noted that 'nearly all the original millwrights in Manchester came from the neighbourhood of the Tweed ... All were Scotchmen – quiet, respectable and mostly middle aged, with experience'.[107] A study of 52 north-west engineers contemporary with Fairbairn found that nine (17 per cent) were Scottish.[108] Besides the Scots, Manchester received many other immigrants in the 19th century. The largest group was the Irish – fifteen per cent of the population in 1851. In comparison the Scots were few, but their influence and enterprise were out of all proportion to their numbers. The importance of the Scottish connection in the Manchester in which Fairbairn thrived will become apparent.

Reformation and Enlightenment

A second network hub was Cross Street Chapel. Historians explain the phenomenon of Scottish success by reference to the cultural effects of the Reformation, with its associated parish schools, and, or, by reference to the Scottish Enlightenment.[109] David Allan links both, viewing the Scottish Enlightenment as the culmination of Scotland's Calvinist experience,[110] or as one of his reviewers puts it, 'any de-Calvinization of the Scottish past ... must inherently misrepresent it'.[111] It is over a century since Weber put forward his thesis that the Protestant ethic influenced the development of capitalism but it continues to generate a historiography.[112] While the Calvinist *doctrines* of Presbyterianism declined among the Scottish emigrants and, as Hobsbawm points out, the characteristic ideology of industrial capitalism became the deist or agnostic rationalism of the professors of the 18th-century 'Scottish Renaissance' – David Hume, Adam Smith, and others – the *culture* of Calvinism, notably its integrity and work ethic, continued to be influential.[113] Indeed most of Manchester's Scots had grown up in that culture, with its advocacy of 'diligence in lawful callings' and in Landes' words, 'the criterion of efficiency on every activity'.[114] It was a culture that epitomised hard work, education, thrift, integrity and systematic thinking. Work was a 'creation ordinance' and this, coupled with the creation mandate to rule over the earth and subdue it, often resulted in work continuing long after there was economic need.[115] In Manchester, Cross Street Chapel, which Fairbairn attended, had its roots in the Puritanism of the Ejection of 1662 and adopted Unitarianism in the 18th century. It espoused both the work ethic of residual Puritanism, and its approach to science.[116] Indeed, Margaret Jacob links the English Puritans with the origins of modern science.[117] By the early 19th century, Cross Street had become a centre of ability and talent, a base for political, educational and philanthropic activities,[118] and had laid 'a cultural foundation ... that wedded science to economic efficiency'.[119] Optimistic about progress and the benefits of useful knowledge,[120] its ethos had much in common with that of the British Association, so important to Fairbairn, which was dominated for at least its first thirty years by liberal Anglican clergy with an 'Enlightenment' approach – broad, accommodating, intellectually tolerant and increasingly Newtonian.[121]

Motivation

DISCUSSION of a Puritan understanding of work raises the associated question of how far industrial development was driven by the simple profit motive, and in the present case how far Fairbairn was driven by that motive. Were there other motivations – the challenge to achieve, fame, the foundation of a dynasty? In the late-19th century and for much of the 20th century there was a tendency to equate entrepreneur and capitalist, 'by whom things were produced only if they would make a profit.' There was nothing to prompt the capitalist to think of anything but making profit as he bought in the cheapest market and sold in the dearest.[122] The founders of many new firms, or their socially ambitious wives, were seen as spurred on by the tangible results of commercial or industrial success – 'the palaces of the cottontot grandees'.[123] Collini writes of historians in the early decades of the 20th century propounding,

> A highly elaborated account of the Industrial Revolution not just as a social catastrophe for certain classes, but also something that now tends to be neglected though it was an even stronger claim – as establishing a quite new form of civilisation, one driven by the narrow and unchecked pursuit of profit.[124]

A study setting out a wide diversity of definitions of an entrepreneur concludes that 'they all boil down to the profit motive'.[125] However, from the 1930s there has been a slow progressive erosion of the orthodox belief in profit-maximisation as the primary motivation of business behaviour.[126] Schumpeter saw motivation as threefold – to succeed, not for the fruits of success, but for the sake of success itself; secondly, the joy of creating, of getting things done, of overcoming difficulties, of delighting in ventures; and thirdly, 'the ultimate motivation of the industrialists ... a dynastic one: to found a family, to endow them splendidly enough to last for ever'.[127] Perkin found 'the limitless pursuit of wealth for its own sake to be a rare phenomenon'.[128] Bolton and Thompson define an entrepreneur as 'a person who habitually creates and innovates to build something of recognised value around perceived opportunities', as distinguished from those who build a business simply to achieve a comfortable lifestyle.[129] Bull, Pitt and Szarka, in a 1993 study, write:

> Traits of personality have an important bearing on entrepreneurial outlook and motivations. For example, the classic portrayal of entrepreneurs suggests that typically they have a high 'need for achievement', they are innovative, oriented to action and change, and above all, they value autonomy and independence.[130]

Their 'artisanal entrepreneur' was likely to believe that product quality should override profit.[131] Mokyr writes – and this is particularly apposite in respect of problem-focussed engineering – of the triumph of getting a problem solved, of 'the thrill of playing and the dream of winning'.[132] Fairbairn's contemporary, Woodcroft, wrote of him, quoting Milton:

> Fame is the spur the pure spirit doth raise
> To scorn delights, and live laborious days.[133]

Thus past and present opinion provides some support that Schumpeter may be right and business behaviour may not be motivated just by pecuniary incentives.[134]

The Family Firm and the Matter of Succession

DYNASTIC motivation raises the issue of how continuity is maintained in a family firm. There is continuing debate about how the handling of succession, as the founder of a family business approaches retirement, affects the firm's future. A widely used definition of a family firm, and the one adopted in this book, is where a family owns enough of the equity to control the strategy and is involved in top management positions. It is generally accepted that the rapid formation of new family firms contributed a vital dynamism during the early phases of industrialisation. Despite the persistence and continuing dynamism of such firms, some historians during the last thirty years have held them responsible for the decline of British industry in the latter part of the 19th-century.[135] Wiener has argued that the decline was caused by widespread gentrification.[136] Some see decline as endorsing the so-called 'Buddenbrook syndrome' (although that was distinguished from Fairbairns in that it was third generation decline whereas Fairbairn's was second generation).[137] In fact the short life and small scale of many 19th-century British family firms meant that gentrification on a scale which might have damaged national business performance was probably not a viable theory. Wiener attracts little support in academia.[138] There were industrialists who bought country estates but they were a minority.[139] Another approach is the 'personal capital' theory of A.D. Chandler, that the dislike of losing personal control over their family firms resulted in entrepreneurs failing to make the three-pronged investment in production, distribution and management necessary to fully exploit the economies of scale,[140] although in the last twenty years there has been growing evidence that Chandler undervalued the resilience and capabilities of the family company.[141] Rose sees succession in family businesses as 'the critical foundation upon which the future prosperity of a family firm rests' and 'since it is possible to plan creatively for leadership succession, such provision should be seen as being as much a part of entrepreneurship as technological or organisational innovation'. She maintains that plans for succession should include an element of meritocracy at family level and almost certainly involve recruitment of people of ability from outside the family.[142] F.M.L Thompson sounds a different note – usually there was only room for one son, not necessarily the eldest, to make a living whilst preparing to take over from his father. The others were to be directed to the professions or civil service to protect the firm from the burden of trying to support 'excessive numbers of functionally superfluous sons'.[143] Neither Rose nor Thompson place emphasis on the creation of a dynastic firm as an entrepreneurial motive, as Schumpeter and Perkin had done. Another potential problem facing family businesses is seen as inter-personal friction, either between generations, or between siblings.[144]

* * *

Chapter	Dates for Wm Fairbairn	Wm Fairbairn	Dates for Firm	Firm: Mode of Trading	Name of Firm
Chapter 3	1789–1804	Childhood			
	1804–1811	Apprenticeship			
	1811–1817	Journeyman			
Chapter 4	1817–1832	Partner	1817–1832	**Partnership (William Fairbairn, James Lillie)**	*Fairbairn & Lillie*
Chapter 5	1832–1841	Sole Principal	1832–1841	**Sole Principal (William Fairbairn)** (1838–43 Andrew Murray and John Hetherington had small shares in the Millwall shipyard)	*Wm Fairbairn (until 1846)*
Chapters 6 and 7	1841–1853	Family Business	1841–1853	**Partnership(s) – Family Business** Wm Fairbairn, John Fairbairn from unknown date until 1844; Thomas Fairbairn from 1841; Wm A. Fairbairn from 1846; George Fairbairn from c.1851	*Wm Fairbairn & Sons (from 1846)*
Chapters 8 and 9	1854–1874	Retired' but gave 'active assistance'	1854–1859	**Partnership(s) – Family Business** (Thomas Fairbairn; George Fairbairn until 1856; Peter Fairbairn until 1859; Wm Andrew Fairbairn until 1859	
			1859–1864	**Sole Principal** (Thomas Fairbairn)	*Fairbairn & Company*
		Director until 1872 but then 'retained interest in the establishment'	1864–1875	**Limited Liability Company** Directors: Thomas Fairbairn, chairman; Augustus Novelli; James McConnell; John Pender; Wm Fairbairn until 1872	*The Fairbairn Engineering Company Limited*
	1874 >	Deceased			
			1875–1899	In voluntary liquidation. Finally wound up in 1899.	

Table 2: Fairbairn Chronology

NOTWITHSTANDING these areas of debate – and there are others – this book is primarily about Fairbairn's engineering achievements and their influence. Three approaches to structuring the book to meet the challenge of integrating so wide a range of work over more than half a century were considered, the thematic, the chronological, and a hybrid between the two. None provides a neat solution, but history does not provide neat solutions. A chronological approach has been chosen because it provides the least compartmentalised and fragmented solution, and because it rectifies the common failure to appreciate the several discrete phases of Fairbairn's career, and also the occasional confusion of Fairbairn with one or other of his sons. The chronological approach endorses Allan Thompson's view that chronology 'helps to relate a number of disparate changes to one another in some consistent way',[145] as well as answering Jennifer Tann's criticism of 'a time- and context-free approach' where the development of engineering products is discussed solely in engineering terms.[146] The emphasis on the discrete and different phases of the Fairbairn business is important – for example the family business is particularly relevant given recent interest in family businesses amongst historians and in business schools. The Fairbairn Chronology is set out in *Table 2*. However, the chronological approach is not followed slavishly and some latitude has been taken where this seems appropriate, particularly in Chapter 1. Fairbairn's influence through education and his trainees and employees was on-going throughout his career and way beyond, and these areas are brought together in Chapter 6, in the chronological sequence when this influence was at its height. Iron and engineering are the uniting features throughout, and the two specific threads which continue through all the phases of Fairbairn's business life are mill-building and experimental engineering. Much engineering history is inevitably bounded by technical parameters, but a study in the way outlined facilitates a synergy between technical aspects of engineering and their business, social and institutional settings. It provides an appropriate context to record Fairbairn's contributions to 19th-century engineering and to question how he, and the engineering company he founded, ascended to success, in what ways he was influential, and what caused the rapid demise of the company.

Notes for the Introduction

1. J Mokyr, *The Lever of Riches: Technological Creativity and Economic Progress,* (1990), p.106; J Mokyr, *Twenty-five Centuries of Technological Change: A Historical Survey,* (1990), p.69.
2. R A Buchanan, *The Engineers: A History of the Engineering Profession in Britain 1750-1914,* (1989), p.38.
3. A W Skempton, 'Introduction' to A W Skempton (ed.), *John Smeaton, FRS,* (1981), p.1.
4. N Rosenberg, 'Review of *The Life of Sir William Fairbairn, Bart.*, Reprint with Introduction by A E Musson', *The Economic History Review,* 25.4, 1972, 712.
5. *Life.*
6. Buchanan, *Engineers,* p.11.
7. C MacLeod, 'Strategies for innovation: the diffusion of new technology in nineteenth-century British industry', *Economic History Review,* 45.2, 1992, 285.
8. M J Wiener, *English Culture and the Decline of the Industrial Spirit, 1850-1980,* (2nd ed. 2004), p.158.
9. C MacLeod, *Heroes of Invention: Technology, Liberalism and British Identity, 1750-1914,* (2007). p.11; S Collini, 'The literary critic and the village labourer: "culture" in twentieth-century Britain'. *Transactions of the Royal Historical Society,* 6th series, 14, 2004, 93-116; D C Coleman, *Myth, History and the Industrial Revolution,* (1962), pp.20-30; E P Thompson, *The Making of the English Working Class,* (1965), p.198.
10. A Herman, *The Scottish Enlightenment: The Scots' Invention of the Modern World,* (2001), pp.326-7, 426.
11. L T C Rolt, 'Introduction to 1967 reprint of the 1863 Edition' of S Smiles, *Industrial Biography: Iron Workers and Tool Makers,* unpaginated. Buchanan says of Smiles volumes that they 'remain excellent biographical studies' (Buchanan, *Engineers,* p.16), whereas Rolt finds he 'was apt to rely too heavily upon ... the fallible recollections of elderly men' (Rolt, 'Introduction').
12. Herman, *Scottish Enlightenment,* pp.326-8; S Heffer, *High Minds: The Victorians and the Birth of Modern Britain* (2013), pp.205-10.
13. For example: A W Skempton et al (eds), *A Biographical Dictionary of Civil Engineers in Great Britain and Ireland, Vol. 1: 1500-1830,* (2002); P S M Cross-Rudkin et al (eds), *A Biographical Dictionary of Civil Engineers in Great Britain and Ireland, Vol. 2: 1830-1900,* (2008); MacLeod, *Heroes of Invention.*
14. The use of the term was popularised by H Butterfield, *The Whig Interpretation of History,* (1931).
15. W Cronon, 'Two Cheers for the Whig Interpretation of History', (American Historical Association, September 2012), on-line at https://www.historians.org/publications-and-directories/perspectives-on-history/september-2012/two-cheers-for-the-whig-interpretation-of-history. For a scholarly discussion on the motivation of Scottish historians see D Allan, *Virtue, Learning, and the Scottish Enlightenment: Ideas of Scholarship in Early Modern History,* (1993). Linear views of history are characteristically optimistic, but this is an optimism with reservations and far from being unqualified.
(D Bebbington, *Patterns in History,* (1979), pp.18, 51).
16. J Mokyr, *The Gifts of Athena: Historical Origins of the Knowledge Economy,* (2002), p.297.
17. Mokyr, *Lever of Riches,* pp.15, 303.
18. Mokyr, *Gifts of Athena,* p.285.
19. Heffer, *High Minds,* p.xiii.
20. E Mayr, 'When is Historiography Whiggish?', *Journal of the History of Ideas,* 51.2, 1990, 309.
21. Thus, the antiquarian is said to be 'the type of man who is interested in historical facts without being interested in history' (A Momigliano, 'Ancient History and the Antiquarian', *Journal of the Warburg and Courtauld Institutes,* 13, 1950, 289).
22. Detailled studies are called for of several important engineers, industrialists and others met with in this book, for example Benjamin Heywood (banker), C Wye Williams (engineer and ship-owner), George Rennie (engineer), Alexander Wilson (engineer), Thomas Bazley (cotton spinner), John Pender (telecommunications entrepreneur), J E Gregan (architect), the Laird family (shipbuilders), the Platt family (textile machine makers), W H Houldsworth (cotton manufacturer), Peter Fairbairn (engineer), W J M Rankine (professor of engineering) and W Pole (mathematician).
23. C P Snow, *The Two Cultures,* ([1959] 1998).
24. J Pickstone, 'Obituary: Professor Donald Cardwell (4 August 1919 – 8 May 1998)', *The British Journal for the History of Science,* 32.4, 1999, 485-8.
25. N S Weyant, *Elizabeth Gaskell: An Annotated Bibliography of English Language Sources, 1976-1991,* (1994) at http://www.lang.nagoya-u.ac.jp/~matsuoka/gaskell/dissertations.html (accessed 15 March 2014). I am only aware of three Masters degrees relating to Fairbairn - Hayward (Hayward, 'Fairbairns'), Tuovinen (T Tuovinen, 'The Life and Work of Sir William Fairbairn: A Case Study in the History of the Engineering Profession in 19th Century Britain', (Pro Gradu Thesis, University of Jyväskylä, 2000)), and my own.
26. W Fairbairn, 'On Public Health', in G W Hastings (ed.), *Transactions of the National Association for the Promotion of Social Science, London Meeting, 1862,* (1863), p.60.
27. P Riden, 'Waterways history research: an alternative view', *Journal of the Railway & Canal Historical Society,* 38.7, 2016, 426-40.
28. M R Bailey (ed.), 'Introduction' to *Robert Stephenson – The Eminent Engineer,* (2003), p.xxii; Mokyr, *Lever of Riches,* p.viii.
29. T A B Corley, 'The entrepreneur: the central issue in business history?' in J Brown and M B Rose (eds), *Entrepreneurship, networks and modern business* (1993), p.15, referring to C H Wilson, *The History of Unilever: A Study of Economic Growth and Social Change,* (1954).
30. W E Bijker, T P Hughes and T Pinch (eds), 'Introduction to *The Social Construction of Technological Systems,* (2nd ed. 2012), pp.3-9.
31. W E Bijker and T Pinch, 'Preface to the Anniversary Edition' of Bijker, Hughes and Pinch, *Social Construction,* p.xxvii.
32. J Appleby, L Hunt and M Jacob, *Telling the Truth about History,* (1994), p.228.
33. R A Buchanan, 'Theory and Narrative in the History of Technology', *Technology and Culture,* 32, 1991, 365-76; Bijker and Pinch, 'Preface, p.xvii.
34. Pickstone, 'Donald Cardwell', 486.

35. T Pinch and W E Bijker, 'The Social Construction of Facts and Artifacts: Or How the Sociology of Science and the Sociology of Technology Might Benefit Each Other', in Bijker, Hughes and Pinch, *Social Construction,* p.16;
P Fridenson, 'Business Failure and the Agenda of Business History', *Enterprise and Society,* 5.4, 2004, 565; R Morris, *Time's Anvil: England, Archaeology and the Imagination,* (2012), pp.19, 319.
36. *London Gazette,* 24 March 1899.
37. R Floud, *The British Machine Tool Industry, 1850-1914,* (1976);
J A Cantrell, *James Nasmyth and the Bridgewater Foundry: A study of entrepreneurship in the early engineering industry,* (1984).
38. W König, 'Design and Engineering' in G Jones and J Zeitlin (eds), *The Oxford Handbook of Business History,* (2007), pp.377-8.
39. J Tann, *The Development of the Factory,* (1970), p.100: 'Fairbairn was prone to exaggerate in order to make his own work look the more impressive'. R S Fitzgerald, 'The Development of the Cast Iron Frame in Textile Mills to 1850', *IAR,* 10.2, 1988, 137: He 'advanced claims that are difficult to substantiate' and is 'equally guilty of the sin of omission'. In neither case are specific instances quoted in support of these generalised criticisms.
40. *Life,* p.v.
41. S Smiles, *Industrial Biography: Iron Workers and Toolmakers,* (1863), pp.v, 299-333. This was translated into several European languages.
42. W J M Rankine, 'William Fairbairn', in *The Imperial Dictionary of Universal Biography, Volume 2,* (1863).
43. *Life,* p.vi; T R Robinson to T Fairbairn, 13 June 1875, Cambridge University Library, Stokes Papers, Add 7342, TR54.
44. *Life,* p.179.
45. Gillian Cookson is incorrect when she refers to Pole's book as 'a hybrid of biography and autobiography, written in the third person yet unmistakably the mouthpiece of the subject' (G Cookson, 'Reconstructing a lost engineer: Fleeming Jenkin and problems of sources: 1 – Biographical Sources', in A Jarvis and K Smith, (eds), *Perceptions of Great Engineers II,* (1998), pp.28, 32n7).
46. *Life,* p.148.
47. J Brittain, 'Review of *The Life of Sir William Fairbairn, Bart.* Reprint with an Introduction by A E Musson', *Technology and Culture,* 12.7, 1971, 505.
48. [Woodcroft] V.
49. *Life,* p.472.
50. [Woodcroft] X, p.309.
51. R A Smith, *A Centenary of Science in Manchester,* (1883), p.258. On Smith see R H Kargon, *Science in Victorian Manchester: Enterprise and Expertise,* (1977), pp.123-34.
52. Fairbairn, *UIfE3,* p.68.
53 . Walker, *Unwin,* p.67.
54. *MPICE,* 39, 1875, 263.
55. J Hewish, *The Indefatigable Mr Woodcroft: The Legacy of Invention,* (1979), pp.9-13.
56. Fairbairn, *UIfE3,* p.17.
57. *Life,* pp.155-6.
58. Hewish, *Woodcroft,* pp.13-4.
59. B Woodcroft, *Sketch of the Origin and Progress of Steam Navigation from authentic documents,* (1848); B Woodcroft, *Brief Biographies of Inventors of Machines for the Manufacture of Textile Fabrics,* (1863).
60. W Fairbairn to W C Unwin, 30 December 1864, London University: Imperial College Archives, Unwin/94.
61. Brittain, 'Review', 506.
62. Smith, *Contribution*; A I Smith, 'William Fairbairn and Mechanical Properties of Materials – 50,000 hours sustained-load tests on iron bars', *The Engineer,* 216, 4 October 1963; A I Smith, 'William Fairbairn and Mechanical Properties of Materials – the Effect of Temperature on Strength', *The Engineer,* 216, 20 December 1963; A I Smith, 'William Fairbairn and Mechanical Properties of Materials: The Effect of Repeated Loading on Strength', *The Engineer,* 217, 26 June 1964, 1133-1136; A I Smith, 'William Fairbairn – Experimental Engineer', in E G Semler, (ed.), *Engineering Heritage Volume 2,* (1966), pp.20-5.
63. Rosenberg & Vincenti, *Britannia,* p.1. On the Britannia Bridge see also S P Timoshenko, *History of Strength of Materials: with a brief account of the history of theory of elasticity and theory of structures,* (1953), pp.156-62; D Beckett, *Stephenson's Britain,* (1984), pp.131-51;
[R] J [M] Sutherland, 'Iron Railway Bridges' in Bailey, *Robert Stephenson,* pp.318-35; J Rapley, *The Britannia & Other Tubular Bridges and the men who built them,* (2003).
64. Hayward, 'Fairbairns' – this has been missing from UMIST library for the last several years and I am very grateful to Ron Hayward for letting me read his copy;
R A Hayward, *The Story and Scandal of HMS Megaera,* (1978). On Fairbairn's shipbuilding see also F C Bowen, 'Shipbuilders of Other Days: No.55 – Sir William Fairbairn (1789-1874), of Manchester and Millwall. A shipbuilding and engineering pioneer of merit', *Shipbuilding and Shipping Record,* 21 August 1952, 245-7; A J Arnold, *Iron Shipbuilding on the Thames, 1832-1915: An economic and business history,* (2000), 24-7, 152;
P Banbury, *Shipbuilders of the Thames and Medway,* (1971), 171-4.
65. Fitzgerald, 'Development of the Cast Iron Frame in Textile Mills to 1850', 127-45; R S Fitzgerald, *Liverpool Road Station, Manchester: An historical and architectural survey,* (1980). On the Ancoats mills see also I Miller and C Wild (eds), *A & G Murray and the Cotton Mills of Ancoats,* (2007).
66. Ahrons, 'Famous Firms'.
67. The Commissioners continued after the Exhibition, distributing the income from the Exhibition's invested profits, as they do to this day.
68. 'Brompton Boilers' was the derogatory name applied to Kensington Gore Museum (G Herbert, *Pioneers of Prefabrication: The British Contribution in the Nineteenth Century,* (1978), p.72.
69. Helen Birley, William Andrew's second wife, was the daughter of Richard Birley and Amelia, the fifth daughter of James Kennedy.
By 1864, William was seventy-five and William Andrew had abandoned engineering and Manchester (*Life,* pp.330-1; G B Smith, *Leaders of Modern Industry, Biographical Sketches,* (1894), p.371).
And it was William Andrew who, from around 1870, was living at an address in Holland Park, which has been wrongly identified as his father's London house – see above.
70. email address at time of publication: richard@byroms.net.
71. J Mokyr, *The Enlightened Economy: Britain and the Industrial Revolution 1700-1850,* (2009), p.9.
72. A E Musson and E Robinson, *Science and Technology in the Industrial Revolution,* (1969); L Roberts, 'The Circulation

of Knowledge in Early Modern Europe: Embodiment, Mobility, Learning and Knowing', *History of Technology,* 31, 2012, 50.
73. A R Hall, 'What did the Industrial Revolution in England Owe to Science', in N McKendrick (ed.), *Historical Perspectives: Studies in English Thought and Society,* (1974), pp.37-56.
74. RV Jenkins, 'Review of *Science and Technology in the Industrial Revolution.* By A E Musson and Eric Robinson', *The Business History Review,* 45.3, 1971, 397; U Wengenroth, 'Science, Technology and Industry' in D Cahan (ed.), *From Natural Philosophy to the Sciences: Writing the History of Nineteenth-Century Science,* (2003), p.227.
75. Rosenberg & Vincenti, *Britannia,* p.71.
76. N Rosenberg, *Exploring the black box: technology, economics, and history,* (1994), p.139.
77. M C Jacob, *Scientific Culture and the Making of the Industrial West,* (1997), pp.3, 177.
78. Mokyr, *Gifts of Athena,* p.30n1. Papers spawned by this book include M Berg, 'The Genesis of "Useful Knowledge"', *History of Science,* 14, 2007, 124-33; K Bruland, 'Technology Selection and Useful Knowledge: A Comment', *History of Science,* 14, 2007, 179-83; J Mokyr, 'Knowledge, Enlightenment, and the Industrial Revolution: Reflections on *The Gifts of Athena*', *History of Science,* 14, 2007, 185-96; K Davids, 'Introduction: "Useful Knowledge" Reconsidered', *History of Technology,* 31, 2012, 1-4; Roberts, 'The Circulation of Knowledge', 47-68; K Davids, 'Gatekeeping. Who Defined "Useful Knowledge" in Early Modern Times?', *History of Technology,* 31, 2012, 69-88.
79. Mokyr, *Enlightened Economy,* pp.40, 44; R C Allen, *The British Industrial Revolution in Global Perspective,* (2009), p.241.
80. L Goldman, 'The Origins of British 'Social Science': Political Economy, Natural Science and Statistics, 1830-1835', *The Historical Journal,* 26.3, 1983, 602; K Olesko, 'Humboldt, Alexander von' and 'Humboldtian Science' in J L Heilbron (ed.), *The Oxford Companion to the History of Modern Science,* (2003), pp.383-7. The term 'Humboldtian Science' was first used by S Cannon, *Science in Culture: The Early Victorian Period,* (1978), pp.104-5.
81. J Morrell and A Thackray, *Gentlemen of Science: Early Years of the British Association for the Advancement of Science,* (1981), pp.271, 426, 512-3; *Life,* pp.220-36, 293-4, 468; A Sachs, *The Humboldt Current: A European explorer and his American disciples,* (2007), p.1; M Dettelbach, 'Humboldtian Science' in N Jardine, J A Secord and E C Spary (eds), *Cultures of Natural History,* (1996), p.287.
82. S F Mason, *A History of the Sciences,* (1956), p.408.
83. B Barnes, 'The science-technology relationship: a model and a query', *Social Studies of Science,* 12, 1982, 166.
84. Jacob, *Scientific Culture,* p.9. The metaphor of twins was also used by E Layton, 'Mirror-Image Twins: `Century America', *Technology and Culture,* 12, 1971, 562-80.
85. Fairbairn, *UIfE,* p.97.
86. Wengenroth, 'Science, Technology and Industry', p.223; Mokyr, *Enlightened Economy,* p.61.
87. S Kuznets, *Economic Growth and Structure,* (1965), pp.85-7.
88. Davids, 'Introduction: "Useful Knowledge" Reconsidered', 1; I Inkster, 'Potentially Global: "Useful and Reliable Knowledge" and Material Progress in Europe, 1474-1914', *International History Review,* 28.2, 2006, 238.
89. V W Ruttan, 'Usher and Schumpeter on Invention, Innovation, and Technological Change', *The Quarterly Journal of Economics,* 73.4, 1959, 597. On the time-lag between invention and innovation see the table reproduced in Rosenberg, *Perspectives,* pp.69-70, from J Enos, 'Invention and Innovation in the Petroleum Refining Industry', in *The Rate and Direction of Inventive Activity: Economic and Social Factors,* (National Bureau of Economic Research Conference Papers, 1962), pp.307-8.
90. N Rosenberg, *Inside the Black Box,* (1982), pp.62-70; N Rosenberg, *Schumpeter and the Endogeneity of Technology: Some American Perspectives,* (2000), p.61; E L Jones, *The European Miracle,* (1981), p.68.
91. A P Usher, *A History of Mechanical Inventions,* (rev. ed., 1954); Ruttan, 'Usher and Schumpeter', 601.
92. Mokyr, *Lever of Riches,* p.12.
93. R A Buchanan, *Brunel: The Life and Times of Isambard Kingdom Brunel,* (2002), p.212.
94. Mokyr, *Lever of Riches.*
95. Allen, *British Industrial Revolution,* pp.239-42. Mokyr describes macroinventions and microinventions in *Lever of Riches,* pp.12-3.
96. F Redlich, 'Innovation in Business: A Systematic Presentation', *American Journal of Economics and Sociology,* 10.3, 1951, 290-1.
97. Rosenberg, *Schumpeter,* p.78; N Rosenberg, 'Economic Development and the Transfer of Technology: Some Historical Perspectives', *Technology and Culture,* 11.4, 1970, 575. See also K Pavitt, 'Innovation Processes' in J Fagerberg, D C Mowery and R R Nelson (eds), *The Oxford Handbook of Innovation,* (2005), p.108; L Marino, P Kreiser and A Robinson, 'Environmental Uncertainty and firm-level entrepreneurship' in H Landström and F Lohrke, *Historical Foundations of Entrepreneurial Research,* (2010), pp.81-97.
98. Pinch and Bijker, 'The Social Construction of Facts and Artifacts', p.14.
99. E M Rogers, *Diffusion of Innovations,* (5th ed. 2003), p.5.
100. D J Jeremy, 'Lancashire and the International Diffusion of Technology' in M B Rose (ed.), *The Lancashire Cotton Industry: A History since 1700,* (1996), pp.210-37; K Bruland, *British Technology and European Industrialization: The Norwegian textile industry in the mid nineteenth century,* (1989); T Nilson, '"The Invisible Hand of Knowledge". Market Forces and the Diffusion of Technology and Knowledge from Britain to Sweden during the Eighteen-fifties: The Case of the Swedish Textile Industry', in *A Case of Identities: Festschrift to Martin Peterson,* (2006).
101. S Jack and M B Rose, 'The historical roots of socio network theory in entrepreneurship research', in Landström and Lohrke, *Historical Foundations of Entrepreneurship Research,* pp.256, 263-6.
102. Brown M B Rose, *Entrepreneurship,* pp.1, 6.
103. M Berman, 'An essay review of A E Musson and Eric Robinson, *Science and Technology in the Industrial Revolution*', *Journal of Social History,* 5.4, 1972, p.525.
104. Mokyr, *Gifts of Athena,* p.66; Mokyr, *Enlightened Economy,* p.48; Allen, *British Industrial Revolution,* p.240; R R Meisenzahl and J K Mokyr, 'The Rate and Direction of Invention in the British Industrial Revolution: Incentives and Institutions', in J Lerner and S Stern (eds), *The Rate and Direction of Inventive Activity Revisited,* (2012), p.469.
105. E J Hobsbawm, *Industry and Empire: An Economic History of Britain since 1750,* (1968), pp.260-1. See also M Weber, *The Protestant Ethic and the Spirit of Capitalism,* (ET 1967). On Calvinism see A Kuyper, *Lectures on*

Calvinism, (1931 ed.), 'Fourth Lecture, Calvinism and Science', pp.110-141; and H R Van Til, *The Calvinistic Concept of Culture,* (1959), Chapter 15, 'Calvinistic Culture and Christian Calling', pp.217-228. For a recent interpretation see: [H] F [R] Catherwood, *The Creation of Wealth: Recovering a Christian Understanding of Money, Work and Ethics,* (2002), pp.17-35.
106. Herman, *The Scottish Enlightenment,* p.ix.
107. D A Farnie and W O Henderson (eds), *Industry and Innovation: Selected Essays, W H Chaloner* (1990), p.100.
108. Unpublished study by the author.
109. Hobsbawm, *Industry and Empire,* pp.260-1; Mokyr, *The Enlightened Economy,* pp.31, 37.
110. D Allan, *Virtue, Learning, and the Scottish Enlightenment: Ideas of Scholarship in Early Modern History,* (1993).
111. A H Williamson, 'Review of *Virtue, Learning, and the Scottish Enlightenment: Ideas of Scholarship in Early Modern History.* By David Allan', *The Journal of Modern History,* 67.3, 1995, 703.
112. D S Landes, *The Wealth and Poverty of Nations: Why Some Are So Rich and Some So Poor,* ([1998] Abacus ed. 1999 pp.175-8). K Lipartito, 'Business Culture' in G Jones and J Zeitlin (eds), *The Oxford Handbook of Business History,* (2007), pp.609-10; S McKinstry and Y Y Ding, 'Alex Cowan & Sons Ltd, Papermakers, Penicuik: a Scottish case of Weber's Protestant Work Ethic', *Business History,* 55.5, 2013, pp.721. K Kininmonth, 'Weber's Protestant Work Ethic: a case study of Scottish entrepreneurs, the Coats Family of Paisley', Business History, 58.8, 2016, pp.1236-61. In 2000 Daniel Treisman produced a paper with robust evidence that the higher the proportion of Protestants in a country, the lower the corruption rating (D Treisman, 'The causes of corruption: a cross-national study', *Journal of Public Economics,* 76, 2000, 399-457).
113. Hobsbawm *Industry and Empire,* p.261. For some criticisms of Weber see C R Trueman, *History and Fallacies: Problems Faced in the Writing of History,* (2010), pp.153-5.
114. McKinstry and Ding, 'Alex Cowan & Sons Ltd', 724, quoting G Marshall, *Presbyteries and Profits: Calvinism and the Development of Capitalism in Scotland, 1560-1707,* (1980), p.235; D S Landes, *The Unbound Prometheus. Technological Change and Industrial Development in Western Europe from 1750 to the Present,* (1969), pp.23-4;
115. Genesis 1. 26-28, 2.15.
116. M C Jacob, *The First Knowledge Economy: Human Capital and the European Economy, 1750-1850,* (2014), pp.100-7.
117. Jacob, *Scientific Culture,* p.52.
118. J Seed, 'Theologies of Power: Unitarianism and the social relations of religious discourse 1800-50' in R J Morris (ed.), *Class, Power and Social Structure in British Nineteenth Century Towns,* (1986), p.142; M J Turner, *Reform and Respectability. The Making of a Middle-Class in Early 19th-century Manchester,* (1995), p.32; Jacob, *First Knowledge Economy,* pp.99-107; A Thackray, 'Natural Knowledge in Cultural Context: The Manchester Model', *The American Historical Review,* 79.3, 1974, 694.
119. Jacob, *First Knowledge Economy,* p.99.
120. Mokyr, *Enlightened Economy,* p.362.
121. Morrell and Thackray, *Gentlemen of Science,* pp.225-9. It was in Oxford's new Natural History Museum, about which Fairbairn had been consulted on structural problems (F O'Dwyer, *The Architecture of Deane and Woodward,* (1997), pp.264-7), that the Association's favoured genre of natural theology was put under great strain, as Samuel Wilberforce and Thomas Huxley debated Darwin's theory of natural selection, before an audience of 750, which must surely have included Fairbairn – it was the year after the publication of Darwin's, *On The Origin of Species,* and the year before Fairbairn became the Association's President (Morrell and Thrackray, *Gentlemen of Science,* p.395).
122. E Fischer, *Marx in His Own Words,* (English Translation, A Bostock, 1973), p.113, quoting *Capital III,* pp.252-4; E J Hobsbawm, *The Age of Capital 1848-1875,* (1975), p.1.
123. P L Payne, *British Entrepreneurship in the Nineteenth Century,* (2nd ed. 1988), p.23.
124. Collini, 'The literary critic', 98.
125. E Chell, J Haworth and S Brearley, *The Entrepreneurial Personality; Concepts, Cases and Categories,* (1991), pp.1-6, 28.
126. D C Coleman, 'Gentlemen and Players', *Economic History Review,* NS26.1, 1973, 96.
127. J A Schumpeter, *The Theory of Economic Development,* (1934), pp.93-4.
128. H Perkin, *The Origins of Modern English Society 1780-1880,* (1969), pp.83-5.
129. W K Bolton and J L Thompson, *Entrepreneurs; Talent, Temperament, Technique,* (2000), p.20.
130. A Bull, M Pitt and J Szarka, *Entrepreneurial Textile Communities: A comparative study of small textile and clothing firms,* (1993), p.67. See also A N Licht and J I Siegel, 'The Social Dimensions of Entrepreneurship', in M Casson et al, *The Oxford Handbook of Entrepreneurship,* (2006), p.514; B H Hamilton, 'Does Entrepreneurship Pay? An Empirical Analysis of the Returns to Self-Employment', *Journal of Political Economy,* 108.3, 2000, 628; Meisenzahl and Mokyr, 'The Rate and Direction', pp.462-70.
131. Bull, Pitt and Szarka, *Entrepreneurial Textile Communities,* p.74.
132. J Mokyr, 'Entrepreneurship and the Industrial Revolution in Britain' in D S Landes, J Mokyr and W J Baumol (eds), *The Invention of Enterprise: Entrepreneurship from Ancient Mesopotamia to Modern Times,* (2010), p.196.
133. [Woodcroft] X, quoting J Milton, *Lycidas. A Lament for a friend drowned in his passage from Chester on the Irish Seas,* 1637.
134. Licht and Siegel, 'The Social Dimensions of Entrepreneurship', p.514; Hamilton, 'Does Entrepreneurship Pay?', 628.
135. A Colli and M B Rose, 'Family Business', in G Jones and J Zeitlin (eds), *The Oxford Handbook of Business History,* (2007), p.194.
136. M J Wiener, *English Culture and the Decline of the Industrial Spirit, 1850-1980,* (1981).
137. T Mann, *Buddenbrooks: Verfall einer Familie,* (1901 – various English translations). This novel won a Nobel Prize for Literature and was banned by Hitler.
138. B Collins and K Robbins (eds), *British Culture and Economic Decline,* (1990); G Jones and M B Rose, 'Family Capitalism', *Business History,* 35.4, 1993, 6-7; M B Rose, 'The family firm in British business, 1780-1914', in M W Kirby and M B Rose (eds), *Business Enterprise in Modern Britain: from the eighteenth to the twentieth century,* (1994), pp.73-4;
139. A Howe, *The Cotton Masters, 1830-1860,* (1984), p.310; W Rubinstein, 'New Men of Wealth and the Purchase of Land in Nineteenth-Century Britain', *Past & Present,* 92, 1981, 125-47.
140. A D Chandler, *Scale and Scope: The Dynamics of Industrial Capitalism,* (1990), p.286.
141. Colli and Rose, 'Family Business', pp.199-200.

142. M B Rose, 'Beyond Buddenbrooks: the family firm and the management of succession in nineteenth-century Britain' in J Brown and M B Rose, *Entrepreneurship, networks and modern business,* (1993), pp.133-5.
143. F M L Thompson, *Gentrification and the Enterprise Culture: Britain 1780-1980,* (2001), p.136
144. Rose, 'Beyond Buddenbrooks', 136; W M Matthew, 'Animus, Absenteeism, and Succession in the Keiller Marmalade Dynasty, 1839-1919', *Journal of Scottish Historical Studies,* 28.1, 2008, pp.51-3.
145. A Thompson, *The Dynamics of the Industrial Revolution,* (1973), p.141. Thompson's illustrations are mainly drawn from developments in agriculture.
146. J Tann, 'Space, Time and Innovation Characteristics: The Contribution of Diffusion Process Theory to the History of Technology', *History of Technology,* 17, 1995,146.

CHAPTER 1

Foundations

Introduction

WILLIAM Fairbairn became one of the great engineers of the 19th century, and did so from a very modest background. What were the foundations of his successful career and were there pointers to the future to be seen there? What were the characteristics of his home and to what extent did these influence his later life? What early education did he receive? What did Fairbairn learn as an apprentice millwright and what features of workshop life influenced the management of his own workshops in the future? What connections made whilst he was an apprentice were influential in his future career? In answering these questions it needs to be appreciated that there were three distinctly separate foundational phases to his success, childhood up to the age of fifteen, followed by seven years as an apprentice millwright, and six years as a journeyman millwright.

From these latter years Fairbairn provides insights into the life of an early 19th-century journeyman millwright. Were there pointers to his future during these years? What was it that drew Fairbairn, like many other leading engineers, to settle in Manchester? Significantly he spent the latter part of his time as a journeyman with Thomas Hewes, one of the leading millwrights in the country at the time, ending with formative time in Hewes' drawing office. What did Fairbairn learn from his time with Hewes?

Fairbairn's experiences from these years help to provide an understanding of some of his future attitudes and actions, and this chapter includes discrete sections looking forward to see the future effects of significant circumstances from these years – paternalism, restrictive practices and domestic stability.

Childhood, Schooling and Early Employment

WHEN Andrew and Margaret Fairbairn's first child, William, was born, into their impoverished home in Roxburghshire, Scotland, the world which he would come to know was beyond their wildest dreams. That William would be an agent in that change, and meet with kings was unimaginable.[1] That Peter, their only other son to survive beyond teenage years, would also build a famous engineering firm which continued until the mid 20th century, become mayor of a great northern city and receive a knighthood, makes history read like fiction.[2]

William's childhood was far from easy, his education was interrupted and his first two jobs were an unfortunate entry into the world of work, but he was clearly influenced by parents who were industrious and 'respectable'. His father, Andrew Fairbairn (1758–1844) was a farm labourer, became a ploughman, and was press-ganged into the navy during the American War of Independence. On his return he married Margaret Henderson, daughter of a Jedburgh tradesman, and worked at Smailholm, near Kelso. William, their first child, was born into their modest home at Coldstream in Roxburghshire, Scotland, on the 19 February 1789.[3] In 1799 Andrew moved his family to Moy, Ross-shire, to take charge of a farm with his brother. Six decades later Mrs Gaskell would write to Fairbairn, 'unless I had directed [my letter] to "Wm. Fairbairn, Esq., Railway Carriage," I don't know where I could have found you',[4] but for the journey in 1799

> my father bought an old horse, and having bent some canvas over one of the carts, he left Kelso in October with a delicate wife and a family of five children, the youngest only six weeks old, on a journey of upwards of two hundred miles through a wild country, at a most inclement season of the year ... with a broken down and restive horse, under rain and sleet ... My mother was so ill as to be lifted in and out of the cart every night and morning.[5]

Things turned out badly at Moy. Crops failed in 1800 and 1801, and Andrew moved to nearby Allangrange as a farm steward. Again there were difficulties and in 1803 he moved his family back to Kelso, by way of 'a tedious voyage' by sailing-ship from Cromarty – its harbour built twenty years earlier by Smeaton – to Leith.[6] Andrew then sought work in Yorkshire but did not remain there, moving in 1803 to became steward of a farm belonging to the owners of Percy Main Colliery, near Newcastle.

William remembered his father as a man of industry and integrity and, perhaps surprisingly, a great reader, affectionate towards his wife, and concerned for the education of his children. Margaret (d.1820) bore at least six children and suffered poor health. She too was imbued with the work ethic. For twenty years she spun the wool and flax for the family's clothes, sent it to the weaver and then made their clothing from it. Both parents were of the Church of Scotland 'from which they never deviated' but the extent to which they attended the kirk is not recorded and Fairbairn thought his father tinged with scepticism. The 'respectability' and strong work ethic of their home can be followed throughout the lives of both William and his brother Peter.[7]

William's education was limited and interrupted. As a four- or five-year-old he attended a school, possibly at Smailholm.[8] From there he moved to the parish school at Kelso where he 'attained some proficiency in reading writing and arithmetic' under a teacher full of enthusiasm who loved his profession.[9] At Moy, William missed two years schooling, but demonstrated an early interest in motive power – a pointer to his future career – by building model wind and water mills, and a four-wheeled wagon to transport his younger brother.[10] Whilst at Allangrange he attended Munlochy School, where the teacher was a severe disciplinarian but otherwise 'an excellent teacher'. When the family moved back to Kelso, William was sent to his uncle, a teacher in Galashields, where he received short courses in book-keeping and land surveying. Enforced memorising led to

Fig.1.1: John Rennie's Kelso Bridge. Built 1803.

distaste for the Psalms. But after three months it was back to Kelso to get a job to help support the family.[11] Thereafter he educated himself. With his very modest background, William had no opportunity of grammar school, university or premium apprenticeship.

William's first two jobs were, to say the least, unfortunate. The first, when he was aged fourteen, was as a labourer on John Rennie's new Kelso Bridge.[12] This job lasted only a few days, for a stone fell on William's leg. He was laid up for three months, during which time his sister, Eliza, died aged two. Andrew was then in Yorkshire.

The family in Kelso was in a state of near destitution with doctors' and funeral bills. Fairbairn records the intensity of his mother's grief 'when she saw her eldest boy, almost a confirmed cripple, take the place of his father in the position of chief mourner'.[13] At the beginning of 1804 William joined his father at Percy Main Colliery and obtained his second job, delivering coal to pitmen's houses by horse and cart. He was just fifteen. His Scottish accent marked him out as different and he was tormented, such that several times he was 'on the point of abandoning the work'. The culmination was a major fight which he won, and the bullying ceased.[14] Here, too, was a pointer to the future where challenges and difficulties would be met and overcome.

Apprentice Millwright

ON 24 March 1804 William Fairbairn was indentured for seven years as apprentice to John Robinson, millwright, who had charge of the machinery and engines at Percy Main Colliery. It is likely that Andrew had been able to use his position to obtain this apprenticeship for his son. Obtaining apprenticeships was not easy as millwrights sought to reserve them for their sons – something William was to experience later in his career. Some millwright craft apprenticeships required the payment of a premium. Evidence from London from the mid-18th century refers to premiums of £5 to £10 for apprentice millwrights.[15] Burgess fails to differentiate between craft apprentices and premium apprentices (or pupils), referring to 'indentures with premiums of £200' – way beyond the resources of the craftsman millwright.[16] A detailed study of the Stephenson locomotive works in the north-east, which discusses craft apprentices, does not refer to any

Fig.1.2: Percy Main Colliery c.1840.

premiums.[17] If his father had made a payment, it is probable that William would have referred to it, as he did to the library subscription which his father paid,[18] but there is no such reference. It is most unlikely that any premium was required.

For his first three years at Percy Main, William was based in the colliery workshop. The work would have involved making, installing, repairing and maintaining the colliery's plant and machinery. For his last four years there, he was in charge of the steam engine and pumps. This opportunity of working with an early steam engine was important, giving him an intimate understanding of steam power, which would be a major feature of his subsequent career. He welcomed the independence that this responsibility gave, although the work could be severe. He records spending hours suspended on a rope in the 900ft shaft, with water descending, and 'every limb numb with cold' – conditions so severe as to 'injure the health of one and destroy the life of another' of his colleagues.[19] The advantage was that, between breakdowns, there was time to pursue his formidable reading programme as well as trying to build a timepiece and an orrery.[20] It was during this time that he struck up a friendship with the then unknown George Stephenson who had charge of an engine at nearby Willington.[21]

William described his wage – five shillings a week, increasing to twelve – as 'good', but he took on extra work making wedges and blocking for a new pit, by which he sometimes doubled his income, enabling him to assist his parents who were struggling with 'a very limited income', and school fees for the younger children.[22]

A feature of William's life at Percy Main, untypical of craft apprentices, was his appetite for self-education through reading. Armed with the ticket, bought by his father, William visited the North Shields subscription library regularly. He read Gibbon's *Decline and Fall of the Roman Empire*; Hume's *History of England*; Robertson's *History of Scotland*; *America*; *Charles the Fifth of Spain*; and *Mary Queen of Scots*. Hume and Robertson were leading authors within the 'Scottish Enlightenment', and Herman describes Gibbon's *Decline and Fall* as 'the great masterpiece of Enlightenment history ... modelled on the Scottish and Edinburgh historical school'.[23] Poets included Milton's *Paradise Lost*, Shakespeare, Cowper, Goldsmith, Burns and Kirke White.

In later years Fairbairn counted William Prescott, Washington Irving, George Eliot, Dinah Craik and Mrs Gaskell among his favourite authors.[24] Critiques of Mrs Gaskell's novels flowed from his pen. 'Your kind and racy critiques both give me pleasure and do me good', she wrote to him on receipt of his comments on *North and South*.[25] He urged the young to 'study the lives of eminent and distinguished men',[26] and recalled that at the difficult time when he commenced business on his own account 'the examples of Franklin, Ferguson, and Watt were always before me'.[27] Again these were men with the values of the 'Enlightenment' – and Ferguson and Watt were Scottish. During his time at Percy Main, despite ridicule, Fairbairn maintained his study programme, frequented the library and attended a place of worship – all aspects of a 'respectable' lifestyle.[28]

In March 1811, his seven-year apprenticeship completed, William Fairbairn was a fully-fledged millwright, and proud to be so.[29] He had acquired skills in working with iron and an understanding of steam power and pumps. Whilst Robinson, to whom he had been indentured was described as a millwright, and Fairbairn described himself as a millwright, he had to all intents and purposes been an apprentice colliery engineer.

Fig.1.3: Inspecting pumps.

Paternalism and the Work Ethic

WORKSHOP culture was one feature of his apprenticeship at Percy Main that made a lasting impression on Fairbairn. Although the Methodists influenced many pitmen, the language of the leading men in the colliery, Fairbairn wrote forty years later, would no longer be tolerated in any society, not even that 'of bargemen or navvies'.[30] In the workshop it was the custom for the men to club together for a keg of ale, resulting in a 'state of intoxication'. He was thankful 'at having escaped the contagion of those irregularities' which had ruined some of his contemporaries.[31] In Manchester he found alcohol abuse endemic. He was not a teetotaller but was not prepared to pay employees to spend time drinking, nor accept the poor work of inebriated men.

> I strictly prohibit on my works the use of beer or fermented liquors of any sort, or of tobacco. I enforce the prohibition of fermented liquors so strongly that, if I found any man transgressing the rule in that respect, I would instantly discharge him without allowing him time to put on his coat ... I wish to have an orderly set of workmen; and ... am decidedly of opinion that it is better for the men themselves and for their families.[32]

This was self-serving, yet beneficent, moral paternalism.[33] Men who showed a 'looseness of character' were, he believed, unlikely to be reliable and conscientious and thus were not the sort of employees he wanted. External indicators were a guide:

> I chiefly judge of their circumstances from seeing them with their wives and families, and their well-dressed and respectable conditions on the Sundays ... if I see any workman in a dirty condition and in his working-clothes in the streets on the Sunday, I do not, perhaps, speak to him then, but on the Monday I tell him I have been looking over the books, that I find that he has had as good wages as other men who dress respectably, and that I do not like to have any one about me who will not dress well on the Sunday. This intimation has generally had the desired effect.[34]

In large towns where high employee turnover was the norm, paternalism was markedly weaker than in small communities where the employer often provided both homes and school, and sometimes influenced the political opinion of employees.[35] It is not widely appreciated that Fairbairn was influential in the shift towards respectability in workshop culture during the first half of the 19th century. The concept of respectability goes back to the earliest days of trade societies,[36] and its values entered the working class 'via the skilled élite'.[37] It was codified by Smiles.[38]

Fairbairn was imbued with the work ethic, not least perhaps through Hugh Blair's *Sermons* with their amalgam of Presbyterianism and the 'Enlightenment' – 'Blair and I are old friends, and I have treasured up his benevolent and homely maxims from early life until they have become almost a part of my existence'[39] :

> Industry is not only the instrument of improvement, but the foundation of pleasure ... He who is a stranger to industry, may possess, but he cannot enjoy. For it is labour only which gives the relish to pleasure.
>
> Think not, that any affluence of fortune, or any elevation of rank, exempt you from the duties of application and industry. Industry is the law of our being; it is the demand of Nature, of Reason, and of God.[40]

Fairbairn diffused his mentor's ethic, advising the members of Bolton Mechanics' Institute that, 'As labour appears to be the lot of all created beings, and as it is essential to health and happiness, we ought to hail it as of inestimable value'.[41]

The Journeyman Years

FAIRBAIRN'S early years as a journeyman show the uncertainty of employment of the traditional itinerant journeyman millwright, moving from job to job. Even with the growth of larger millwrighting firms, it was common practice for a millwright to move on following his apprenticeship, sometimes because there was no opening in the firm in which he had trained, more often to gain experience by working for short periods of time with leading firms.[42] The *London Review* in 1862 noted, with reference to Fairbairn, 'There was at that time a good old custom, less in vogue now, among the superior class of mechanics, of making, once in their lives, a tour of the country in search of new ideas and varied experience in their business'.[43] In any event he 'left reluctantly' what had become 'the scene of my trials and many friendships'. His first engagement was erecting a sawmill at Newcastle. Six months at Bedlington followed, probably repairing waterwheels at Bedlington Iron Works. It was a 'most agreeable time', where Fairbairn met his future wife. 'Respectable' pursuits continued: he subscribed to a circulating library, became leader of a Discussion Society and 'patronised the players'. After that, with no further work to be found, Fairbairn with another journeyman, David Hogg, also from Coldstream, boarded a collier bound for London on 11 December 1811. His belongings comprised 'half-a-dozen new shirts, a new suit of clothes, a watch and four pounds in his pocket', and presumably some tools of his trade. It was a hair-raising twelve-day voyage.[44]

Fig.1.4: Cheshunt Windmill, 1811, on which Fairbairn worked as a journeyman.

In London, Fairbairn and Hogg obtained an interview with John Rennie. He offered them employment, referring them to a foreman, who said they would have to be cleared by the Millwrights' Society, whose next meeting was a month away. Starving in a freezing garret, 'we spent one of the most uncomfortable months of our existence'. Attending on the Millwrights Society they were refused permission to work for Rennie. Hearing there was work at Hertford, they walked there in foul weather, but there was none. They struggled on to Cheshunt where they got a fortnight's work on a new windmill.

Returning to London, they determined that, if still unsuccessful, they would work their passage to America. However Hogg met a schoolfellow from Coldstream who explained there were three millwrights' societies in London. He was secretary of the Independent Millwrights, 'a Society' he explained, 'greatly superior to the vagabonds at Little Eastcheap'. Presumably Fairbairn and Hogg joined his Society, for they obtained eighteen months' employment building Grundy's rope-works at Shadwell. Then Fairbairn moved to work on Stratford Mill for three months for John Penn, who had established his famous works at Greenwich in 1799.[45]

Once established in London, where he was paid between two and three pounds a week, there was opportunity for 'respectable' social pursuits. He subscribed to a library, went to theatres, and visited the Westminster Forum, a debating society (where he heard the radicals John Cartwright and Gale Jones). He had Sunday lunches with a cousin of his mother and her joiner husband, who also hosted an eccentric Scottish clergyman, Revd Mr Hall, who wrote for Tilloch's *Philosophical Magazine*. Hall introduced Fairbairn to Alexander Tilloch who had founded *The Philosophical Magazine*, one of the earliest scientific journals, in 1798, and to the Society of Arts. This, Fairbairn wrote, was more than he could have expected 'as a common workman'. Encouraged by Hall, who was to meet half the cost, Fairbairn lost 'upwards of £20' – his entire savings – in making a model of a digging machine which Hall arranged to be exhibited at the Society of Arts, believing it would make their fortunes. It was not well received, and Hall turned out to be unable to pay his share of the costs. This pattern of rising to an engineering challenge, with meagre concern for the financial implications, would be repeated in later life. Undeterred Fairbairn took on the first commission obtained on his own account – the challenge of building a sausage machine. This was successful and he was paid thirty-three pounds.[46]

Finding no more work in London and with seven pounds in his pocket, Fairbairn moved to Bath, where he obtained six weeks' work. In Bath he 'entered into all the reminiscences' of the days of Beau Nash, went to the opera, and visited the Pump Room, hiring a sedan chair to ensure entry. He felt 'quite at home', having read the whole of Smollett's works and Fielding's *Tom Jones*, suggesting his 'respectability' may have been spiced with the risqué.[47] He went sightseeing in Bristol, moved on to Newport and Cardiff, visited Llandaff, and took a sloop to Dublin. Arriving penniless, he obtained four months' work at Phoenix Foundry, constructing a nail-making machine. He lodged with a young man 'who had received a liberal education', and with his assistance endeavoured to improve himself by reading and discussing 'some of the best authors'. Then, via Liverpool, he travelled to Manchester, finding employment with Adam Parkinson, erecting calico-printing machinery.[48]

Fairbairn would not have recognised Carlyle's description of Manchester as a place of loneliness, isolation and anomie.[49] A short time after arriving he caught scarlet fever and was laid up for nearly three months. During this time he was frequently visited by James Houston, his foreman at work, who remained a friend for the rest of Houston's life. At Houston's home, Fairbairn attended a weekly meeting which engaged in literary discussions followed by supper and singing.[50] Among those attending were Dr Hardie, a Scottish physician at the Infirmary, and Leo Schuster, two years younger than Fairbairn, who had recently arrived from Germany, was Jewish, became a Unitarian, and was to have a very successful commercial career.[51] This all resonates with Hewitt's identification of respectability as 'a powerful structuring distinction that cut across class'.[52] It also says much about Fairbairn and the type of company he was drawn towards – on this occasion his foreman, a doctor, and a very able German immigrant.

After two years with Parkinson, Fairbairn moved to Thomas Hewes, initially working on unidentified waterwheels at Macclesfield, and in 1817 moving into Hewes' drawing office.[53] Hewes had come to Manchester from Kent in 1792.[54] By the early 1820s his engineering works was amongst the biggest in the country employing 40 men on heavy millwork and about 110 on machine-making.[55] He built mills, waterwheels, gearing and shafting, and spinning machinery for mills throughout the British Isles. He probably used wrought-iron for shafting before Fairbairn, to whom its introduction was often credited.[56] What Fairbairn introduced was more slender shafting, revolving faster and more smoothly, leading to considerable savings in power.[57] Hewes told an 1824 Select Committee that his best men were deployed on site.[58] This is consistent with one of his time-sheets relating to a Macclesfield waterwheel, listing eight journeymen, six of whom – and Fairbairn was one of them – had by 1845 become 'heads of important mechanical concerns in the neighbourhood of Manchester'.[59]

Fairbairn's time with Hewes was formative. He acknowledged Hewes' introduction of 'an entirely new system in the construction of waterwheels', in which wheel and axle were held together with light wrought-iron rods – the so-called 'suspension wheel'. Hewes was the leading builder of waterwheels in the first two decades of the 19th century, and Fairbairn was engaged in the erection of such wheels. It was a form of waterwheel construction that Fairbairn would use and develop during the 1820s, constructing some of the most powerful industrial waterwheels ever built.[60] Working in Hewes' drawing office was even more important. Here Fairbairn was within the hub of Hewes' firm, able to observe the design and detailing of waterwheels, millwork and textile machinery, and, probably more importantly, to witness at first hand the organisation and working methods of one of the largest and most successful millwrights. It was valuable experience for the next step. This came very suddenly towards the end of 1817, when tension arose between Hewes and Fairbairn about a competition, which both men had entered, for a new Blackfriars Bridge over the Irwell.[61] Neither man won the competition, but for Fairbairn, now twenty-eight, the time had come to launch out. He handed in his notice.[62]

Manchester

WHAT was it that drew Fairbairn to Manchester and then caused him to settle there? Unlike Nasmyth twenty years later, when Fairbairn came to Manchester in 1813 it was not specifically to set up a business. He came looking for employment as a journeyman.[63] Arriving by coach from Liverpool, seventeen years before Stephenson's railway, and before Manchester's streets were paved or policed, Fairbairn instinctively appreciated that wide opportunities were open for him.[64] He could not have found a more exhilarating place to pursue his calling. Until the 1820s, engineering was largely centred in London – Bramah, Maudsley, Rennie, Penn, with Boulton & Watt in Birmingham as an exception. But the hub of engineering was moving north, drawn by the opportunities provided by the burgeoning cotton industry, bringing to Manchester in the second quarter of the 19th century 'a higher profile than any other [city] in the world'.[65] By the start of the century there were numerous engineers and ironfounders, notable among them Bateman & Sherratt who had been building steam-engines since the 1780s, Thomas Hewes, Peel Williams & Co., and Galloway, Bowman & Glasgow.[66] Scots were prominent amongst the early engineers drawn to Manchester.[67] Later immigration included six outstanding engineers who had been employed at the famous Maudslay 'nursery' in London, and several outstanding engineers from abroad.[68]

Up to the 1960s historians centred Manchester history almost exclusively on cotton manufacture and an impression of social and geographical cleavage stemming from Cooke Taylor and Engels.[69] Since then the importance of the cotton factory has been balanced by the warehouses of the mercantile sector, and attention given to Manchester's metal, chemical and finishing trades.[70] By 1856 Manchester had 120 engineering works compared with 123 textile mills.[71] It had become not only a centre of textile machine-making but also of locomotive building and machine-tool manufacture, and a terminus of the first main-line railway – with Fairbairn's Water Street Bridge.

Manchester Infirmary had been established back in 1752, but it was in the run up to the mid-19th century and beyond that Manchester became a city of transformation, as environmental problems, highlighted by Kay and Engels,[72] were tackled – streets were paved, and J.F. Bateman, son-in-law of Fairbairn, engineered Manchester's water supply.[73] It was a city at the centre of social movements: indeed Fairbairn was an eye-witness of the Peterloo tragedy in 1819.[74] There were immense Chartist meetings at Kersal Moor in 1838–9.[75] In the alcove of the reading room at Chetham's Library, Engels and Marx worked on the *Communist Manifesto*.[76] Manchester was the city of the Anti-Corn-Law League, and of free trade.[77] Nor was Manchester without culture, as the Art Treasures Exhibition, driven by William's son, Thomas, showed. Rubinstein concluded that 'entrepreneurial *effort* is not as important as *place in the total economy* in determining entrepreneurial rewards'.[78] Fairbairn's instinct was right. He had arrived in the right place at the right time, with the right skills.

Restrictive Practices

FAIRBAIRN'S initial experiences in London influenced him throughout his career. He never forgot the cold and hunger he suffered as a result of the actions of the London Millwrights' Society. Nearly fifty years later he told young men in Derby about it, 'Laws of a most arbitrary character were enforced ... by cliques of self-appointed officers, who never failed to take care of their own interests'. He looked back further to the exclusive restrictions of the trade guilds which he saw as having retarded British mechanical progress. Their removal in 1815 'left open to all classes *a fair field and no favour* in the race of national progress'.[79] But change did not occur overnight. The years 1815–1852 witnessed a struggle in the engineering industry as millwrights sought to retain their traditional control of many of the workshops, including Fairbairn's, with reference to standard wages, employment restricted to their members, and control over who could be apprenticed. They faced increasing pressure from machine tools which worked faster, more accurately, and at a fraction of the cost of manual work – and the new machines could be operated 'by boys and labourers'. In 1824 Hewes complained about the difficulty in engaging sufficient skilled men because those 'engaged in that sort of work will not allow other workmen to come into the trade.' He and others had 'completely given up on this point'.[80]

By the time Fairbairn had been in partnership with James Lillie for around six years, and when they were employing 60–70 men, their millwrights demanded the dismissal of John Brown, an apprentice, because his father was not a millwright. Fairbairn & Lillie refused, and a strike ensued.[81] The *Manchester Guardian* supported the employers: to prevent a young man becoming a mill-wright because his father was a weaver or a joiner was anything but just and reasonable. The newspaper also carried an advertisement from Fairbairn & Lillie,

'Millwrights Wanted – From forty to fifty good workmen will meet with liberal encouragement and constant employ ... Persons unconnected with the Manchester Millwrights' Club will be preferred'.[82] There was a similar situation in 1837 when about fifty boilermakers struck because of the apprentice issue. It featured in Fairbairn's last address, a few months before his death:

> Some 40 years ago ... I had large orders on hand; and being unwilling to allow the men to dictate the terms on which I should engage apprentices and conduct the work, I received notice of a turn-out, which immediately took place, and the works were suspended for a number of weeks.
> In the dilemma, with impatient customers, I was driven to the necessity of supplying the place of the riveters, and resuming the works by a passive and unerring workman, which, from that day to this, has never complained, and did as much work in one day as was formerly accomplished by 12 of our best riveters and assistants in the same time.[83]

This illustrates Marx's reference to inventions made to 'supply capital with weapons against the revolts of the working-class' – his footnote specifically refers to Fairbairn.[84] Within twelve months machine-made boilers were the boilers of first choice.[85]

The Millwrights' Society admitted only time-served millwrights. In 1826 the Friendly Society of Mechanics was formed in Manchester, also including fitters and turners. In 1838 it became the Journeymen Steam Engine and Machine Makers' Society and in 1851 part of the Amalgamated Society of Engineers,[86] as did the Millwrights. From 1844 there was competition from the United Machine Workers' Association which was in constant conflict with the Millwrights' Society and its successors, including the Amalgamated Society of Engineers, which the Machine Workers did not join until seventy years later in 1920. It seems that Fairbairn may have given support to the Machine Workers as, at the opening of the new Manchester Town Hall, its members marched behind their banner with portraits of Fairbairn and Whitworth above the caption, 'The men whom we delight to honour'.[87] Without further evidence it is not possible to say what part the United Machine Workers' Association played in the 1851–2 engineering dispute.[88] Strikes in the engineering industry over the apprentice and 'illegal men' issues, were endemic for over three decades, coming to a head with the1851–2 dispute, which forms part of Chapter 4. While Fairbairn had a concern for the well-being of operative millwrights, amongst whom he had lived and worked, his experiences in London left him bitterly opposed to any restrictive practices. It was a position from which he never deviated.

Domestic Stability

FAIRBAIRN'S Ancoats works and his home, always within walking distance of each other, were the two hubs around which his life and career revolved. His marriage provided a stable domestic base for fifty-six years, a base that played a large part, not just in his family life, but in entertaining many engineers, scientists, and those involved in education – friends, clients and colleagues, many of whom contributed to Fairbairn's ascent to fame.

Fairbairn married in June 1816 during his journeyman years. He had no thoughts of gentrified living. His dream of domestic bliss was a small house, with 'a neat parlour, every corner filled with books'.[89] Having corresponded for five years, and having amassed the largest sum he had ever possessed – nearly thirty pounds –William took a coach to Bedlington, married Dorothy Mar, and brought her back to Manchester. Initially they found lodgings in Macclesfield where he was engaged in erecting Hewes' waterwheels. With the advancement to Hewes' drawing office, the couple moved to a cottage at the poorer end of Ancoats, where their first child, Anne, was born. It was an inauspicious start, with Dorothy ill and curtains catching fire. Doctor's fees and the fire-damage took all their money. When Fairbairn left Hewes, things got worse. They were forced to borrow five pounds.[90] Woodcroft wrote:

> The writer has heard, in years now long gone by, from Mrs. Fairbairn's own lips, how her own and her husband's food consisted largely of 'water porridge', and the noble simplicity with which she at an after period narrated the facts of this time to the writer's mother, made an impression upon him – then a boy – which he remembers, and has ever viewed as one of the lessons of his life.[91]

Their subsequent moves, from Marshall Street (1817–23), to Great Ancoats Street (1824–7), to Ancoats Crescent (1828–34), to Medlock Bank (1835–9), and finally to the Polygon (1840–74), demonstrate a classic 'Ancoats to Ardwick' story. Howe has shown that for the Lancashire cotton master, his factory, his relationship with his operatives, and his contribution to the economy, were all-important, such that 'suburban mansions, within easy access of the mill, were preferred to country retreats'.[92] This was also true for William Fairbairn. These moves illustrate, in this case at least, that Cooke Taylor was wrong when he said that 'Ardwick knows less about Ancoats than it does about China', suggesting that employers living in fashionable Ardwick knew nothing of the conditions in which their employees lived in Ancoats.[93] Having moved to the Polygon, William and Dorothy lived there for their remaining thirty-four years together. It was a good address. Next to them was the Rector of Ardwick. The Kennedys were nearby at Ardwick Hall and the George Murrays at Ancoats Hall.[94]

There were eight children from the Fairbairn marriage, Anne (1817–94), John (1821–67), Thomas (1823–91), William Andrew (1824–10), Margaret (1826), Peter (1828–59), George (1830–68) and Adam (1836–88). The older boys had some tutoring from Eaton Hodgkinson,[95] and attended Karl Voelker's school at Everton, where the roll – Ashton, Heywood, Fielden, Rathbone, Worthington – reflected leading north-west Unitarian families. Voelker, a German refugee, was a Unitarian and ran the school on Pestalozzian lines.[96] By the time the Fairbairns' youngest son was of school age, Voelker's school had closed,[97] Adam was sent to Rugby,[98] a school favoured by the cotton masters.[99] Its famed headmaster Dr Arnold had died a few years earlier but his influence continued: Adam went on to Trinity, and became an Anglican clergyman[100] (as did Fairbairn's grandson, William Bateman – via Harrow and Trinity – whom his grandfather financed on the 'grand tour').[101] The influence of Cross Street and Unitarianism had waned by the mid-century.

While there were children still at home, there were family holidays in

Southport.[102] Subsequently William and Dorothy spent their holidays at the homes of friends, or on short trips abroad, or revisiting the Scotland of William's youth. He had no taste for field sports, but enjoyed walking.[103] In later years he and Dorothy spent many holidays with their daughter, Anne, and her husband, the water engineer, J.F. Bateman, Fairbairn's protégé and friend.[104]

The census returns suggest a comfortable standard of living at the Polygon – in 1851 there was a cook plus five other living-in servants; the 1861 census recorded six living-in staff – butler, cook, lady's maid, housemaid and two kitchen maids; and the 1871 return, a staff of five.[105] Little is recorded about Dorothy Fairbairn, but with the eight children, at least thirty-five grandchildren,[106] the home to manage, and her husband's incessant flow of influential visitors, she cannot have been short of things to do. All the indications are that she was the ideal complement to her ever-active husband.

At Medlock Bank in the mid-1830s, Fairbairn convened discussion evenings, attended by Eaton Hodgkinson, Bennet Woodcroft, James Nasmyth and John Elliot. Their intended quarterly, *The Workshop,* never got off the ground, but the meetings did generate Fairbairn's *Observations on Improvements of the Town of Manchester,* with two plates by Nasmyth.[107] This envisaged a new Exchange and Quadrant at the junction of Market Street and Piccadilly; plus statues of the Duke of Bridgewater, whose coal-carrying canal from Worsley to Manchester in 1761 heralded the canal era, Fairbairn's hero James Watt, Sir Richard Arkwright, and Hygæia, goddess of health, opposite the Infirmary in Piccadilly. The publication also included a paper by Nasmyth, 'The Art of Design as applied to Steam Engine Chimneys'.[108] Fairbairn described John Elliot as 'exceedingly intelligent' and 'foreman of the millwrights'.[109] This example of meetings with employer and foreman is indicative of peer approval amongst engineers being no respecter of social position. It reflects Fairbairn's meetings at James Houston's, where foreman and journeyman had met.[110] When Shaftesbury stayed at the Polygon, Thomas Wright, the prison visitor and a foreman with engineers, Richard Ormerod & Sons, joined them for dinner.[111] As Asa Briggs found, rather than middle and working class arrayed in hostility, 'the town springs to life when the interplay of the groups and the relations within them are studied in something of their intricate detail'.[112]

An insight into the high standing of the social circle into which the Fairbairns moved can be gleaned by reference to an 1852 letter of Mrs Gaskell where she refers to the Fairbairns together with the Rathbones, Caton & Norcliffe Gregs, Murrays, Mellys, Worthingtons, Potters '(the fat Sir John)', Wm Woods, Philips, Horners and Schwabes. Apart from the visiting Horners, these were some of the North-west's leading Unitarian families, of considerable wealth and influence. However, she added that a dinner at the Polygon was 'rather flat because there were too many of the Fairbairn family'.[113] Guests at the Polygon from further afield included Robert Stephenson, Sir Marc Brunel, Baron von Bunsen, Sir David Brewster, Wm Hopkins, Lord Wrottesley, Vernon Harcourt, T. Romney Robinson, Lord Rosse, Sir Edward Sabine, Earl Granville, Lord Brougham, and many others.[114]

Most families endure times of sorrow and some surviving letters from Fairbairn to his friend T. Romney Robinson of Armagh Observatory show the

Fairbairns to have been no exception. These letters provide an insight into Fairbairn's humanity. In 1859 he wrote about the family's great affliction on the death of his daughter-in-law, William Andrew's wife, leaving four young children; and of his fears about his son Peter's 'mental as well as physical health'. Peter died later in the year. In 1861 the death of Fairbairn's brother Peter caused much sorrow – 'he was my only brother, to whom I was particularly attached'.[115] In 1867 his son John returned home with an illness from which he never recovered – 'his gentle disposition and great goodness of heart endeared him to his mother and myself beyond that of any other member of the family. He only returned from India a few months before his death'.[116]

Fairbairn's home was the secure domestic base from which, when in Manchester, he set out like clockwork every morning to walk to the works at Canal Street. It was where he had his library, in which, when the family had retired to bed, he wrote his prolific letters, papers and books, 'beneath a picture of von Humboldt in his *arbeitszimmer*'.[117]

Conclusion

WILLIAM Fairbairn inherited the 'respectable' and industrious qualities of his parents. His schooling was only basic – and two years of it were missed. However, his father was able to arrange an apprenticeship for William at Percy Main and after that his experience was widened as a journeyman. From early days Fairbairn showed an innate fascination with machines, and as a journeyman there were indications that in later life engineering achievement would assume priority over financial prudence.

Fairbairn's thirst for knowledge, stimulated partly by a library ticket, and drawing ridicule from his shop-mates, was unusual for an apprentice millwright. It continued throughout his career, finding expression in his engineering experiments and in a lifelong commitment to education, particularly of young engineers.

From his apprenticeship and journeyman years Fairbairn obtained wide experience – wider than traditional millwrighting. Apart from a lack of theoretical and mathematical knowledge, and not having worked on a cotton mill, it provided a useful grounding for his future career. Two disagreeable experiences from these years had lasting effects. It is clear William had little respect for John Robinson to whom he was indentured, and the ethos of the colliery workshop was unpalatable, influencing the discipline in his own workshops in later years. Initial harsh experiences as a journeyman in London provoked life-long opposition to restrictive practices. Whilst William faced some trials at Percy Main there were also good friendships, and the apprenticeship system provided secure employment. As a journeyman, in contrast, employment was unpredictable and uncertain, although overall the journeyman years were enjoyable, especially in London.

Fairbairn seemed to be drawn to people who would become, or who already were, influential. He would become a consummate networker. His early friendship with the then unknown George Stephenson had unforeseen consequences. Bringing his millwrighting skills to Manchester, at the same time that the centre of engineering was moving there from London, was opportune, not least because it was also the time when steam power and machine tools were on the cusp of rapid development. That he obtained employment with the leading millwright of the day was significant, particularly so the move into Hewes' drawing office giving Fairbairn valuable knowledge and insights into the running of a leading millwrighting business.

Martin Hewitt has shown that in 19th-century Britain 'respectability' cut across class, its distinctions no respecters of occupation, income or residence, but emphasising character, self-reliance and individual improvement.[118] Fairbairn, during his apprenticeship and journeyman years, provides a useful illustration of this. He was a workman, never able to save much money, on some occasions having none, and sometimes no work. Yet he made extensive use of libraries, and partook of various 'respectable' activities, untypical of journeyman millwrights. One of Fairbairn's pupils wrote, 'his manners might have put to shame many who had been born in a far higher rank of life'.[119]

In youth Fairbairn showed a determination and resoluteness which were to feature in later years. Trials were overcome and self-education pursued. It was characteristic of Fairbairn to make an impulsive decision to leave Hewes when tensions arose, notwithstanding poor trading conditions at that juncture, and that by then he had a wife and young child. When he left Hewes it was not with thoughts of fame or wealth, but with confidence in his own abilities, resolute that he would no longer be the servant of another. He would 'endeavour to be useful' and was 'determined to excel'.[120]

Notes for Chapter 1

1. Queen Victoria (1857), Prince Albert (several occasions), William IV of Prussia (1849), Louise of the Netherlands who became Queen of Sweden in 1850 (1849), Emperor Nicholas I of Russia (1850), Princess Charlotte, Empress of Russia (1850), Grand Duke Alexander of Russia (1850), Emperor Louis Napoleon of France (1854).
2. *Fortunes Made in Business*, pp.252-79; G Cookson, 'Fairbairn, Sir Peter (1799-1861)', *Oxford DNB on-line* (2004), (accessed 9 January 2014).
3. R A Smith, *A Century of Science in Manchester,* (1883), p.257, 'His life by Dr Pole says he was born in Kelso, but the present writer had Coldstream from Mr Fairbairn himself'; *Life,* p.53.
4. *Life,* p.462.
5. *Life,* p.59.
6. *Life,* pp.59-64; A W Skempton, 'Harbours' in A W Skempton (ed.), *John Smeaton, FRS,* (1981), pp.204-6.
7. *Life,* pp.54-70, 192. The children were William (1789-1874), two girls – names unknown, Thomas (drowned at sea 1812), Peter (1799-1861), Eliza (1801-1803).
8. *Fortunes Made in Business*, p.237.
9. *Life,* pp.54-5.
10. *Life,* pp.60-1.
11. *Life,* pp.62-4.
12. For Fairbairn on Rennie, see Fairbairn, *UIfE2,* pp.215-6.
13. *Life,* p.65.
14. *Life,* pp.69-71. Tristram Hunt's description of Fairbairn as a 'pugilistic pit lad' gives a completely inaccurate impression. (T Hunt, *Building Jerusalem. The Rise and Fall of the Victorian City* (2004), p.73.)
15. R Campbell, *A General Description of All Trades,* (1747), quoted in J Tann, *The Development of the Factory,* (1970), p.95; There are records of such premiums being paid to London millwrights John and Thomas Cooper in the years 1760-80 (J G Moher, 'The London Millwrights and Engineers 1775-1825', (University of London PhD Thesis, 1988), pp.253, 75n2, 116, referring to Records of the Worshipful Company of Turners).
16. K Burgess, *The Origins of British Industrial Relations,* (1975), p.7.
17. M R Bailey, 'Decision-making Processes in the Manufacturing Sector: The Independent Locomotive Industry in the 19th Century', (University of York DPhil, 1999), pp.265-6. A later report (1909) found only nine engineering firms out of eighty-five requiring premiums – 'small sums of five pounds or so' (C Jackson, *Report on Boy Labour in London and certain other Towns,* Parliamentary Papers, 1909, XLIV, pp.1078-82, cited by C More, *Skill and the English Working Class, 1870-1914,* (1980), p.69).
18. *Life,* p.73.
19. *Life,* p.78.

20. *Life*, pp.73, 75. Monday: Arithmetic, Mensuration. Tuesday: Reading History and Poetry. Wednesday: Recreation, Reading Novels and Romances. Thursday: Mathematics. Friday: Eculid, Trigonometry. Saturday: Recreation and Sundries. Sunday: Church, Milton, and Recreation.
21. S Smiles, *Lives of the Engineers with an Account of their Principal Works*, (1862), Vol.3, pp.41-2.
22. *Life*, p.72.
23. A Herman, *The Scottish Enlightenment: The Scots' Invention of the Modern World*, (2001), pp.99. 215.
24. *Life*, pp.468-9.
25. *Life*, pp.460-1.
26. Fairbairn, *UIfE*, pp.64-5, at the opening of the Southport Athenaeum.
27. *Life*, p.197. Benjamin Franklin (1706-1790), American , astronomer, from humble background. James Watt, FRS, FRSE, (1736-1819), steam-engine builder.
28. *Life*, pp.73-4; Fairbairn, *UIfE3*, p.64. The denomination of the 'place of worship' is not known.
29. *Life*, pp.72, 78. .
30. *Life*, p.70. The roots of the cultural shift during these years are compellingly set out in H Schlossberg, *The Silent Revolution and the Making of Victorian England*, (2000).
31. *Life*, p.79. In 1817 it was said that the collieries neighbouring Newcastle were 'enwrapt in profound and opake [sic] moral darkness' (Newcastle Religious Tract Society, quoted in R Colls, *The Pitmen of the Northern Coalfield: Work, Culture and Protest*, (1987), pp.1-2).
32. M W Flinn, (ed.), *Report on the Sanitary Condition of the Labouring Population of Great Britain by Edwin Chadwick 1842*, (1965), p.314.
33. S Pollard, 'Factory Discipline in the Industrial Revolution', *Economic History Review*, NS16.2, 1963, 269-70. For a recent discussion from a different discipline see A Dawson and E Garrard, 'In defence of moral imperialism: four equal and universal prima facie principles', *Journal of Medical Ethics*, 2006, 32.4, 200-4.
34. Flinn, *Report*, p.323.
35. P Joyce, 'The Factory Politics of Lancashire in the Later Nineteenth Century', *The Historical Journal*, 18.3, 1975, 525-53.
36. A E Musson, *British Trade Unions, 1800-1875*, (1972), p.18.
37. G Crossick, 'The Labour Aristocracy and its Values: A Study of Mid-Victorian Kentish London', *Victorian Studies*, 19.3, 1976, 314.
38. M Hewitt, 'Introduction: Victorian Milestones, in M Hewitt (ed.), *The Victorian World*, (2012), p.18. 'The poor man ... may be honest, truthful, upright, polite, temperate, courageous, self-respecting, and self-helping - that is, be a true gentleman' (S Smiles, *Self-help, with illustrations of Conduct and Perseverance*, (new ed. 1884), pp.399-400).
39. *Life*, p.458.
40. H Blair, *Sermons*, Vol.I (1820 ed.), pp.266-8.
41. Fairbairn, *UIfE3*, pp.50, 52.
42. In due course ambitious young millwrights would gain experience with Fairbairn - See Chapter 6.
43. Quoted in *MG*, 6 January 1862.
44. *Life*, pp.83-7; *MPICE*, 32 Part 1, 1871, 203.
45. *Life*, pp.92-3. Moher is incorrect in saying that Fairbairn's indentures were 'fraudulent' (Moher, 'London Millwrights', p.8); R Hartree, *John Penn and Sons of Greenwich* (2008). I am grateful to Richard Hartree for the reference to Stratford Mill. J W Roe is incorrect in stating that William worked for George Rennie (who was two years younger than Fairbairn) (J W Roe, *English and American Tool Builders*, (1916), p.54).
46. *Life*, pp.93-4.
47. *Life*, p.99.
48. *Life*, pp.83-101.
49. T Carlyle, *Past and Present* ([1843], 1965 ed.), p.148. The Carlyles met Fairbairn when they were guests of Lord Ashburton at The Grange in the first week of 1856, and Jane wrote of him. 'There was too, an old Scotch engineer for two days, called Fairbairn, whose *broad Scotch* pleased me so much that I sat next to him at dinner both days' (J W Carlyle to Margaret Welsh 10 January 1856 in 'The Carlyle Letters on-line' at http://carlyleletters.dukejournals.org/cgi/content/full/31/1/lt-18560110-JWC-MW-01?... (accessed 18 November 2008). Carlyle and Fairbairn did not normally move in the same circles.
50. *Life*, pp.101-3.
51. W D Rubinstein, M Jolles and H L Rubinstein, *The Palgrave Dictionary of Anglo-Jewish History*, (2011), p.882; S Chapman, *Merchant Enterprise in Britain, From the Industrial Revolution to World War I*, (1992), p.146. It is tempting to link Fairbairn's supplying of locomotives to the London & Brighton Railway with Schuster's role as its chairman, but the dates suggest this was not the case.
52. M Hewitt, 'Introduction: Victorian Milestones' in M Hewitt (ed.) *The Victorian World* (2012), p.18.
53. On Hewes see S B Smith, 'Thomas Cheek Hewes (1768-1832) An Ingenious Engineer and Mechanic of Manchester', (MSc Thesis, University of Manchester, 1969).
54. *MG*, 11 February 1832.
55. 'Report of Select Committee on Artizans, Machinery and Combinations', *Parliamentary Papers*, (1824), V, p.340.
56. Ure, *Philosophy*, pp.34-5; R Buchanan, *Practical Essays on Mill Work and other Machinery*, ('revised into a third edition with additions by George Rennie, 1841), p.xix; E Leigh, *The Science of Modern Cotton Spinning*, (5th ed. 1882), Vol.2, p.273; See Chapter 4.
57. Ure, pp.34-7; W J M Rankine, 'William Fairbairn' in *The Imperial Dictionary of Universal Biography*, Vol.2 (1863).
58. 'Report of Select Committee on Artizans and Machinery', pp.340-3.
59. 'Mr. Alderman John Hopkinson, J.P.', *Manchester Faces and Places*, 4, 1893, 155. Thomas Hewes' successors were Hewes & Wren, then Wren & Bennett, and then Wren & Hopkinson.
60. Fairbairn, *M&MWI*, p.120; W Fairbairn, 'The Rise and Progress of Manufactures and Commerce, and of Civil and Mechanical Engineering in Lancashire and Cheshire' in T Baines, *Lancashire and Cheshire, Past and Present* (1869), Vol.2, pp.cxliv-cxlv.
61. No details of the entries are known except that neither was successful.
62. *Life*, pp.105-7. Fairbairn's journeyman years are summarised in tabular form at Appendix 1.1.
63. S Smiles (ed.), *James Nasmyth, Engineer: An Autobiography*, (1895 ed.), pp.176-80.
64. *Life*, p.101.
65. M Hewitt, *The Emergence of Stability in the Industrial City: Manchester, 1832-67* (1996), p.23.
66. Musson and Robinson, *Science and Technology*, pp.393-472.
67. D A Farnie and W O Henderson (eds), *Industry and Innovation: Selected Essays, W H Chaloner*, (1990), p.100.
68. See Chapter 2.
69. Hewitt, *Emergence of Stability*, pp.26-7.

70. R Lloyd-Jones and M J Lewis, *Manchester and the Age of the Factory: The Business Structure of Cottonopolis in the Industrial Revolution,* (1988); A E Musson and E Robinson, *Science and Technology in the Industrial Revolution,* (1969), p.455.
71. Hewitt, *Emergence of Stability,* p.33.
72. J P Kay, *The Moral and Physical Condition of the Working Classes employed in the cotton manufacture in Manchester,* (2nd ed. 1832); F Engels, *The Condition of the Working Class in England from Personal Observations and Authentic Sources,* (translated and edited by W O Henderson and W H Chaloner, 1958). For a criticism of Engels' *Condition of the Working Class,* see Hunt, *Building Jerusalem,* p.33, quoting G Stedman Jones, 'The First Industrial City? Engels Account of Manchester in 1844', (unpublished paper), 'The proletariat became the Irish: they were portrayed as a race of subhuman beings reduced to utter pauperism by the process of urban living and industrial capitalism ... the reason for this sleight of hand had little to do with the working class in Manchester and everything to do with Engel's developing creed of communism'.
73. H L Platt, *Shock Cities. The Environmental Transformation and Reform of Manchester and Chicago,* (2005), pp.201-7; P Russell, 'John Frederic La Trobe-Bateman (1810-1889) Water Engineer', *TNS,* 52, 1980-81, pp.119-38; T Quale, *Resevoirs in the Hills – the story of the construction of the reservoirs in the Longdendale Valley which feed Manchester with water,* (1988).
74. [Woodcroft] III.
75. R G Gammage, *History of the Chartist Movement 1837-1854,* (2nd ed. 1894), pp.59-63, 113.
76. R and E Frow, *Karl Marx in Manchester* (1985), p.7.
77. W D Grampp, *The Manchester School of Economics,* (1960).
78. W D Rubinstein, *Men of Property: The very wealthy in Britain since the Industrial Revolution,* (1981), p.163 [Rubinstein's italics].
79. Fairbairn, *UIfE2,* pp.211-4 [Fairbairn's italics].
80. 'Report from the Select Committee on Artizans, Machinery and Combinations', p.341.
81. 'Report from the Select Committee on Artizans, Machinery and Combinations', p.566; *MG,* 26 February, 1825.
82. *MG,*19 February, 1825.
83. W Fairbairn in *Manchester Scientific and Mechanical Society, Opening Meeting, October 1873,* (1873), p.6.
84. K Marx, *Capital: A Critique of Political Economy,* (1967 ed.), Vol.1, p.411; see also K Marx, *The Poverty of Philosophy,* (1910 ed.) p.161.
85. *Life,* p.164. See Chapter 3.
86. K Burgess, *The Origins of British Industrial Relations,* (1975), pp.11, 16.
87. *MG,* 17 September 1877.
88. 'A Catalogue of the Papers of The United Machine Workers' Association' at http://web.warwick.ac.uk/services /library/mrc/ead/259MW.htm (accessed 8 February 2013). Neither Burgess nor Murphy mention the UMWA in their papers on the 1851-2 engineering dispute (see Chapter 4). While Jefferys mentions the UMWA. he says nothing about it in relation to the 1851-2 dispute. (J B Jefferys, *The Story of the Engineers 1800-1945,* (1945), p.19).
89. *Life,* pp.83-5.
90. *Life,* pp.103-5.
91. [Woodcroft] III.
92. A Howe, *The Cotton Masters 1830-1860,* (1984), pp.252-3.
93. W Cooke Taylor, *Notes of a Tour in the Manufacturing Districts of Lancashire,* (2nd edition 1842), p.164.
94. J McConnel, *James McConnel (of Carsriggan): His Forbears & Descendants,* (1931), p.32.
95. *Proceedings of the Manchester Literary and Philosophical Society,* 17, 1877-8,145-6; D Cardwell, 'Hodgkinson, Eaton (1789-1861)', *Oxford DNB on-line* (accessed 10 January 2012).
96. There were about sixty boys. Most of the masters were from continental Europe. William Rathbone (1819-1902), head boy and later an MP, speaks of Voelker as generous, courageous, enthusiastic, and 'his moral influence over the boys did much to make them the useful men they were in after life.' The school's influence was out of all proportion to its numbers and short life. At least three alumni became MPs. (Pole, *Life,* p.340, 445; *Liverpool Mercury* 15 October 1884; *Plan for the Education & Instruction of the Pupils Under Mr Chas Voelker's Care at St Domingo House, Everton, near Liverpool,* quoted in S E Koss, *Sir John Brunner, radical plutocrat, 1842-1919* (2008) p.5; [I M Heywood (ed.)], *Reminiscences, Letters, and Journals of Thomas Percival Heywood, Baronet,* (1899), pp.4-5; E F Rathbone, *William Rathbone: A Memoir,* (1905), pp.60-1).
97. B R Law, *Fieldens of Todmorden: A Nineteenth Century Business Dynasty,* (1995), p.53.
98. J Chapple and A Shelston (eds), *Further Letters of Mrs Gaskell,* (2003 edition), p.177.
99. Howe, *Cotton Masters,* p.57.
100. *Life,* p.480; *Crockford's Clerical Directory,* (1868 ed.). In 1857 the Rev William Gaskell lent his wife's copy of Hughes' newly- published *Tom Brown's School Days* to 21-year-old Adam, whose grandfather took it to Scotland and there 'wickedly lent it to grandsons' such that it came back 'dirty & dog-eared' (Chapple and Shelston, *Further Letters,* p.177). After curacies, Adam became Vicar of Walton St Lawrence, but died aged only 52, leaving his widow with eleven children.
101. M La Trobe-Bateman (ed.), *Memories Grave and Gay of William Fairbairn La Trobe-Bateman,* (1927), pp.3-6.
102. *UIfE3,* p.60.
103. *Life,* p.469.
104. *Life,* pp.191-2, 430-1.
105. Howe, *Cotton Masters,* p.87, gives the average number of servants per cotton master's household in 1851 as 3.5 based on a sample of 122. Presumably these were living-in servants.
106. See Appendix 1.2.
107. *Life,* pp.155-6; W Fairbairn, *Observations on Improvements of the Town of Manchester particularly as regards the importance of blending in those Improvements, the Chaste and the Beautiful with the Ornamental and Useful,* (1836). There is a suggestion (pp.iii-iv) that the book was written by 'another William Fairbairn', but it is now accepted that this was not the case.
108. J Nasmyth, 'The Art of Design as applied to Steam Engine Chimneys' in Fairbairn, *Observations,* pp.39-44.
109. *Life,* p.155
110. *Life,* pp.101-3; See above.
111. E Hodder, *The Life and Work of the Seventh Earl of Shaftesbury, K.G.,* (1886), Vol.2, p.376; 'Thomas Wright, the Prisoners' Friend', *The Ragged School Union Magazine,* 1852, 173-4. Wright spent his working life in Ormerod's foundry, starting as an apprentice at five shillings a week at the age of fifteen, and remaining there until his sixty-third year when he was a foreman earning £3.10s per week.

He had 19 children. See also J A V Chapple and A Shelston (eds), *Further Letters of Mrs Gaskell,* (2003 ed.), p.69.
112. A Briggs, *Victorian Cities*, (1968 ed.), p.109.
113. J A V Chapple and A Pollard (eds), *The Letters of Mrs Gaskell,* (1966), pp.847-9. The Horners were the Leonard Horners – he is known for his work in factory inspection and was an occasional guest of Fairbairn at the Polygon.
114. Fairbairn to Stephenson, 21 August 1846, No.105, Stephenson Papers, ICE, London; *MG,* 22 June 1842; *Life,* p.451.
115. He was the only brother to survive beyond teenage years. Their other brother, Thomas, was drowned in 1812.
116. Cambridge University Library, Stokes Papers, Add 7342, TR22, 26, 38.
117. *Life,* pp.468-9.
118. M Hewitt, 'Class and Classes' in C Williams (ed.), *A Companion to Nineteenth-Century Britain,* (2004), p.318; Hewitt, 'Introduction', p.18. On 'respectablity' see also Smiles, *Self-help,* pp.399-400; Musson, *British Trade Unions,* (1972), p.18; Crossick, 'The Labour Aristocracy', 314; T W Laqueur, *Religion and Respectability. Sunday Schools and Working-class Culture, 1780-1850,* (1976), p.239.
119. E J Molesworth, (ed.), *Life of Sir Guilford L. Molesworth, K.C.I.E. 'The Nestor of the Engineering Profession',* (1922), pp.19-20.
120. *Life,* pp.106-7.

Sources, Credits and Acknowledgements for Illustrations in Chapter 1

Fig.1.1 Old print
Fig.1.2 H. Hair, *A Series of Views of the Collieries of Northumberland and Durham,* (1844), p.16.
Fig.1.3 J. Parkinson, reproduced in D. Anderson, *The Orrell Coalfield, Lancashire 1740–1850,* (1975), p.99. (Pemberton Colliery, c.1915).
Fig.1.4 C. Moore, *Hertfordshire Windmills and Windmillers,* (1999), p.59.

CHAPTER 2

Fairbairn & Lillie: Partnership

1817–1832

Introduction

THE first phase of William Fairbairn's business career – his first experience in business – was a very successful partnership with James Lillie, which lasted fifteen years. They undertook predominantly millwrighting work – line-shafting with its gearing, waterwheels, and iron columns, beams and roofs – in Britain and Continental Europe. The partnership, having commenced in the short fortuitous moment of opportunity when millwrights' skills were in demand but little capital was needed, grew steadily in number of employees and in assets, with expansion largely financed from profits. The initial growth of the business illustrates how important one key commission can be. Obtained within weeks of the partnership's commencement, this was an unexpected and remarkable appointment, achieved by Fairbairn convincing one of the largest cotton spinners that Fairbairn & Lillie could renew the power transmission in his mill to give substantial savings in running costs. This can be seen as an example of innovation in the sense of setting up a new function which was more than an incremental improvement. It provided the pattern for future power transmission, and was the critical event leading to Fairbairn & Lillie becoming celebrated and successful millwrights.

The partners also built the most powerful industrial waterwheels, but unlike the new function in power transmission, these demonstrate optimisation by way of assembling the best components available, sometimes with improvements. They have been used to illustrate the 'persistence of old technology'. Fairbairn & Lillie's patterns for parts for power transmission and waterwheels provide confirmatory evidence that interchangeability was well established in Manchester in the 1820s.

The growth of Fairbairn & Lillie's business also emphasises the role of networks based on places of origin, membership of organisations and shared interests. Fairbairn's commitment to education, particularly of young engineers, found early expression in the Manchester Mechanics Institution, at the same time widening his network of influential friends. During these partnership years Fairbairn

met, and in many cases came to know well, many leaders of Manchester's industry, commerce and science, enhancing his standing and status in the area. His election to the Institution of Civil Engineers reflected that standing and introduced him to many more of the leading engineers of the day, and hence to a national profile.

Fairbairn's role as the leading experimental engineer of his time dates from his testing of beams in 1824, using apparatus he built himself. This became widely known as the 'Fairbairn lever'. Then from 1827 he provided facilities and finance for Eaton Hodgkinson's experiments on the strength of cast iron which provided the basis for much subsequent 19th-century cast-iron structural design. Fairbairn continued this work with other investigations. One outcome was the safer and more economical 'Hodgkinson beam', designed for mills, but first used by way of a technology transfer, for an important bridge.

Fig.2.1: Fairbairn & Lillie's Workshop, 20 Canal Street, Ancoats. Built 1823.

The partnership illustrated division of responsibilities between the partners, but there were also differences of temperament. After fifteen years this renowned partnership, with its constant flow of profitable work, came to an unhappy end, as Fairbairn sought to enlarge the firm's range of products to encompass ship-building and cotton manufacture. Always excited by engineering challenges, he seemed to have little concern fortheir financial implications, whereas Lillie was more prudent and cautious. It was this that led to the dissolution of their partnership.

Starting in Business

November 1817, when Fairbairn left Thomas Hewes, appeared an inauspicious time to commence a new business. As Woodcroft points out, the peace of 1815 had swelled the ranks of pauperism as soldiers returned, and factories equipping the army no longer had work, with the situation made worse by poor harvests in 1816 and 1817. At the same time steam power and machinery were bringing distress to hand-workers, leading to machine-breaking and bread riots.[1] In March 1817 the Blanketeers marched from Ardwick Green.[2] The economy had entered a state of depression from which there would be no real recovery until 1821.[3] In July 1818 there was a strike, lasting several weeks, of 15,000 Manchester cotton spinners.[4] On 16 August 1819 Fairbairn witnessed Peterloo and in later years 'gave some graphic accounts of the terrible scenes of disorder'.[5]

Whilst the partnership started at a time when 'business was slack',[6] these were nevertheless potentially prosperous times for innovative millwrights with a driving work ethic, as steam power was spreading exponentially. There were already approaching 200 steam engines in Manchester;[7] and in the decade following the Napoleonic Wars, 400 new factories were built in Lancashire.[8]

Mechanisation was progressing at a faster pace than ever before, bringing opportunities with it – mills, millwork, machinery, steam-engines, boiler-making. There were few machine tools, requiring minimal capital outlay.[9] Thus the partners, with their skills, were able to start in business without capital. Ten years later, when the necessary outlay on capital-intensive machinery had increased, they would almost certainly have required a partner with capital,[10] in the way that the Nasmyths, commencing in 1836, were initially assisted by the Birley family.[11]

Fairbairn had made no plans for the future when he left Hewes other than that he was determined that he would work for himself.[12] He sought work from Otto Hulme, a calico-printer in Clayton, for whom he had erected machinery whilst with Parkinson. Hulme offered the job of building a conservatory, which Fairbairn accepted. He sought a 'clever, active partner' to join him and found James Lillie, also from Coldstream, with whom he had worked at Parkinson's. After some weeks, work on the conservatory stopped abruptly, on the grounds of infringement of a patent. With no money, insufficient food and fruitless attempts to get work, Lillie suggested abandoning their venture. Fairbairn urged firmness. Two small jobs, erecting machines, enabled the partners to hire a shed and build a large manually-operated lathe, powered by 'James Murphy, a muscular Irishman', at a time when there were only a few small lathes in Manchester.[13] At this point the partnership was transformed by unexpected major millwork commissions, followed by an increasing flow of work, requiring premises, finance, machines and men.

Premises, Finance, Machines and Men

ON completing their first major commission at Murrays' Mill, Fairbairn & Lillie moved from their shed to an old building at 14 Mather Street. The partners worked long hours, with division of responsibilities, Fairbairn doing the books and the visits to clients, travelling 'by hired hack', and Lillie managing the workshops. With maintenance work to be done when mills were shut on Sundays, it was seven-day working. By the end of 1822 they were able to buy a 16hp second-hand Boulton & Watt engine, and build a four-storey workshop at 20 Canal Street. In 1824 land was bought on the other side of Canal Street and in 1825 a three-storey building was constructed on Pott Street, incorporating a fire-proof pattern store.[14] In 1827 a yard was hired for work on the Catrine waterwheels.[15]

By the end of 1822, whilst still living from hand to mouth and without ready money, Fairbairn & Lillie were worth £5,000 in material, stock and tools – around £350,000 at 2015 figures. Their new building and steam engine were financed with occasional help from Heywood's Bank. Besides the bank, the partners had 'good friends and considerate customers' and found 'no difficulty in obtaining money on account for work … in progress'.[16] It is unclear if Fairbairn's on-going links with Benjamin Heywood started through the bank or through Cross Street

Chapel where Heywood was involved, or through John Kennedy. At the end of 1827, Fairbairn wrote to Lillie from London:

> Mr. Cooke's wheel, Mr. Potter's work, and all the rest, we shall talk over together. In the meantime I shall look in at the watchmaker's and order both yourself and me a gold watch, *but on the condition that it is not to be delivered* until we have *paid for our buildings*, and are FAIRLY OUT OF DEBT.[17]

In 1828 Fairbairn was in a position to advance £500 to assist his younger brother, Peter.[18] By 1830 Fairbairn & Lillie's stock-book showed a balance of nearly £40,000 – equivalent to around £2.8M in 2015.[19] All the indications are that apart from occasional bank loans, their business was self-financing. Without doubt, it was very successful.

Machine tools became increasingly crucial for Fairbairn & Lillie's business but records of them are sparse. At the beginning there was the large lathe. For their second major commission – Sedgwick Mill – this was supplemented by two other lathes turned by three more labourers. Another lathe for large shafts was erected temporarily at the mill.[20] At first Fairbairn & Lillie were unable to make their own castings, but it is likely that a foundry was incorporated into their new building in 1823 as J.C. Fischer from Switzerland, visiting their works in 1825, noted two cupola furnaces. This was an important step forward for Fairbairn & Lillie, a form of vertical integration, giving them full control of the millwork they were installing – casting, machining and erecting. Fischer also noted a circular saw and large grinding wheel.[21] In the mid-1820s millwrights held that key-ways must be cut by them with chisel and file. Fairbairn, in response to a strike of millwrights at that time, developed the existing technology of the planing machine to build a slotting machine to cut key-ways.[22] The exact date of this is uncertain but it is likely that it was during the strike at the beginning of 1825. Richard Roberts claims to have built the first slotting machine in 1824.[23] In any event this is a very early example of a machine being developed to resist a strike. In 1828 two French engineers visiting the works noted two horizontal boring machines.[24] Around 1830 a new foundry was constructed.[25] By then Fairbairn & Lillie's machine tools 'embraced a considerable array of lathes, drills and other tools'.[26] The firm had also built up a large and valuable stock of toothed-wheel patterns, spur and bevel, of many sizes, from which identical castings could be made. From 1825 they had a 'fire-proof' pattern store.[27] These patterns had been paid for by the owners of the mills in which they were first used, but remained the property of Fairbairn & Lillie. This was of commercial benefit in providing on-going work as it encouraged mill-owners to return to the partnership – on grounds of speed and economy – whenever a breakdown occurred requiring the replacement of any gearing, or in the event of alteration or extension.[28] Standardisation and interchangeability were established practices by the 1820s. Another factor in the early decades of the 19th century was the increasing availability of malleable wrought-iron, by 'decarburisation' and 'puddling'.[29] It was much tougher and more ductile than brittle cast iron, and was used by Fairbairn & Lillie for shafting, waterwheel spokes, boilers and iron ships.

By 1824 Fairbairn & Lillie employed between sixty and seventy men.[30] There were difficulties getting enough men because existing Millwrights Societies would 'not allow other workmen to come into the trade'.[31] Nevertheless by the end of the decade the firm was employing 300 men.[32] Records of their various trades do not exist but it is possible to surmise that these comprised mainly millwrights both within the works and outside erectors, with foremen and apprentices. It is unclear how successful the Millwrights' Societies were in keeping out semi-skilled machine operators during the 1820s but the evidence indicates that in subsequent decades some were engaged, and the opposition to their employment ended with the 1851–2 dispute. The foundry required pattern-makers, moulders and other foundrymen. Skilled tradesmen were assisted by labourers and there would be a small office staff of draughtsmen and clerks, and probably also a storekeeper. The steam-engine would require an engineman and stoker. There is a disappointing gap in the evidence as to how the undertaking of the work was organised. Transport of goods to sites, which included those in Scotland, Ireland and Continental Europe, was sub-contracted to carriers. The workforce in engineering in the first half of the century was mobile and the total number of men who spent some time employed by Fairbairn & Lillie during their fifteen years must have been several times the complement at any one time.

Networks and Networking

NETWORKS based on common values and attitude have had a profound influence in business activity.[33] In her 1876 novel, *The Manchester Man,* Isabella Banks captured the spirit of mutual support among emigrant Scots when she wrote of John McConnel, 'who had already given a lift to his rising young countryman, Fairbairn the engineer'.[34] The most important event in the establishment of Fairbairn & Lillie came unexpectedly when Fairbairn called on Adam & George Murray at their Ancoats mill, and had astounding success, obtaining a commission to renew the line-shafting and gearing at the mill.[35] How, without money, without an established works or labour force, without cotton mill experience and with no track record, did Fairbairn & Lillie obtain this commission from one of the world's largest cotton spinners;[36] and in the face of established local millwrights such as Thomas Hewes, and Peel, Williams & Co., at a time when 'business was slack'? Their commission may have been facilitated by a common ethnic origin. George Murray (1772–1866) and his brother Adam (1766–1818) shared Fairbairn & Lillie's Scottish ancestry. They too had served trade apprenticeships, as machine-makers, before moving into cotton spinning.[37] Thus they were skilled in a branch of engineering. There is every reason to believe it was these common bonds that gave Fairbairn the entrée, without which he would not have had opportunity to convince the Murrays to accept the innovational renewal of the power-transmission in their mill, which set Fairbairn & Lillie on the road to success.

The Murrays were delighted with Fairbairn & Lillie's work and recommended the new firm to their neighbouring cotton spinners, M'Connel & Kennedy, both of whom were also Scots. Like the Murrays, they too had served apprenticeships as machine makers with William Cannan, M'Connel's uncle, at Chowbent in Lancashire. Moving to Manchester, they initially built textile machinery before expanding into cotton spinning, and thus they too were skilled in a branch of engineering.[38] The Murrays' recommendation came just at the opportune time, when M'Connel & Kennedy were about to build a new mill. Fairbairn & Lillie were employed.

John Kennedy (1769–1855) was an outstanding inventive engineer and cotton spinner who is widely lauded by historians today.[39] R.C. Allen includes him in his list of seventy-nine important inventors of the 17th and 18th centuries, citing his invention of 'double speed' which enabled the mule to spin fine yarn.[40] Fairbairn came to know Kennedy well, had great respect for his abilities and, following Kennedy's death, wrote a memoir of him.

Through Kennedy, Fairbairn became acquainted with William Murdoch (1754–1839), another Scot (and also one of Allen's seventy-nine), when Murdoch came to commission the Boulton & Watt engine at the Sedgwick Mill in 1818.[41] Murdoch, over seventy, was described by Fairbairn as 'the right hand of the illustrious Watt'. They clearly struck up a rapport and Fairbairn visited Murdoch to see his model of a 'locomotive-engine' which travelled at five miles per hour.[42] It is likely that Fairbairn also met Peter Ewart (1767–1842) – yet another Scot – around the same time. He was a friend of John Dalton and Eaton Hodgkinson, and an advocate of the application of scientific knowledge to engineering.[43] For a time he represented Boulton & Watt in Manchester but then became a cotton spinner, at the same time acting as a consulting engineer. It is likely that he was involved with the Boulton & Watt engine at the Sedgwick Mill.[44] In any event he was involved with Fairbairn in the start of the Mechanics' Institution.[45] It was Ewart who persuaded Fairbairn to make the far-reaching decision to provide facilities for Hodgkinson – see below. Network connections with Murdoch and, probably, with Ewart, almost certainly led to Benjamin Gott appointing Fairbairn & Lillie in connection with the extension of Bean Ing Mill, Leeds, in 1824. Gott was a friend of Murdoch, and his extension was to be powered by a Boulton & Watt engine. Ewart was also involved. It is reasonable to assume that they recommended the partners.[46]

Kennedy was of major importance in the ascent of Fairbairn & Lillie.[47] He was widely respected and had many business contacts. It was he who passed their name to fellow Scot, Archibald Buchanan, partner in James Finlay & Co., and manager of their Catrine cotton mills in Ayrshire. Buchanan had been apprenticed to Arkwright and was himself no mean engineer.[48] This recommendation led to Fairbairn & Lillie's most famous waterwheels, at Catrine, followed by waterwheels at Finlay's Deanston factory, where they worked with Buchanan's nephew, James Smith, a talented engineer and agricultural inventor.[49] The centre of mechanical experiment in Scotland at this time has been identified as in these mills of James Finlay & Co., where Buchanan and Smith exercised their remarkable engineering abilities.[50] The work at Finlay's Catrine and Deanston mills led to Fairbairn's involvement with Thomas Grahame and steamboat experiments on the Forth & Clyde Canal. The links appear to have been Kirkman Finlay, who was Senior Partner of James Finlay & Co. and Chairman of the Canal Company, and James Smith of Deanston – 'my much esteemed friend' – who was also involved in the Forth & Clyde experiments. A major significance of these experiments was that they led directly to commissions to Fairbairn & Lillie to build steamboats for the Canal Company, and hence to Fairbairn becoming a shipbuilder.[51]

These Scottish network connections were important in the growth of

Fairbairn & Lillie in that they brought them into contact with some of the leading Scottish engineers and manufacturers who placed major work with them and recommended them to others. In addition Ewart was the important link to the prestigious experiments with Hodgkinson. Later Fairbairn would help other Scots, such as Nasmyth who arrived in Manchester in 1834.[52]

The second important network for Fairbairn centred on Cross Street Unitarian Chapel, of which he became a member 'on his settlement in Manchester'.[53] M.C. Jacob sees a marriage of enlightened piety and science-based rationalism in the ethos of Unitarianism,[54] and Mokyr describes Unitarianism as the Enlightenment religion *par excellence*.[55] However, in stressing the roots of Unitarianism in Socinianism, Jacob diminishes the influence of Puritanism, with its work ethic, which is where Cross Street chapel had its roots.[56] Gatrell points to the many members of Cross Street who could trace their ancestral commitment to the Presbyterian and Unitarian ministry back to the 17th century.[57] In viewing the Enlightenment as the culmination of the Reformation, rather than a departure from it, David Allan has sought to change the historical canon of alienation between the Scottish Reformation and the Scottish Enlightenment.[58] In the espousal of the work ethic *and* an enlightenment approach to science, both Fairbairn and Cross Street exemplified this link between Reformation and Enlightenment.[59] In Scotland, as Allan points out, 'Virtue had to be evinced ... by a sincere and committed orientation towards the service of the public interest', and Scottish historians sought to 'stigmatise the fatal consequences of luxury and ignorance'.[60] This is all consistent with Fairbairn's warm approval of the *Sermons* of Hugh Blair (1718–1800), the latitudinarian minister of the High Church of St Giles in Edinburgh,[61] and fellow member of the Select Society with William Robertson, David Hume, Adam Smith and others of the 'Scottish Enlightenment'.[62] Commenting on Blair's sermon on Matthew 20 v.6, John Hill writes that Blair 'proposes to prove, and does so incontrovertibly, that the idle man fails in duty to God, to the world, and to himself'.[63] The extent to which Fairbairn actually held Unitarian beliefs is unclear. Blair's sermons major on moral conduct, but orthodox protestant doctrine is to be found within them, and there is no hint of Unitarianism,[64] whereas Fairbairn's writings, whilst seeing his achievements to be 'under the blessing of Divine Providence', suggest some form of theism.[65] Proposing a toast at a Manchester dinner in 1868 in honour of the former Secretary of State for India and subsequent Prime Minister, the Marquis of Salisbury, Fairbairn, with Unitarian scepticism and pragmatism, 'was convinced that if we were to develop India it must be done by mechanical science, irrigation and cheap transit. ... [We] might send out missionaries, and do other things; but his opinion was that the locomotive engine and mechanical science would do ten times more than anything else'.[66]

Unitarianism probably had no relevance to Fairbairn & Lillie's initial major commission for the Murrays, as Slugg does not include them in his list of prominent Unitarians.[67] It is difficult to track specific Fairbairn & Lillie commissions to Cross Street members, although the Mr Potter, for whom they were working in 1827, was almost certainly one of the Potter family which was associated with the Chapel.[68]

Importantly Cross Street was part of the network introducing Fairbairn to

Fig.2.2: Cross Street Chapel. Built 1694.

Manchester's civic and intellectual life, much of which was centred there in the first half of the 19th century. Indeed it was often seen as a launching-pad for a public career, a base for political activity, educational and philanthropic work and cultural advancement.[69] In this respect a major Cross Street link for Fairbairn was his banker Benjamin Heywood (1793–1865). Again Fairbairn appears drawn to those who would go on to achieve – four years younger than Fairbairn, Heywood was a prominent local banker, a pillar of the Literary and Philosophical Society, and a founder of the Royal Manchester Institution (1823). He went on to be a founder of the Manchester Statistical Society (1833), long-time treasurer of the Manchester & District Provident Society, a Fellow of the Royal Society, Member of Parliament and a Baronet.[70] From 1840 to 1842 he was President of Cross Street Chapel (although he later embraced the Trinity, joining the Church of England and building St John's, Miles Platting – designed by J.E. Gregan, but demolished in 1973).[71] It was the association with Heywood in the establishment of the Manchester Mechanics' Institution – Heywood was President and Fairbairn was Secretary – that thrust Fairbairn into civic prominence.[72] Half the committee of the Institution were Unitarians, including Heywood and Fairbairn from Cross Street and Kennedy, Ewart, Wood, William Henry, R.H. Greg and Henry Houldsworth from the Mosley Street Chapel.[73] Engineers on the committee, besides Kennedy, Ewart and Fairbairn, were James M'Connel, Thomas Sharp, Richard Roberts, William Williams, and also building contractor David Bellhouse.[74] In 1993 David Allan suggested that 'Manchester Mechanics' Institute ... would perhaps bear further examination for Scottish influence'[75] – Ewart, Kennedy, M'Connel and Fairbairn were Scots. Significantly, James Lillie was not on the committee, an indication of the different characteristics of the two partners. Whereas Fairbairn moved comfortably in these circles, Lillie was less at ease.[76]

Many of the members of the Unitarian chapels and of the Mechanics' Institution committee were also members of the Manchester Literary and Philosophical Society which, by the end of the 18th century had become 'one of the most vivid expressions of English provincial culture', deriving its tone from its Unitarian membership.[77] At the time Fairbairn joined in 1824, Ewart, Henry and Wood were vice-presidents, under John Dalton's Presidency.[78] Fairbairn was

Fig.2.3: Mechanics' Institution. Built 1827.

probably introduced by Kennedy, or possibly Ewart or Benjamin Heywood, who were all members.[79] These were three socially very powerful groups. Of the membership of Cross Street in 1829, nine became Members of Parliament and five Mayors of Manchester.[80] Of the Mechanics' Institution committee, four became MPs – Heywood, Wood, Greg and Brotherton. These interlocking networks of chapels, the Mechanics' Institution and the Literary & Philosophical Society brought Fairbairn into the company of Manchester's political, manufacturing and commercial leaders, and its intellectual and scientific elite. His earlier self-education and his engineering prowess enabled him to integrate into these circles. The former journeyman millwright had moved into the middle class.

William Fairbairn was a 'first generation' north-west engineer. He did not have an affluent father, nor did he marry into wealth. His father came from a working-class background as an agricultural labourer, but by rising to farm steward was bridging to the middle class, from where the vast majority of the north-west's successful engineers were drawn. A study of forty leading north-west engineers contemporary with Fairbairn, showed only nine (22.5%) from working-class backgrounds.[81] Daunton sees 'the most striking instances of social mobility ... in sectors where technical skill provided the basis for entry, such as William and Peter Fairbairn'.[82]

National networks were also important for Fairbairn, bringing him into contact with leading engineers and scientists beyond Manchester. First there was the Institution of Civil Engineers to which he was elected in 1830, proposed by James Walker, Robert Stephenson and John Farey, Jnr. There he met the President, Thomas Telford, one of the greatest of engineers, who congratulated Fairbairn on his first book.[83] He was now moving amongst the leading engineers nationally, as he would continue to do in the engineering institutions and, more particularly from the mid 1830s, in the British Association for the Advancement of Science, providing a conduit to a national profile. That profile was already prominent when, in 1861, he became the first engineer to preside over the British Association, raising his national, and indeed international, standing still higher. For Fairbairn the British Association was very markedly 'a vehicle for upward social mobility'.[84]

Mills

Cotton Mills: The Transmission of Power

THE shared Scottish and technical backgrounds were important in gaining an entrée with the Murrays, but Fairbairn must have convinced them that he and Lillie could deliver significantly improved efficiency by replacing their heavy, ponderous, part-timber line-shafting with light, fast and smooth-running wrought-iron shafts. It may be that it helped that Fairbairn & Lillie could turn to the work immediately, but men with the Murrays' experience do not place work with a firm unless reasonably convinced that what is promised will be achieved. Two facets of entrepreneurship met here. The Murrays took the risk – not without some misgiving, says Woodcroft – of engaging men who could tick none of the boxes, but with whose skills and background they resonated.[85] Their money followed their instincts, in a way that is impossible in a large public company. Fairbairn had the vision of what could be achieved and the confidence that it would be. He had appreciated that the principle of using a low-speed main drive and then having to speed it up by means of countershafts in order to transmit the power to the fast-running machines was wrong. For the transmission of a given amount of power the weight and strength of shafts and wheels could be reduced in direct proportion to the increase in the speed of rotation. Fairbairn & Lillie's introduction of high-speed, light, iron shafting effected a saving in capital cost, and significant savings in running costs.[86]

Fairbairn & Lillie's role in connection with wrought-iron shafting is questioned by Musson who credits Peel, Williams & Co. with its introduction.[87] This is probably correct, but it was Fairbairn & Lillie who integrated lighter wrought-iron shafts into a very much faster and smoother composite system. There may have been a precedent at a flour mill in Zurich designed by J.G. Bodmer around 1815, but if there was, Fairbairn could not have seen it until he visited Zurich in 1824.[88] 19th-century writers including Andrew Ure, George Rennie, Evan Leigh, W.J.M. Rankine and Bennet Woodcroft endorsed Fairbairn's claims to have introduced the faster and smoother system. Rankine and Woodcroft also attributed to him improved lap couplings forged in one piece with the shafting, 'which have long become common everywhere'.[89] The endorsement from George Rennie is important because Musson attributes the commencement of 'this revolution' to John Rennie, George's father.[90]

Fairbairn was also well aware of friction and the importance of lubrication. In *Mills and Millwork* he notes work done by Morin in France in the 1830s and by George Rennie, and refers to the effects of different lubricants, systems of lubrication and temperature. He had known large cotton mills 'where as much as ten to fifteen horses' power was absorbed by a change in the quality of oil used for lubrication'.[91] Landes observes that few manufactures gave lubrication the attention it deserved.[92] Indeed it was a century after *Mills and Millwork* that the subject came to the fore with the research of Bowden and Tabor,[93] and the 'Jost Report' of 1966 which introduced the now widely-used term, 'tribology', to describe inter-disciplinary studies in friction, wear and lubrication.[94]

The partners worked extremely hard on Murrays' Mill – some days working from 5am to 9pm. They succeeded – a great saving of power was achieved.

Fig.2.4: Murrays' Mill as it was in 1831.

The shafting at Murray's Mill was the tinder that ignited Fairbairn & Lillie's ascendency. Fairbairn & Lillie did not invent wrought-iron shafts, and, as Woodcroft pointed out, 'It must have been seen by this time by very many mechanics and mill proprietors ... that great advantages would arise by the substitution of much lighter shafting driven at higher velocities'. They might have seen it, but if they did, they did nothing about it. Fairbairn saw it and gave effect to it – a classic example of entrepreneurial 'innovation' – with the result that

> to him appears to be ascribable the entire suppression of square shafting, the substitution for it of cylindrical shafting truly turned by the slide lathe ... [such that] ... an almost boundless field of work of nearly one common character was thus presented to Fairbairn and his partner ... after they had once completed the job at Mr Murray's mill.[95]

This lends credence to Fairbairn's ascription of his own and Lillie's success in life 'to the saving of power effected by increasing threefold the velocity of the shafting' at Murrays' Mill.[96] It concurs with the other 19th-century writers referred to above. The matter is put beyond reasonable doubt by Murrays' neighbours, M'Connel & Kennedy. Engineers and cotton spinners of their standing would not have engaged this small partnership to undertake the millwork in their new Sedgwick Mill if they had not seen something wholly exceptional in their neighbours' mill – they had, a twenty per cent increase in efficiency with the associated savings in running costs.[97] What was achieved with the line-shafting at Murrays' Mill became the norm for the rest of the steam age.[98] In 1836 Ure recorded that 'Fairbairn's millwright establishment'

> turns out from three hundred to four hundred yards of shaft-geering [*sic*] every week, finely finished, at a very moderate price, because almost every tool is now more or less automatic, and performs its work more cheaply and with greater precision than the hand could possibly do.[99]

Fig.2.5: Sedgwick and Murray's Mills, in 1962.

Fairbairn & Lillie's success at Murrays' Mill came at the opportune moment, as M'Connel & Kennedy were about to build the Sedgwick Mill, which would make them, like the Murrays, one of the largest cotton spinners in the world.[100] Kennedy commissioned Fairbairn & Lillie, and Fairbairn records that he

> laid down all his plans for the new mill to a scale, calculated the proportions and strength of the parts, fixed the position and arrangement of the different machines, and introduced, under that gentleman's direction, a new system of double speeds, which, I believe, was an original invention of his own, for giving an increased quality of twist to the finer description of mule yarn.[101]

This is less than clear. It appears that Kennedy and Fairbairn worked closely together, with Fairbairn as draughtsman. Fairbairn would have designed the details of the power-transmission which was installed by Fairbairn & Lillie.[102] As they did not have a foundry at this stage, the shafts and gears must have been cast elsewhere, probably paid for direct to the foundry by M'Connel & Kennedy, prior to machining by Fairbairn & Lillie. The 'double speeds' refers to the mules not the shafting.[103] The building work was undertaken by James Lowe with M'Connel & Kennedy providing the materials, and the cast-iron frame was by J. & P. Sherratt.[104] There is a query as to how advanced the project was when Fairbairn was first engaged as work was in progress on site by mid 1818, yet Fairbairn states that his engagement followed the completion of Murrays' work, which did not start until the beginning of 1818.[105] The probable explanation is that it was the installation of the millwork that did not take place until after Murrays' work was finished. This is consistent with the payments made to Fairbairn & Lillie which did not commence until September 1818, with around £3,600 having been paid by the end of 1822 – equivalent to around £250,000 in 2015.[106] The power transmission was by a primary horizontal shaft in a duct beneath the ground floor, driving seven vertical shafts.[107] This arrangement, apart from the sub-floor duct,[108] was not repeated in Fairbairn's later mills. His norm became the single vertical shaft with long line-shafting on each floor.[109] Fairbairn gave safety as one of the reasons for this.[110] The duct was not novel as a similar duct is shown on Boulton & Watt's drawing of Phillips & Lee's Salford mill of 1801[111] – a building whose designer Fairbairn praised, incorrectly believing

Fig.2.6: K.F. Schinkel's sketch of 1826 showing the Sedgwick and Murray Mills.

it to be the first iron-framed mill. It was the model for the eight-storey Sedgwick mill.[112] It is clear that M'Connel & Kennedy were well satisfied as Fairbairn & Lillie continued to be engaged on further work, probably repairs and renewals at their adjoining older mills, up to the dissolution of the partnership, by which time M'Connel & Kennedy had paid them £8,166.[113] Kennedy retired in 1826 to devote himself to technical pursuits in much the same way as Fairbairn would do.[114]

The state-of-the-art mill immediately became 'one of the sights of Manchester',[115] drawing visitors from home and abroad, and diffusing knowledge of Fairbairn & Lillie's gearing and shafting. Peter Beuth, main mover in Prussia's industrial renewal, visited Ancoats in 1823 and wrote to the prominent Prussian architect, K.F. Schinkel, that it is a place where 'machinery and buildings can be found commensurate with the miracles of modern times – they are called factories'.[116] Beuth and Schinkel visited Manchester together in 1826, when they toured the Sedgwick and Murray mills, which Schinkel sketched.[117]

For Fairbairn & Lillie to be associated with a mill of this importance was a powerful advertisement for them, at a time when most new firms worked their way up by way of small commissions.[118] The good word of the well-connected Kennedy was 'a passport to fame and business'.[119]

The pattern of association with large and prestigious mill projects extended to the largest cotton mill in Ireland, built at Portlaw by David Malcolmson, a Quaker flour-miller from Clonmel.[120] Fairbairn & Lillie's connection was almost certainly through their Quaker clients, Henry and Edmund Ashworth.[121] James Cropper, a successful Quaker merchant from Liverpool, was much concerned about poverty in Ireland. He proposed the building of cotton mills, but the only outcome was the 1826 Portlaw factory.[122] It was a six-storey mill, driven by two waterwheels on the Clodagh, with a canal link to the river Suir, and a model village.[123] Fairbairn & Lillie were involved with machinery and shafting, and with the waterwheels, for which the components were cast in Cork, possibly to provide work in Ireland.[124]

As a sequel to the previous paragraph, whilst there are exceptions,[125] there is a distinct correlation between those for whom Fairbairn provided high-quality mill buildings and plant, and those who, by the standards of their day, were amongst the best employers. This is true of M'Connel & Kennedy,[126] the Ashworths at Turton and Egerton,[127] and the Malcolmsons at Portlaw,[128] as it is of John Wood of Bradford,[129] and Archibald Buchanan at Catrine.[130] On the Continent it was true of Gros, Davillier, Roman & Co. in Alsace,[131] and F.A. and W. Jung at Wuppertal in Germany.[132] Later examples were Herdman's at Sion Mills in Northern Ireland,[133] and Thomas Ainsworth at Cleator,[134] followed by Saltaire,[135] and the linen factory for Fairbairn's friend, the Irish railway contractor and philanthropist William Dargan at Chapelizod, Dublin.[136] Portlaw, Sion Mills, New Eagley, Egerton, Saltaire and possibly Cleator had 'model' factory villages – very much more than the mere provision of industrial housing which was widespread and could exhibit some of the worst aspects of the 19th century,[137] and which Pollard includes in his understanding of a factory village.[138] Opinions are divided as to whether industrial villages were primarily a managerial tool to obtain, retain and control employees, or models of social progress.[139] However the provision of a resident surgeon and a school at Portlaw; and of almshouses, hospital, school, library, reading rooms, baths and wash-houses, chapels and a large park at Saltaire – way beyond the mere provision of housing – indicate that, at least in these cases, they were built by men with a concern for the well-being of their employees.[140] Salt was also a significant benefactor to many charities.[141] The number of complete 'model' factory villages built in the British Isles during Fairbairn's lifetime, as opposed to just the provision of industrial housing, was quite small.[142] The point being made here is that Fairbairn undertook work at the factories which were at the centre of a significant proportion of those 'model' factory villages. However this is secondary to the primary argument that his work illustrates that there is often (but not invariably) a correlation between the best mill buildings and plant, such as he built, and the best employers.

Fairbairn had an on-going concern for safety and for the working environment in mill buildings. One expression of this was ventilation. Dust and fluff were problems in cotton mills, particularly in the carding rooms. In Elizabeth Gaskell's *North and South,* the dying Bessy Higgins tells Margaret Hale, 'there's many a one as works in a carding-room, that falls into a waste, coughing and spitting blood, because they're just poisoned by the fluff'.[143] Mechanical ventilation in textile mills was first contrived by Henry Houldsworth, and 'executed under his direction' by Fairbairn & Lillie who then went on to manufacture fans for textile mills. As well as removing dust 'disengaged in cleaning the fibrous materials', they were used to ventilate the toilet towers, 'so as to cause a breeze into every seat'.[144] This is one of the earliest examples of attempts to improve conditions in the workplace and a forerunner of what would become mandatory. The carding room at Portlaw was ventilated, but it is unclear if the ventilation was installed when the mill was built or later.[145]

Records from the 1820s are scarce. For some mills there are records that are tantalisingly inadequate. One such, from 1824–6, and probably by Fairbairn & Lillie, was Gorton Mill, a cotton mill important because similarities suggest it may have been a precedent for Fairbairn's famous Orrell's Mill of 1836.[146] Known mills with which Fairbairn was involved are listed in Appendix 2.1 but, given the size of the firm, there must have been much textile mill work undertaken by Fairbairn & Lillie which has been swept away without trace.

On occasions mill work led to work in other fields in which mill-owners were active. Thus it was that Fairbairn & Lillie became architects for Stalybridge Town Hall![147]

Experiments made at Leeds in 1824.

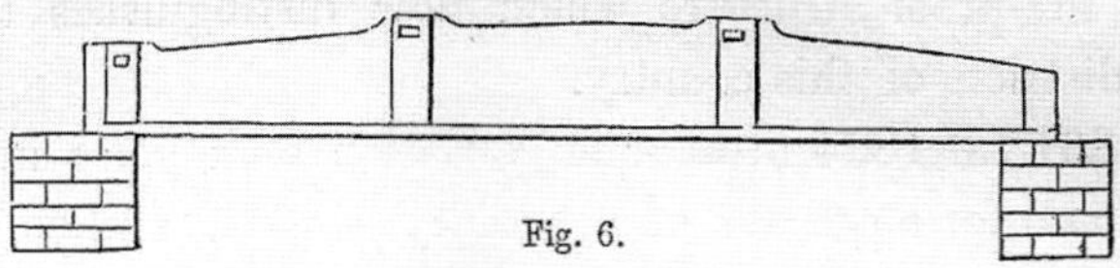

Fig. 6.

EXPERIMENT I.

Distance between the supports, 14 feet.
Depth of beam in the middle, 15 inches.
Depth of beam near the ends, $9\frac{1}{2}$ „

Fig. 7.

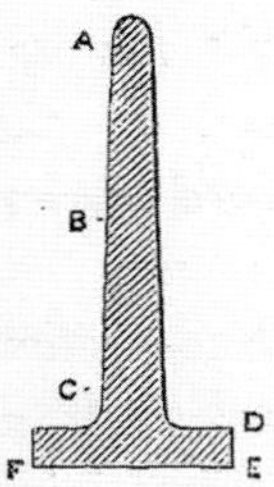

Weight of beam (taken from the average weights of several beams from the same model), 7 cwt. 3 qrs. 20 lbs.

Dimensions of section.

Thickness at A = $\frac{5}{8}$ inch.
„ C = 1 „
„ DE = 1 „
„ FE = 5 inches.

Weights.	Deflections in parts of an inch.	Remarks.
4 tons. 10 cwt.	0·21	
6 „ 0 „	·278	
10 „ 0 „	·48	A little warped.
11 „ 4 „	·537	The top edge of beam pressed considerably out of perpendicular.
12 „ 10 „	·665	The pressure outward much increased, and danger of breaking.

The experiments on this beam were not performed with the same accuracy as those subsequently made at Leeds and Bradford; it was nevertheless tested with considerable care.

Fig.2.7: Testing of beams, Leeds, 1824.

Woollen Mills: The Testing of Beams

FAIRBAIRN'S first experimental work – load testing beams and measuring deflection – took place when Fairbairn & Lillie's work expanded into the woollen districts of Yorkshire with a four-storey extension to Gott's Bean Ing Mill. Fairbairn & Lillie provided the structural ironwork, shafting and gearing; and Boulton & Watt provided the steam-engine and gas lighting.[148] Again Fairbairn put the main horizontal shaft in a sub-floor duct but the layout of the rest of the shafting is unclear.[149] The circumstance that led to the beam testing was a failure, due to a blemish in the cast-iron.[150] Fairbairn's response was to load-test other beams in order to reassure Gott.[151] The work at Bean Ing must have turned out well as five years later there was repeat business.[152]

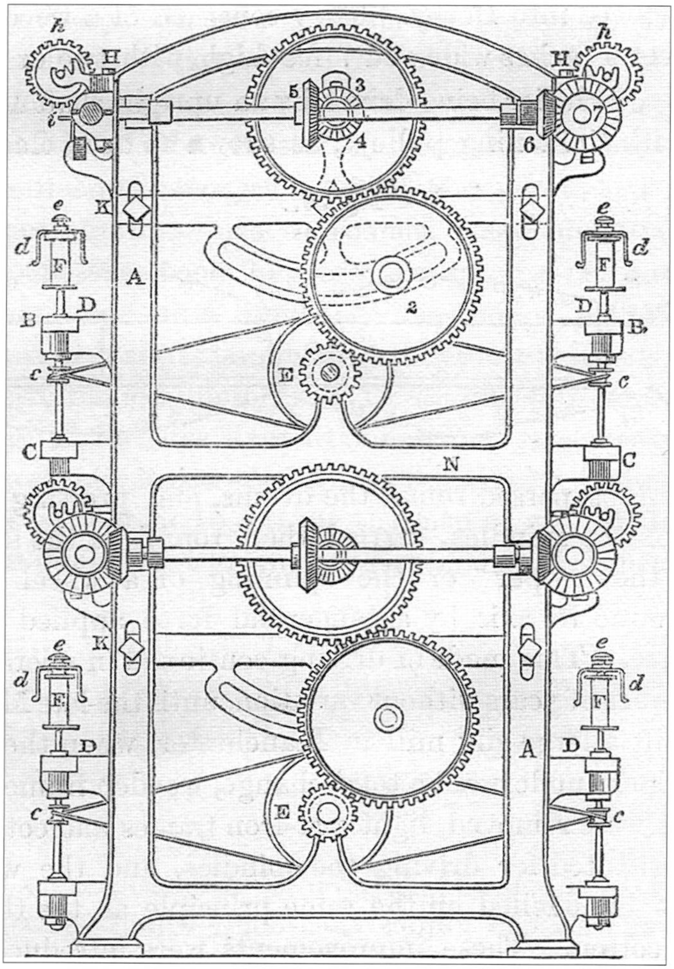

Fig.2.8: End View of Fairbairn & Lillie's Silk Spinning-mill.

The work at Bean Ing was followed by a mill for the large worsted spinner, John Wood – the first 'fire-proof' mill in Bradford.[153] Here Fairbairn load-tested a beam to failure.[154] The pattern was established and he was still testing beams thirty years later at Saltaire.[155]

Silk Mills: Spinning Machinery

FAIRBAIRN & Lillie's textile mill work extended from cotton and wool to silk, where, exceptionally, they built spinning machinery. Manufacture of silk was introduced into Manchester on an increased scale in 1815 and was developed by Vernon Royle (1784–1854) who, in 1825, with Thomas Crompton, built the seven-storey Havelock Mill, to spin silk on a larger scale than any of the existing mills in the East Cheshire silk district. Williams concludes that Havelock Mill is so important that it may 'be placed within a sequence of exceptional silk mills which includes: Lombe's Mill in Derby of 1718–21, Old Mill in Congleton of c.1752, and Lister's Manningham Mill in Bradford of 1871–3'.[156] Havelock Mill was furnished with Fairbairn & Lillie's new spinning machines. These were modelled on the throstle used for cotton spinning.[157] Heavy wooden frames were replaced by light cast iron and the speed was greatly increased.[158] This was derivative development, but was certainly innovative. Once again Fairbairn & Lillie took existing technology and developed it, in this case to give a fifty per cent increase in productivity.[159] Ure wrote:

> Ever since the throwing-mills for doubling and twisting silk were re-modelled by Messrs Fairbairn and Lillie, upon the cotton-throstle plan, they are incomparably superior in convenience, precision, and speed of performance, to what they formerly were in this country, and still are upon the continent. When these mechanicians took the silk-mill in hand, the spindles moved at the rate of only 1200 revolutions per minute; they forthwith raised it to 3000.[160]

The machinery was illustrated in 1836 by Ure and by G.S. White in America, and later by Fairbairn in *Mills and Millwork.*[161] Similar machines continued to be used into the 20th century.[162] In view of this initial success, it is curious that Fairbairn

& Lillie ceased to build silk spinning machinery. Ure records that they 'resigned this branch of engineering' in favour of 'Mr Rison, a cotton-mill mechanic'.[163] There is no recorded date when this occurred, nor reasons for it, but textile machinery making was a complex specialist branch of engineering, distinct from traditional millwrighting.[164]

Corn Mills

IN parallel with work for the textile industry, the partnership undertook corn-mill work. They supplied the millwork and milling machinery to the six-storey Albion Corn Mill, Tib Street, Manchester in 1829, but the engine was by Peel, Williams & Peel.[165] In 1831 they built the Macclesfield Steam Corn Mill.[166] Fairbairn strongly favoured the linear layout of millstones rather than clusters. In *Mills and Millwork* (1863) he wrote:

> It is now upwards of forty years since the new system of corn mills, having the millstones in a continuous line, was first adopted. For many years it was called into question, and it met with a determined opposition from the old millers and millwrights, who stoutly maintained that the bevel-wheel principle was decidedly inferior to the old plan of stones ranged round a large spur wheel. Time, however, showed the advantages of the new plan.[167]

In *Useful Information for Engineers: Third Series* (1866) he claimed that 'the new system of placing the stones in a line, and driving them by bevel gear along one side of the mill' was first introduced 'by Fairbairn and Lillie, of Manchester, and some others in Yorkshire'.[168] The Macclesfield mill is the first known mill to have this layout. It was four stories in height, plus attic. The main stone room was on the first floor. It was rectangular in plan, with twelve pairs of mill-stones – five along one wall and seven along an adjacent wall, driven by shafting and gearing on the ground floor.[169] This mill is also believed to have been the first mill building with Hodgkinson beams.[170] The mill was modernised by others in 1854, including the introduction of a ventilation system. A major explosion in 1881, caused by the ignition of the dust in this ventilation system, demolished the mill.[171]

Fig.2.9: Alternative layouts of millstone – clusters or linear.

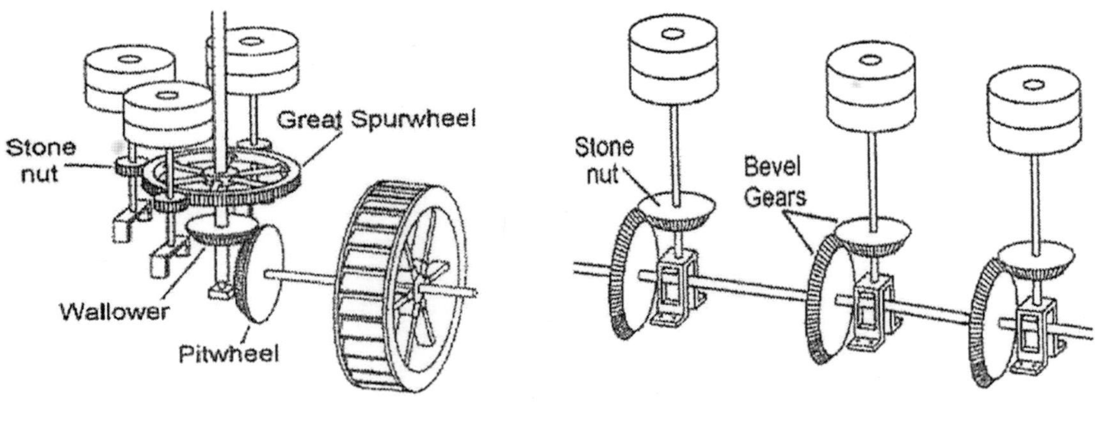

Fig.2.10: Milford Mills, Carlow, Ireland.

There was an extensive corn-mill at Milford Mills, Carlow, in Ireland, although the exact date is unclear – it could have been after the dissolution of the Fairbairn & Lillie partnership. In 1836 Milford Mills was described as 'unrivalled',[172] and in 1840 Mr and Mrs Hall's *Ireland* stated:

> The establishment at Milford is one of the most extensive and celebrated in Ireland. ... The Milford works have been constructed under the superintendence of Mr William Fairbairn of Manchester; and the chief water-wheel made by him, of iron ... In the two establishments for producing flour and oatmeal, there are twenty-two pair of millstones in constant work; thirteen of which, with all the attendant machinery, are driven by one wheel.[173]

Waterwheels: Into Europe

FAIRBAIRN and Lillie became the most renowned builders of industrial vertical waterwheels during the 1820s, and Fairbairn carried this on during the next twenty years, during which waterwheels continued to be widely used.[174] Fairbairn's approach was that of the optimiser, bringing together the best components available from whatever source, and introducing incremental improvements. The success of Fairbairn's waterwheels validates this approach, at least in respect of an old technology such as waterwheels. There are four elements to industrial water power – the water supply, the pentrough which delivers the water to the wheel, the wheel itself, and the transmission of power from the wheel to the machinery. Where water supply was intermittent, reservoirs were interposed. The fall could be increased by a tail-race discharging some distance down-stream – on occasion by means of a tunnel forming an inverted siphon.[175] Fairbairn preferred the high-breast wheel, that is one where water is fed above the axis.[176] Smeaton had realised 'the vast potentialities of the breast wheel',[177] and the high-breast wheel became practicable after John Rennie's introduction of the sliding hatch in the 1780s.[178] From Hewes Fairbairn adopted the 'suspension wheel' with slender wrought-iron spokes like a bicycle wheel and with the transmission of power from the inside of the circumference of the wheel, rather than from the axle.[179] Hewes' spokes had threaded ends, secured to flanges on the wheel hub by nuts, which tended to work loose causing the wheel to go out of balance. Fairbairn squared the ends of the spokes, securing them in sockets in

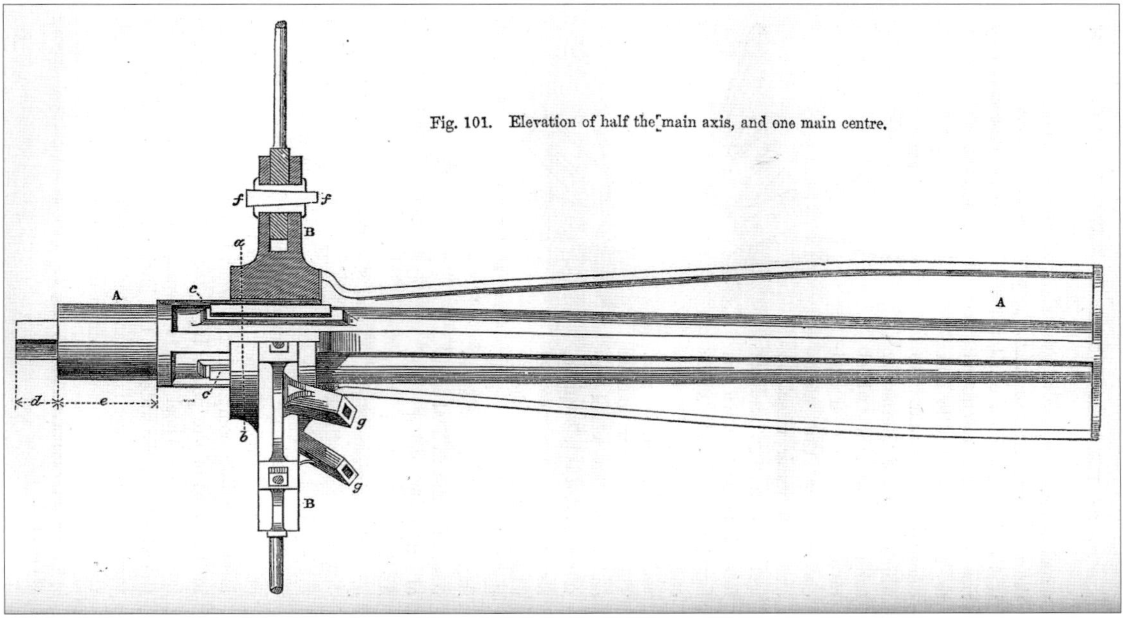

Fig.2.11: One hub and half of main cast-iron axle of a Fairbairn waterwheel.

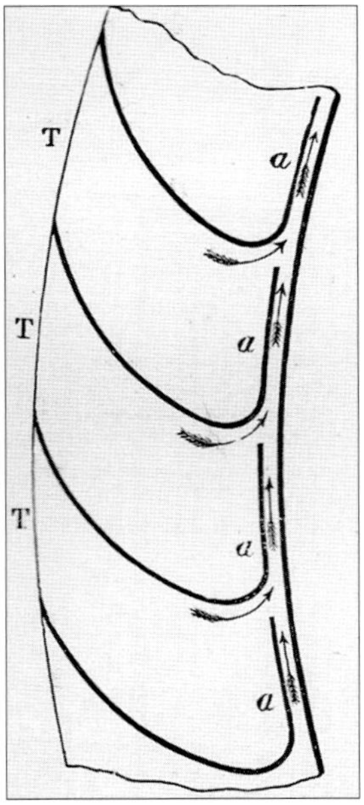

Fig.2.12 : Ventilated waterwheel buckets.

a cast-iron hub by means of gibs and cotters, instead of nuts.[180] Fairbairn preferred to deliver the water to the wheel via an iron pentrough. From John Rennie he adopted, and improved, the curved moveable pentrough shuttle which controlled the flow of water.[181] As textile machinery became more sophisticated it was important that it was driven at a constant speed. Primitive centrifugal fly-ball governors were applied to windmills around 1780 and to steam engines around 1790. Hewes used one for a waterwheel at Belper but it was cumbersome.[182] Fairbairn improved both the governor and the mechanism which it operated to raise or lower the pentrough shuttle. This controlled the flow of water to the wheel, maintaining a constant speed, essential for the operation of more sophisticated machinery.[183]

In 1828 Fairbairn made an innovation with significant economic consequences – the provision of 'ventilated buckets'. When unventilated buckets were filling, air could not escape fast enough, forcing spray to around 6ft above the shuttle, causing a nuisance and, more seriously, reducing efficiency.[184] He overcame this by introducing airways at the tops of the buckets, which stopped the spray and increased the power by nearly a quarter. However this may be a case of Fairbairn adopting an existing feature as his own. In 1816 John Sutcliffe had written of a wheel at Sowerby Bridge with 'a very ingenious but simple plan for supplying the buckets with air'.[185] Bodmer's obituary claimed that the Egerton waterwheel, designed by Bodmer and completed by Fairbairn & Lillie, incorporated a novel feature allowing the free

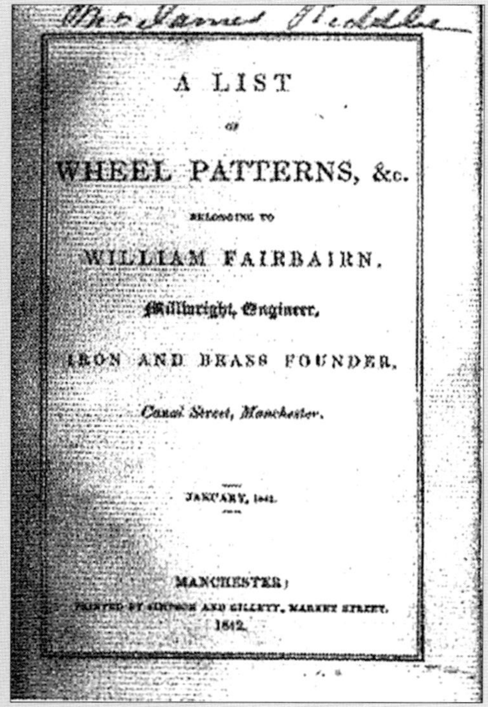

A LIST
OF
WHEEL PATTERNS, &c.
BELONGING TO
WILLIAM FAIRBAIRN,
Millwright, Engineer,
IRON AND BRASS FOUNDER,
Canal Street, Manchester.
JANUARY, 1842.
MANCHESTER:
PRINTED BY SIMPSON AND GILLETT, MARKET STREET.
1842.

Fig.2.13:
Fairbairn's
List of Wheel Patterns &c.,
1842 Edition.

WATER WHEEL SHROUDS,

No. of Pattern.	Diameter of Shrouds. Feet	Inches	Depth of Shroud. Inches.	No. of Buckets.	Opening in buckets. Inches.
436	60	4	14	144	3¾
313	50	0	15	120	6
306	46	0	10	114	4½
305	39	9	12	112	5
380	36	0	18	80	7
308	30	0	12	96	5½
401	26	0	10	70	5
307	24	0	14	70	6
303	22	9	8	80	3¾
340	22	3	14	70	6
301	22	0	10	70	4½
338	22	0	10	48	7
304	19	9	10	60	4½
391	19	0	12	48	5
334	18	0½	15	48	6
400	18	0	20	40	13
310	16	0	floats 18	40	15
319	16	0	15	40	9
339	16	0	16	40	7½
484	16	0	24	32	10½
493	40	0	10	112	4¾
502	22	6	14	60	6
525	18	0	20	32	11½
530	36	0	16	80	7
540	16	0	21	32	17
559	26	0	18	60	7
564	26	3	14	70	6
566	18	0	12	48	6½
577	18	0	22½	32	21

54

AND SEGMENTS.

No. of Pattern.	Diameter Feet	Inches	No. of Cogs.	Pitch. Inches.	Breadth. Inches.
430	69	11¾	912	2½	10
Inter. 313	48	6	560	3¼	15
Inter. 306	44	6	560	3	10
305	39	4	592	2½	8
Inter. 380	33	8½	384	3¼	15
Inter. 308	28	9	360	3	10
Inter. 307	22	4	280	3	10
303	21	6½	390	2⅜	6
343	21	6	270	3	8
301	21	10	330	2½	6½
353	21	8	272	3	12
Inter. 304	18	6½	280	2½	8
391	17	7½	204	2½	10
336	17	2½	216	3	12
416	16	1	200	3	12
Inter. 339	14	0¾	176	3	12
310	15	4½	192	3	12
Inter. 492	20	6½	310	2½	8
Inter. 493	38	5¼	480	3	9
Inter. 503	20	8	260	3	10
Inter. 525	16	0	200	3	14
Inter. 530	33	6½	384	3½	14
Inter. 540	14	0¾	176	3	9
555	15	4½	192	3	8
Inter. 559	23	8	270	3¼	12
562	60	2¾	696	3¼	12
Inter. 564	24	8	310	3¼	10
Inter. 566	16	6	248	2½	8
Inter. 573	15	11¼	200	3	10

55

escape of air from the buckets.[186] A letter in *The Engineer* in 1877 about this wheel referred to its '144 ventilating buckets' and found it 'rather remarkable that Mr Fairbairn should have forgotten to mention the wheel' and 'the name of its designer'.[187] It could be that ventilated buckets were Sutcliffe or Bodmer's idea but Fairbairn was the 'innovator' who developed them and made them known.

Mass-production technology, dependent upon the assembly of interchangeable parts, has been claimed as 'a distinctively American achievement',[188] although its origins were in 18th-century France.[189] In 1812, when exporting steam engines, Maudslay sent them with 'all necessary duplicates ... of the wearing parts to ensure their perfect success in countries where mechanical assistance cannot easily be procured'.[190] Fairbairn and Lillie's 'extensive stock of patterns' confirms that standardisation and interchangeability of parts was well-established in Manchester by the mid-1820s.[191] Its origins there went back to as early as 1808 when engineers Peel Williams were distributing catalogues of standard castings.[192] Fairbairn published a *List of Wheel Patterns* which included standard waterwheel components. The earliest edition seen is that of 1842 but there were almost certainly earlier editions.[193]

Fairbairn's waterwheels are examples of evolving and developing technology. They exhibit cumulative improvements – securing of spokes, ventilated buckets, automatic control of the pentrough shuttle, standardisation of parts – thus supporting the view that inventive activity is 'a gradual process of accretion, a cumulation of events'.[194] There is no record of how many waterwheels Fairbairn built. In *Mills and Millwork* he lists the dimensions of sixty-two of his waterwheels, but without locations.[195] He almost certainly built many more than this and there may have been multiple wheels of the same size, using standard castings. To date approaching fifty locations have been identified, but the information from varied sources in Appendix 2.2 does not correlate with the dimensions in *Mills and Millwork*. The probability is that he built several hundred and there is little doubt that he was the leading manufacturer of vertical industrial waterwheels in the second quarter of the 19th century.

The importance and quality of these wheels can be gauged from some examples. The first known Fairbairn & Lillie waterwheel was for the Ashworths at New Eagley Mill, Bolton, in the early 1820s.[196] In 1825–7 Fairbairn & Lillie built the famous waterwheels at Catrine – two 50ft diameter wheels, with provision for a further two. Archibald Buchanan visited Manchester, interviewed Fairbairn and engaged the firm. Fairbairn spent a fortnight at Catrine, surveying, and designing the water supply.[197] He claimed these wheels were the first erected on the principle of concentrating all the power for a mill complex in one wheel-house.[198] A tunnel led from the sluice gates on the Ayr to a reservoir. From there the water flowed along a 12ft wide arched-over conduit to the wheelhouse, 600yds from the river.[199] They were described as housed in 'a great dripping stone mansion' with 'a noise like thunder everlastingly'.[200] The evidence suggests that when it was built this installation of two wheels in tandem was the most powerful hydraulic prime-mover of the time. It is a testimony to Fairbairn & Lillie's work that these wheels drove the Catrine mills for 120 years, until 1945.[201] It also endorses Woodcroft's assessment that 'the construction exhibited in all

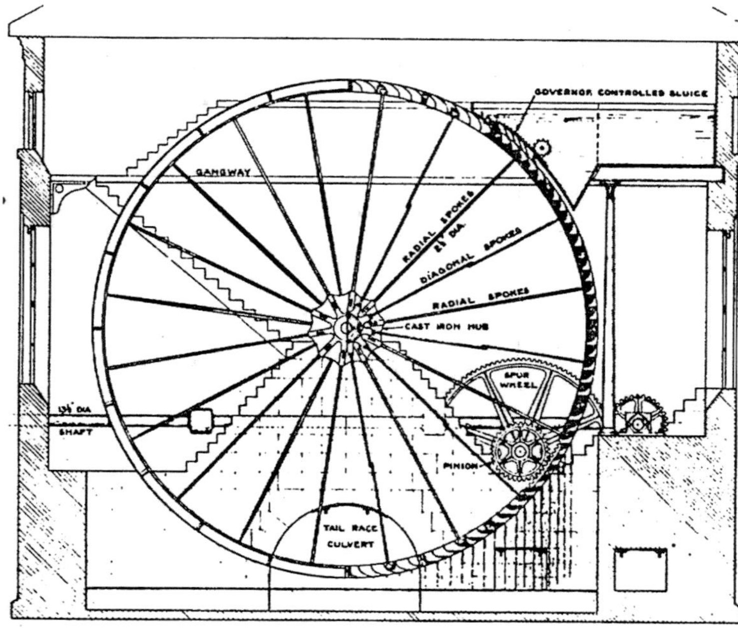

Fig.2.14: Catrine Waterwheels.

(above right) **Fig.2.15: Section/ Elevation of Catrine Waterwheels.**

the examples which issued from the Canal-Street works was masterly'.[202] There is a rebuilt Fairbairn waterwheel at the National Trust mill at Styal, Cheshire, moved there from Glasshouses, Yorkshire.

Success at Catrine led to waterwheels for another Finlay mill, at Deanston, Perthshire. Here the proposal was for eight 36ft diameter wheels, four on each side of a central iron pentrough supported on cast-iron columns. Only four wheels were built and only two of those by Fairbairn & Lillie. Deanston's manager, the engineer and inventor James Smith, observed the construction of the first two wheels and copied them, using local labour.[203] Here is diffusion of technology in its simplest form, a classic example of 'imitation'. Smith then developed what he had learnt, building a massive 70ft diameter wheel at Greenock.[204]

Fairbairn & Lillie's line-shafting and waterwheels quickly disseminated into mainland Europe. Just as their first major clients in Manchester had engineering backgrounds, it was through contact with one of the most brilliant engineers of the day that a chain of European clients was established. In the first decade of the century the Swiss engineer and inventor Johann Georg Bodmer had helped J.C. Escher to establish a cotton mill in Zurich, which 'was among the largest cotton manufactories on the continent'.[205] Around 1823 Bodmer was involved with the reconstruction of the Schinznach Hydro in the Swiss Canton of Aarau. Bodmer, like many leading engineers of that time, was drawn to Manchester, where he found employment with Sharp, Roberts & Co. He arranged for the waterwheels and pumps for Schinznach to be constructed by Fairbairn & Lillie.[206] Escher, with his son Albert, visited Bodmer in Manchester in 1824. It was arranged that Albert would study mechanics under Bodmer, plus a pupillage with Fairbairn.[207] Escher returned to Zurich, accompanied by Fairbairn, who was

Fig.2.16:
Johann Georg Bodmer
(1786–1864).

Fig.2.17:
Hans Caspar Escher
(1775–1859).

Fig.2.18:
Albert Escher
(1807–45).

there for three weeks, 'devising the alterations and improvements' but also fitting in some sightseeing, including walking up the Righi. He received orders for waterwheels and millwork for Escher's new mill, and for renovating Escher's existing mill, on the Limmat, the fast-flowing river draining the Zurichsee. There was no fall, so Fairbairn designed waterwheels that could be raised or lowered by cast-iron levers, so that they were always at the right depth in the water.[208] In this he followed a precedent set by Bodmer at a flour mill on the Limmat, built around 1816.[209] Subsequently Albert Escher returned to Zurich to run the engineering department of Escher Wyss.[210]

On route to Zurich, Escher and Fairbairn visited the Vosges and Alsace, where

Fairbairn found the name of Fairbairn & Lillie was already known; and in the next three years the firm received orders for several new waterwheels in France.[211] The visit to Gros, Davillier, Roman & Co. at Wesserling was particularly opportune, as they were planning a new cotton mill. Fairbairn obtained the order for the waterwheel and millwork, and subsequently attended the opening of the mill. He believed this was the first waterwheel 'on the suspension principle' in France.[212] This may be incorrect as a 'roue anglaise' was built at Gisors in 1816.[213] In any event there was a technological time-lag from Britain, where Hewes had built suspension wheels since 1802.[214] Further orders were received from France until 'the French were able to construct the improved wheels for themselves', as when Gros, Davillier, needing a second wheel, had a copy of Fairbairn & Lillie's wheel built locally, another instance of 'imitation'.[215] A.L. Dunham writes that the adoption of these waterwheels 'by the leading textile mills and machine shops of Alsace' was one of the factors which increased the utilisation of water-power in France during the Restoration.[216]

Old technologies persisted alongside new technologies well into the 19th-century. In the UK textile industry, in 1838 there were 3,053 steam engines and 2,230 waterwheels (although the average horse-power for the engines was 24.3 compared with 12.6 for the waterwheels).[217] Rosenberg, using Fairbairn waterwheels as his illustration, has identified the reason for this persistence as 'continued improvements'.[218] This may have been so in the 1820s, but beyond 1830 there were few improvements and, with little financial advantage in water-power over steam,[219] technological inertia was probably the major reason. *Table 2.1* shows the number of *known* Fairbairn waterwheels built in each five-year period.[220] This must be treated with caution as relatively few of the total wheels built are known. Nevertheless it clearly suggests a slow decline in the 1830s, levelling off in the 1840s and '50s, and falling away in the 1860s.

The decline was primarily due to the increasing use of steam power, but also to

Table 2.1: Numbers of known Fairbairn waterwheels with dates.

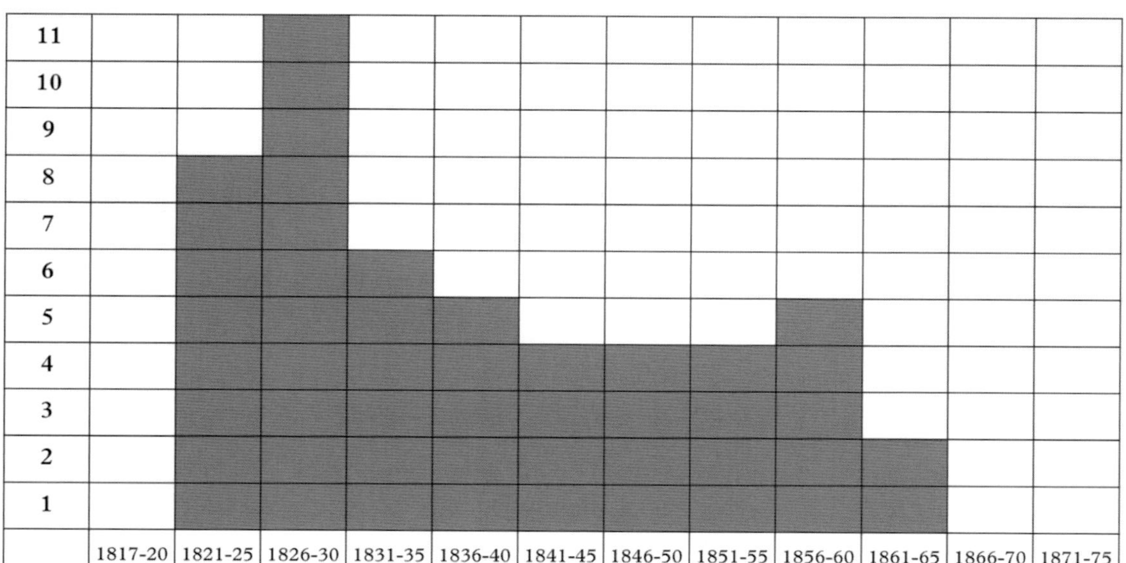

the gradual replacement of the vertical waterwheel by the more efficient horizontal waterwheel or turbine, particularly on the Continent. There was inertia on Fairbairn's part in moving to turbines. Recent interest in technological inertia prompts the question why this was so.[221] He was well aware of turbines. In 1824 he had met Benoît Fourneyron, a major figure in their development, and inspected one of his turbines.[222] The probable explanation is that Fairbairn had a high reputation and profitable trade in a machine which was tried and tested, simple and understandable, whereas the turbine required knowledge of theoretical dynamics which he did not have. Furthermore the turbine was more appropriate to the higher falls more generally found on the Continent. Thus there was little incentive to change, and by mid-century when the case for the turbine was clear, Fairbairn was taken up with a plethora of other work. This explanation is consistent with the difference between the empirical, experimental Baconian science found in Britain, and the theoretical and abstract science of France. In Britain research was conducted by practical engineers – Smeaton, Banks, John Rennie, Hewes and Fairbairn – in search of a better water mill.[223] On the Continent work on water power was largely theoretical and carried out by mathematicians.[224] Woodcroft's observations, made in 1877, are apposite to this, as to the wider science and technology debate:

> Abstract principles, purely scientific rules, never took a first place in his mind, which was intensely practical in its conceptions, and never wandered far into the ideal. In saying this, we derogate nothing from Fairbairn's real merits; for has not Bacon himself declared that the end of science, as of all knowledge, is 'fruit'.[225]

Fairbairn's inertia in respect of turbines was in parallel with his on-going cumulative improvements to the vertical waterwheel, primarily during the 1820s. Well before mid-century Fairbairn had brought the vertical waterwheel to the zenith of its 2,000 years of development – increasingly it was giving way to steam or turbine. The lasting quality of Fairbairn's engineering of weirs and leats is seen in the adoption of several of them to drive hydro-electric turbines in the 21st century.[226]

An unusual transfer of technology took place in the 1860s when Fairbairn was engaged to apply his expertise with waterwheels to treadwheels for Walton gaol in Liverpool. There had been treadwheels in over fifty prisons since the 1820s. In some areas they were seen as punishment and rarely put to any useful purpose, but this was not so in Lancashire where magistrates were keen that power from the wheels should be applied to practical objects. For example the Kirkdale wheel ground grain and the Lancaster wheel drove power looms.[227] What Fairbairn did in the 1860s was to increase the scale of the installations. Thus at Walton six lines of wheels, capable of accommodating 216 men, powered, via typical line-shafting, a weaving shed and mat-dressing machinery for the manufacture of cocoa-nut fibre matting. In addition the mill pumped all the water for the prison from an 80ft well to storage tanks.[228] Designed by Fairbairn, the machinery was built by C.J. Appleby & Associates. A similar installation, built by Fairbairns, followed at Belle Vue prison in Manchester.[229] Treadwheels were abolished in British prisons in 1902.

Fig.2.19: Eaton Hodgkinson (1789–1861).

Eaton Hodgkinson, Fairbairn's Lever and the Water Street Bridge

FAIRBAIRN'S experimental work was quickened for two decades, from the mid-1820s, by links with Eaton Hodgkinson (1789–1861). These links brought reward by way of prestige, and the opportunity for Fairbairn & Lillie to use 'Hodgkinson beams' in a high profile bridge and elsewhere. Hodgkinson, from a modest background although with a grammar school education, studied the works of continental mathematicians under John Dalton.[230] In 1820 he joined the Manchester Literary & Philosophical Society, of which Dalton was President. In 1822 he read his first paper, 'The Transverse Strength of Materials'.[231] Then he undertook experiments on the strength of cast-iron beams at Hatton's Salford foundry. Ewart noted Hodgkinson's work and realised its importance. He prevailed upon Fairbairn & Lillie to provide facilities for Hodgkinson's experiments, and attended during many of them.[232] When Hodgkinson married, at Cross Street Chapel in 1841, Fairbairn and John Dalton were witnesses.[233]

Equipped with the best resources and freed from financial constraints, Hodgkinson undertook experiments which found that the moduli of elasticity, elastic limits, and ultimate strength of cast iron have different values for tension and compression.[234] With this information, he produced a form of cast-iron beam with a twenty-five per cent saving in metal for the same strength, thus reducing self-weight and cost.[235] As the beams in a typical mill could weigh 100 tons, the savings were substantial. He presented his results in a second paper in 1830, 'Theoretical and Experimental Researches to ascertain the Strength and Best Form for Iron Beams'.[236] This investigation, more than any other, illustrated the benefits of bringing Hodgkinson and Fairbairn together, in an extremely productive working arrangement that lasted for twenty years,[237] an illustration of the 'communication between those who knew things and those who made things' as highlighted by Mokyr.[238] Fitzgerald sees Hodgkinson's paper as the 'basis of all subsequent 19th-century cast-iron [structural] design' and 'a clearly defined turning point in the history of cast iron'.[239] Fairbairn's *Application of Cast and Wrought Iron to Building Purposes* includes a fold-out chart of the strengths and other properties of fifty-one different cast-iron bars which he and Hodgkinson had tested.[240] In 1840 Hodgkinson wrote to Fairbairn about his works, that 'more experiments, of a really useful character have been made there, either by yourself or me, than have been made at any one place in Europe in the time'.[241] Fairbairn & Lillie's role in connection with Hodgkinson's work was the provision of premises, apparatus – the 'Fairbairn lever' – and finance.[241] This did not go unnoticed: in 1836 John Dalton wrote a letter of introduction for Fairbairn to the Director of the Paris Observatory, Francois Arago – mathematician, scientist, biographer, politician and close friend of Alexander von Humbold:

> [He is] a very large manufacturer of iron machinery in this place, such as steam-engines, boats, boilers, mill wheels, etc ... He is a member of our

> Philosophical Society and has a great fund of practical mechanical knowledge. He has been liberal in furnishing models for one of our members, in order to ascertain the best form of strength.[242]

Fig.2.20: George Stephenson (1781–1848).

The 'Fairbairn lever' was a mechanical device to apply loading to beams or samples in order to measure such properties as deflection or crushing strength. It was used for many investigations throughout Fairbairn's career, sometimes with modifications.[243] It was also used for many years by Hodgkinson.[244] As Fitzgerald notes, this 'understanding of cast-iron beam behaviour was ... unequivocally the result of scientific investigation'.[245] As late as the 1867 edition of Peter Barlow's standard work on the strength of materials, Hodgkinson's results still occupied eleven pages.[246]

Fairbairn was responsible for the first use of Hodgkinson beams, in the construction of a railway bridge in 1830. Again a peer engineer, George Stephenson, who was now Engineer to the Liverpool & Manchester Railway, provided the patronage.[247] Stephenson visited Fairbairn's works during some of Hodgkinson's experiments. Adjacent to its Manchester terminus, the railway had to bridge Water Street, for which the Act imposed onerous headroom requirements. Stephenson consulted Fairbairn who designed a bridge with five Hodgkinson beams spanning 24ft 6in. The beams, cast by Fairbairn & Lillie, were supported by nine columns –

Fig.2.21: Fairbairn Lever as used by Hodgkinson in determining the strength of cast-iron columns.

Fig.2.22: Water Street Bridge: the first iron bridge on a main-line railway.

probably cast-iron – on each side of the roadway.[248] Water Street Bridge was a highly influential structure. Bennet Woodcroft wrote:

> The rapidity with which Hodgkinson's results were adopted in every part of the world was largely due to the courage and sagacity with which Fairbairn from the first employed them ... [Water Street Bridge] remained for some years a shrine to which engineers from every part of the world resorted for instruction.[249]

The significance of this bridge is that, not only was it the first structure using Hodgkinson beams, it was the first main-line iron railway bridge. It was also an example of the technology designed for one application – multi-storey mills – being transferred to another – railway bridges. In the mid-1830s Robert Stephenson used twenty-one bridges with Hodgkinson beams on the London & Birmingham Railway, and many more were used in bridges elsewhere.[250] The beams were also widely used in mill buildings, for which they were originally intended – the first is believed to have been Fairbairn & Lillie's Steam Corn Mill at Macclesfield in 1831,[251] and the best documented is Orrell's Mill at Stockport.[252]

Dissolution of the Partnership

BY 1830 the partners considered themselves rich, but storms were brewing.[253] In August 1831 there was a serious fire at the works, affecting the three-storey building on Pott Street. Only the fire-proof pattern-store survived. The building was insured but not the contents, on which the partners suffered a loss of £5,000.[254] More seriously, dissention was creeping in between the partners. They had a steady flow of very profitable millwork and waterwheels, in which they were market leaders. Why expand, as Fairbairn was pressing to do, into new, uncharted and hazardous territory? Why take unnecessary risks? Two ventures brought matters to a head, shipbuilding and cotton manufacturing.

Fig.2.23: Experiments with the Swift using a mercury dynamometer and a stopwatch.

The Forth & Clyde Canal

THE first venture to put strain on the partnership was shipbuilding. The Forth & Clyde Canal Company was moving towards iron steamboats, following pressure from one of its committee members, Thomas Grahame, who was convinced that the application of steam power to canal boats would fend off any challenge from the railways. By 1828 he was running a steamboat service, but only at three miles per hour.[255] In 1830 Fairbairn was commissioned by Grahame, on behalf of the Canal Company, to further investigate the application of steam power to canal boats.[256] The initial experiments were with the horse-drawn *Swift,* using a mercury dynamometer and a stopwatch to determine power required at various speeds. Fairbairn published his first book, setting out the results of the experiments which showed a great increase of force necessary to overcome the effect of surge as speed increased.[257] The experiments were a combination of empirical trials, without prior theoretical knowledge, in an attempt to increase the speed of canal boats, and the systematic measuring of force and speed in order to determine a mathematical relationship between them. Who designed and made the dynamometer is unclear. Fairbairn was typically optimistic, indicating to the Company that he anticipated achieving nine to ten miles per hour.[258] In consequence the Company placed orders for iron steamboats with Fairbairn & Lillie. For the engine of the first, Fairbairn approached Boulton & Watt, through Peter Ewart, and Stephensons. He gave 'young Stephenson' the order for the boiler and two cylinders, for which Fairbairn & Lillie did not have the necessary machine tools.[259] The boat had a single central paddle wheel with the engine immediately in front of it.[260] In this there was some affinity with John Laird's first iron steamboat, the *Lady Dunally* of 1829, the design intended to reduce damage to canal banks which was the principal objection to

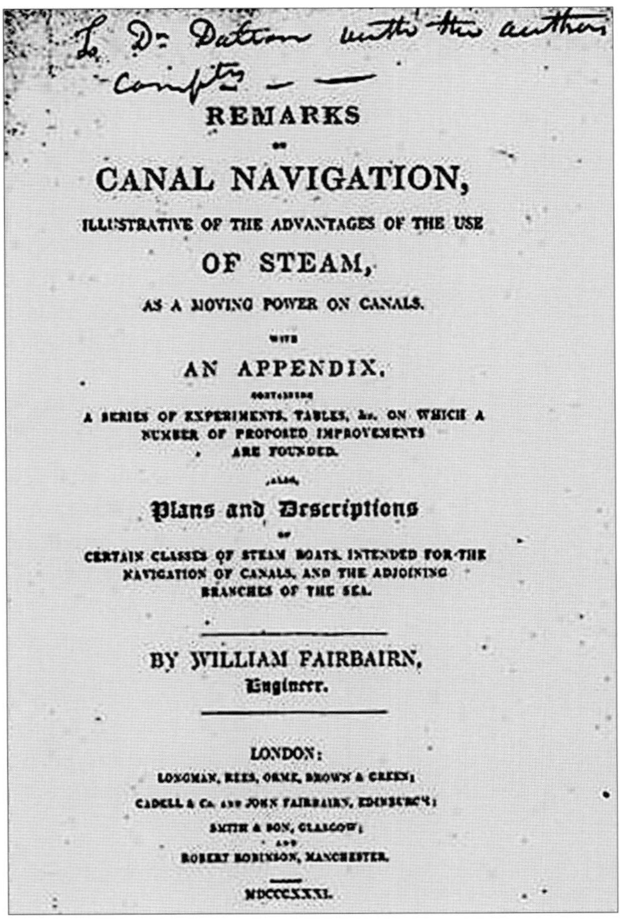

To Dr Dalton with the author's compts — —

REMARKS
ON
CANAL NAVIGATION,
ILLUSTRATIVE OF THE ADVANTAGES OF THE USE
OF STEAM,
AS A MOVING POWER ON CANALS.
WITH
AN APPENDIX,
CONTAINING
A SERIES OF EXPERIMENTS, TABLES, &c. ON WHICH A NUMBER OF PROPOSED IMPROVEMENTS ARE FOUNDED.
ALSO,
Plans and Descriptions
OF
CERTAIN CLASSES OF STEAM BOATS, INTENDED FOR THE NAVIGATION OF CANALS, AND THE ADJOINING BRANCHES OF THE SEA.

BY WILLIAM FAIRBAIRN,
Engineer.

LONDON:
LONGMAN, REES, ORME, BROWN & GREEN;
CADELL & Co. AND JOHN FAIRBAIRN, EDINBURGH;
SMITH & SON, GLASGOW;
AND
ROBERT ROBINSON, MANCHESTER.
MDCCCXXXI.

Fig.2.24: Fairbairn's first book. Presentation copy to John Dalton.

steamboats on canals.[261] As the launch day approached, Fairbairn, having typically, unhesitatingly and confidently taken on the challenge, gave an insight into his state of mind: 'my doubts and misgivings continued to increase ... if I did not succeed, I must fail, and a failure was, of all things, to my mind, the most obnoxious and disagreeable'.[262]

The trials on the Irwell were not successful. He despaired of achieving above eight miles an hour.[263] However the boat was sent, in parts, for trials on the Forth & Clyde. She realised eight miles an hour without passengers, but when laden this fell to six. Grahame reported to the committee, counselling a sympathetic approach because:

> Not much more than nine months have elapsed since any one proposing to move a body or vessel weighing fourteen or fifteen tons thro' a Canal at the rate of eight miles an hour would have been pronounced to be insane.[264]

Notwithstanding his failure to achieve the speed he anticipated, that he achieved a speed in excess of any other canal steamboat at that time indicates how advanced Fairbairn's work was in this field. A compromise was reached. Fairbairn & Lillie would substitute a lighter engine, and the company would make a deduction for the first engine which would be returned, and a further deduction of £164 for the 'total failure' of the boat.[265]

The second boat for the company caused Fairbairn much stress but clarified issues between canals and railways. She was a paddle-steamer with wheels at each side of her stern. Trials were held on the Mersey, with George Rennie aboard. Fairbairn decided she should steam to Douglas on the Isle of Man where he would join her for the rest of the voyage to the Clyde. When he arrived in Douglas, she was not there. He took a vessel to the Clyde and after frantic fruitless searching returned to Douglas to be told she was at Ramsey. The iron hull had affected the compass and she had ended up off Cumberland. She reached Glasgow with Fairbairn on board but after trials on the canal he concluded there was no chance of canal boats matching the speed of railway locomotives and advised the company to concentrate on freight.[266]

Fairbairn also advised constructing iron vessels designed for both canal and sea navigation, and the third vessel for the Forth & Clyde, with two side paddles at the stern, was designed for this dual use.[267] In November 1831 the boat, on drags, was pulled across Manchester from Ancoats to the Irwell, by fourteen horses, taking several days and causing much inconvenience.[268] In February 1832 she sailed from Liverpool to the Clyde, meeting a severe gale through which she 'behaved to the admiration of all on board'.[269]

Showing Bennet Woodcroft over a partially constructed boat in his Ancoats yard, Fairbairn told him:

> I am too old but you may yet live to see almost every ship afloat made of iron; the wooden ships will be the exception, and the iron ships of the future will possess a lightness, strength, and durability unknown to wooden vessels.[270]

It was a visionary and prophetic observation from a man whose main concern seemed to be technological achievement, regardless of the risks in sending

Fig. 1.

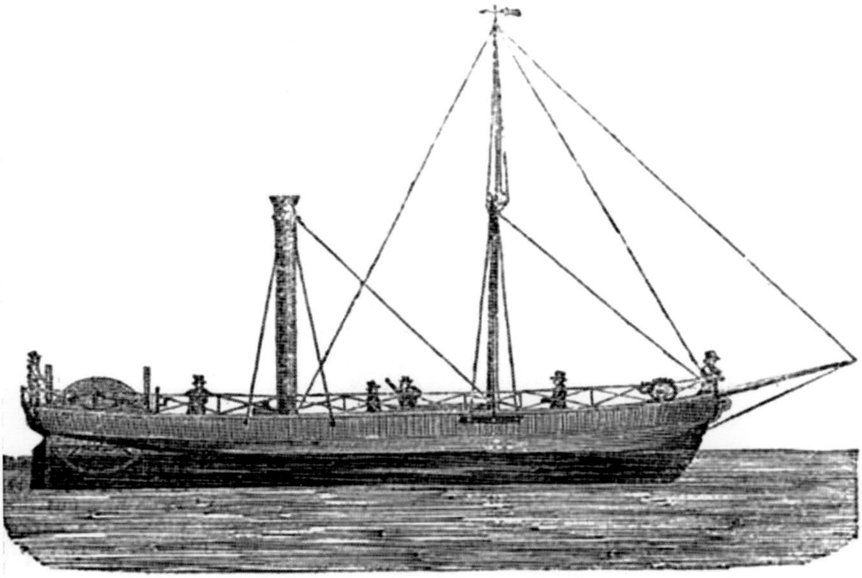

Fig.2.25: Fairbairn's first sea-going iron steamship.

Fig. 2.

these pioneering boats to sea and the risks of financial loss. The partners did lose money on the first Forth & Clyde boat, and probably on the others. Lillie, more cautious and prudent than his business partner, had reservations – in Fairbairn's words, 'although he did not oppose, he did not cordially join with me in the undertaking'.[271] Lillie's reservations were with good reason and such as any prudent businessman might have had. Future events would show them to have been well founded. But, whilst shipbuilding might have provoked tension, Lillie had a much larger concern.

Egerton Cotton Mill

THE final straw that broke the Fairbairn & Lillie partnership arose from the purchase, in partnership with their clients Henry and Edmund Ashworth, of a partly-built cotton mill and dye-works at Egerton, Bolton. J.G. Bodmer, for whom Fairbairn & Lillie had worked at Schinznach, was one of the most inventive of engineers, who is widely recognised today as a pioneer of automation.[272] Like Fairbairn, Bodmer ranged over a wide span of engineering, apparently, at least on the basis of this project, more interested in engineering solutions than money.

Bodmer, with a Swiss chemist and an Italian merchant resident in Manchester, set out to build a cotton mill and dyeworks at Egerton, Bolton. Bodmer was to design the cotton mill and machinery, his Swiss colleague the dye-works, with Novelli, the Italian, providing the finance.[273] It was no ordinary mill as Bodmer proposed great savings by automation. By 1828, £40,000 had been spent with no completion in sight. Novelli took fright. Bodmer, suffering ill-health, returned to Switzerland. The part-built mill was offered for sale well below cost. It was bought for £13,000 with the two Ashworths, Fairbairn, and Lillie taking a quarter-share each. Fairbairn knew Bodmer well, having undertaken work for him, and it may be that Bodmer's proposals for the automation of the Egerton mill were particularly attractive to Fairbairn. The building was doubled in size, the weir and 62ft diameter water-wheel were completed, millwork and machinery were erected and commissioned. But the drain on capital was crippling Fairbairn & Lillie's core business.[274] This suggests that Fairbairn had typically grasped at a venture that excited him, without adequate assessment, albeit his instinct had been right as the mill went on to have a successful future.[275]

Misunderstanding and mistrust deepened between the partners and Fairbairn made the decision to end the partnership, a decision with which Lillie was not in accord. Fairbairn offered to pay Lillie out, or for Lillie to have the business and pay him out. On 29 September 1832 the partnership was dissolved. Lillie opted for the money, which Fairbairn did not have to hand. It was therefore agreed that Fairbairn would transfer his Egerton shares to Lillie, and that the Ashworths would buy the shares. In the event Lillie loaned money to the Ashworths, and was paid out over several years.[276] In addition to handing over his Egerton shares Fairbairn made an undisclosed lump sum payment to Lillie. To do this he was obliged to mortgage part of the Canal Street property for £4,500.[277] The various departments at Canal Street were broken up as most of the foremen joined Lillie.[278] Established clients were lost, such as Wood at Bradford and Gott at Leeds who turned to Wren & Bennett (successors to Hewes & Wren from 1832).[279] These were difficult days for Fairbairn, but his response was typically entrepreneurial:

> I entertained the utmost confidence in my own powers; and ... came to the conclusion that I had nothing to fear, and that ultimate success was sure to follow.[280]

Fairbairn recorded at some length 'the causes and circumstances of the disagreement' but they were edited out by Pole. Similarly Woodcroft did not discuss 'the motives or their justifiableness' which caused the dissolution. The general impression was that Lillie was forced out by Fairbairn and, opening a new business in Store Street, he received sympathy and support.[281] The reasons for the decision of the foremen are unrecorded and it is only possible to speculate. As Lillie had run the workshops he would have established closer working relationships with them. Some might also have been concerned about their own security given Fairbairn's costly and largely unplanned ventures. Others may have moved because of sympathy for Lillie.

Tensions subsided over time. By 1847, Fairbairn wrote of 'My friend Mr Lillie', and later that he found Lillie 'as an opponent the same honourable, kind-hearted man that I found him as a partner'.[282] Lillie's entirely understandable caution about entering new and risky fields of business appears to have been at the root of the friction, and it is typical that Fairbairn was frustrated by such caution. It is also possible that Fairbairn, now with five sons, might have been thinking about a dynastic business. The two men were very different. Whereas Fairbairn's commitment to engineering was lifelong and led to fame, Lillie's firm closed after a fire in 1849, and he retired to the obscurity of a farm in Cheshire.[283]

Fig.2.26: The older part of Egerton Mill as it was in 1883.

Conclusion

THE key factor, from which the success of Fairbairn & Lillie's fifteen-year partnership stemmed, was Fairbairn's engineering insight and confidence, convincing the Murrays to engage an untried partnership to do something that had not been done before. Wrought-iron shafts, of reduced diameters, at increased velocities, transformed factory power distribution. This developed existing technology, but it was so far beyond the scope of an incremental development that it is better classed as an 'innovation'. The technology diffused rapidly as the Ancoats cotton mills, with their revolutionary power transmission, drew visitors from Britain and the Continent. Satisfied clients and Kennedy's network connections ensured continuing success for Fairbairn & Lillie. This provides correctives to Tann's statements that Fairbairn's important millwork took place in the 1830s and 1840s, that it is misleading to give him credit for the innovation of improved millwork, and that the adoption of improved millwork transmission systems was slow.[284]

In contrast waterwheels provide an example of incremental and cumulative development. Fairbairn optimised by bringing together and building on the elemental work of others, producing outstanding wheels whose technology was further diffused by copying or 'imitation'. The persistence of the old technology of the waterwheel is seen to result from improvements by Fairbairn and others, but there were no major improvements after the 1820s.[285]

The link with Ewart led to experimental work with Hodgkinson. It is in character that Fairbairn was prepared to accommodate Hodgkinson – at considerable cost – with no surety of future benefit. Yet benefit there was, as Fairbairn's own early experimental work was widened through this patronage – an amalgam of science and technology generating reliable knowledge and useful artefacts. In this case it was the Hodgkinson beam, which meshed with the link with George Stephenson, providing the catalyst for the transfer of beams developed for mill buildings to railway bridges; and demonstrating the recurring phenomenon within technological history of the interaction of innovations in separate spheres, as highlighted by Rosenberg and Vincenti with reference to tubular structures.[286]

The commissioning of Fairbairn, then primarily a millwright, to undertake canal experiments indicates the regard in which he was held by peer engineers. The aim was to develop a faster canal steamboat and the experiments included systematic measurements to determine the relationship between force and speed. In that they did not produce a steamboat of the planned speed, they failed, but they did provide the company with the knowledge, not previously appreciated, that canals would be unable to compete with railways for passenger traffic, and should concentrate on freight. It was through this unexpected commission that Fairbairn became a shipbuilder, exhibiting the opportunism and the unpredictability of Redlich's creative entrepreneur.[287] Yet as the range of work widened to embrace shipbuilding and steam-engine building as well as mill building, there was no clear indication of what Fairbairn's goals were, other than McClelland's classic entrepreneurial 'need to achieve'.[288]

The importance in business activity of networks related to places of origin, religious affiliation, educational, business or familial contacts, highlighted by Brown and Rose, is well illustrated by Fairbairn, in all except the familial.[289] His commitment to education found an early expression in the Mechanics' Institution where there was the significant side-effect of bringing him into contact with many influential people in the entwined networks of the Institution's committee, Cross Street and Mosley Street Chapels and the Literary & Philosophical Society. These networks certainly assisted Fairbairn's upward social mobility.[290] In becoming the Mechanics' Institution's first Secretary, as in providing facilities for Hodgkinson, Fairbairn firmly placed himself in the Industrial Enlightenment of Mokyr and Allen.[291]

Fairbairn and Lillie were very different in temperament. Fairbairn, the restless entrepreneur, ambitious and innovative, was excited by large and challenging engineering projects, but with little concern for their financial outcome.[292] Lillie, a competent engineer, was much more cautious and prudent, less ambitious, less socially adept, and understandably averse to taking risks which might bring calamitous consequences. A very successful partnership came to a painful end as these very different men went their separate ways.

Postscript to Chapter 2

From Millwright to Engineer

WHEN Fairbairn commenced business in partnership with Lillie in 1817 he was a millwright, and proud to be one. When the partnership came to an end in 1832, and he commenced business as a sole proprietor, he was an engineer.

Millwrights

FAIRBAIRN began his career as an apprentice colliery-millwright in 1804. The term 'millwright' is widely associated with him but its meaning is not always clear. Around the beginning of the 19th century significant changes were in train affecting millwrights and these continued, with the development of millwrighting firms and the transition of master-millwrights to engineers. For generations windmills, water mills and horse wheels were built and repaired by local country millwrights and itinerant journeyman millwrights. Their work involved the application of power to an industrial process and the transmission of that power to machinery. In milling they were also involved with the machinery and millstones. The first stages of the transfer to the factory system were achieved with the aid of traditional millwrights. They designed and built the shafting and gearing, and often designed the factory itself. They recommended the prime mover, and unless it was a steam-engine, they designed and built it.[293] As the number and size of factories increased in the latter part of the 18th century, master-millwrights engaged other millwrights, so that millwrighting firms emerged with some, such as Thomas Lowe of Nottingham, undertaking work throughout the country.[294]

By this time the division between master-millwrights and tradesmen millwrights was widening. Important among the former group were Andre Meikle (1719–1811) to whom John Rennie (1761–1821) was apprenticed, John Penn (1770–1843), and Bryan Donkin (1768–1855). Meikle apart, these master-millwrights gravitated to London and set up significant firms. From around 1777 they formed a society to protect their position. In 1799 there were eighteen members, with Rennie and Donkin much involved. In 1805, now styled 'The Society of Master Millwrights', they met regularly during a protracted dispute with journeymen.[295] In parallel, by 1799 there was also an organisation of journeyman millwrights, with the purposes of obtaining increased wages and limiting who was employed.[296] By the end of 1811 there were at least three competing London societies of employee millwrights.[297] In Manchester there were important master-millwrights, including Peter Ewart (1767–1842) and Thomas Hewes (1768–1832), and there was also the employees' Manchester Millwrights' Society.

These master millwrights worked in areas wider than traditional millwrighting. Meikle also invented the threshing machine. Rennie, besides traditional millwork, built rolling mills for mints, applied steam power to pile-driving and dredging, and built canals, harbours and bridges. Penn specialised in milling machinery, moving to marine engines in the mid-1820s. Donkin built papermaking machines, and was involved in canning, machine tools and instrument making. Ewart worked as agent for Boulton & Watt and subsequently became chief engineer to the navy's steamships. Hewes combined traditional millwrighting with machine-making. These master millwrights were exceptional men. However, the majority of millwrights were artisans and many had only limited skills. Some were much more akin to carpenters and concentrated on simple work – in the mid-18th century millwrighting had been seen as a branch of carpentry.[298]

At the start of the 19th century, wooden transmission systems and waterwheels were beginning to give way to iron, with components manufactured in foundries and workshops. At the same time steam engines were replacing some waterwheels. As more factory-made iron components were introduced, facilitated by an increasing range of capital-intensive machine tools – notably slide lathes, drilling, boring, planing and slotting machines – there were increasing challenges to the dominance of the millwrights' societies in the workshops. In the 1830s Nasmyth was selling standardized machine tools by catalogue. They did not require a skilled millwright to operate them, but a 'well-educated labourer'.[299] Traditional millwrighting was gradually being reduced to site erection of factory-made gearing and shafting, with a diminishing number of waterwheels. In 1841 William's brother, Peter Fairbairn (1797–1861) gave evidence to a Parliamentary Committee that it was only repair work that now required a millwright's all-round skills.[300] Increasing specialisation was reflected in new trade societies, including the Steam Engine Makers in 1824, the Journeymen Steam Engine and Machine Makers and Millwrights in 1826 and the Boilermakers in 1837.[301] Around 1851 many of the societies in the industry joined together to form the Amalgamated Society of Engineers, Machinists, Millwrights, Smiths and Pattern-makers, generally known as the Amalgamated Society of Engineers, bringing further loss of status to the millwrights and confusion about the designation 'engineer'. A decisive change came through the 1852 engineering dispute, after which the ASE lost control of the workshops, and employers were freely able to employ semi-skilled men to operate machine tools.[302] In parallel with improved machine tools in planned workshops, was the growth of the drawing office. This transferred considerable responsibility from the millwright to the draughtsman and, coupled with the increasing division of labour in the workshops, tended to restrict the opportunities for traditional general millwrights, notably to the roles of 'foremen and out-door superintendents'.[303]

Engineers

ALONGSIDE the growth of millwrighting firms, the civil engineer was emerging. In the early part of the 18th century, an engineer was a military man dealing with fortifications and strategic roads and bridges. In 1761 Smeaton described himself as a 'civil engineer'.[304] He, as Brindley and Telford, is best known for transport engineering, but all were also involved in millwrighting work.[305] In 1771 Smeaton founded the Society of Civil Engineers with a small membership mainly engaged in designing and supervising such works as roads, canals and harbours. Rennie was involved in the Society, but it did not welcome younger engineers.[306] This led to the founding of the Institution of Civil Engineers in 1818 'for the general advancement of mechanical science'. The Royal Charter given to the Institution in 1828 declared civil engineering to be: 'The art of directing the great sources of power in nature for the use and convenience of man', and one of the several descriptions to which this was applied was 'the construction and adaptation of machinery'.[307] The Charter also provided official recognition that civil engineering had achieved professional standing.[308]

In 1818, when the Institution was founded, there was no clear differentiation between civil and mechanical engineering and the two types of activity were commonly combined. In fact all the founding members of the Institution of Civil Engineers undertook primarily mechanical engineering work.[309] It was commonplace for engineers to be active in many areas of work – the Stephensons, Brunels, Rennies and Penns all exemplify this. Later there was increasing specialisation. As demands and opportunities arose, such as those brought by the railways, new products were introduced, whilst others declined or were eclipsed. In the second quarter of the 19th century there was a growing divergence between civil and mechanical engineering as they are understood today, with the 'civils' majoring on railways, water supply, drainage, roads, bridges and harbour works. Engineers undertaking these classes of work also tended to divide into consulting engineers, who designed and certified work, and contractors, who actually built the projects.

The description 'mechanical engineering' came into being mainly as a result of the development of the rotative steam engine.[310] In 1841 the Manchester engineer William Jenkinson described mechanical engineering as divided into three classes: first, steam-engines, mill-gearing, hydraulic presses and other heavy machinery; second, tool-making; and third, machine-making. Amongst the third group were those he described as 'mechanics', engaged in textile factories 'not only in the making of machinery, but in the repairing of it'.[311] At the beginning of the 19th century, whilst steam-engine building centred on Boulton & Watt in Birmingham, London was the centre of mechanical engineering, with Joseph Bramah (1748–1814), and Henry Maudslay (1771–1831) who had worked for Bramah for nine years. Bramah manufactured water closets, locks and hydraulic presses, straddling Jenkinson's second and third categories, whereas Maudslay was famous for his machine tools – and the men who gained experience with him.[312] The separate Institution of Mechanical Engineers was not founded until 1847.[313]

The divisions between civil and mechanical engineers, between consulting engineers and contractors, and between machine-makers and millwrights were not always clear-cut. Some engineers and some firms worked in more than one sector or manner. Robert Stephenson acted as consulting engineer for much of his railway work but also contracted for the supply of locomotives.[314]

By the mid-1820s the hub of mechanical engineering was moving from London to Manchester, drawn by the opportunities provided by the burgeoning cotton industry. There were already well-established firms of millwrights, machine-builders and iron-founders in Manchester, such as Bateman & Sherratt who had been building steam-engines since the 1780s, Peel Williams & Co., Ormrod & Son, Galloway Bowman & Glasgow, W.&J. Mather, Adam Parkinson and Thomas Hewes. Scots were prominent amongst the early millwrights and machine-builders drawn to Manchester.[315] Later immigration included six outstanding mechanical engineers who had been employed at the Maudslay 'nursery' – Richard Roberts (arrived 1816), Francis Lewis (1816), Joseph Whitworth (1833), James Nasmyth (1834), George Nasmyth (1834) and William Muir (1842).[316] Those from abroad included Joseph Dyer (1816) from Connecticut,[317] the Swiss inventor J.G. Bodmer (1824)[318] and C.F. Beyer from Saxony (1834).[319]

William Fairbairn: from Millwright to Engineer

HOW did William Fairbairn fit into this picture? His training, he tells us, was as a millwright – he was bound apprentice to John Robinson at Percy Main Colliery, described in the indenture as 'Millwright'.[320] However Fairbairn's concept of a millwright was wider than traditional millwork. His millwright also had 'knowledge of mill machinery, pumps and cranes'.[321] His own apprenticeship illustrated this, for his work included repairing pumps and looking after a steam engine, and, despite the description in the indenture, he variously described Robinson as an 'engine-wright' and a 'colliery engineer'.[322] Similarly his journeyman years illustrated a wide breadth of work. It included waterwheels and millwork but also involved machine-making in Dublin and erecting textile printing machinery in Manchester.[323] His second employer in Manchester was Thomas Hewes who had built up one of the largest and most outstanding firms of millwrights and machine-makers.[324] Over the ensuing twenty-five years Hewes and his successor firms continued to be very important millwrights.[325] Initially Fairbairn was engaged in the site erection of Hewes' waterwheels, notably in Macclesfield, moving from there to his drawing office.[326]

Fairbairn arrived in Manchester in 1813 and, in a widely quoted passage, recorded that:

> When I first entered this city, the whole of the machinery was executed by hand. There were neither planing, slotting, nor shaping machines, and, with the exception of very imperfect lathes and a few drills, the preparatory operations of construction were effected entirely by the hands of the workmen.[327]
>
> There were only three or four establishments that could make a steam-engine, and those were Boulton and Watt of Soho, Fenton, Murray and Wood of Leeds, and Messrs. Sherratts of this town.[328]

An 1824 Manchester directory listed ten firms as 'Millwrights', including Fairbairn & Lillie, which was the sole firm with the added designation '& engineers' – the directory did not include a separate list of Engineers. The firm was also one of twenty-four listed as 'Iron Founders'. Under 'Machine Makers' there were thirty-eight firms whose owners would have been drawn from a variety of backgrounds. These firms included four of the ten listed as millwrights, but not Fairbairn & Lillie. Under their home addresses both Fairbairn and Lillie were individually described as 'millwrights'.[329] For its first decade almost all the firm's work was that of the traditional millwright – transmission shafting, gearing, waterwheels and structural ironwork (an exception was the silk spinning machinery). In 1824 they were employing between sixty and seventy men, mainly artisan millwrights, members of the Manchester Millwrights' Club.[330] In the late 1820s, as Fairbairn & Lillie's work increased, some of it beyond traditional millwrighting, so the industry's perception of a millwright was changing to that of an artisan employee. There is little doubt that Fairbairn was aware of this at the time, and later he wrote about it:

> The introduction of the steam engine, and the rapidity with which it created new trades, proved a heavy blow to the distinctive position of the

> millwright, by bringing into the field a new class of competitors in the shape of turners, fitters, machine-makers, and mechanical engineers; ... [which] lowered the profession of the millwright, and levelled it in a great degree with that of the ordinary mechanic.[331]

He also saw the decline of the millwright stemming from the 'excesses' of the Millwright Societies, leading to 'the almost ultimate extinction of the name of millwright as a distinct profession'.[332]

It is clear that despite the elite description given by Fairbairn, the typical 18th-century millwright was 'vastly remote from the paradigm of the gentleman, his upbringing, and his doings!'[333] With the declining status of the millwright, it was an obvious move for Fairbairn to ally himself with the ascending civil engineers whose status tended to be managerial and professional. They were employers rather than employees. By 1830 he saw himself as a civil engineer, a designation sealed by his election to the Institution of Civil Engineers, after which he used the suffix 'C.E.'.[334] It is in this sense that the description 'engineer' is used of Fairbairn in this book.

After 1830, like many other engineers, Fairbairn's range of work was wide, changing as circumstances and opportunities changed, as has been illustrated in *Table 1*. Much of his post-1830 work was within the category of what is now known as mechanical engineering but he always had a foot in civil engineering – his factory structures, water supplies to waterwheels, bridges and drainage schemes were all civil engineering as it is now understood. Most of Fairbairn's mechanical engineering work fell within Jenkinson's first class of that work – steam engines and mill gearing – to which were added iron shipbuilding, locomotives and iron structures. These were constructed in the foundry, workshop and yard by millwrights and other artisans, increasingly using standardised parts, and with increasingly sophisticated machine tools. The products were then erected on site and, or, commissioned, worldwide by peripatetic erectors and fitters who retained an affinity with the traditional itinerant millwright. For the half century 1820–70, Fairbairn was one of the foremost of millwrights and engineers. He provides a useful example of the transition of master millwright to civil engineer.

Fairbairn on the 18th-century Millwright

IT is somewhat ironic that Fairbairn, with a vast portfolio of engineering projects, is best known in the academic world for his description of an 18th-century millwright, whom he refers to as a 'jack-of-all-trades', and then proceeds to describe as a master-of-all-trades! His description, or part of it, has been quoted by historians on at least twenty-two occasions between 1945 and the present.[335] It also appears on the Science Museum's website.[336] Most quote it without comment, but in an accepting manner. Some are more discerning, pointing out that this 'does not represent the whole story of millwrighting' – some millwrights were 'much more akin to a carpenter';[337] or that the account is 'perhaps favourably biased';[338] and one describes it as 'a pervasive myth ... that the ancestor of the engineering worker was a millwright'.[339] Yet in 2014 a historian referred to Fairbairn's piece as 'a good description of a millwright of the 18th century'.[340] Others consider that 'the truth of this statement is clearly evident', supporting their view by references to Smeaton, Telford and Rennie.[341] Indeed, Fairbairn himself referred to millwrights as 'generally men of talent', and he 'ranked amongst them the celebrated names of Brindley, Smeaton and Rennie'.[342] But Smeaton, probably the greatest of 18th-century engineers, 'did not learn these skills by being an apprentice millwright'.[343] What Fairbairn actually wrote was:

> The millwright of the last century was an itinerant engineer and mechanic of high reputation. He could handle the axe, the hammer, and the plane with equal skill and precision; he could turn, bore or forge with the ease and despatch of one brought up to these trades, and he could set out and cut in the furrows of a millstone with an accuracy equal or superior to that of the miller himself. These various duties he was called upon to exercise, and seldom in vain, as in the practice of his profession he had mainly to depend upon his own resources. Generally, he was a fair arithmetician, knew something of geometry, levelling and mensuration, and in some cases possessed a very competent knowledge of practical mathematics. He could calculate the velocities, strength and power of machines: could draw in plan and section, and could construct buildings, conduits or watercourses, in all the forms and under all the conditions required in his professional practice; he could build bridges, cut canals, and perform a variety of work now done by civil engineers. Such was the character and condition of the men who designed and carried out most of the mechanical work of this country, up to the middle and end of the last century.[344]

None of those citing Fairbairn's description point out that this was not a universal view amongst his contemporaries. For example *The Mechanics' Magazine*'s reviewer of Fairbairn's book in which the passage appeared wrote:

> It does not take a very extended retrospect to recall the days when everything connected with prime movers, wheel-work or mechanical appliances, dwelt in the hands of a class called 'millwrights' ... Illiterate and conceited, drunken and unmannerly, their presence was tolerated only because nothing better was to be had.[345]

Fairbairn had eulogized the roots from which he had sprung but he was misleading because he blurred the distinctions between the variable levels of skills of millwrights.[346] From the 1760s there emerged 'specialist millwright/engineers', such as Brindley, Smeaton, Jessop, Telford and others – the recognition of whose group identity brought about the establishment of the Society of Civil Engineers.[347] By the middle of the 19th century this had led to a clear division, with master-millwrights having become professional engineers – members of the engineering institutions – whilst tradesmen millwrights became members of the Amalgamated Society of Engineers (and subsequently the Amalgamated Engineering Union in 1920). The millwrights' name was thus largely lost in Britain, where today it tends to be limited to those involved in the conservation of heritage machinery.[348] In North America the term is more widely used, to describe tradesmen who erect machinery.[349] Judging by its repetition, and the few historians who have raised doubts about it, Fairbairn's description of the 18th-century millwright is influential, but it is a misleading influence, and a reminder that however many times something is repeated is no guarantee that it is correct.

Notes for Chapter 2

1. [Woodcroft] III.
2. T S Ashton, *The Industrial Revolution 1760-1830,* (1968 ed.), p.124.
3. Ashton, *Industrial Revolution,* p.122; M Casson and A Godley, 'Entrepreneurship in Britain, 1830-1900', in D S Landes, J Mokyr and V J Baumol (eds), *The Invention of Enterprise: Entrepreneurship from Ancient Mesopotamia to Modern Times,* (2012), p.213; [Woodcroft] III.
4. Ure, *Philosophy,* pp.287-8. Strikers included men from Fairbairn & Lillie's clients at the time, A&G Murray, for whom the Irish trade union leader John Doherty worked (J Saville, 'Doherty, John (1797/8-1854)', *Oxford DNB online, (*2004), (accessed 8 April 2014); http://radicalmanchester.wordpress.com/2009/08/john-doherty/ (accessed 8 October 2011)).
5. [Woodcroft] III.
6. *Life,* p.111.
7. Farey gives the figure at 240 in 1825 and Baines 212 in the same year. By the mid-1820s Manchester had approaching one-fifth of the steam horsepower in the country. Between 1835 and 1856 horsepower in the Lancashire cotton industry almost trebled. (G N von Tunzelmann, *Steam Power and British Industrialization to 1860,* (1978), pp.32, 236.)
8. D Bindman and G Riemann (eds), *K F Schinkel, The English Journeys: Journal of a Visit to France and Britain in 1826,* (ET 1993), p.175.
9. W Fairbairn, 'Presidential Address', *BAAS1861,* p.lxiv; M Berg, *The machinery question and the making of political economy 1815-1848,* (1980), p.21.
10. J Mokyr, *The Enlightened Economy: Britain and the Industrial Revolution 1700-1850,* (2009), p.263.
11. J A Cantrell, 'James Nasmyth and the Bridgewater Foundry: Partners and Partnerships', *Business History,* 23.3, 1981, 348-50.
12. *Life,* p.106; A Bull, M Pitt and J Szarka, *Entrepreneurial Textile Communities: A comparative study of small textile and clothing firms,* (1993), p.67.
13. *Life,* pp.111-2.
14. Hayward, 'Fairbairns', pp.1.13-15.
15. *Life,* p.129.
16. *Life,* pp.114-7.
17. *Life,* p.452. [Fairbairn's italics and capitals].
18. *Fortunes Made in Business,* pp.253-6. Peter Fairbairn had served an apprenticeship with a millwright in Newcastle and then worked for several years as a journeyman, starting with Holdsworth in Glasgow and including two spells with Fairbairn & Lillie, some time with the Rennies in London and at the famous Charenton Engineering Works in Paris. He accepted a partnership at Holdsworths in 1824, but was not happy there. In 1828 he consulted William as to what he should do, resulting in the decision to commence his own business in Leeds The Glasgow partnership had left him in debt to the extent of £500. William lent him the money to pay it off.
19. *Life,* p.129.
20. *Life,* p.112; 115. At this time Roberts was employing three men, at a rate of 11s. per week, to turn his machines. (L T C Rolt, *Tools for the Job. A History of Machine Tools to 1950,* (1965), p.106).
21. K Schib (ed.), *Johann Conrad Fischer: Tagebücher,* (1951), pp.262, 274-6, quoted in W O Henderson, *J C Fischer and his Diary of Industrial England 1814-1851,* (1966), p.63. Fischer wrote that Fairbairn was 'one of the first and cleverest millwrights in England'.
22. [Woodcroft] IV. The planing machine originated in France ([Woodcroft] IV). See also *Life,* pp.44-5.
23. *The Engineer,* 11, Jan-June 1861, 121; R L Hills, *Life and Inventions of Richard Roberts 1789-1864,* (2002), pp.87-9.
24. *Annales des Mines,* Second Series 6, 1829, 94-6; E A Forward, 'The Early History of the Cylinder Boring Machine, *TNS,* 5, 1924, 34.
25. *Life,* p.129.
26. [Woodcroft] IV.
27. Hayward, 'Fairbairns', pp.1.13-15.
28. [Woodcroft] IV; [B Love], *Manchester As It Is,* (1839), p.212.
29. W Fairbairn, *Iron, Its History, Properties, and Processes of Manufacture,* (1861), 88-101.
30. 'Report from the Select Committee on Artizans, Machinery and Combinations', *Parliamentary Papers,* (1824), V, p.566.
31. 'Report from the Select Committee on Artizans, Machinery and Combinations', p.341.
32. *Life,* p.129.
33. J Brown and M B Rose (eds), *Entrepreneurship, networks and modern business,* (1993), pp.1, 6.
34. [I] (Mrs G L) Banks, *The Manchester Man,* ([1876], 1970 ed.), p.207.
35. *Life,* pp.111-4.
36. *Life,* p.113; R Lloyd-Jones and M J Lewis, *Manchester and the Age of the Factory: The Business Structure of Cottonopolis in the Industrial Revolution,* (1988), p.25.
37. I Miller, 'The Rise and Progress of an Iconic Cotton-Spinning Firm: The History of A & G Murray' in I Miller and C Wild (eds), *A & G Murray and the Cotton Mills of Ancoats,* (2007), p.61.
38. C H Lee, *A Cotton Enterprise 1795-1840: A History of M'Connel & Kennedy, Fine Cotton Spinners,* (1972), pp.10-22.
39. W Fairbairn, *A Brief Memoir of the Late John Kennedy, Esq.,* (1861); M C Jacob and L Stewart, *Practical Matter: Newton's Science in the Service of Industry and Empire, 1687-1851,* (2004), pp.127-38; M C Jacob, *The First Knowledge Economy: Human Capital and the European Economy, 1750-1850,* pp.85-109; Mokyr, *The Enlightened Economy,* p.111.
40. R C Allen, *The British Industrial Revolution in Global Perspective,* (2009), pp.251, 269-71.
41. *Life,* p.116.
42. Fairbairn, *UIfE2,* p.239. Fairbairn is also said to have recalled Murdoch filling a bladder with gas, placing it under his arm like a bagpipe, and discharging the gas through the stem of an old tobacco pipe to provide a torch (W Matthews, *A Historical Sketch of the Origins and Progress of Gas Lighting,* (1832), quoted in J Griffiths, *The Third Man: The Life and Times of William Murdoch 1754-1839, the Inventor of Gas Lighting,* (1992), p.248). This refers to walking to Medlock Bank, but when, and who lived at Medlock Bank, is unclear. Fairbairn did not move to Medlock Bank until the mid-1830s, by which time Murdoch was incapacitated.
43. Ewart was knowledgeable in theoretical mechanics, as witnessed by his best known paper (P Ewart, 'On the Measure of Moving Force', *Manchester Literary and Philosophical Society Memoirs,* 2nd series, 2, 1813, 105-258).
44. Jacob, *First Knowledge Economy,* p.92.
45. *MG,* 10 April 1824.
46. H Heaton, 'Benjamin Gott and the Industrial Revolution in Yorkshire', *Economic History Review,* 3.1, 1931, 52-3;

Fairbairn to Boulton & Watt, 2 October 1824, Fairbairn to Gott, 10 February 1825, MS193/150 and 153, Gott Papers, Brotherton Library, Leeds
47. *Life*, p.115; S Smiles, *Industrial Biography: Iron Workers and Tool Makers*, (1863), p.324.
48. *Life*, p.121; 'Biographical Notice : Archibald Buchanan of the Catrine Cotton Works, Ayrshire', *Journal of the Society of Arts*, 6.168, 1856, 205; [C Brogan], *James Finlay & Company Limited, Manufacturers and East India Merchants 1750-1950*, (1951), pp.60-2; E Leigh, *The Science of Modern Cotton Spinning*, (3rd ed. 1875), Vol.1, pp.106-10, which illustrates Buchanan's self-stripping carding engine.
49. [Brogan], *James Finlay*, pp.69-71; Leigh, *Modern Cotton Spinning*, Vol.1, pp.110-12.
50. G M Mitchell, 'The English and Scottish Cotton Industries', *The Scottish Historical Review*, 22, 1924-5, 104.
51. W Fairbairn, *Remarks on Canal Navigation, illustrative of the advantages of the use of Steam as a Moving Force on Canals* (1831), pp.30-2, 71.
52. S Smiles (ed.), *James Nasmyth, Engineer: An Autobiography*, (1895 ed.), p.178.
53. *Life*, p.456.
54. Jacob, *First Knowledge Economy*, pp.100-7; M C Jacob, 'Commerce, Industry, and the Laws of Newtonian Science: Weber Revisited and Revised', *Canadian Journal of History*, 35, 2000, 277.
55. Mokyr, *Enlightened Economy*, p.362.
56. Thomas Percival Heywood, son of Benjamin Heywood and brought up at Cross Street Chapel wrote, 'I resent their [Unitarians] being called Socinians' ([I M Heywood, ed.], *Reminiscences, Letters, and Journals of Thomas Percival Heywood, Baronet*, (1899), pp.4-5).
57. V A C Gatrell, 'Incorporation and the pursuit of Liberal Hegemony in Manchester 1790-1839', in D Fraser (ed.), *Municipal reform and the industrial city*, (1982), p.25.
58. D Allan, *Virtue, Learning, and the Scottish Enlightenment: Ideas of Scholarship in Early Modern History*, (1993). For a useful précis see 'Review' by A H Williamson, *The Journal of Modern History*, 67.3, 1995, 700-3. See also T Devine, 'A Puzzle from Scotland's Past: Why did the Scottish Enlightenment Happen?' (The Royal Society of Edinburgh, 25 April 2013) at www.royalsoced.org.uk/cms/files/.../A_Puzzle_from_Scotlands_Past.pdf. (accessed 18 August 2014), 'Of all the forces relevant to the Scottish Enlightenment, Calvinism is at the heart of the matter'.
59. The Calvinist theologian would attribute this resilience of Calvinist culture to 'common grace'. See H R Van Til, *The Calvinist Concept of Culture*, (1959).
60. Allan, *Virtue, Learning, and the Scottish Enlightenment*, pp.95, 203.
61. *Life*, p.458; R B Sher, 'Blair, Hugh (1718-1800)' *Oxford DNB on-line*, (2009), (accessed 16 November 2012); H Blair, *Sermons*, (5 vols. 1777-1801, five editions)
62. A Herman, *The Scottish Enlightenment: The Scots' Invention of the Modern World*, (2001), p.184.
63. J Hill, *An Account of the Life and Writings of Hugh Blair DD*, (1808), p.92.
64. Blair, *Sermons*, (1820 ed.), Vol.I pp.5, 30, 35, 44, 98; Vol.2 pp.99, 224-5, 225, 227, 349; and many more references.
65. *Life*, p.72; *UIfE3*, pp.81-2; and many more.
66. *MG*, 28 October 1868.
67. J T Slugg, *Reminiscences of Manchester Fifty Years Ago*, (1881), pp.170-6.
68. *Life*, p.452; Slugg, *Reminiscences*, p.173; W A Shaw, *Manchester Old and New*, (1894), Vol.3, p.8.
69. M J Turner, *Reform and Respectability. The Making of a Middle-Class in Early 19th-century Manchester* (1995), p.32. See also J Seed, 'Theologies of Power: Unitarianism and the social relations of religious discourse 1800-50' in R J Morris (ed.), *Class, Power and Social Structure in British Nineteenth-Century Towns* (1986), p.142.
70. A J Kidd, 'Introduction: The Middle Class in Nineteenth-century Manchester', in A J Kidd and K W Roberts (eds), *City, class and culture: Studies of social policy and cultural production in Victorian Manchester*, (1985), pp.10-11. As Kidd notes, Benjamin Heywood was in many ways a key Manchester figure in the second quarter of the century and deserves closer study (*City, class and culture*, p.22n68).
71. [Heywood], *Reminiscences of Thomas Percival Heywood*, pp.5, 139; A J Dobb, *Like a Mighty Tortoise: A history of the Diocese of Manchester*, (1978), p.258.
72. R H Kargon, *Science in Victorian Manchester: Enterprise and Expertise*, (1977), p.20.
73. J Seed, 'Unitarianism, Political Economy and the Antinomies of Liberal Culture in Manchester, 1830-1850', *Social History*, 7.1, 1982, 12; Slugg, *Reminiscences*, pp.170-6.
74. *MG*, 10 April 1824.
75. Allan, *Virtue, Learning, and the Scottish Enlightenment*, p.237.
76. Lillie did join the Literary and Philosophical Society in 1830, but does not appear to have taken any active part in it.
77. Gatrell, 'Incorporation', pp.28-9.
78. *Life*, p.383. Fairbairn was President 1855-60.
79. *Complete List of the Members & Officers of the Manchester Literary and Philosophical Society*, (1896).
80. Slugg, *Reminiscences*, p.173.
81. Unpublished study by the author.
82. M J Daunton, *Progress and Poverty: An Economic and Social History of Britain 1700-1850*, (1995), pp.197-8.
83. *Life*, pp.130, 134-5.
84. M Berman, 'An essay review of A E Musson and Eric Robinson, *Science and Technology in the Industrial Revolution*', *Journal of Social History*, 5.4, 1972, p.525.
85. [Woodcroft] III.
86. Fairbairn, *M&MWII*, pp.72-3; Rolt, *Tools for the Job*, p.125.
87. A E Musson and E Robinson, *Science and Technology in the Industrial Revolution*, (1969), pp.462-3.
88. *MPICE*, 28, 1868-9, 579.
89. Ure, *Philosophy*, pp.34-5; G Rennie (ed.), *Practical Essays on Millwork and other Machinery. By Robertson Buchanan, Engineer*, (3rd ed. 1841), p.xix; Leigh, *Modern Cotton Spinning*, Vol.2, (5th ed. 1882), p.273; W J M Rankine, 'William Fairbairn' in *The Imperial Dictionary of Universal Biography*, Vol.2, (1963); [Woodcroft] III.
90. Musson and Robinson, *Science and Technology*, pp.462-3.
91. Fairbairn, *M&MWII*, pp.74-8.
92. D S Landes, *The Unbound Prometheus: Technological change and industrial development in Western Europe from 1750 to the present*, (1969), pp.298-9.
93. F P Bowden and D Tabor, *The Friction and Lubrication of Solids*, (1950).
94. P Jost (ed.), *Lubrication (Tribology): Education and Research*, (1966).
95. [Woodcroft] III. Cf. Schumpeter: Innovation involves seeing old things in new ways and moving 'outside the ruts of established practice' (J Schumpeter, 'The Creative Response and Entrepreneurial History', *Journal of Economic History*, 7, 1947, pp.149-59).

96. Fairbairn, *M&MWII*, pp.72-3.
97. Ure, *Philosophy*, pp.34-5.
98. Rolt, *Tools for the Job*, p.125. There were some inroads made by belt and rope drives, but they did not displace shafting.
99. Ure, *Philosophy*, p.37 read with A Ure, *A Dictionary of Arts, Manufactures, and Mines: containing a clear exposition of their principles and practice*, (4th ed. 1853), Vol.II, p.165.
100. A & G Murray and M'Connel & Kennedy were the largest cotton spinners in Manchester (R Lloyd-Jones and M J Lewis, *Manchester and the Age of the Factory: The Business Structure of Cottonopolis in the Industrial Revolution*, (1988), p.25), but Peter Marsland in Stockport is said to have been the largest in the north-west around this time (M Williams with D A Farnie, *Cottton Mills in Greater Manchester*, (1992), p.25).
101. *Life*, p.115.
102. *Life*, p.115.
103. Tann implies that the 'double speeds' refers to the shafting. (J Tann, *The Development of the Factory*, (1970), p.103). This is not the case - 'Fairbairn is referring to mules with two speeds, not that the second speed was double the first. As the carriage commenced its outward run then the spindles would be driven at the first, lower speed, and at some point in the run would be switched to the higher speed ... Double speeds were used in spinning fine yarns which needed large amounts of twist put in them.' (R Holden, email to R Byrom 14 August 2014); Fairbairn, *Memoir of John Kennedy*, p.4; Lee, *Cotton Enterprise*, p.151.
104. Lee, *Cotton Enterprise*, p.102; R Fitzgerald, 'The Development of the Cast Iron Frame in Textile Mills to 1850', *IAR*, 10.2, 1988, 137-8.
105. *Life*, pp.114-5; Lee, *Cotton Enterprise*, p.102; Fitzgerald, 'Development of the Cast iron Frame', 137-8.
106. Hayward, 'Fairbairns', p.1.10. Hayward lists M'Connel and Kennedy's payments to Fairbairn & Lillie, from M'Connel & Kennedy Papers, University of Manchester Rylands Library, Cash Books Vols. 2 and 3, Ledgers 1 and 2.
107. I Miller, C Wild and S Little, 'The Development of Ancoats: Archaeological Case Studies', in I Miller and C Wild, *A & G Murray and the Cotton Mills of Ancoats*, (2007), pp.52-3. At Arkwright's first mill in Nottingham, Thomas Lowe had used a primitive form of this arrangement with vertical shafts (J Tann, 'The Textile Millwright in the Early Industrial Revolution', *Textile History*, 5, 1974, 83).
108. For example at Gott's: Fairbairn to Gott, 18 September 1824, MS 193/149, Gott Papers, Brotherton Library, Leeds
109. *M&MWII*, p.104.
110. Fairbairn to Heathcote, 20 April 1841, in 'Report on the Practicality of Legislative Interference, to diminish the Frequency of Accidents to the Children and Young People engaged in Factories', *Parliamentary Papers*, 1841, X[311], p.24.
111. J Gloag and D Bridgwater, *A History of Cast Iron in Architecture*, (1948), Fig.115, p.112.
112. *Application*, pp.2-5, Fairbairn illustrates Phillips & Lee's mill but does not show a sub-floor duct; *Life*, p.115.
113. Hayward, 'Fairbairns', p.1.10
114. Lee, *Cotton Enterprise*, p.151.
115. A Kidd, *Manchester*, (1993), p.26.
116. A Von Wolzogen, *Aut Schinkels Nachlass*, (1862-4), Vol.3, p.141,
117. D Bindman and G Riemann (eds), *Karl Friedrich Schinkel, 'The English Journey', Journal of a Visit to France and Britain in 1826*, (1993), pp.175-7.
118. A similar case was Peter Fairbairn, in Leeds, who obtained commissions and financial support from the largest flax spinner of the day, John Marshall (*Fortunes Made in Business*, pp.262-3).
119. Smiles, *Industrial Biography*, p.324; *Life*, p.115.
120. A Bielenberg and J M Hearne, 'Malcolmsons of Portlaw and Clonmel: some new evidence on the Irish cotton industry 1825-50', *Proceedings of the Royal Irish Academy*, 106C, 2006, 339-66.
121. R Boyson, *The Ashworth Cotton Enterprise: The Rise and Fall of a Family Firm 1818-1880*, (1970), p.8.
122. K Charlton, 'The State of Ireland in the 1820s: James Cropper's Plan', *Irish Historical Studies*, 17.67, 1971, 320-39.
123. Bielenberg and Hearne, 'Malcolmsons', 339-66; T Hunt, *Portlaw, County Waterford 1825-1876: Portrait of an Industrial Village and its Cotton Industry*, (2000).
124. Bielenberg and Hearne, 'Malcolmsons',342, 346-7. The exact nature of Fairbairn & Lillie's involvement remains unclear. When the mill was extended in 1833, Fairbairn had further involvement.
125. Fitzgerald, 'Development of the Cast Iron Frame', p.139.
126. Lee, *Cotton Enterprise*, p.123.
127. Boyson, *Ashworth Cotton Enterprise*, pp.84-140.
128. Hunt, *Portlaw*, pp.29-45.
129. J T Ward, *The Factory System. Volume I: Birth and Growth*, (1970), pp.131-8.
130. [Brogan], pp.60, 63.
131. A Ure, *The Cotton Manufacture of Britain, Volume I*, (1836), pp.lxvii-lxix.
132. T C Banfield, *Industry of the Rhine. Series II: Manufactures embracing a view of the social condition of the rural & manufacturing population of that district.* (1848), p.145.
133. 'Sion Mills Village & Its People', at http://www.sionstables.com/the-history/sion-mills-stories/village-people/ .
134. C Caine, Cleator *and Cleator Moor: Past and Present*, (1916), pp.378-9.
135. R Balgarnie, *Sir Titus Salt, Baronet*, (1877), pp.143-6; J Reynolds, *The Great Paternalist: Titus Salt and the Growth of Nineteenth-century Bradford*, (1983), pp. 265-86; 'Nomination of Saltaire Village for Inclusion in the World Heritage List', (2001), pp.40-52.
136. *Amicus* [Thomas Fairbairn], 'The Irish Exhibition', *The Times*, 20 June 1853; B MacAongusa, 'William Dargan: Great Railway Builder and Patriot', *Journal of the Irish Railway Record Society*, 24.171, 2010, 4-12; W Fairbairn to Jane Dargan, 10 February 1867, Jane Dargan Papers, National Archives, Dublin, quoted in F Mulligan, *William Dargan: An Honourable Life 1799-1867*, (2013), p.202.
137. G Darley, *Villages of Vision*, (1975), pp.122-3.
138. S Pollard, 'The Factory Village in the Industrial Revolution', *The English Historical Review*, 79.312, 1964, 513-31.
139. Pollard, 'The Factory Village', 513-4; Reynolds, *The Great Paternalist*, pp.256-65.
140. Hunt, *Portlaw*, pp.29-42, Reynolds, *The Great Paternalist*, pp.274-9.
141. Reynolds, *The Great Paternalist*, pp.75-6.
142. Darley, *Villages of Vision*, pp.275-307. Darley lists eighteen in Britain, but does not mention New Eagley or Egerton, or Cleator, and perhaps others. In addition there were around five in Ireland. This suggests a total of perhaps thirty or more in the British Isles.
143. E Gaskell, *North and South*, ([1855], 1970 ed.), p.146.

144. Ure, *Philosophy*, pp.380-4; A Ure, *The Cotton Manufacture of Great Britain systematically investigated...*, Vol.2, (1836), pp.23-5.
145. Hunt, *Portlaw*, p.56.
146. D A Farnie, *John Rylands of Manchester*, (1993), Fig.3 opposite p.12. Gorton Mills was sold in 1841 following the bankruptcy of A Lees, when Lillie was one of the assignees of the estate. (Advertisment of 1841, M95, Chapman MSS, Manchester Central Reference Library). The date of 1825 given by Farnie appears to be too early for the extensive weaving shed shown in his Fig.3, but not for the main building. Roger Holden suggests that only the central block of the mill was built in 1825 (e-mail 2 April 2016). On Orrell's Mill see Chapter 3.
147. F R Stott, 'Town Hall Buildings in Tameside', (MA Thesis, University of Manchester 1983), pp.22-40
148. Fairbairn to Boulton & Watt, 2 October 1824; Fairbairn to Gott, 10 February 1825; MS193/150 and 153, Gott Papers, Brotherton Library, Leeds
149. Fairbairn to Gott, 18 September 1824, MS193/149, Gott Papers, Brotherton Library, Leeds,
150. Fairbairn to Gott, 21 October 1824, MS193/152.
151. Fairbairn, *Application*, pp.6-7; Fairbairn to Gott, 21 October 1824.
152. Fairbairn to Gott 19 February 1829, MS193/170, Gott Papers. However from 1832 Wren & Bennett received several commissions from Gott – see below.
153. Tann, *Development of the Factory*, p.40; D T Jenkins, *The West Riding Wool Textile Industry, 1770-1835: A Study of Fixed Capital Formation*, (1975), pp.106-8. This mill was very similar to one, built in Stalybridge in 1822 for Thomas Harrison, in which Boulton & Watt were involved, and, on balance of probabilities, so were Fairbairn & Lillie (Tann, p.24.). Tann claims that the millwork for Wood's mill was by Wren & Bennett, but she appears to be confusing work in 1833 with the building of the mill in 1825 which was prior to Wren & Bennett's formation (Tann, p.103); Fitzgerald, 'Development of the Cast Iron Frame', 139; Fairbairn, *Application*, pp.2-3. On Wood, a supporter and financier of factory reform, see J T Ward, *The Factory System, Volume 1: Birth and Growth*, (1970), pp.131-8.
154. Fairbairn, *Application*, p.7.
155. See Chapter 4.
156. M Williams, 'Havelock Mill, Manchester: A case-study in the emergency recording of a large urban mill complex', *IAR*, 16.1, 1993, 100-3, 110.
157. The throstle frame was a descendant of the water frame and differed from the mule by having a continuous action.
158. N Cossons (ed.), *Rees's Manufacturing Industry (1819-20): A selection from 'The Cyclopaedia; or, Universal Dictionary of Arts, Sciences and Literature' by Abraham Rees*, (1972), p.194; *M&MWII*, pp.213-4.
159. Fairbairn, *M&MWII*, p.213.
160. Ure, *Philosophy*, p.266.
161. Ure, *Philosophy*, pp.272-5; G S White, *Memoir of Samuel Slater, the Father of American Manufactures, connected with a History of the Rise and Progress of the Cotton Manufacture in England and America*, (1836), Vol.2, p.408; Fairbairn, *M&MWII*, p.214.
162. W S Murphy, *The Textile Industries. A Practical Guide to Fibres, Yarns & Fabrics in every branch of Textile Manufacture*, (1910), Volume 3, Plate opposite p.174.
163. Ure, *Philosophy*, p.266; Fairbairn, *M&MWII*, pp.213-4.
164. However Peter Fairbairn, having trained as a millwright, specialised in flax machinery, and Hewes was both a millwright and a textile machine maker.
165. 'Notice of Auction' on 29 October 1833, displayed in Capes Dunn's Auction Rooms, Manchester. I am grateful to Terry Wyke for this reference.
166. T Bonson, 'Macclesfield Mill: From Doomsday to Global Trade Name', *Wind and Water Mills: The Occasional Journal of the Midland Water Mills Group*, 23, 2004, 32-5; N Brindley, 'Francis Brindley and the Marple Brindleys', at http://www.marple-uk.com/marple_brindleys.htm (accessed 26 March 2013).
167. Fairbairn, *M&MWII*, p.118. Moher states that Fairbairn attributed the change to John Rennie (J G Moher, 'The London Millwrights and Engineers 1775-1825', (University of London PhD Thesis 1988), p.162). The evidence does not support this.
168. Fairbairn, *UIfE3*, p.142. This is in a paper 'On the Comparative Merits of the Machinery of the Paris Universal Exhibition, 1855', said to be 'extracted from Mr Fairbairn's report ... to the Board of Trade'. Again he refers to 'upwards of *forty years*' [his italics]. Presumably the reason for the italics is that this relates to the 1866 publication date rather than to the 1855 report (forty years before which would have been before the foundation of Fairbairn & Lillie). Fairbairn is open to criticism in respect of this paper in that, notwithstanding its title, his section on 'Flourmills' is largely a description, with illustrations, of his own mills, which were not exhibited but which he clearly considered superior to what he saw.
169. Bonson, 'Macclesfield Mill', 32-5.
170. See below.
171. Bonson, 'Macclesfield Mill', 32-6.
172. *The Carlow Sentinel*, 8 October 1836.
173. S C Hall and A M Hall, *Ireland: Its Scenery, Character, &c.* (1841), p.405. There is an affinity with Portlaw in that both mills had castellated parapets.
174. G N von Tunzelmann, *Steam Power and British Industrialization to 1860*, (1978), pp.138-9; A E Musson, 'Industrial Motive Power in the United Kingdom,!800-70', *The Economic History Review*, NS29.3, 1976, 424. The comparative costs of water *v.* steam are discussed, based on Fairbairn's figures, by D T Jenkins (Jenkins, *West Riding Wool Textile Industry*, pp.159-60, citing *M&MWI*, (3rd ed. 1871), pp.91-3); and in more detail by von Tunzelmann who refers to Fairbairn waterwheels (von Tunzelmann, *Steam Power*, pp.125ff).
175. Fairbairn, *M&MW1*, pp.114-5.
176. In some nineteenth century writings high-breast wheels are described as 'overshot'. In modern works the term 'over-shot' refers to wheels where the water is introduced just past the top of the wheel.
177. J Mokyr, *The Gifts of Athena: Historical Origins of the Knowledge Economy*, (2002), p.47n24.
178. T S Reynolds, *Stronger than a Hundred Men. A History of the Vertical Water Wheel*, (1983), pp.282-3.
179. C Rynne, 'Technological Innovation in the Early 19th Century Irish Cotton Industry: Overton Cotton Mills, County Cork, Thomas Cheek Hewes and the Origins of the Suspension Water Wheel' in E C Casella and J Symonds (eds), *Industrial Archaeology: Future Directions*, (2006), pp.205-15; *MPICE*, 8, 1849, 62-3. The first known suspension wheel was by Hewes at Overton Mill in 1802.
180. Reynolds, *Vertical Water Wheel*, pp.315-7; Fairbairn, *M&MW I*, pp.120-1.
181. One of John Rennie's sons, probably George, said, 'The late Mr. Rennie introduced the system of laying the

water on to the wheel in a thin stream, not exceeding ten inches in depth. In addition to this ... he used the curved moveable shuttle, and at the same time tried various curves for the buckets. It appeared that Mr. Fairbairn had directed his attention to nearly the same points, as there was great similarity in the machines, he having apparently taken the subject up where Mr. Rennie had left it'. (*MPICE*, 8, 1849, 59-60).
182. Reynolds, *Vertical Water Wheel*, pp.294-6.
183. Fairbairn, *M&MWI*, p.147.
184. Fairbairn, *M&MWI*, p.136. Note however that Ure describes Fairbairn's Wesserling wheel, of about 1826, as 'on the ventilating plan'. (Ure, *Cotton Manufacture*, Vol.I, p.lxx).
185. J Sutcliffe, *A Treatise on Canals and Reservoirs ...*, (1816), p.255.
186. *MPICE*, 28, 1868-9, 581.
187. *The Engineer*, 44, July-Dec.1877, 240.
188. N Rosenberg, *Perspectives on Technology*, (1976), pp.286-7. Rosenberg and Ames cite the construction of the Royal Small Arms Factory at Enfield in 1854 as 'the beginning of the movement of mass-production techniques from the United States to Europe'. (E Ames and N Rosenberg, 'The Enfield Arsenal in Theory and History', *The Economic Journal*, 78.312, 1968, 827); J W Roe, 'Interchangeable Manufacture', *TNS*, 17, 1936-7, 165-74. Fairbairn was deeply involved in the Enfield factory – see Chapter 4.
189. J Mokyr, 'Technological Inertia in Economic History', *The Journal of Economic History*, 52.2, 1992, 332, citing K Alder, 'Forging the New Order: The Origins of French Mass Production and Two Languages of the Machine Age, 1763-1815', (Harvard University PhD thesis 1991),pp.36-9;
190. F T Evans, 'The Maudslay Touch', *TNS*, 66, 1994-5, p.157.
191. [Woodcroft] IV.
192. Musson and Robinson, *Science and Technology*, pp.461-3. See also J A Cantrell, *James Nasmyth and the Bridgewater Foundry: A study of entrepreneurship in the early engineering industry*, (1984), pp.72-3.
193 [W Fairbairn], *A List of Wheel Patterns, &c. belonging to William Fairbairn, Millwright, Engineer, Iron and Brass Founder, Canal Street, Manchester*, (1842).
194. Rosenberg, *Perspectives*, p.192.
195. Fairbairn, *M&MWI*, pp.148-9.
196. Boyson, *Ashworth*, p.15.
197. *Life*, pp.121-3.
198. Fairbairn, *M&MWI*, p.132.
199. Fairbairn, *M&MWI*, pp.129-33.
200. [Brogan], *Finlay*, p.62.
201. P N Wilson, 'British Industrial Waterwheels', *Transactions of the Third Symposium of the International Molinological Society*, 1973, 22-4. When they were demolished in 1947 they were found to be true within 3mm on a diameter of 15m.
202. [Woodcroft] IV.
203. Fairbairn, *M&MWI*, pp.133-7.
204. Reynolds, *Vertical Water Wheel*, pp.315-7.
205. W O Henderson, *Industrial Britain under the Regency: The Diaries of Escher, Bodmer, May and de Gallois 1814-18*, (1968), p.8; Smiles, *Industrial Biography*, p.325.
206. Henderson, *Industrial Britain*, p.11. The thermal baths at Schinznach are still flourishing and the leat which served the Fairbairn waterwheel still exists (visited 2016).
207. *MG*, 24 September, 1845; *Life*, pp.124-5, 129.
208. *Life*, pp.127-8.
209. *MPICE*, 28, 1868-9, 579.
210. *Life*, p.129; H Sitterding (ed.), *Escher Wyss 1805-1955. 150 Years of Development*, (1955).
211. *Life*, pp.125, 129. Fairbairn was not impressed by the French engineering works he visited: they appeared 'very deficient in arrangements and methods' (Report from the Select Committee on Artisans, Machinery and Combinations, *Parliamentary Papers*, 1841, (51) V, p.568).
212. *Life*, p.77; Association <Wesserling, un Patrimoine pour l'Avenir>, *Une toiture neuve pour l'ancienne chaufferie de la friche industrielle de Wesserling*, (2006), pp.5,17; J-A Haan and J-M Bobrienrieth, *Gros-Roman: 130 ans d'industrie textile a Wesserling et dans la haute vallée de la Thur* (2nd ed. 2008), p.113 - 'Le premier établissement du Haut-Rhin'.
213. P Fluck, 'Forcopar – Module 6: Technologies de production, d'énergies, de transports. Unité 1: technologies d'énergie', (Université de Haute-Alsace, 2005), p.5.
214. Rynne, 'Technological Innovation in the Irish Cotton Industry', p.212.
215. *Life*, p.129; Ure, *Cotton*, Vol.1, p.xx. Probably by a emigrant Welshman, David Lloyd (J-M Schmitt, 'Relations between England and the Mulhouse Textile Industry in the Nineteenth Century : An Outline Explanation', *Textile History*, 17.1, 1986, p.35.
216. A L Dunham, *The Industrial Revolution in France 1815-1848*, (1955), p.112.
217. Musson, 'Industrial Motive Power in the United Kingdom', 424.
218. Rosenberg, *Perspectives*, p.203; N Rosenberg, *Inside the Black Box*, (1982), pp.62-70;
219. von Tunzelman, *Steam Power*, p.138.
220. See Appendix 2.2. A D George states that Fairbairn built over 300 waterwheels at Ancoats, but gives no source for this information (A D George, 'Industrial Archaeology in Manchester', *Industrial Archaeology*, 8.4, 1971, p.367).
221. Mokyr, 'Technological Inertia', pp.325-38.
222. *Life*, pp.126-7.
223. J Mokyr,'Editor's Introduction: The New Economic History and the Industrial Revolution' in J Mokyr (ed.), *The British Industrial Revolution. An Economic Perspective*, (2nd ed.1999), p.78.
224. Reynolds, *Vertical Water Wheel*, pp.258-65.
225. [Woodcroft] V.
226. e.g. Milford, Hazelbank and Zion Mills, all in Ireland.
227. M DeLacy, *Prison Reform in Lancashire, 1700-1850: A Study in Local Administration*, (1986), pp.207-9.
228. C J Appleby, 'Mechanism and Convict Labour', *Engineering*, 6, 1868, 181 and 209.
229. J A Bennison, 'Prize Essay' (1869, Lancashire Record Office, DDX/184).
230. B Warburton, 'Eaton Hodgkinson (1789-1861) and the Science of Strength of Materials (PhD Thesis UMIST 1971), pp.2.1-2.14.
231. E Hodgkinson, 'On the Transverse Strain, and Strength of Materials', *Memoirs of the Literary and Philosophical Society of Manchester*, 4, 1824, 225-84.
232. E Hodgkinson, 'Some Account of the late Mr Ewart's paper on the Measure of Moving Forces and of the recent applications of principle of Living Forces to estimate the effects of Machines and Movers', *Memoirs of the Manchester Literary and Philosophical Society*, Second Series 7, 1846, 156.
233. Certified Copy of an Entry of Marriage, Manchester.
234. S P Timoshenko, *History of Strength of Materials, with a brief account of the history of theory of elasticity and theory of structures*, (1953), p.127.
235. Fitzgerald, 'Development of the Cast iron Frame', 136.

236. E Hodgkinson, 'Theoretical and Experimental Researches to ascertain the Strength and Best Form for Iron Beams', *Memoirs of the Manchester Literary and Philosophical Society,* Second Series 5, 1831, 407-544.
237. Warburton, 'Hodgkinson', p.2.24.
238. Mokyr, *The Enlightened Economy*, p.87.
239. Fitzgerald, 'Development of the Cast iron Frame', 136.
240. Fairbairn, *Application,* between pages 74 and 75.
241. Hodgkinson to Fairbairn, 11 December 1840, (*Life,* p.181).
242. A L Smyth, *John Dalton 1766-1844. A bibliography of works by and about him with an annotated list of his surviving apparatus and personal effects,* (rev. ed. 1977), p.21.
243. See Chapter 7.
244. Hodgkinson, 'Theoretical and Experimental Researches', pp.450-1.
245. Fitzgerald, 'Development of the Cast iron Frame', 136.
246. P Barlow, *A Treatise on the Strength of Materials,* (6th ed. 1867), pp.173-83.
247. Fairbairn & Lillie also supplied the winding mechanisms for the tunnels at the Liverpool end of the Liverpool and Manchester Railway (R H G Thomas, *The Liverpool & Manchester Railway,* (1980), pp.108-15). The Rainhill trials were held to determine who should build the first locomotives for the line. Fairbairn did not enter but was there to witness Stephenson's success(*UIfE2,* p.241). Fairbairn's patron, John Kennedy, was one of the three judges at the Rainhill trials (Smiles, *Industrial Biography,* p.322).
248. R Fitzgerald, *Liverpool Road Station, Manchester: An historical and architectural survey,* (1980), pp.21-7.
249. [Woodcroft] IX.
250. [R] J [M] Sutherland, 'Iron Railway Bridges' in M R Bailey (ed.), *Robert Stephenson – The Eminent Engineer,* (2003), p.304; 'Returns and Plans of Iron Bridges, 1847', MT8/1, National Archive, refers to many cast-iron girder bridges, but it is not clear which used Hodgkinson beams.
251. Bonson, 'Macclesfield Mill', 32-5; *MG,* 22 May 1830.
252. A Ure, *The Cotton Manufacture of Great Britain,* (1836), Vol.1, pp.296-304, 311-2 and Plates 1 and 2; Fitzgerald, 'Development of the Cast iron Frame', 139-41.
253. *Life,* p.129.
254. *MG,* 13 August 1831.
255. C J A Robertson, *The Origins of the Scottish Railway System, 1722-1844,* (1983), p.50.
256. *Life,* p.133.
257. W Fairbairn, *Remarks on Canal Navigation illustrative of the Advantages of the use of Steam, as a Moving Power on Canals,* (1831), pp.53, 59. Three years later on an adjoining canal Scott Russell, conducting experiments to determine the most efficient design for canal boats, discovered what he described as 'the wave of translation', now known as Russell's solitary wave. In the 1960s the significance of his discovery in physics, electronics, biology and especially fibre optics started to be understood, leading to the modern general theory of solitons (J Scott Russell, 'Report on Waves', *BAAS1844,* pp.311-90; O Darrigol, *Worlds of Flow: A History of Hydrodynamics from the Bernoullis to Prandtl,* (2005)).
258. Fairbairn, *Canal Navigation,* p.32.
259. Forth & Clyde Canal Navigation Archive 1/70, p.152, Scottish Record Office.
260. Fairbairn, *Canal Navigation,* Plate I. Some confusion is caused by the fact that both the first two boats were both called *Lord Dundas* and Pole does not differentiate. Fairbairn refers to the first as 'experimental'.
261. *Liverpool Albion,* 14 September 1829.
262. *Life,* p.137.
263. *Life,* p.138.
264. Forth & Clyde Archive, 1/71, pp.55-9.
265. Forth & Clyde Archive, 1/72, pp.88-9.
266. *Life,* pp.138-42.
267. *Life,* p.142; Fairbairn, *Canal Navigation,* Plate IV. There is uncertainty as to whether there were three or four iron steamboats for the Forth & Clyde Canal Co. The *CE&AJ,* 1841, 147, lists three including the *Manchester* with a length of 70ft and a beam of 15ft. Hayward, having examined the Forth and Clyde archives, agrees. However, Fairbairn refers to four, including one of 84ft length and 14ft beam, stating, 'This vessel (the *Manchester*) was for several years employed in carrying goods and passengers between Port-Dundas, Grangemouth and Dundee' (W Fairbairn, *Treatise on Iron Ship Building: Its history and Progress,* (1865), p.3; W Fairbairn, 'The Rise and Progress of Manufactures and Commerce and of Civil and Mechanical Engineering in Lancashire and Cheshire' in T Baines, *Lancashire and Cheshire, Past and Present: A History and a Description of the Palatine Counties of Lancaster and Chester, forming the North-Western Division of England. From the Earliest Ages to the Present Time (1867),* (1869), p.cxxiii). The increased size could explain why there were problems transporting this vessel across Manchester, whereas there do not appear to have been similar problems with the earlier vessel. (*MG* 26 February 1831). It is possible that the earlier vessel was transported in parts. However, it is possible that Fairbairn's memory may be faulty as it is unlikely that the locks on the Forth & Clyde Canal in 1832 would have accommodated an 84ft boat.
268. *MG,* 12 November 1831.
269. *CE&AJ,* 4, 1841, 147.
270. [Woodcroft] VI.
271. *Life,* p.134.
272. S Giedion, *Mechanization Takes Command: a contribution to anonymous history,* (1948), pp.78, 87; J W Roe, *English and American Tool Builders,* (1916), pp.75-80; *MPICE,* 28, 1868-9, 573-608; Henderson, *Industrial Britain,* pp.7-16, 73-127. Rolt is incorrect in describing this as 'a small factory for making textile machinery' and the primary source does not refer to 'an *overhead* travelling crane' (my italics), (L T C Rolt, *Tools for the Job,* (1965), p.126).
273. Boyson, *Ashworth,* p.19.
274. *Life,* pp.146-7; Boyson, *Ashworth,* pp.20-1
275. Boyson, *Ashworth,* p.65. The mill became part of the English Sewing Cotton Co in 1898.
276. *Life,* p.148; *MG,* 6 October, 1832; Boyson, *Ashworth,* p.35. Lillie appears to have built a weaving shed for the Ashworths in 1839 (Boyson, *Ashworth,* p.59).
277. Hayward, 'Fairbairns, p.1.42.
278. *Life,* p.149.
279. Tann, *Development of the Factory,* p.103 – see above; J Tann, 'The Textile Millwright in the Early Industrial Revolution', *Textile History,* 5, 1974, 86.
280. *Life,* p.149.
281. *Life,* pp.147-9; [Woodcroft] V.
282. *B&CTB,* pp.158-60; *Life,* pp.148-9.
283. Hayward, 'Fairbairns', pp. 6.2, 6.5-6; *Manchester Examiner & Times,* 6 April 1850. Lillie is remembered for a conical willow machine, used for cleaning both cotton and wool (Ure, *Philosophy,* pp.160-4; Ure, *Cotton,* Vol.2, pp.8-15) and used at Orrell's Mill (Ure, *Cotton,* Vol.1, pp.311-2), and a warp sizing machine. (Ure, *Philosophy,* p.370; Ure, *Cotton,* Vol.2, pp.249-53). From 1842 he was a member of John

Shaw's Club, (F S Stancliffe, *John Shaw's 1738-1938,* (1938), pp.221-2).
284. Tann, *Development of the Factory,* p.105.
285. Rosenberg, *Perspectives,* p.203.
286. Rosenberg and Vincenti, *Britannia,* pp.70-1.
287. F Redlich, 'Innovation in Business. A Systematic Presentation', *American Journal of Economics and Sociology,* 10.3, 1951, 291.
288. D C McClelland, 'Need for achievement and entrepreneurship: a longitudinal study', *Journal of Personality and Social Psychology,* 1, 1965, 389-92; Bull, Pitt and Szarka, *Entrepreneurial Textile Communities,* p.67.
289. J Brown and M B Rose (eds), *Entrepreneurship, networks and modern business,* (1993), pp.1, 6.
290. Berman, 'An essay review of *Science and Technology in the Industrial Revolution*', p.525.
291. Mokyr, *Enlightened Economy,* pp.86-91: Allen, *British Industrial Revolution,* pp.239-42.
292. *Life,* p.466.
293. Tann, *Development of the Factory,* p.95.
294. Tann, 'The Textile Millwright', 83-4; Fairbairn, *Application,* pp.2-5; *Life,* p.122. The millwork at Arkwright's Nottingham factory was designed and executed by Thomas Lowe, a local millwright, with power from a nine horse animal wheel. Lowe built up a sizeable business, not least for Manchester cotton spinners. In 1801 he executed the millwork for G A Lee's Salford Twist Mill, the first fireproof factory in Lancashire, and one which influenced Fairbairn who, incorrectly, believed it to have been the first fireproof mill anywhere.
295. J G Moher, 'The London Millwrights and Engineers 1775-1825', (London University PhD Thesis 1988), pp.70-1.
296 J B Jefferys, *The Story of the Engineers 1800-1945,* (1945), pp.10-11.
297. *Life,* p.92.
298. R Campbell, *A General Description of All Trades,* (1747).
299. W H G Armytage, *A Social History of Engineering,* (1976), p.126.
300. Report of the Select Committee on the Operation of Laws affecting the Exportation of Machinery, *Parliamentary Papers,* (1841), VII, p.217.
301. Jefferys, *Story of the Engineers,* pp.27-31.
302. See Chapter 4.
303. *Life,* p.47.
304. *Life,* p.17.
305. Tann, 'Textile Millwright', 81-2.
306. R A Buchanan, *The Engineers: A History of the Engineering Profession in Britain 1750-1914,* (1989), pp.54, 60-1.
307. *Life,* p.23.
308. W J Reader, *Professional Men: the rise of the professional classes in nineteenth century England,* (1966), p.164.
309. Buchanan, *The Engineers,* pp.61-2.
310. R H Parsons, *A History of the Institution of Mechanical Engineers 1847-1947),* (1947), p.1.
311. Report of Select Committee on the Operation of Laws affecting the Exportation of Machinery', p.107.
312. J Cantrell and G Cookson (eds), *Henry Maudslay & the Pioneers of the Machine Age,* (2002).
313. Buchanan, *The Engineers,* p.80
314. M R Bailey, 'The Mechanical Business' in Bailey, *Robert Stephenson,* p.164.
315. D A Farnie and W O Henderson (eds), *Industry and Innovation: Selected Essays, W H Chaloner,* (1990), p.100.
316. Cantrell and Cookson, *Henry Maudsley and the Pioneers of the Machine Age*; J A Cantrell, 'Two Maudslay Protégés: Francis Lewis and George Nasmyth', *TNS,* 73, 2003, 257-74.
317. C W Sutton, rev. A McConnell, 'Dyer, Joseph Chessborough (1780-1871)', *DNB on-line* (2004).
318. *MPICE,* 28, 1864, 573-602.
319. R L Hills, *Life and Inventions of Richard Roberts 1789-1864,* (2002), pp.192-3.
320. *Life,* p.71.
321. Fairbairn, *UIfE2,* p.212.
322. *Life,* pp.71-2.
323. *Life,* pp.100-1, 111.
324. 'Report from the Select Committee on Artisans, Machinery and Combinations', *Parliamentary Papers,* (1824), V, p.340.
325. Tann, *Development of the Factory,* p.103.
326. *Life,* pp.101, 105; 'Mr. Alderman John Hopkinson, J.P.', *Manchester Faces and Places,* 4, 1893, 155.
327. W Fairbairn, 'Presidential Address', *BAAS1861,* p.lxiv.
328. Fairbairn, *UIfE,* p.112.
329. E Baines, *History and Directory of Manchester,* (1824).
330. 'Report from the Select Committee on Artisans, Machinery and Combinations', p.566; *MG,* 19 and 26 February 1825.
331. Fairbairn, *M&MWI,* p.vii.
332. *Life,* p.93.
333. D C Coleman, 'Gentlemen and Players', *Economic History Review,* NS26.1, 1973, 104.
334. *Life,* p.130.
335. Jefferys, *Story of the Engineers,* pp.9-10; S B Smith, 'Thomas Cheek Hewes (1768-1832) An Ingenious Engineer and Mechanic of Manchester', (University of Manchester MSc thesis 1969), p.16; A E Musson, 'The Diffusion of Technology in Great Britain during the Industrial Revolution' in Musson and Robinson, *Science and Technology,* p.73; D S Landes, *The Unbound Prometheus,* (1969) p.63; A E Musson, 'Introduction' to 1970 ed. of *Life,* p.ix; Coleman, 'Gentlemen and Players', 104; Jenkins, *West Riding Wool Textile Industry,* pp.101-2; C More, *Skill and the English Working Class, 1870-1914,* (1984), p.195; J Wrigley, 'The Division between Mental and Manual Labour: Artisan Education in Science in Nineteenth Century Britain', *American Journal of Sociology,* 885, 1982,538; Reynolds, *Stronger than a Hundred Men,* pp.191-2; J A Cantrell, *James Nasmyth and the Bridgewater Foundry: A study of entrepreneurship in the early engineering industry,* (1985), p.225; M Berg, *The Age of Manufactures: Industry, Innovation and Work in Britain 1700-1820,* (1985), pp.269-70; Moher, 'London Millwrights and Engineers', p.29; Buchanan, *The Engineers,* p.35; Griffiths, *The Third Man,* p.4; G Cookson, 'The West Yorkshire Textile Engineering Industry, 1780-1850', (University of York PhD 1994), pp.48-9; T King, 'Millwrights to Mechatronics: The Merits of Multi-disciplinary engineering', *Mechatronics,* 5.2/3, 1995, 96; G N von Tunzelmann, 'Engineering and Innovation in the industrial revolution', *Interdisciplinary Science Reviews,* 22.1, 1997, 69; L Purbrick, Ideologically Technical: Illustration, Automation and Spinning Cotton around the Middle of the Nineteenth Century', *Journal of Design History,* 11/4. 1998, 281; I Inkster, 'Potentially Global: "Useful and Reliable Knowledge" and Material Progress in Europe, 1474-1914', *The International History Review,* 28.2, 2006, 275; J Mokyr, 'Entrepreneurship and the Industrial Revolution in Britain' in D S Landes, J Mokyr and W J Baumol (eds), *The Invention*

of Enterprise: Entrepreneurship from Ancient Mesopotamia to Modern Times, (2010), pp.20-4; Mokyr, *The Enlightened Economy,* p.109; B Trinder, *Britain's Industrial Revolution: The making of a manufacturing people 1700-1870,* (2013), p.68.
336. http://www.makingthemodernworld.org.uk/stories/manufacture_by_machine/02.ST0... (accessed 4 June 2015).
337. Cookson, 'West Yorkshire Textile Engineering', p.49.
338. Cantrell, *Nasmyth,* p.225.
339. More, *Skill and the English Working Class,* p.195.
340. Jacob, *First Knowledge Economy,* p.4n6.
341. Musson, 'Diffusion of Technology', p.73; Inkster, 'Potentially Global', p.275.
342. *Life,* p.93.
343. Allen, *British Industrial Revolution,* p.242; T Turner and A W Skempton, 'John Smeaton' in A W Skempton (ed.), *John Smeaton, FRS,* (1981), p.7-16.
344. Fairbairn, *M&MWI,* pp.v-vi. There is a similar description in Fairbairn, *UIfE2,* p.212, which has been largely overlooked, although it was quoted by J W Roe, *English and American Tool Builders,* (1916), p.72.
345. *MM,* 10(NS), 1863, 700.
346. Moher, 'London Millwrights', p.29.
347. Tann, 'Textile Millwright', 81-2.
348. Thus the main UK list of millwrights is that published by the Society for the Protection of Ancient Buildings (SPAB) at https://www.spab.org.uk/spab-mills/millwrights-directory/ . The 2014 update contains fourteen entries
349. http://www.unionmillwright.com/ (accessed 5 June 2015). Fairbairn's description of the eighteenth century millwright is also quoted on this website.

Sources, Credits and Acknowledgements for Illustrations in Chapter 2

Fig.2.1 R.J. Byrom, photograph 1962.

Fig.2.2 G. Head, *Cross Street Chapel – Unitarian – Manchester: A Brief History,* (2009) at http://cross-street-chapel.org.uk/history (accessed 8 August 2010).

Fig.2.3 M. Tylecote, 'The Manchester Mechanics' Institution, 1824–50' in D.S.L. Cardwell (ed.), *Artisan to Graduate: Essays to Commemorate the Founding in 1824 of the Manchester Mechanics' Institution* (1974), p.56.

Fig.2.4 G. Pyne, *Lancashire Illustrated in a Series of Views* (1831), reproduced in E Baines, *History of the Cotton Manufacture of Great Britain,* (1836), p.395, and widely elsewhere.

Fig.2.5 R.J. Byrom, Photograph 1962.

Fig.2.6 Widely reproduced e.g. in D. Bindman and G. Riemann (eds), *Karl Friedrich Schinkel, 'The English Journey', Journal of a Visit to France and Britain in 1826* (1993), p.177; I. Miller and C. Wild (eds), *A & G Murray and the Cotton Mills of Ancoats* (2007), p.80; C. Hartwell, *Manchester* (2001), p.279.

Fig.2.7 Fairbairn, *Application,* p.6.

Fig.2.8 Fairbairn, *M&MWII,* p.214.

Fig.2.9 T. Bonson, *Driven by the Dane: Nine Centuries of Waterpower in South Cheshire and North Staffordshire,* (2003), pp.246–7. I am grateful to Tony Bonson for permission to reproduce this diagram.

Fig.2.10 S.C. Hall and A.M. Hall, *Ireland: Its Scenery, Character, &c.* (1841), p.405. Reproduced in W. Ellis, 'Milford Mills' in *Carloviana: Journal of the Old Carlow Society,* 1962, 12. There is an affinity with Portlaw in that both mills had castellated parapets.

Fig.2.11 Fairbairn, *M&MWI,* p.118.

Fig.2.12 W. Fairbairn, 'The Rise and Progress of Manufactures and Commerce and of Civil and Mechanical Engineering in Lancashire and Cheshire' in T. Baines, *Lancashire and Cheshire, Past and Present* (1869), p.cxlvii.

Fig.2.13 [W. Fairbairn], *A List of Wheel Patterns, &c. belonging to William Fairbairn, Millwright, Engineer, Iron and Brass Founder, Canal Street, Manchester* (1842), Title page and pp.54–5.

Fig.2.14 http://yesterdayrules.wordpress.com/ (accessed 27 July 2011).

Fig.2.15 J. Williamson, 'Measured Drawing' (c.1947, prior to dismantling). Photocopy held by author – location of original not known.

Fig.2.16 W.O. Henderson, *Industrial Britain under the Regency: The Diaries of Escher, Bodmer, May and de Gallois 1814–18* (1968), opp. p.92.

Fig.2.17 Henderson, *Industrial Britain,* Frontispiece.

Fig.2.18 H. Sitterding (ed.), *Escher Wyss 1805–1955. 150 Years of Development* (1955), p.15.

Fig.2.19 *MPIME,* 1929, 645. Reproduced in B. Warburton, *Eaton Hodgkinson (1789–1861) Mathematician and Engineer,* (nd but c.1970), p.1.

Fig.2.20 S. Smiles, *Lives of the Engineers, with an account of their principal works* (1862), Vol.3, Frontispiece.

Fig.2.21 E. Hodgkinson, 'Experimental Researches on the Strength of Pillars of Cast Iron and Other Materials', *Philosophical Transactions of the Royal Society of London,* 130, 1840, plate XIII, fig.1.

Fig.2.22 Crane's lithograph (detail). Reproduced in R.S. Fitzgerald, *Liverpool Road Station, Manchester: An historical and architectural survey* (1980), p.23

Fig.2.23 As with his later books, *Remarks on Canal Navigation was* widely distributed, including (as illustrated here) to John Dalton, and to Thomas Telford (*Life,* pp.134–5).

Fig.2.24 W. Fairbairn, *Remarks on Canal Navigation illustrative of the Advantages of the use of Steam, as a Moving Power on Canals* (1831), Plate II.

Fig.2.25 *MM,* 16.427, 1831–2, 33.

Fig.2.26 *British Trade Journal,* 1 February 1883, pp.95–6, (Detail).

CHAPTER 3

Sole Proprietor

1832–1841

Fig.3.1:
B.R. Faulkner, *William Fairbairn*, c.1840, (detail).

Introduction

WHAT would happen in the second phase of Fairbairn's business career – his sole proprietorship – which started with the crisis of the dissolution of the partnership in September 1832? He soon recovered from the dissolution and, no longer restrained by his more cautious partner, exploited his independence. To the core business of mills and millwork were added increasing shipbuilding and steam engine manufacture, while experimental engineering and papers to the British Association helped to make him more widely known.

In this phase, as throughout his career, it is argued that Fairbairn was in the vanguard of mill-building, as he built complete new mills for entrepreneurs and foreign governments, and replaced steam engines and line-shafting for existing mill-owners. Expertise in one branch of engineering was transferred to another. There were innovations such as the riveting machine and Lancashire boiler. By the mid-1830s Fairbairn's sole proprietorship was well-established with his reputation bringing clients to him, such that it is more difficult to trace network links from one commission to another than in the Fairbairn & Lillie phase – Alexander Wilson in Russia provides an exception. Personal contacts with clients, whether by travelling to visit them, or by welcoming them to his works, are seen to have been important, as the geographical client base widened with the most advanced mills and millwork diffused into Germany, Belgium, Austria, Italy, Russia, Turkey and Egypt.

Fairbairn's contributions to the early development of iron steamships are traced, initially in Ancoats and then from 1835 to 1844 – the most important decade in the establishment of modern shipbuilding – at Millwall on the Thames.[1] What was his standing as a builder of iron ships and marine engines sent to Europe, the Baltic, the Black Sea, India, Australia and South America? His commitment to research is shown to have meshed with his shipbuilding as he investigated the strength of wrought-iron plates and riveted joints, providing knowledge that was relied upon by boiler-makers and iron shipbuilders for many years. Fairbairn ceased shipbuilding in the mid 1840s, but his interest continued. Papers were written over the next thirty years – referred to, out of chronological sequence, in this chapter. From 1859 to 1867 he was a member of the British Association's Committee on Steamship Performance.[2] Research into the properties of cast-iron continued, including some of the first tests on long-continued loading, or creep, and on the effects of temperature on strength, again widely publicised, and relied upon.

In 1838 Fairbairn built his first railway engine, but locomotive building was a minor area of work until the 1840s. It more logically forms part of Chapter 4.

In spite of an increasing national and international reputation, this phase of Fairbairn's career contained difficult times, sometimes when opportunities were grasped, perhaps without full consideration of their implications – such as his acceptance of invitations to visit Turkey without having adequate management in place at Ancoats and Millwall. The 1837 recession affected cash flow, and the Millwall shipyard, described by Smiles as 'the first great iron shipbuilding yard in Britain'[3] – and hence in the world – turned into a financial disaster, clouding the early years of the ensuing family business.

This was all against the backdrop of local and national events. In 1832 the passing of the Reform Bill and enfranchisement of Manchester led to celebrations; but there was also cholera in the town.[4] Victoria acceded to the throne in 1837, and in 1841 she married Albert. The Grand Junction Railway linked Manchester and Birmingham in 1837 and the London and Birmingham Railway completed the link from Manchester to London. Prior to this, in the setting up of the Millwall shipyard, and en route to the Continent, Fairbairn travelled by coach. In 1838 Richard Cobden and others set up the Anti-Corn-Law League in Manchester.

The economic background is particularly relevant. 1832 was not a good year for trade but there was improvement in 1833, continuing to the spring of 1836, after which there was a fall. In 1837 – the year of financial panic in the United States – recession became sharp. The next years saw some recovery, but 1841 and 1842 were depressed.[5] Axon records distress in Manchester in 1837 and in 1842.[6] The General Strike of 1842 was fuelled by a reduction of wages in the cotton industry, support for the Charter, want of employment and the high cost of food. The millwrights were not involved.[7] Nevertheless the strikers forced Fairbairn to shut down his boilers for ten days.[8] Economic recovery came in 1843, although it did not immediately lift the economy to any very high level of activity.[9]

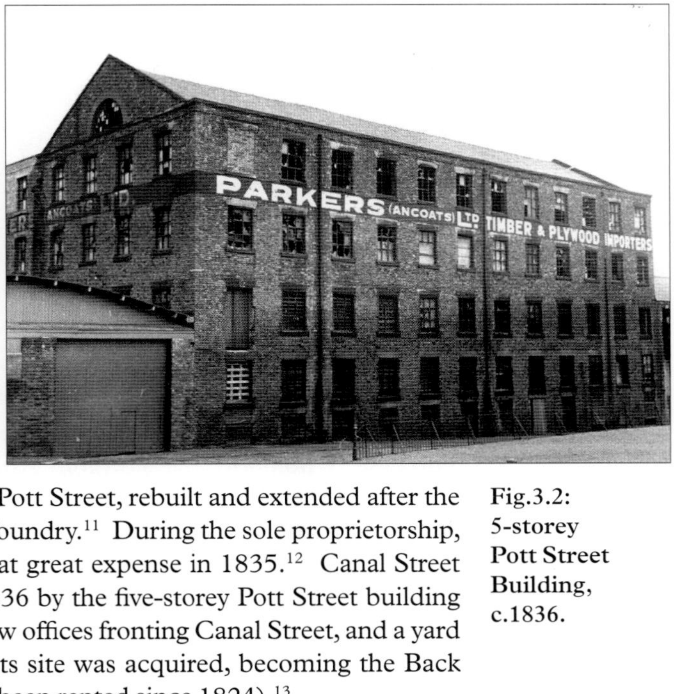

Fig.3.2: 5-storey Pott Street Building, c.1836.

Resources

An engineering business needed premises, capital and personnel. On the dissolution of Fairbairn & Lillie, Fairbairn retained the partnership's premises, comprising the original three-storey building, No.20, Canal Street, and the Canal Street Works across the street.[10] By 1832 the latter comprised the three-storey building on Pott Street, rebuilt and extended after the fire of 1831 and the recently-built new foundry.[11] During the sole proprietorship, the Millwall shipyard was constructed at great expense in 1835.[12] Canal Street Works was further extended, around 1836 by the five-storey Pott Street building and new stables; and around 1839 by new offices fronting Canal Street, and a yard on Pott Street. In 1839 another Ancoats site was acquired, becoming the Back Mather Street Works (where a yard had been rented since 1824).[13]

In 1830 Fairbairn & Lillie had a sound balance sheet.[14] Apart from the short-term loan for their premises, they appear to have financed the business out of income, but the Egerton Mill venture strained their finances. At the dissolution, the Ancoats buildings remained with Fairbairn but he was obliged to mortgage some of them.[15] The Millwall shipyard was built at a cost in excess of £50,000, over twice the cost of a typical new iron shipyard.[16] A lot of money was spent on the foundations in the Thames mud.[17] All the money for it was borrowed but there is no indication where from.[18] Around 1837 rumours circulated that Fairbairn was in financial difficulties. This affected his mercantile credit.[19] There was no

Fig.3.3: 3-storey Pott Street. building, rebuilt c.1831, after a fire.

shortage of work, but with large sums tied up in on-going contracts, there were cash-flow problems. In some cases payers were slow, and money could be held back if a client was dissatisfied, both of which circumstances affected payments for Fairbairn's pumping engine at the Water Grove Mine in Derbyshire.[20] Wages at Ancoats alone were running at £1,000 per week.[21] Cash was raised on the security of the fixed assets,[22] and employees John Elliot and Robert Smith put modest sums into the business.[23] With 'a numerous family' entirely dependent on him, Fairbairn faced 'fits of melancholy'.[24] It was his own fault, as a 'friend who knew him well in business' wrote:

> He liked to secure a great order, and his one anxiety, when such an opportunity presented itself, was to 'do the work', thinking little of the result, whether for profit or the reverse ... He was never, in the ordinary acceptation of the term, a 'good man of business'.[25]

Fig.3.4: Fairbairn's Offices, Canal Street, Ancoats, built c.1839.

After the problems of 1837 there was sufficient improvement in the firm's finances to enable Fairbairn to build new offices on Canal Street and acquire the Back Mather Street site in 1839;[26] and in 1840 to allow him to purchase a home at the Polygon, Ardwick. The serious problems affecting Fairbairn's shipyard in the early 1840s, form part of the next chapter.

Throughout the Fairbairn & Lillie years the number of men employed had grown steadily until it was above 300 by the end of the 1820s. At the dissolution when most of the foremen and many of the men went with Lillie, Fairbairn, hardly surprisingly, 'felt alone'. For neither the first nor the last time, his entrepreneurial perseverance and determination drove him on.[27] The dip in employee numbers at that time is impossible to quantify but was soon reversed by rapid expansion, and Fairbairn's figure of 2,000 employees by the end of the decade, including perhaps 600 at Millwall, may be correct. However Love & Barton indicate much lower figures, of 550–600 employees at Ancoats and 400 at Milwall.[28]

The Ancoats employees in the late 1830s included 'smiths, strikers, moulders, millwrights, mechanics, boiler makers and pattern makers' – the patternmakers numbering around fifty and the boilermakers fifty, reducing to around five following the introduction of the riveting machine.[29] Financial difficulties in 1837 and in the early 1840s must have led to some reduction in numbers and the uncertainties around 1840 in *Table 3.1* may reflect this.

Mill Building

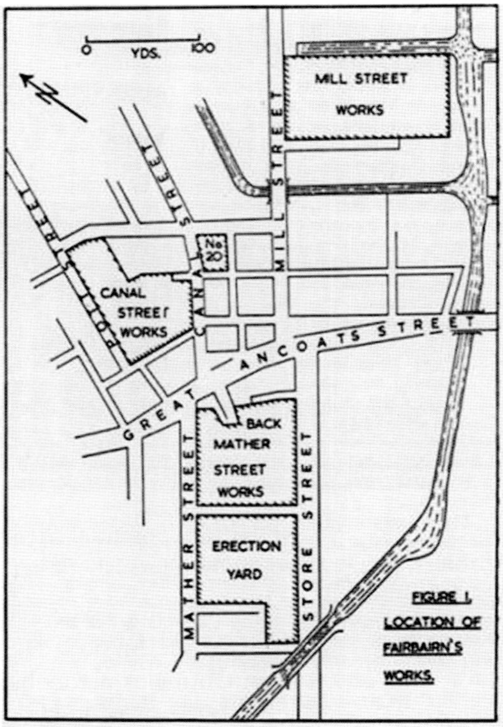

Fig.3.5: Locations of Fairbairn's Ancoats Works.

WILLIAM Fairbairn was the leading figure in the building of mills, with their prime-movers and power transmission, for half-a-century. In *Mills and Millwork* he discusses and illustrates the progression of mill architecture from the brick boxes of the functional tradition of the 1820s, via buildings with corner pilasters and cornices, to factory buildings 'vieing [*sic*] with ... public buildings as works of art', designed in collaboration with architects. In *The Works: The Industrial Architecture of the United States,* Betsy Bradley reproduces an illustration from *Mills and Millwork* showing the extremely modest adornment of the simple brick box mills with distinctive corner pilasters and modest cornice, which Fairbairn claims to have introduced in 1827 and was 'speedily copied in all directions'.[30] She then illustrates Talbott & Brother's Machine Works at Richmond, Virginia (1853) which repeats the detail.[31] The use of this apparently trivial detail throughout the textile districts was remarkable, extending into the first decade of the 20th century. Tann is incorrect to suggest this form was less common than Fairbairn indicates.[32] It is doubtful if most of those using this minor detail had any idea of its origin but its widespread, perhaps unthinking, imitation is an indication of the respect in which Fairbairn mills were held. However, Fairbairn's assessment of the progression of mill architecture was flawed in that he failed to recognise several fine 18th-century mills, in some of which architects were involved (referred to below). In the textile districts of Lancashire architects were not involved in mill

Year	Ancoats	On Site	Millwall	Total	Source
1824	60–70	incl.	-	60–70	Select Com.
1826	142	incl.	-	142	*MG*
1830	300+	incl.	-	300+	Fairbairn
c.1836	incl.	incl.	incl.	2,000+	Fairbairn
c.1839	incl.	incl.	incl.	2,000+	Pole
1839	550–600	not incl.	400+	c.1,000	Love
1840	not incl.	not incl.	600	?	*CE&AJ*
1841 or 1846	600	probably not incl.	not incl.	?	Butterworth
1842	incl.	incl.	incl.	1,000–2,000	Chadwick
1851	c.2,500	incl.	-	c.2,500	Hayward
1871	350	incl.	-	350	*M CityNews*

Table 3.1: Fairbairn complement at various dates.

Fig.3.6: Fairbairn's pilasters and cornice, 1827.

building until into the mid 1860s, although the building contractor, David Bellhouse, who built mills, was referred to as an 'architect' in the 1820s.[33]

There were also changes in methods of building procurement, from the traditional system of tradesmen organised by, for example, a millwright, and paid by the employer, to the fixed-price contract for the total works undertaken by a 'general contractor', with hybrids in between.[34] There is evidence of general contracting in the north-west from the late-1820s.[35] Andrew Ure wrote in 1835 that a potential client had only to provide a budget and a brief and Fairbairn would then furnish 'designs, estimates and offers', becoming responsible for

> the masonry, carpentry, and other work of the building, for the erection of a sufficient power … to drive every machine it is to contain, and for the mounting of all the shafts and great wheels by which the power of the first mover is distributed.[36]

This indicates a threefold division, between the building, the power and its transmission, and the machinery, rather than a single contract for everything – a 'turnkey' contract as it would be called today. What Ure does not say is whether Fairbairn acted as main contractor for all this work on a gross lump sum basis, or in a hybrid capacity whereby he contracted only for the iron frame, prime-mover, gearing and shafting; and in a consultant role for the building work – walls, floors, roof, staircases, flues, chimney. The latter is almost certainly the case, but if he did contract to build the complete factory, the traditional building work was all sub-let as there is no evidence that Fairbairn employed any building trade operatives. At

Fig.3.7: Pilasters and cornice, Whittakers' Mill, Ashton-under-Lyne, 1847.

Brunswick Mill, Manchester, (1839) Fairbairn apparently had only a minor role, limited to approving the structural ironwork, which involved Hodgkinson beams.[37]

Although not for a mill, another procurement method is illustrated by Manchester City Council's Smithfield Market of 1853–4, for which Fairbairn was appointed 'consulting engineer'.[38] The roof was the subject of a competition which drew entries from London, Dublin, Birmingham, Glasgow and Derby. The adjudication of the submissions and the superintendence of the work were entrusted to Fairbairn. The premium went to a Mr Wheeldon of James Haywood's Phœnix Foundry, Derby.[39] Haywood was appointed as contractor for the roof, to be 'tested and proved' by Fairbairn.[40] In addition there appear to have been separate contracts for the groundworks, the foundations and the shops.[41] The probability is that these contracts were coordinated by Henry Bake who, on Fairbairn's recommendation, was appointed 'clerk of the works', at a salary of £200 a year, 'acting, so far as the roof of the market is concerned, under the direction of Mr Fairbairn'.[42] Bake's role was clearly wider than the modern clerk of works, which is limited to that of site inspector, as 'the original design … was revised by Mr Bake'.[43] Manchester's Smithfield was 'the largest market roof of its day'.[44]

Fairbairn's standing drew clients from throughout Europe, in most cases because there was insufficient expertise in their own country. In each case something of his knowledge was diffused through his involvement in a factory which was typically the first, or the largest, or the most advanced in its region. These commissions emphasise the importance of clients visiting Britain, evident in respect of Fairbairn mills in Germany, Turkey, Russia, and later, Norway and Sweden. Travelling to see potential clients was also important – Fairbairn obtained Russian and Turkish contracts during or following visits. Some of the commissions demonstrate the role in the diffusion of technology of Britons who had established themselves abroad.

Whereas in the 1820s links from one client to another helped to build the business, in the 1830s an established reputation, Ure's books, and personal meetings in Ancoats or on-site secured the work. Orrell's Travis Brook Mill in Stockport was state-of-the-art and its influence can be followed to Germany.

Fig.3.8: Sections, Orrell's Travis Brook Mill, Stockport.

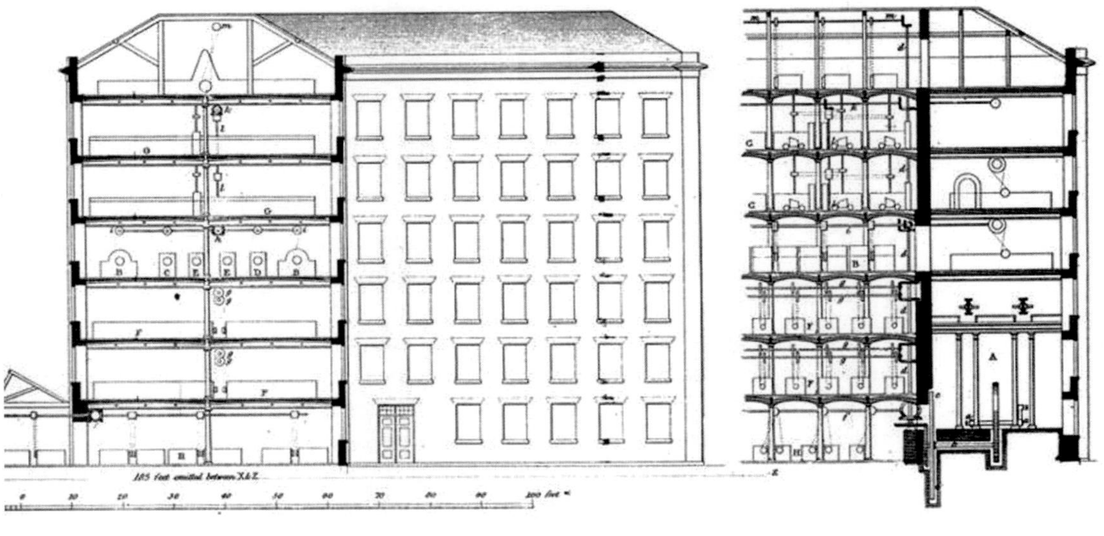

Fig.3.9: Orrell's Travis Brook Mill, Stockport.

It was chosen by Ure for description, and for the frontispiece of his *Philosophy of Manufactures* (1835).[45] The illustration, with an expanded description and detailed plans and sections, appeared the following year in his *Cotton Manufacture of Great Britain.*[46] The earlier book was immediately translated into French, and both books into German.[47] Sections of Ure's books were also published in America, including part of the description of Orrell's Mill.[48] Gatrell claims that, far from representing S.D. Chapman's 'mill of the future', large fireproof mills like Orrell's were rare in Britain.[49] There is no statistical evidence, but the impression given by more recent publications suggests otherwise.[50]

At Orrell's Mill – a combined spinning and weaving mill built 1833–4 – Fairbairn was responsible for the design of the building, the layout of the machinery, the construction of the engines and shafting and the supervision of the whole of the works. It was described by Ure as 'a model of factory architecture'.[51] The cast-iron beams were of the Hodgkinson type, with 'fire-proof' brick-arched floors.[52] In plan the factory occupied three sides of a rectangular mill-yard. The two shorter wings were used as storerooms, winding rooms and counting-house. The main body of the mill, 280ft by 48ft, had a single row of columns. There were seven floors including the attic storey. An extensive single-storey weaving shed, 280ft by 80ft, was attached to the rear of the main block. The 1,100 or so looms, because of the desirability for stability and humidity, occupied the ground floor of the main block and the adjoining weaving shed (which had a saw-tooth roof, albeit, according to Ure's section, facing south). The throstle frames occupied the first and second floors of the main block. The mules were on the fourth and fifth floors. In between, the third floor was allocated to the preparation of raw materials – cleaning, blowing and lapping machines, carding engines, drawing machines, bobbin and fly frames, and roving machines. Altogether there were 45,850 spindles. It is unlikely that the full complement of machinery was installed at the outset. The attic storey was used for warping and dressing the yarn for the looms. The engine house extended through three stories, housing a pair of beam engines with common flywheel between them. The boilers were in a single-storey boiler house alongside the engine house, with an underground flue to the chimney some distance away on a knoll. Staircases, semi-circular in plan, were located in each of the inner corners of the projecting wings, surrounding lift shafts. The corner pilasters incorporated toilets. There was space heating and fresh air was provided by ample sliding-sash windows.

Fig.3.10: Scharfenstein Mill, 1837.

Fairbairn's Travis Brook Mill was the pattern for the largest and most modern cotton factories in the Rhineland and in Saxony. When F.A. and W. Jung planned a new factory in Hammerstein, Wuppertal, in 1835, they visited Manchester to consult Fairbairn, who provided detailed plans and costings.[53] In July 1838 Fairbairn visited Hammerstein to view the completed mill, which another visitor considered 'might have been transported overnight from Lancashire'.[54] In Saxony the Fiedler & Lechla cotton spinning factory at Scharfenstein, built in 1837 with the assistance of a German architect, was almost a replica of Travis Book Mill – see *figs 3.8–3.10*.[55] The influence of Orrell's Travis Brook Mill was on-going. The German travel writer, J.G. Kohl, wrote about his visit there in the early 1840s. It had been recommended to him 'as one in which all the newest improve ments in machinery had been adopted'.[56]

Fig.3.11: Dixon's Mill, Carlisle, built 1836.

The first examples of architects and engineers working together on major industrial buildings include William Wyatt II working with Matthew Boulton at the Soho Manufactory, Birmingham, 1764–6, and the architect Samuel Wyatt working with Boulton & Watt and John Rennie at the Albion Mill on the Thames in 1784–8.[57] In the late 18th century, architects Samuel and William Stretton worked on textile mills in the Nottingham area. In the early decades of the 19th century, architects were involved with gasworks, ironworks and breweries.[58] The first major 19th-century textile factory on which an architect and engineer worked together appears to have been Peter Dixon's Shaddon Mill in Carlisle in 1836. Here Fairbairn

Fig.3.12: Dixon's Mill: Typical Upper Floor.

Fig.3.13: Robinwood Mill, Todmorden.

worked with the Manchester-based architect, Richard Tattersall, who, with a restrained design in dressed red sandstone, produced a fine mill.[59] The main block, 224ft by 58ft externally, differs from Travis Brook in that each floor has two rows of columns. This mill also had what is believed to have been the tallest mill chimney of its day in Britain, at 305ft – an octagonal brick tapering shaft without plinth, reduced to 270ft in 1950.[60]

The engagement of architects for mill buildings was the exception in the first half of the century. The six-storey Robinwood Mill, Todmorden, is a fine building from the late 1830s for which Fairbairn prepared the drawings and superintended the work.[61] It is however an atypical Fairbairn mill in that it is built in coursed dressed stone without pilasters, although it has a central pediment and there are band courses at cill levels; and its floors are of timber rather than of 'fire-proof' construction (other than above the engine house). This is a reminder that until the fourth quarter of the century timber floor construction continued to be used in many industrial buildings. Cost was the main determinant – Jennifer Tann has estimated that in 1836 a cotton mill with traditional timber floors cost 1*s*.8*d* per square foot, whereas an advanced Fairbairn iron-framed factory cost 4*s*.8*d* per square foot.[62] Robinwood shows that Fairbairn was involved in both and, as Richard Morris points out, is a corrective to the notion that industrial archaeology must always focus upon technical innovation'.[63]

A magnificent flax mill from 1838 is Hunslet Mill which, on balance of probabilities, is by Fairbairn for the following reasons:

1. The beams are of the Hodgkinson type and similar to those in other Fairbairn mills.[64]
2. The plan and massing, the projecting stair tower and location of the boiler house, the pilasters and cornice are all typically Fairbairn, and the overall character is that of a Lancashire mill. However, documentary evidence is to date lacking.[65]

This building features in the Victorian Society's top ten most endangered Victorian buildings (in 2016).[66]

An illustration of the role of emigrants in the diffusion of industrial technology is provided by the Scot, Alexander Wilson, at St Petersburg who became responsible for Government engineering and textile works at Colpino and Alexandrovsk, where he built mills 'containing all the improvements known at the time'. With the Empress Maria Fedorovna (widow of Tsar Paul I and mother of Tsar Nicholas I) he had also built an institution for 800 foundling children with the twofold aim of training them to gain their own livelihoods, and to carry their knowledge of manufacturing processes in the latest mills to the remotest parts of Russia.[67] Fairbairn visited Russia around 1837 when he was 'principally engaged in connection with the Government cotton mills under the direction of General Alexander Wilson, MInstCE'.[68] Wilson was a good friend of Nicholas I, and Fairbairn's building of the iron steam-yacht, *Nevka,* for the royal family comes from this link.[69] Wilson also had associations with the banker and philanthropist Alexander von Steiglitz, which probably led to Fairbairn building the water-driven flax mill at Invangorod for von Steiglitz.[70]

This pattern of diffusion was reversed when Fairbairn obtained work through an Italian emigrant who settled in Britain. Benedetto Albano (1796–1881), a graduate of the Royal Polytechnic School of Naples, had to flee for his life from the restored Bourbon dynasty. He was initially employed by J. & G. Rennie, subsequently becoming a consulting engineer and member of the Institution of Civil Engineers.[71] After political changes in Italy he undertook various commissions there, including the flax and hemp works of Battaglia & Co. at Cassano d'Adda, c.1841, for which Fairbairn supplied the waterwheel and millwork. This was only the second mechanised factory in Lombardy and 'excited a good deal of attention in Italy'.[72]

In 1838 Fairbairn was commissioned in connection with a locomotive works for the Wien-Raaber-Eisenbahn in Vienna. He sent John Haswell to set it up.[73] Haswell settled there, managing the works for forty years – the founder of Austrian locomotive building, who introduced industrialisation into Austria.[74]

Turkey proved to be a fruitful source of work but Fairbairn's absence whilst visiting there proved disastrous for his shipyard. Turkish diplomats visited Fairbairn's works followed by a call to him to visit Turkey. This he did with his eldest son, 18-year-old John, in 1839. At Constantinople they were looked after by an Armenian officer, Ohanes Dardian. An afternoon audience with the Sultan was organised, but the Sultan died in the morning of the day of the audience and it never took place. Five weeks were spent inspecting and making recommendations. The latter included a steam-engine and pumps to empty a dry-dock, to replace mules and horses. In the small-arms manufactory he found recent machinery from Rennie and from Maudslay & Field, but little use was being made of it due to 'the apathy of the Turks and their aversion to new things'.[75] Ohanes Dardian led a second Turkish delegation to Manchester in 1843, with the Turkish Ambassador, Ali Effendi, in attendance, and Fairbairn made a second visit to Turkey.[76] Large orders followed – a corn mill after the first visit, and a woollen mill the second – but it is unclear which visit led to furnaces, forges and rolling mills, or to the silk and cotton mills for which the ironwork was sent out.[77] Fairbairn also investigated Turkish iron ores from Samakoff, leading to a paper to the Institution of Civil Engineers,[78] which received a silver Telford medal.[79]

Fairbairn also worked for the Ottomans in Egypt. In 1843 he submitted an estimate for work at the arsenal at Alexandria, comprising forging equipment,

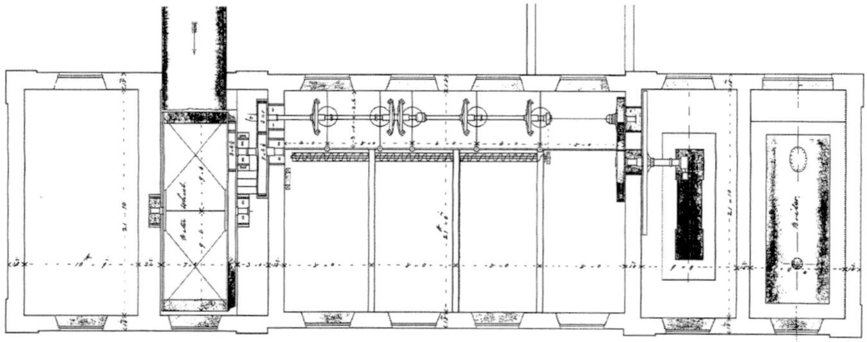

Fig.3.14: Lower floor plan of Park Mill, Brereton, Cheshire, c.1833.

forty-three machine tools, millwork, a steam engine and boiler, for the sum of £8,333. This was followed by other estimates. They were subject to the 'usual terms of payment' – 'one third when the order is given, one third when the work is reputed to be three parts finished, & the remainder when delivered'. It is not known which of these estimates was accepted but in 1845 Fairbairn wrote that 'the first order is about finished'.[80]

Two small corn-mills from this period are significant – one because of the layout of the millstones and antiquity of the existing millwork, and the other because of its place in the history of architecture. Brereton Mill in Cheshire, dating from 1833, had its millstones laid out in line, rather than in clusters around a great spur wheel. John Howard was the owner of Greenfield Mill, Hyde, a town where Fairbairn's millwork was well-known.[81] He purchased the Brereton Estate and commissioned Fairbairn to design a corn mill to replace the existing run-down mill. The surviving drawings, signed by Fairbairn, show four sets of millstones in line on the first floor, driven by a main shaft on the lower floor, with disengaging mechanisms for each pair of stones. At one end of the main shaft there was a waterwheel and at the other end a steam engine.[82] Brereton is believed to be the oldest remaining corn-mill with a linear layout. The waterwheel, gearing and shafting remain and, on the assumption that this millwork is by Fairbairn, it is believed to be the only surviving example of his millwork.[83] It was a forerunner of Fairbairn's greatest corn-mill at Taganrog, Russia, with its line of thirty-six pairs of millstones.[84]

The significance of the other small corn-mill, in Constantinople, is that it was the first iron-framed three-storey building. It also illustrates the transfer of technology from boiler-making, via prefabricated ships to a prefabricated building. Built of iron, it had a row of three sets of millstones on the first floor, driven by a main shaft on the ground floor, powered by a high-pressure columnar steam engine, served by two boilers.[85] No elevations exist, although Bannister has attempted to reconstruct them.[86] For forty years textile mills had been constructed with internal cast-iron framed structures, surrounded by outer walls of heavy masonry. Fairbairn, in a bold, creative move, constructed this three-storey corn-mill, in which structural cast-iron pilasters and 'sheet-iron' panels – probably wrought-iron – were used for the external walls in lieu of masonry. This is best described as part of a cumulative development, as it had decades of development

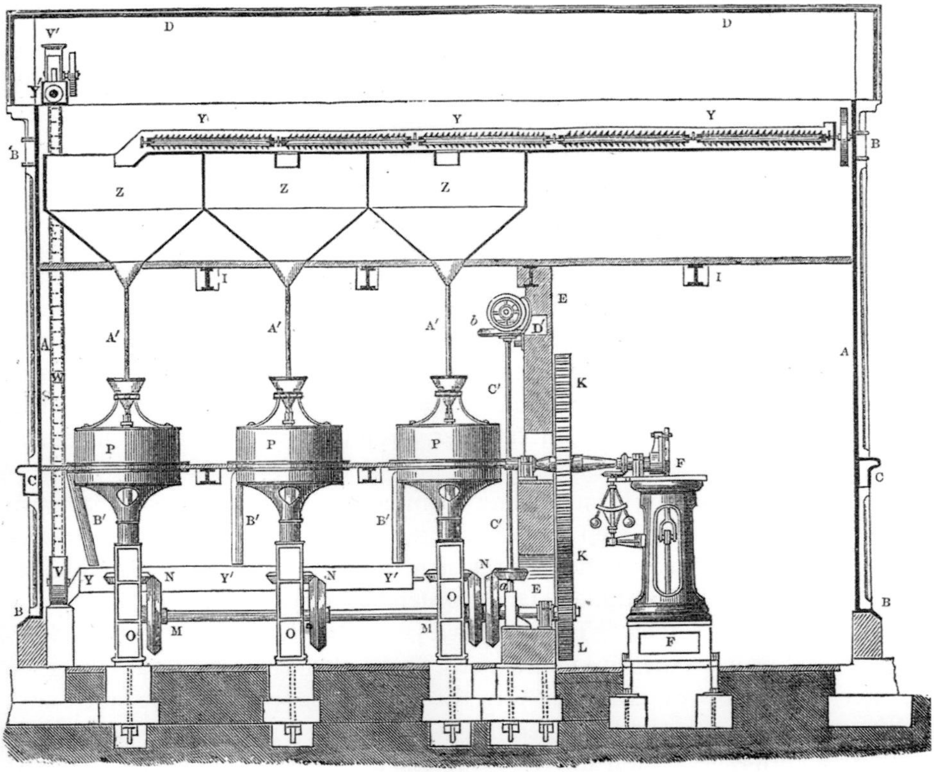

Fig.3.15: Section through prefabricated corn-mill, Constantinople, 1840.

leading to it, although to what extent Fairbairn was aware of this remains unclear.[87] But he did have the precedents of internally-framed textile mills and of prefabricated iron ships.[88] In 1833 he had sent out, in sections, the 108 ton, 98ft long, *Minerva* to the Zurichsee, followed by larger boats, also in sections, to the Bodensee, the Upper Rhine and, around the same time as this corn mill, to the Ganges.[89] The latter significantly included four 'accommodation boats' – 125ft long wrought-iron travelling house-boats to be towed by a steamer.[90]

The prefabricated corn-mill was exhibited at Fairbairn's Millwall yard, prior to being shipped to Turkey.[91] It was almost certainly seen there by James Bogardus, the American who claimed to have built the first iron building, and who left London for New York in the autumn of 1840.[92] Even if he did not see the building in London, Bannister believes he must have been aware of it from the American press.[93] Henry-Russell Hitchcock's assertion that the building Bogardus erected in New York in 1848–50, with an exterior frame as well as an interior skeleton of cast-iron, 'was undoubtedly the first storeyed urban structure to be built', is unsustainable.[94] One can only speculate why Fairbairn, having constructed this early prefabricated iron building, did not construct further such buildings. In 1843 he did design a single-storey prefabricated iron woollen mill, also for Turkey, but in the event the external walls were built of stone, perhaps because of the costs of iron.[95] The probable reason why Fairbairn built no more prefabricated buildings was the financial difficulties at Millwall in the early 1840s.[96] In any event this building, unique amongst Fairbairn's works, is of significance in the history of construction, as the first three-storey iron building without load-bearing external walls, a forerunner of the steel-framed buildings of the 20th century.[97] Having returned to America, in 1856 Bogardus published an advertising booklet claiming to be the builder of 'the first complete cast-iron building ... in the world'.[98] Plates affixed to his buildings read 'James Bogardus Originator and Constructor of Iron Buildings'.[99] He was not their originator.

Steam-Engines

FAIRBAIRN, as sole proprietor, continued to build stationary steam engines, an activity which had commenced shortly before the end of the Fairbairn & Lillie partnership.[100] By the mid-1830s he had also built at least five marine engines. These would have been side-lever engines – more compact with a lower centre of gravity – following Boulton & Watt, the market leaders in early marine engines. At the end of the 1830s Fairbairn's works was inundated with orders for steam engines – 'upwards of sixty ... many of them of great size, were in hand'.[101]

Around 1835, having been building steam-engines for only six years, Fairbairn took the unprecedented step of installing two side-lever marine engines, rather than traditional beam engines, in the new cotton mill of Bailey Brothers at Stalybridge, another illustration of the transference of technology from one sphere of engineering to another.[102] The advantages were that these engines occupied less space and had less expensive foundations.[103] Another innovation was taking the power, as with his waterwheels, from the toothed periphery of the single 20ft diameter flywheel, located between the two engines.[104] A single flywheel ironed out irregularities in motion and the peripheral toothing drove the main shaft fast

Fig.3.16: Side-lever marine steam engine, Bailey Bros Mill, Stalybridge, 1836.

enough to eliminate intermediate trains of gear wheels.[105] There were prognostications of failure for peripheral toothing from some engineers, but Baileys' engine dispelled these, and driving from the rim of the flywheel became widely used.[106] A measure of his peers' response to this engine can be gauged by Scott Russell illustrating it in his article on the 'Steam Engine' in the 1841 edition of *Encyclopaedia Britannica*.[107] It is not known how many of these engines were built. In 1861 Fairbairn wrote that he had introduced them 'on an extensive scale' and that 'numbers now at work ... are ... giving entire satisfaction'.[108] However, records of mid 19th-century mill engines are sparse and the only other known example of a side-lever mill engine by Fairbairn is Kingston Mill Stockport.[109] William McNaught of Glasgow is known to have designed similar engines.[110] The window of opportunity for these engines was only open until the horizontal engine came into prominence in the 1860s. The evidence suggests that side-lever engines were never widely adopted for mill use.[111] This seems surprising, given their advantages, and may be a case of technical inertia.

By the late 1830s Fairbairn was one of the leading manufacturers of steam engines. In 1839 Love & Barton wrote of his steam engine department:

> All sizes and dimensions are frequently under hand, from the diminutive size of 8 horses' power, to the enormous magnitude of 400 horses' power. One of this latter size contains the vast amount of 200 tons or upwards of metal, and is worth, in round numbers, from £5,000 to £6,000.[112]

Woodcroft, visiting the works between 1839 and 1844, referred to 'such constant crops of orders as have seldom fallen to the lot of any other engineering establishment'

> Such was the pressure against time, that he found the patterns for some engines of marine construction being proceeded with from full sized chalk drawings on the floor, and without any scale drawings on paper having been made.[113]

Elsewhere there is another description of this practice, suggesting that it was more than an expediency:

> It was ... a characteristic practice with him to order many of his mechanical contrivances to be drawn out full size on a large surface. For this purpose the floor of one large room nearly seventy feet long was kept free as a huge drawing-board; to these full-sized drawings the wooden patterns, when completed, were brought down and adjusted.[114]

Substantial profits were being made, sufficient to offset serious losses at Millwall.[115] The firm was exporting to all parts of the world, '*this* article is for Calcutta, *that* for the West Indies; this for St. Petersburgh, that for New South Wales'.[116]

The development of the high pressure expansive engine was a watershed in the evolution of steam power technology, bringing economies of fuel. Established in Cornwall by the 1820s, these engines were not generally adopted elsewhere until the mid 1840s.[117] In the textile districts Fairbairn was an early advocate of the advantages of high pressure steam worked expansively, as in his paper to the Manchester Geological Society in 1840 *(fig.3.17)*.[118] In 1849 he wrote that 'for

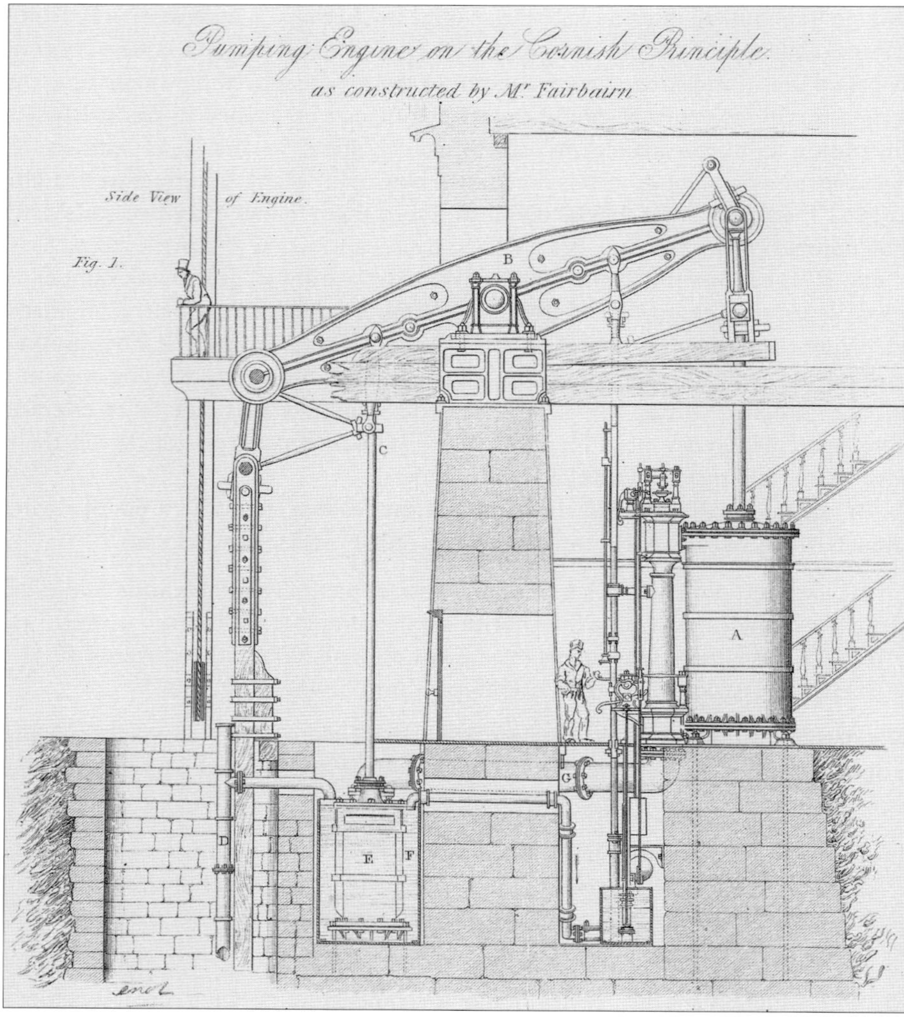

Fig.3.17: Fairbairn high pressure 'Cornish' Pumping Engine for Biolly et Fils, Verviers, Belgium 1839–40.

many years [I] had to contend with the fears and the prejudices of the manufacturer, before the present system ... was adopted'.[119]

Murdoch's D-type or slide valves in use at that time were unsuitable for working steam engines expansively.[120] Fairbairn introduced a remarkably simple solution of drop valves worked by cam wheels. In this he was the innovator, 'greatly modifying and perfecting' the invention of Robert Brownhill'.[121] The valves were described in a widely publicised paper in 1849 and illustrated in *The Engineer and Machinist's Assistant* the following year.[122] They were not perfect. At Saltaire they were replaced by Corliss valves, pioneered in America, after only ten years.[123] However Fairbairn's valve-gear continued to be used into the 1870s.[124] In 1869 *The Engineer* illustrated it, stating:

> It is certain that neither the Corliss, Allen, or any other gear ... can do more [for beam engines] than has already been done years and years ago to obtain a good diagram, or, in other words, a good distribution of steam. ... It is an old device now, a device which has borne the test of years of

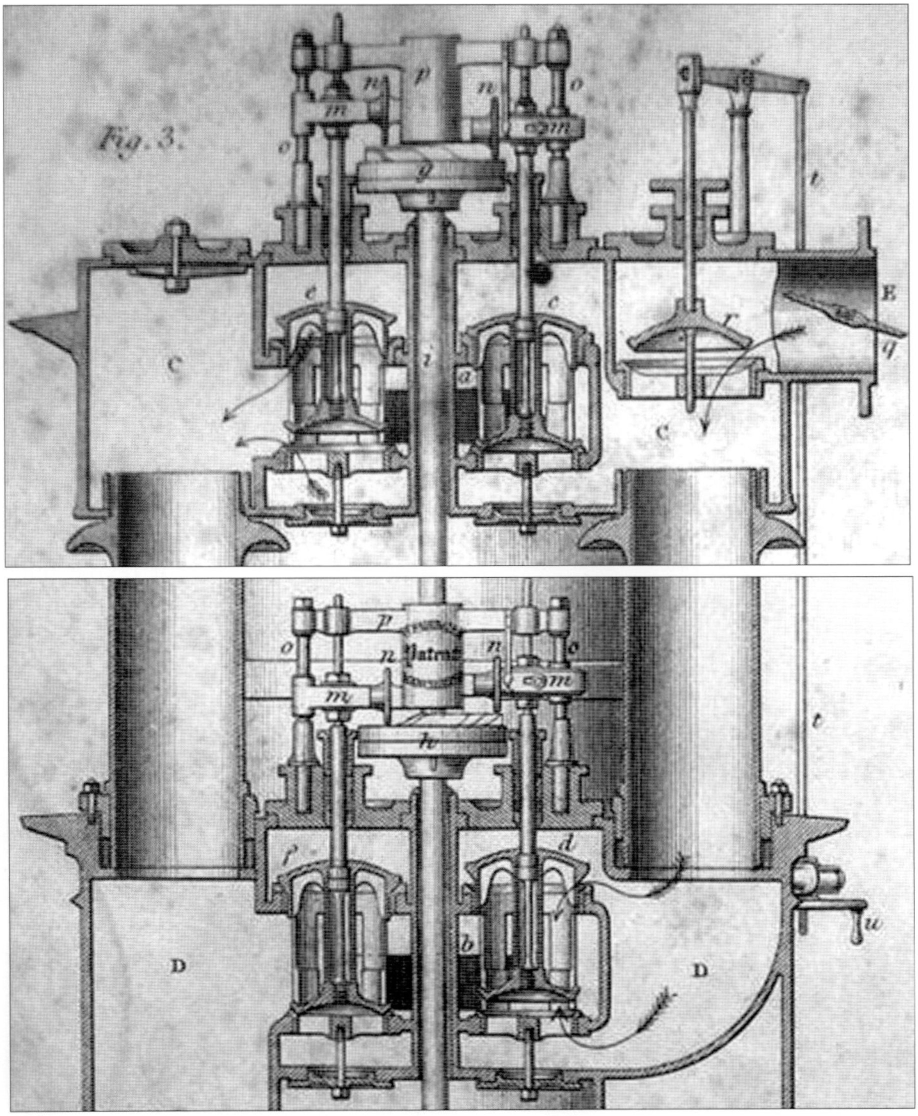

Fig.3.18: Fairbairn's Valve-gear.

> trial. When it can be shown that any modern valve-gear produces a better diagram, and is not more complex or expensive, we will admit that the Fairbairn valve-gear has been fairly beaten, but not before.[125]

Drop valves came to the fore again in the 1890s when Sulzer and others developed them, leading to economies and their wide use, and providing an example of Fairbairn's engineering foresight.[126]

In contrast Fairbairn was not an early advocate of compounding (that is introducing a second cylinder to maximise the use of the steam). Writing as late as 1861, he pointed out the higher initial cost and the greater complexity of compound engines, such that he had 'no hesitation in recommending the single engine worked expansively, as an efficient competitor of the compound engine'. This would change as more reliable compounding was developed by others.[127]

Fig.3.19: Columnar engine. Paris Exhibition, 1855.

Construction of steam engines took place at the Fairbairn works from 1828 into the 1870s. One can only make an educated guess of how many steam-engines Fairbairns built during those forty-five years, but it would seem to be well over 1,000 stationary steam engines, of which it has been possible to identify only around 50,[128] plus around 500 locomotive engines and approximately 100 ships' engines. The locomotives were built from 1838 to 1862 and the ships' engines from 1830 to 1845 – side-lever, oscillating and direct-action – see *Table 3.2* (p.125). The Fairbairn range of stationary steam-engines for industry was wide. Marine-type engines for textile mills have been referred to. For his most powerful installations Fairbairn used high-pressure condensing beam engines, as at Saltaire,[129] and at the Nevsky Cotton Mill at St Petersburg, one of the largest cotton spinning factories in Russia.[130] For smaller installations his high-pressure columnar engines, such as at the Constantinople corn mill, were well-regarded. One was exhibited at the 1855 Paris Exhibition.[131] There were innovative winding and pumping engines for coal mines.[132] In the early 1860s Fairbairns were building horizontal engines, notably hauling engines for the São Paulo Railway, of which a pair is preserved at Paranappiacaba – the only known surviving stationary steam engines manufactured by Fairbairns.[133] In the late 1860s the firm was building engines that had

Fig 3.20: Horizontal engine at Ancoats works, 1862.

not been designed in-house, notably three-cylinder triple expansion engines designed by J. Crosland.[134] During the four decades 1830–1870 Fairbairn's built a wider variety of important steam engines than any other firm. There is scope for a mechanical engineer to undertake a detailed study of Fairbairn's engines illustrating the development of steam engine design over these years.

Boilers

STEAM-ENGINES require boilers, and from around 1830 Fairbairn built boilers of the 'Cornish' type, cylindrical with a single cylindrical furnace or flue tube within them. In 1844, with John Hetherington, he patented a boiler using *two* internal flue tubes – the Lancashire boiler as it was known – which became the most important boiler in the textile districts.[135] It was safer and more efficient than the single-tube Cornish boiler. The alternate firing of the tubes reduced the smoke.[136] One of these boilers was introduced to a mill in Ancoats that year.[137] The boiler received wide publicity and was an outstanding success.[138] Diffusion was rapid. In 1859 Henry Harman, the Manchester Steam Users' Association's chief inspector (and soon to become manager of the Fairbairn works) wrote that he 'knew of no better construction than the cylindrical two-flued boiler, which was the one in general use'.[139]

By no means were all 'Lancashire' boilers built by Fairbairns. Some are likely to have been built under licence, others after the patent had expired. Nor was their use limited to Lancashire – one by Fairbairn was for sale in Australia in 1858.[140] There were many modifications by well-known boilermakers such as W. & J. Galloway and Daniel Adamson & Co., yet Henry Powles could write in 1905 that 'the boiler of today has practically returned to its original form, and, with the exception of the rings in the furnace flues, differs in no material way from Fairbairn's earliest designs'.[141] The rings on the boiler flues were the direct result of Fairbairn's later research.[142] Fairbairn maintained that the principle of the Lancashire boiler, that is the double furnaces within the same boiler, 'was first introduced by myself'.[143] Lavington Fletcher, in what remains the definitive text on the Lancashire boiler, attributes it to Fairbairn *and* Hetherington, as does Evan Leigh.[144] Daniel Adamson and D.K. Clark give the credit for its introduction to Fairbairn,[145] and both Powles and Dickinson refer to Fairbairn as its 'inventor'.[146] But, as with waterwheels, there were rumblings that Fairbairn was not the 'inventor' of the Lancashire boiler, and recent research has endorsed this. Hetherington's obituary stated that the two-flued boiler was his invention.[147] R.J. Law has provided evidence that two-flued boilers were in use in Cornwall from at least 1830 under Woolf, and that in 1835 Fairbairn's assistant, Robert Smith, visited the mine where Woolf had been engineer.[148] In 1838 Robert Armstrong referred to boilers with 'two or more flues placed low down in the boiler'.[149] In 1846 John Bourne illustrated three two-flued boilers by Maudslay Sons & Field for the Blackwall Railway.[150] W. & J. Galloway dated these Blackwall boilers around 1840, and said that 'in 1844 an exact copy of these boilers was first introduced into Manchester at a mill in Ancoats'.[151]

Paul Quinn sees a paper by C. Wye Williams at the Manchester Royal Victoria Gallery in 1842 as a catalyst for the Lancashire boiler.[152] Williams had studied chemistry in Manchester under John Dalton and become the very successful owner of a fleet of steamboats.[153] (These included the *Lady Burgoyne,* built by Fairbairns and which it is just possible remains sunk in the mud of Loch Derg in Ireland).[154] His paper followed a lethal explosion in a Manchester mill. He propounded the view that failure to maintain water in contact with the iron exposed to the furnace was the important factor in causing over-heating and consequent rupture of a boiler. Thus he held that accidents or damage to boilers were accompanied by inadequate water levels.[155] His views prevailed as Fairbairn and Hetherington 'further advanced the idea of ensuring adequate water coverage of the iron' within boilers, by the use of two smaller fire tubes in place of one larger one 'in their development of their patented Lancashire boiler'.[156] Another feature of the Lancashire boiler patent was that it described how to stoke the boiler so as to avoid smoke:

> By feeding each furnace alternately with a fresh supply of fuel, the superfluous carbon from the fire last fed is burnt in the main flue on meeting with the current from the other flue in the boiler, in consequence of such current being at a higher temperature from having a clear or glowing fire, whereby the products of combustion are almost invisible, and whereby also less fuel is consumed.[157]

The prevention of smoke had been a concern of the Society of Arts for sixty years.[158] In 1839 Williams patented the Argand boiler in which jets of air were introduced into the furnace, in a manner having affinity to the Argand lamp.[159] Publication of Williams' book, *The Combustion of Coals and the Prevention of Smoke Chemically Considered,* and his paper to the British Association, both in 1840, generated much interest.[160] The paper resulted in the formation of a BAAS committee on the subject.[161] In 1841 a large wagon-boiler with Williams' Argand arrangement was erected in Manchester and visited by 'several eminent engineers'.[162] On 26 May 1842, the same year in which Edwin Chadwick's *Report* drew attention to the drastic mortality rates in Manchester,[163] one of Britain's first smoke abatement societies, the Manchester Association for the Prevention of Smoke, was founded under the chairmanship of the Revd J.E.N. Molesworth, Vicar of Rochdale (and father of a subsequent pupil of Fairbairn), 'for the purpose of directing attention to plans for the prevention of smoke'. The Committee included Roberts, Houldsworth, Fairbairn, William Sturgeon and Peter Clare.[164] In a paper to the BAAS meeting in Manchester later in the year, Fairbairn referred to experimental observations on ten steam engines working in Manchester, including Williams' arrangement, attached to Hetherington's boiler, which had showed a four per cent saving in fuel, and from which the abatement of 'the nuisance of smoke' would no doubt follow.[165] The paper initiated a long discussion in which, among others, Wye Williams, Richard Roberts, Henry Houldsworth and Edmund Ashworth took part.[166] The following year Government set up a Select Committee to which Molesworth gave evidence of the Manchester Association's aims.[167] In 1844 Molesworth and his Committee used their influence to get a smoke-control clause inserted into the Manchester Improvement Act.[168] In the BAAS Report for 1844 – the year of the Lancashire boiler patent – there was a long Report by Fairbairn, 'On the Consumption of Fuel and Prevention of Smoke'.[169] Hills suggests that 'in all probability Fairbairn derived the idea of firing the twin fires from Wye Williams, who had been experimenting with smoke consumption on board ships by dividing large furnaces with an internal partition and firing each side alternately at regular intervals'.[170] Thus the patent for the Lancashire Boiler did not occur in a void. Five years on the *Mining Journal* reported:

> All who have, for a few years, been acquainted with the towns of Manchester and Salford must have noticed a marked change in the purity of the atmosphere. Instead of the dense plumes of black smoke vomited forth by a thousand chimneys of steam engines until the air was thoroughly impregnated with soot, begriming clothes, goods in the shops, and penetrating every room, the atmosphere is now comparatively clean.[171]

This is perhaps over-optimistic, but the Lancashire boiler undoubtedly made some contribution to this improvement.

The factual evidence about the origin of the Lancashire boiler presents an enigma. From 1840 Fairbairn had been deeply involved in the discussions at the British Association and elsewhere on smoke prevention, which brought him into contact with others at the cutting-edge of boiler technology, making it hard to believe he was unaware of developments elsewhere. When in 1861 he made the claim to have introduced double flues, he cannot have been unaware of the challenges to this claim. Was the patent a shrewd entrepreneurial move? Was it that having spotted an opportunity, he exploited it? Did he believe he had patented something novel or did he turn a blind eye to what he did not want to see? Alternatively did he equate 'first introducing' with obtaining the patent? Hayward sees the patent as successfully bringing together the admission of air behind the bridge (which was not new) with the alternate firing of the tubes such that the fire from one consumed the smoke from the other.[172] Hills is of the view that we are unable to discover who developed the Lancashire boiler with its twin fire-tubes. He sees what Fairbairn and Hetherington patented as 'not so much the type as the

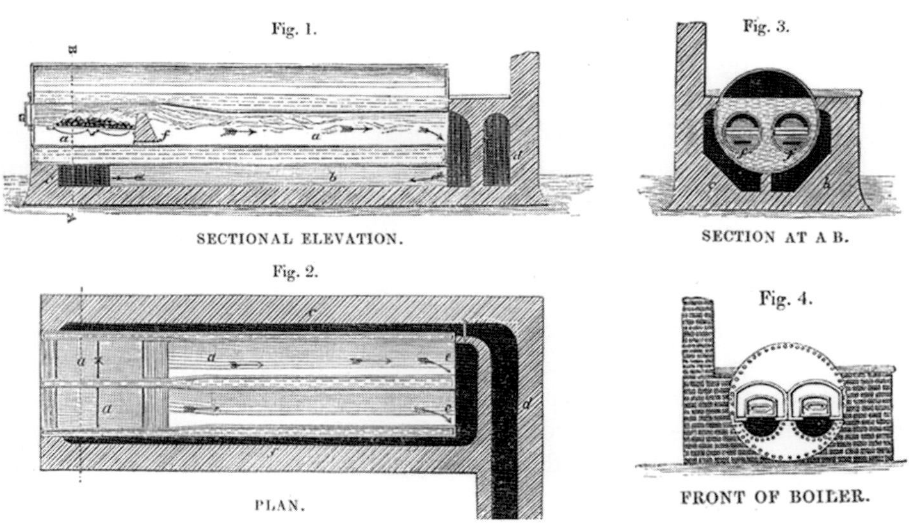

Fig.3.21: Improved Lancashire Boiler, c.1854.

way the boiler was to be fired'.[173] Neither explanation satisfies. Fairbairn comes out of this venture, at least in the present state of knowledge, somewhat tarnished. Yet here was the entrepreneur *par excellence*. He saw the opportunity which others did not appreciate, he developed the technology, patented the result, and publicised the patented boiler. Others could have done it, but none did. It brought prestige and profit, with the Lancashire boiler remaining the boiler of choice in the textile districts, albeit with improvements, for the remaining century of the steam era.[174] In the years after its introduction there was plenty of competition in boiler construction, notably in the 1850s and 1860s from W. & J. Galloway. Their most important patent was in 1851 for conical water tubes which became widely used.[175] In 1854, a month after William's 'retirement', Galloways warned Fairbairns:

> We are informed that you are putting into the boilers you are making the conical vertical waterpipes. If this is correct we beg to intimate to you that these pipes are patented by us, and that your use of them is an infringement.
>
> Many other boiler makers are introducing them, but of course under our licence, which may also be granted to you if you desire it.[176]

There is no record of any reply.[177] In the 1870s Fairbairn introduced a safer 'five-tube' boiler, but it never displaced the Lancashire boiler.[178]

Shipbuilding

HAVING commenced shipbuilding at Ancoats, Fairbairn made a decision to build a shipyard at Millwall on the Thames. Here he became the leading builder of iron steamships, in terms of numbers of ships and marine engines built, during the formative decade 1835–44. His major contributions were research into the strength of wrought-iron plates and riveted joints and the transfer of cellular construction from the tubular bridges to ships' hulls. However, before Millwall, he was responsible for the diffusion of iron ships and shipbuilding to the Central European Lakes, a classic example of the spread of technology.

Shipbuilding had continued in Manchester with *La Reine de Belge* to ply between Ostend and Bruges; and the *Railway* and *L'Hirondelle* to ply on the Humber.[179] When Albert Escher, who had been an informal pupil of Fairbairn ten years earlier,[180] was drawn back to Manchester in the early 1830s by the youngest daughter of Fairbairn's patron John Kennedy, Annie, whom he married, he saw iron ships under construction at Fairbairn's works.[181] In the intervening period, he had built up the engineering department of Esher Wyss to employ 400 people.[182] There had been wooden-hulled steamships from 1823 on Lake Geneva, and from 1826 on Lakes Neuchâtel and Maggiore.[183] There was little progress with steamships on the Zürichsee, until 1834 when Franz Caspar and Johann Lämmlin, with Albert Escher's guidance, ordered an *iron* steamship from Fairbairn.[184] The *Minerva,* originally known as *Vulcan,* was 108ft long with a 15ft 6in. beam and was powered by two high-pressure engines.[185] On completion at Ancoats she was taken apart, sent to Selby, reassembled, and steamed to Rotterdam and up the Rhine to Basel. There she was taken apart again and

Fig.3.22: *Minerva* – the first iron steamship on the Central European Lakes, 1835.

transported on ox-carts to Zurich, the larger sections crossing the Aar, Reuss and Limmat on wagon-ferries.[186] At Zurich Esher reassembled her yet again. She made her maiden voyage in July 1835 – the first iron steamship on the Central European Lakes.[187] A year later Esher built his own iron steamship, the *Linth-Escher.*[188] The technology had transferred from Manchester to Central Europe. Within eight years Escher had built nineteen steamships, some with Fairbairn engines, and had opened a shipyard on the Danube.[189] During the next 100 years Escher Wyss built over 600 steamships and many marine engines.[190]

Iron shipbuilding in the first half of the 1830s was centred in Northern England. There was a friendly rivalry between Lairds in Liverpool and Fairbairn in Manchester. John and McGregor Laird attended the trials of Fairbairn's first *Lord Dundas,* on the Irwell.[191] Within days of the voyage of his second *Lord Dundas* to Glasgow in June 1831, Lairds launched the *Alburkah* which made the epic voyage to the coast of Africa.[192] There was cooperation rather than industrial secrecy: in April 1834 Fairbairn wrote to John Laird:

> I have taken the liberty of sending our Foreman Boiler Maker ... to look at the manner in which you form the skeletons of the iron boats: perhaps you will have no objection to allow your foreman to show him the way in which it is done; with a liberal desire to accommodate your people in the same way.[193]

There was clearly some business between the two firms as on 29 October 1834 Fairbairn wrote to John Laird acknowledging a draft for £29.9*s*.4*d*. In the same letter he accepted an invitation to attend the 'starting' of Laird's *Garyowen,* and invited Laird to the 'start' of the *Vulcan of Zurich* at Selby.[194] By May 1841 Fairbairn had built at least 43 iron ships,[195] and Lairds 40.[196] By 1845, when Fairbairn had ceased taking orders for ships, he had built over a hundred,[197] whereas Lairds had built at least 65,[198] but Lairds did not commence building their own marine engines until 1854.

Ancoats was not the place to build ships. Fairbairn faced abandoning shipbuilding or moving elsewhere. He decided on the latter and, perhaps spurred by a desire to make himself known in the capital, he chose London rather than near-by Liverpool.[199] It was an unwise decision. The chosen site, on the Isle of Dogs at Millwall, was unhealthy, marshy and difficult of access.[200] The ground

Fig.3.23: Nicholas I reviews the Baltic fleet of the Russian navy in 1848 from the *Nevka*, the first iron steam royal yacht (built by Fairbairn in 1838).

conditions – Thames mud – made it expensive to develop. More was expended 'below the surface of the ground, than upon all the buildings and plant which showed above it'.[201] It was a fine yard with the latest technology including a railway system to facilitate the movement of heavy items. There was an octagonal chimney with subterranean flues, all typically Fairbairn. Buildings included a two-storey boat and boiler manufactory, with its ground floor open to the yard, a pattern-making shop, a foundry, a fitting and turning shop, smithies, ship joiners' shed, engine house and offices.[202] Speaking in 1859 he said he was the first iron shipbuilder in London, and that he had built 'upwards of a hundred-and-twenty iron vessels', of which nine were built in sections at Manchester and the rest at Millwall.[203] Seventy-five vessels have been identified, as in Appendix 3.2.[204]

The Millwall yard got off to a good start. The *Ludwig* was the first iron steamer built there: she was for the Bodensee.[205] In 1837 the *Sirius,*[206] built to ply the Rhone from Marseilles, was a triple first for Fairbairn. She was 'the largest yet made of iron', the first to be *launched* on the Thames, and the first to be classified by Lloyd's Register. The ratio of her length to breadth at 10:1 was 'greatly in excess of anything then usual'. She had three watertight bulkheads, and high-pressure tubular boilers with copper tubes.[207] In 1838 Fairbairn's 21-year-old daughter, Anne, launched the first iron steam-yacht, the *Nevka*, for the Emperor of Russia, an occasion witnessed by 'thousands of spectators'.[208] A second Royal Yacht, the *Aegir,* was built for Christian VIII of Denmark in 1841.[209] At the start of that year nearly 600 were employed at the Millwall yard, and thirty-one iron vessels had been built.[210] These included at least eleven for the East India Company.

Fig.3.24: The *Rose*, the first iron steamship to reach Australia (built by Fairbairn in 1841).

There were ships to ply the Elbe, the Neva, the Weser, and the Black Sea,[211] and the first two iron steamships to reach Australia, the *Rose* and the *Thistle*.[212]

The decade, 1835–1844, saw remarkable and revolutionary changes which started an ever-increasing expansion of marine technology and design that is still with us today.[213] This was the decade during which Fairbairn was building ships at Millwall and contributing to those changes. He did not 'invent' steamships or iron ships. There had been wooden steamers operating on the East Coast since 1814, when Fairbairn witnessed the *Caledonia* entering South Shields.[214] The first iron steamer was the *Aaron Manby*, built around 1824 by the Horsley Company. She steamed across the Channel to Paris. Several others were built in the 1820s.[215] So what did Fairbairn contribute to the development of iron steamships?

Firstly, there were engines. In 1841 Fairbairn took out a patent for 'Certain Improvements in the Construction and Arrangement of Steam-engines'. It referred to making the engines direct acting, and bringing the working parts into a smaller compass,[216] but it is unclear exactly what was included. In 1845 *The Practical Mechanic and Engineer's Magazine* published a Table of 95 leading marine engines.[217] It included twelve Fairbairn engines from the six years 1839–45, the last years of the dominance of the paddle-wheel before it was replaced by the propeller.[218] Several features are apparent from these Fairbairn engines – the move from side-lever (up to 1842), to oscillating (1842–3) to direct-action (from 1843); the increasing horsepower, from 21 to 290; the increasing cylinder diameter and length of stroke; and the increasing paddle-wheel diameters, from 10ft 4in. to 27ft. These engines are listed in *Table 3.2*. Information is limited: illustrations are currently known of only four Fairbairn marine engines. One is the side-lever engine of the *Nevka*,[219] and the other three, for the wooden-hulled ships *Vulture*,[220] *Dragon*,[221] and *Odin*,[222] built by the Admiralty, are very similar direct-action engines, with 'grasshopper' motion. These were amongst the most powerful of their day and widely admired. It was the side-lever principle of marine engine which Fairbairn transferred to textile mills, and the 'grasshopper' motion that he transferred to the winding engine at Astley Pit.[223]

Table 3.2. Marine Steam Engines by Fairbairn as listed in *The Practical Mechanic and Engineer's Magazine*, 1845.

Type	Year	Ship	HP each engine	No. of engines	Cylinder diameter	Cylinder stroke	No. per minute	P-wheel diameter
Side-lever	1839	Dolphin	21	2	28"	2'0"	35	10'4"
	1839	Enterprise	21	2	27 1/4"	2'6"	34	12'6"
	1840	Rose	53	2	40 3/4"	3'6"	29	17'0"
	1840	Thistle	53	2	40 3/4"	3'6"	29	17'0"
	1841	Agir	41	2	36 3/4"	3'0"	30	16'0"
	1842	Hamlet	21	2	27 1/2"	2'3"	38	5'6"
Oscillating	1842	Rocket	10	2	19"	2'0"	40	9'8"
	1843	L.Burgoyne	46	2	39"	3'0"	32	16'0"
Direct-action	1843	Cormorant	158	2	65 1/2"	5'3"	22	26'8"
	1844	Vulture	248	2	80 5/8"	5'9"	19	26'6"
	1845	Dragon	290	2	87 1/2"	5'9"	19	27'0"
	1845	Odin	290	2	87 1/2"	5'9"	19	26'6"

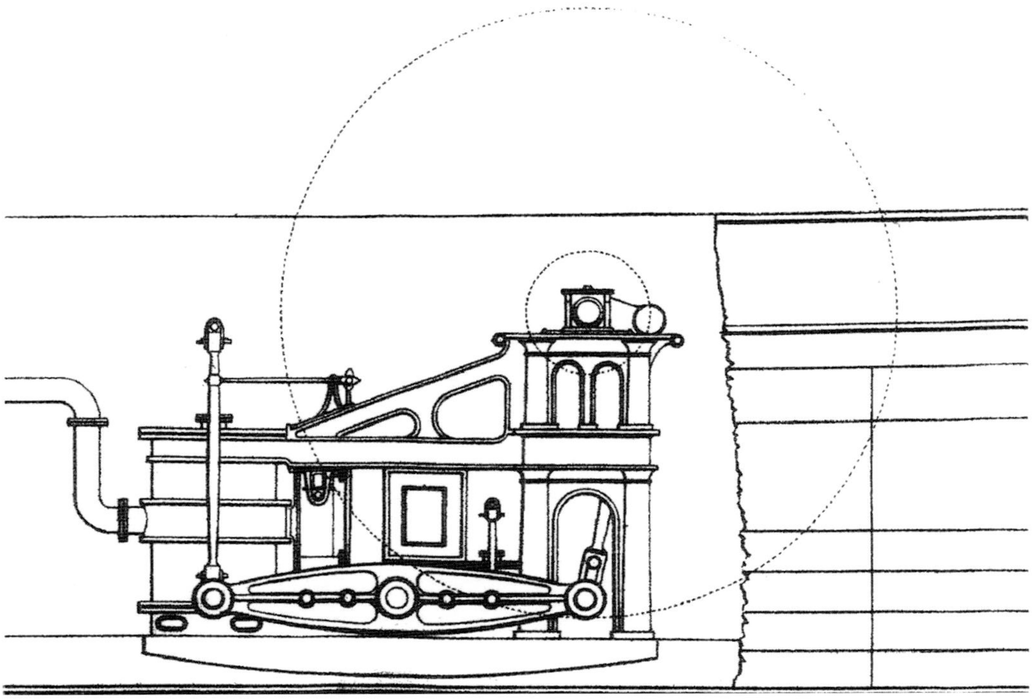

Fig.3.25: Side-lever engine, the *Nevka*, 1838.

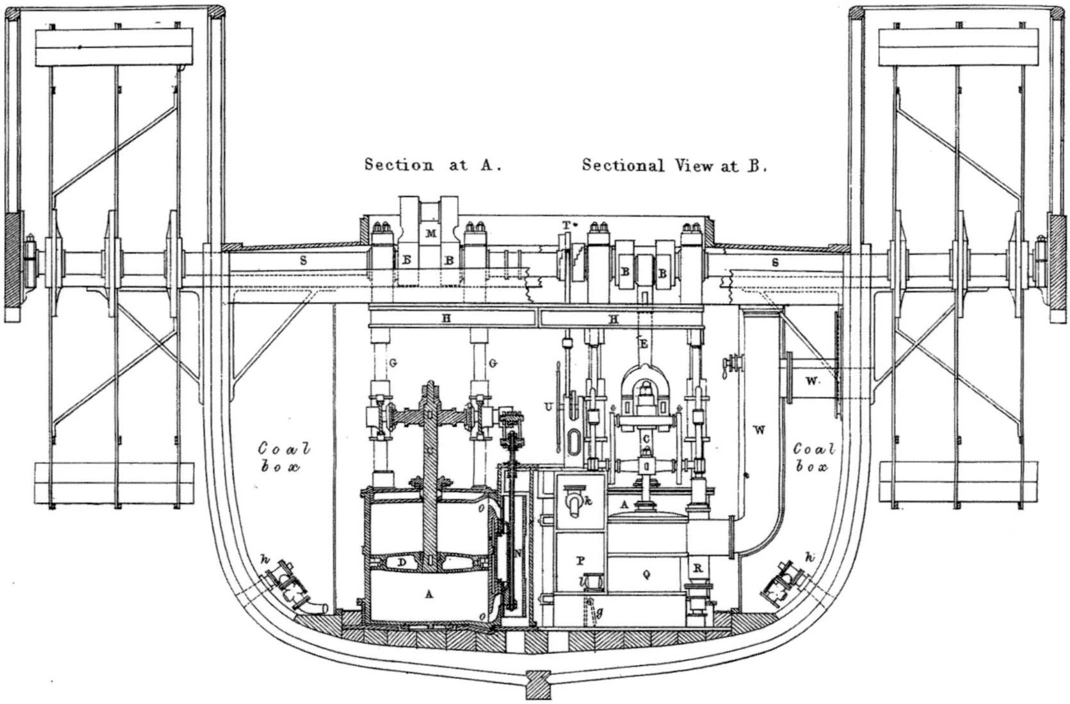

Fig.3.26: Engines of HMS *Vulture*, 1844.

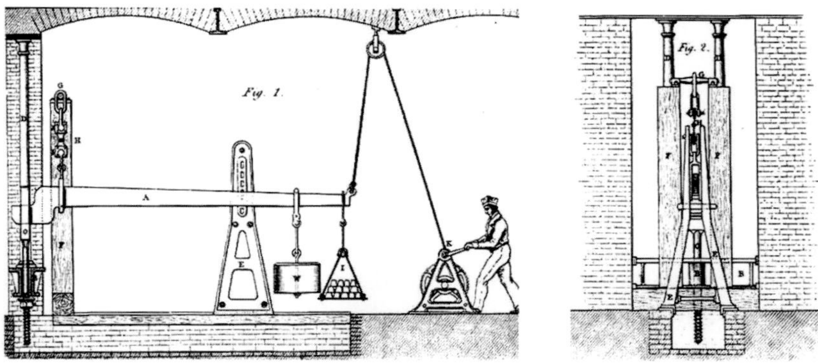

Fig.3.27: Testing apparatus re strength of iron plates and riveted joints.

Fairbairn's position as a marine engine builder can be appreciated by an analysis of the *Practical Mechanic*'s 1845 Table of leading marine engines. The firms with more than five listed engines were Maudslay and Fairbairn (12 each), Robert Napier (9) and Seaward (7).

Secondly, there were boilers. Fairbairn was involved in the early transfer of multi-tube boilers from locomotive to marine use and their subsequent development. Multi-tube locomotive boilers originated from Marc Seguin in France in 1828 and Robert Stephenson's *Rocket* the following year.[224] The first 'horizontal tubular locomotive boiler' for a steamboat appears to have been by William Gravatt in 1829 for an iron steamboat built by Fenton & Murray of Leeds.[225] After this Robert Stephenson supplied a *Planet*-type multi-tubular boiler for Fairbairn & Lillie's first steamboat for the Forth & Clyde Canal in 1830–1.[226] Fairbairn's *Sirius* of 1837 had 'locomotive tubular boilers' and in the same year he provided boilers for two French-built ships, the *Castor* and the *Pollux,* each with three boilers containing 51 tubes.[227] In the table of 95 engines referred to above, eight engines are noted as having tubular boilers, including two by Fairbairn. All these eight boilers are later than the *Sirius*, but further research is necessary to ascertain the early history of multi-tube marine boilers. The speed of development of marine boiler technology can be appreciated from Fairbairn's three boilers for HMS *Dragon* in 1845. Only fifteen years on from the introduction of small multi-tube marine boilers, these each contained 1,200 tubes of 3in. diameter.[228]

Thirdly, there were iron hulls. Fairbairn records being hampered in the early 1830s by a lack of data on the properties of iron for shipbuilding. To overcome this, he and Hodgkinson undertook very extensive experiments in 1838–9, with the intention of showing 'that the iron ship, when properly constructed, is not only more buoyant, but safer, and more durable than vessels built of the strongest English oak'.[229] The initial results were published in 1840, but the full results not until 1850, in the *Philosophical Transactions of the Royal Society.*[230] This delay may have been for commercial advantage but this would be out of character for Fairbairn. More likely reasons were the financial pressures and difficulties faced at Millwall in the 1840s. Fairbairn's four-part paper to the Royal Society was his primary contribution to the development of iron shipbuilding, providing shipbuilders with definitive information on the strength of iron plates and, more importantly, on the strength of riveted joints, leading to stronger, but lighter, ships. In addition it advocated cellular hull construction.

Part 1 of his paper to the Royal Society sets out a systematic evaluation of the tensile strength of wrought-iron plates from various sources. Here, and in Part 2, the Fairbairn Lever was used. The results were painstakingly recorded, including drawings of the fractures.[231] He obtained from Peter Barlow and others the tensile strengths of various timbers and compared them with the results for iron, concluding that a ½in. iron plate equalled a 2½in.-thick oak plank.[232] Fairbairn went on to examine the effect of stretching of iron, establishing (in association with Hodgkinson) that, once stretched, the iron did not yield when next loaded until the previously applied force was exceeded.[233] This is thought to be the first description of the phenomenon of work hardening.[234] The variations found in the properties of wrought irons from various sources provoked ongoing discussion. To keep costs down many owners and shipbuilders used poor grade wrought iron. There were, as Quinn points out, exceptions – 'Laird, Ditchburn, Samuda, Fairbairn and a few others persistently built to a high quality'.[235]

The quality of cast iron was also an ongoing issue and in 1853 Fairbairn worked on this with the Manchester chemist F. Crace Calvert.[236] They sought increased strength by using 'purified coke' in smelting. A joint paper was presented to the Institution of Civil Engineers.[237] In the ensuing discussion Locke noted that 'Engineers only wanted to ascertain where good iron was to be obtained' and Brunel that:

> every Engineer of eminence ... used all their influence to obtain the production of good quality iron, at any price. They knew the ... fatal effects of using materials whose quality could not be depended upon.[238]

To his *Useful Information for Engineers: Second Series* (1860), Fairbairn appended two letters from his son Thomas to *The Times,* following the wreck of the *Royal Charter* off Anglesey on 26 October 1859 with the loss of around 450 lives:

> Is not any kind of iron thought good enough to build a ship with? What is the meaning of 'boat plates' being the lowest priced in any iron-maker's list? If we pay £25 or £30 a ton for the plates of which a locomotive boiler is made, why should we give only £8.10s or £9 per ton for those of which a ship is built. If safety can only be bought at the high price in the one case, are we not courting disaster with the low price in the other? ... built of the 'boat plates' of the present day, God help the human freight of the ship that strikes upon a rocky shore![239]

Part 2, 'On the Strength of Iron Plates united by Rivets, and the best mode of Riveting', was the major part of the paper. Riveting had been used in boiler-making from 1810 and in shipbuilding from the 1820s. It continued to be used until the mid 20th century by which time wrought iron had been replaced by steel, and riveting was being replaced by welding. Up to Fairbairn's work in 1838, no research of any consequence had been done on this subject. There was a widespread, and wholly erroneous, assumption that a well-riveted joint was stronger than the plate itself.[240] Fairbairn undertook tests with pieces of iron riveted together with different arrangements of rivets. These showed that a joint made with a single row of rivets had only a little over half the strength of the plate. Even double rows of rivets only reached 70 per cent of the strength of the plates.[241]

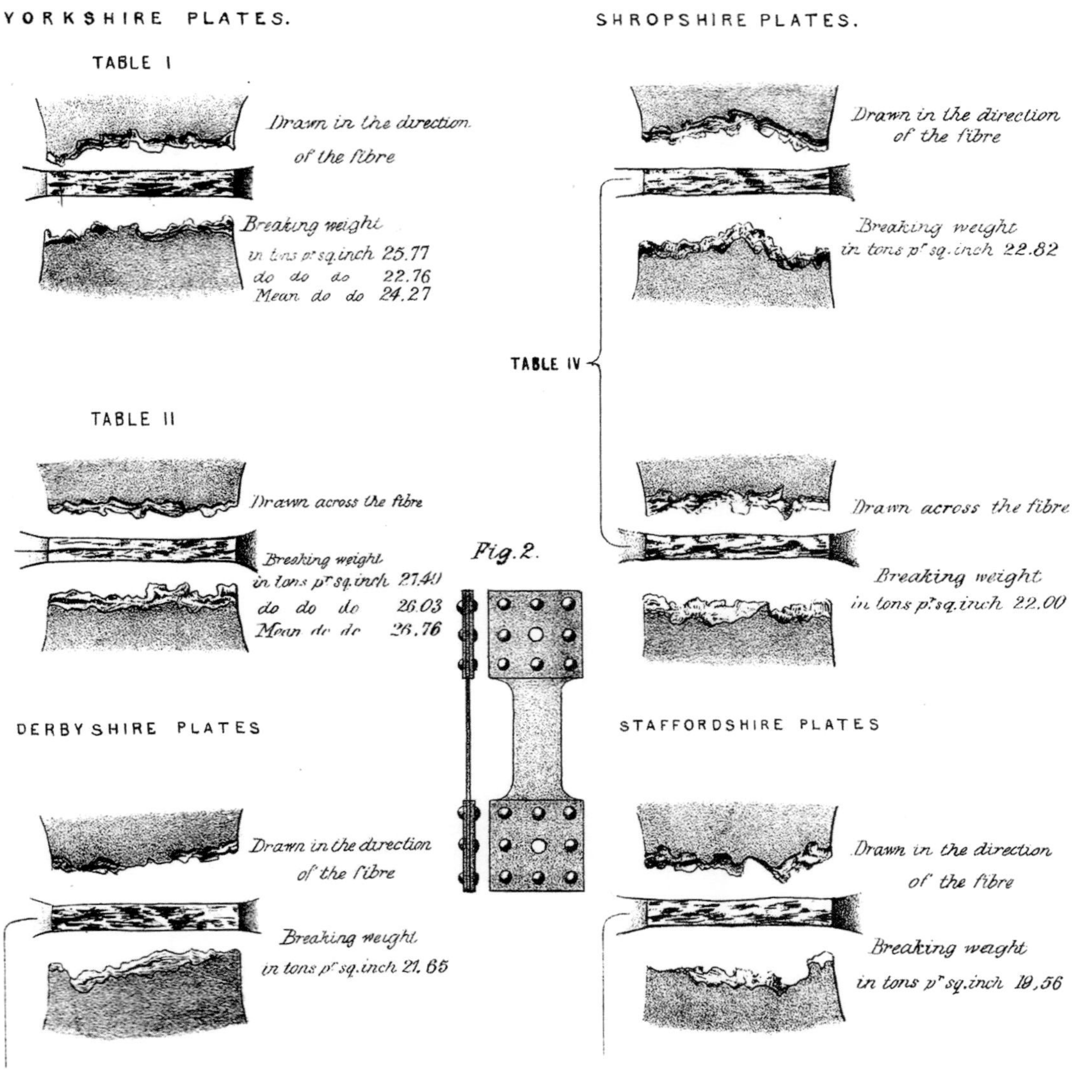

Fig. 3.28: Failure of various wrought-iron plates.

A significant outcome of the initial results presented in 1840 was Brunel's adoption of double riveting for the *Great Britain,* built 1840–3.[242] Indeed Quinn writes that 'the initial credit for introducing … double riveting must go to Brunel'.[243] No evidence has been found of an earlier application of double riveting to a *whole* ship. However between Fairbairn's experiments in 1838–9 and the construction of the hull of the *Great Britain* in 1841, Fairbairn built at least twenty ships and in two cases, the *Rose* and the *Thistle,* there is evidence, at the very least, of *some* double riveting.[244] Other Fairbairn ships from these two years included the *Coquette,* 150ft long, for the Neva, the *Telegraph,* 136ft long, for the Weser, and the *Iron Duke,* 104ft long, for Demerara.[245] It would be surprising if some of these ships were not double riveted, as was the last ship that Fairbairn built, the screw frigate, HMS *Megaera,* designed in 1843 but, due to Admiralty delays, not completed until 1849.[246]

Three decades after Fairbairn's paper to the Royal Society, the Institution of

Mechanical Engineers formed a Committee on Riveted Joints. In its First Report to Council in 1881, its Reporter, William Unwin, referred to Fairbairn's paper as 'the earliest published experiments on riveted joints', and probably the most influential in practice.[247] In a 2011 paper by Belgian scholars on the 'Evolution of historical riveted connections', Fairbairn is the only 19th-century researcher specifically referred to.[248]

Part 3 sought to establish 'the resistance of wrought iron to a blunt instrument at right angles to the surface of the plate'. Fairbairn's intention was not to establish its resistance to shot but the relative ability of wood and iron to resist penetration by rocks should the vessel ground. The test plates were clamped to a cast-iron frame to restrain them from dishing and to replicate what happened to plates between frames in a grounding on a rocky shore. A 3in.-diameter hemisphere, mounted on the end of a strong bar, was applied with increasing force until the plate ruptured. Oak planks of various thicknesses were subjected to the same test. He found that dry wood was better than wet, that wood tended to burst along the grain, and that the force required to cause the timber to fail varied as the square of its thickness, but that iron obeyed a different law, the force needed being proportional to its thickness.[249] This result appears to have been unexpected, but is correct.[250]

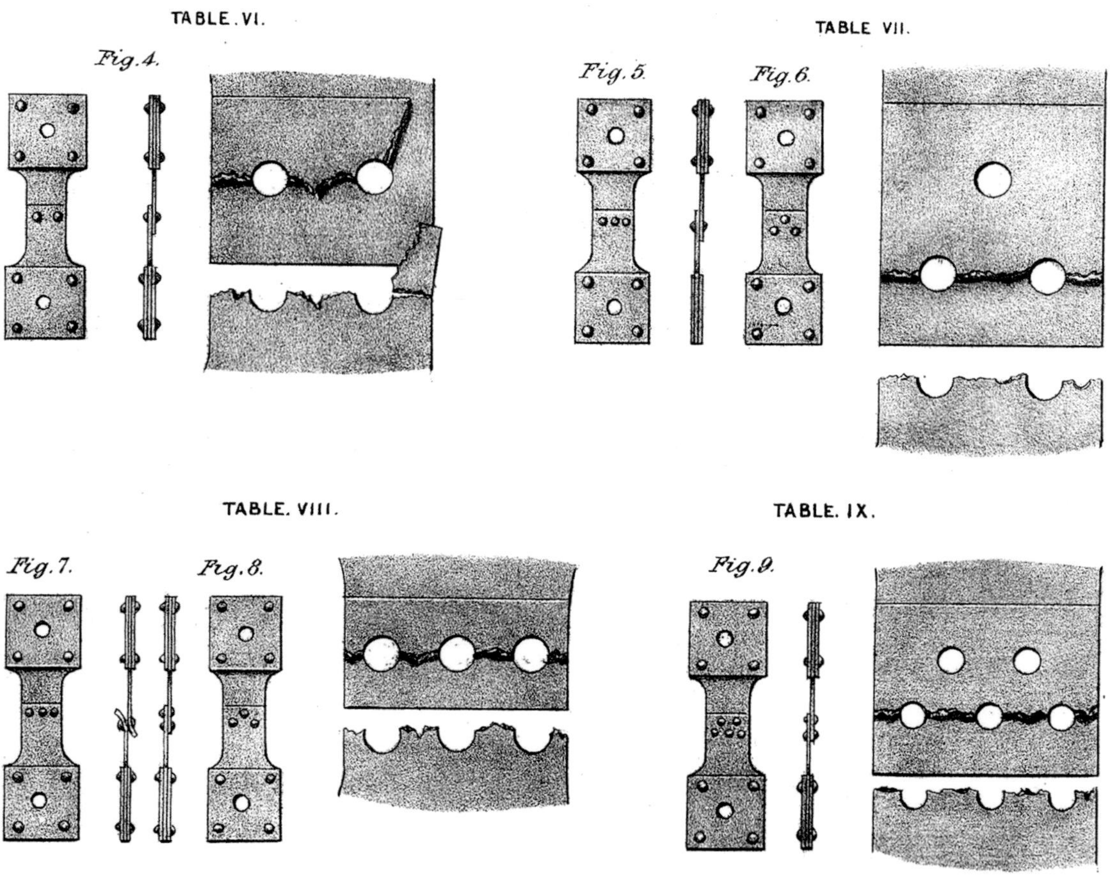

Fig.3.29: Failure of Riveted Joints.

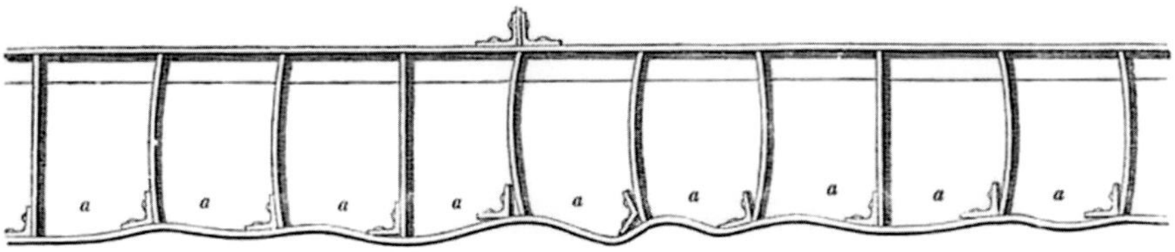

Fig.3.30: Damage to the pounded cellular hull of the Vanguard.

Part 4 covered experiments on rolled and built-up wrought-iron beams. Fairbairn illustrated and recommended forming the sheathing of ships' hulls into a cellular form.[251] In a lecture to the Manchester Mechanics' Institution in March 1852 he referred to a survey of the cellular hull of the *Vanguard* after she had been pounded for several days on a reef of rocks off the west coast of Ireland, sustaining 'comparatively little injury'.[252]

Cellular hulls had some affinity with the longitudinally framed double bottom of Brunel's *Great Britain*,[253] but it was not until the Britannia Bridge experiments that the truth dawned that vessels floating on water and subjected to the swell of a rolling sea – to say nothing of their being stranded or beaten upon the rocks of a lee shore – are governed by the same laws of transverse strain as simple hollow beams like the tubes of the Conway and Britannia Tubular Bridges, and the deck should approach the strength of the bottom for the strongest box girder result.[254] This was a case of reciprocal transference – the concept of the hull of a ship as a beam generating the great girder bridges, and then the cellular construction from the bridge experiments transferring back to strengthen ships' hulls. In his lecture in March 1852 Fairbairn referred to a ship as a 'huge hollow beam or girder' like the tubular bridges, and envisaged calculating, 'from some formula yet to be ascertained by experiment, the ultimate strength of the ship'.[255]

Six years before the building of the *Great Eastern* commenced, Brunel had been at Conway for the floating of the first tube of the Conway Bridge and the following year he had visited the Britannia Bridge under construction.[256] He adopted cellular construction for his *Great Eastern,* begun in 1854. Its builder, Scott Russell, noted:

> The introduction by Mr Brunel of a double bottom … and the doubling of the upper deck – an addition which closely identifies the structure with the cellular bridge system of Mr Robert Stephenson, Mr Fairbairn, Mr Hodgkinson and Mr Clark.[257]

In the discussion following a paper by Scott Russell at the British Association in 1854, referring to the *Great Eastern,*[258] Fairbairn said that Brunel had shown him the plans and:

> though he had at one time thought a ship of that size would be too large for strength, he had, after examination of the plans, arrived at the opposite opinion. He had now no doubt that the ship would be perfectly strong, and able to bear a gale of wind without bending. It was built on the same principle as the Britannia Tubular Bridge; and when it was perceived that that mode of structure is able to sustain a bridge without any support in

Fig.3.31: Bow and stern suspended – ship breaks in two.

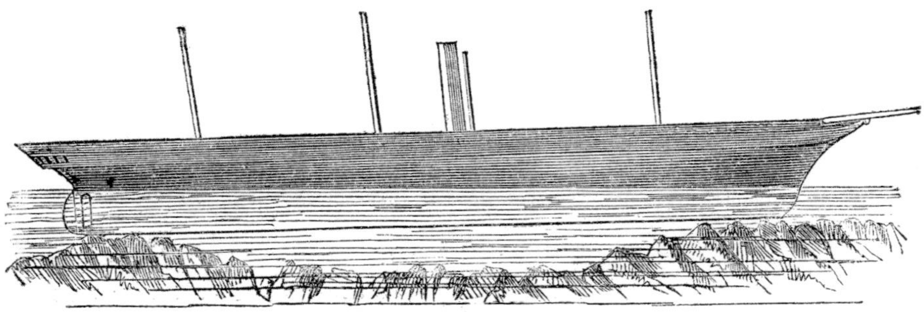

Fig.3.32: Bow and stern supported – tensile strain along the keel.

> the middle, there could be no doubt that supported as the ship would be by the water, it would under all circumstances be able to bear the strains to which it might be subjected.[259]

Fairbairn continued to advocate longitudinal cellular beam construction for ships in his widely reported paper on 'The Strength of Iron Ships', delivered to the Institute of Naval Architects at its opening session in 1860,[260] and in his *Treatise on Iron Ship Building* (1865) where Tate's calculations, treating a ship as a beam, are set out.[261] Notwithstanding, there was a reluctance to adopt longitudinal framing and cellular construction because of the greater expense, and reduced interior space of the ship.[262] Fairbairn's paper to the naval architects was the first paper of scientific note on the overall strength of ships. He believed that an initial appraisal of the necessary strength of a ship was possible if one considered the vessel to be constrained at the mid-ship and, in a second load-case, at each of the ends. Fairbairn's paper was received with lively interest and heralded the era of strength research in shipbuilding. Its approach was described in 2014 as close to 'a very modern ultimate strength calculation'.[263]

Fairbairn was far-sighted. His definitive study on the strength of ships confirmed the longitudinal frame as superior to the traditional transverse frame and extended the idea to the deck. As ships grew larger this approach became essential and large modern welded ships exclusively use the longitudinal system.[264] And in the 21st century, double skin v. single skin is again a subject of debate – in respect of bulk carriers where double skin provides greater safety, but at increased capital and running costs.[265]

Part 2 of the paper to the Royal Society also makes reference to Fairbairn's riveting machine.[266] He brought steam-driven riveting to engineering in 1837, the

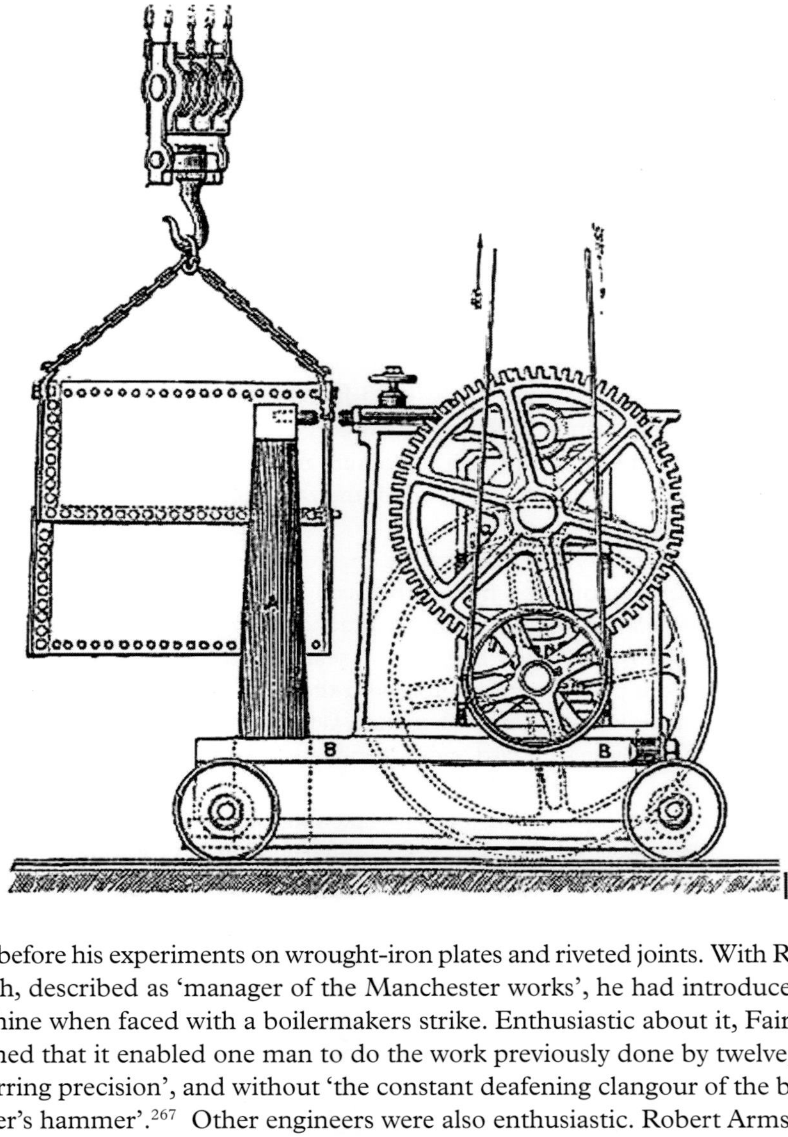

Fig.3.33: Fairbairn riveting machine as exhibited at the Great Exhibition 1851.

year before his experiments on wrought-iron plates and riveted joints. With Robert Smith, described as 'manager of the Manchester works', he had introduced the machine when faced with a boilermakers strike. Enthusiastic about it, Fairbairn claimed that it enabled one man to do the work previously done by twelve, with 'unerring precision', and without 'the constant deafening clangour of the boiler-maker's hammer'.[267] Other engineers were also enthusiastic. Robert Armstrong wrote, in 1839, of the 'impossibility of overrating the importance of this invention'. If it had been predicted three years earlier that an establishment would be able to turn out a boiler a day, it 'would have been treated as the wildest romance'.[268] By October 1838 the machine was advertised as 'ready to supply'.[269] The first sale was to peer engineer Benjamin Hick at Bolton.[270] Fairbairn read a paper about it at the British Association meeting in 1838, and it was exhibited at an exhibition held in conjunction with the Association's meeting in Manchester in 1842.[271] An improved version, on rails, was exhibited at the Great Exhibition in 1851.[272] It dramatically changed boiler-making, including marine boilers.

Fairbairn drove the development of the riveting machine and his name is associated with it, but the indications are that he received a 'useful hint' from elsewhere. Thirty years later John Bourne wrote that he used machine-riveting in

Dublin in 1836, and that it was only after his foreman moved to Fairbairn's 'that the riveting machine made its appearance in Manchester'.[273] In June 1836 Smith filed a patent, which was never enrolled.[274] In February 1837 another patent was enrolled, at Fairbairn's expense, but in Smith's name, because he 'was the person first to accomplish the task'.[275] Fairbairn may have received a 'useful hint' from Dublin, but it was he who had the entrepreneurial drive to develop, patent and market the machine. The initial machine was cumbersome and around 1840 Fairbairn brought out an improved version which was more compact, and on rails.[276] The speed, economy, lack of noise, and quality of work of the machine ensured its rapid spread by way of sales to other engineering firms. J.J. Mayer of Mulhouse purchased one and modified it by adding a clutch to stop the machine if a rivet was poorly positioned.[277] In 1844 Schneiders at Le Creusot produced a modification of the machine.[278] In 1845 James Garforth patented a machine using the direct application of the expansive force of steam.[279] This was followed by hydraulic riveting. Yet Fairbairn's machine remained in use for over twenty years: one was ordered by the contractor for the Tarradale Viaduct near Melbourne in 1860.[280]

Fairbairn's research and the riveting machine highlighted major problems affecting the building of ships' hulls – the need to double-rivet and the fact that the use of the riveting machine was made practically impossible by every plate in a hull being cut to the curved lines of the seams from stem to stern. In 1842 Fairbairn lodged a patent to overcome both these problems.[281] The framing of the ship would be longitudinal rather than vertical. The plates would then be vertical and of the same standard width, which would enable them to be fixed by machine riveting. In addition the plates would have thickened edges to obviate double riveting by compensating for the weakening caused by the punching of the rivet holes. The often critical Woodcroft described this as 'amongst the most brilliant creations of Fairbairn's mind'. It appears to be the first reference to longitudinal framing for ships. In 1844–5 HMS *Grappler* was built by this method but it failed at an unexpected critical point. The thickened-edge plates came off the rollers buckled and twisted, to the extent that great difficulty was experienced in obtaining enough plates to complete the ship.[282] *The Practical Mechanic and Engineer's Magazine* discussed the application of plates to the *Grappler*:

> The breadth of the plates is made to correspond exactly with the distance from centre to centre of frames, those for the 'Grappler' being 15 inches broad by 10 feet long, and they are arranged vertically like hoops around the ship, the joint being formed upon the back of the T iron frames, with each of which they make a watertight joint [283]

The use of vertical plating subsequently became the norm, although thickened-edge plates never did, and ultimately welding replaced rivets. Here were brilliant and far-sighted ideas that failed through a completely unforeseen technical difficulty. It seems likely that it was Fairbairn's wider financial problems that extinguished any motivation to overcome these technical difficulties.

The subsequent history of HMS *Grappler* is instructive. In the three years, 1846–9, she liberated over 2,000 slaves from thirteen ships during a tour of duty patrolling the West African coast. On the voyage home she sprang a leak and continual pumping was needed to prevent her sinking. Dry-docked at Woolwich, her hull, unsurprisingly after three years in the tropics, was found to be badly fouled with barnacles and live oysters. Furthermore a plate in the bottom of the hull had completely worn away where a copper pipe from the engine room passed through the hull: the two metals immersed in salt water had caused galvanic action. Elsewhere many of the plates were badly pitted, indicating that the paint used to prevent corrosion was incapable of providing protection for three years in the tropics. Embroiled in the Admiralty dispute about the suitability of iron for the construction of warships, the *Grappler,* having cost £33,314, was sold for £550. Around the same time some experiments took place at Woolwich to determine the effect of shot on iron hulls. Fairbairn assisted with these. The results were 'unexpected and perhaps discouraging'. The case for iron was lost. All five of the Admiralty's iron frigates were converted to troopships, including Fairbairn's *Megaera.* Fairbairn voiced his disapproval in no uncertain terms. The Admiralty had arrived at a hasty conclusion after only a limited number of tests – 'in action, the chances were in favour of the iron ship, as it was not only secure from fire, but was much stronger'.[284] The matter was resolved in November 1853 when the Russians destroyed a squadron of Turkish wooden-hulled frigates in the first naval engagement of the Crimean War, at Sinope. By 1859 the French, who, like the British, had decided iron was unfit for the hulls of warships, had built a wooden-hulled frigate, *La Gloire,* protected by iron plates over four inches thick. Alarmed by such a vessel the Admiralty set to and within two years the larger and faster HMS *Warrior* was launched, protected by iron plates but also with an iron hull.[285] Fairbairn included a photograph of the *Warrior* as the frontispiece of his *History and Progress of Shipbuilding* but was critical that she did not 'contain the elementary strength ... as exhibited in the longitudinal keelson and cellular form of construction'[286] of the later HMS *Bellerophon* (1865),[287] or HMS *Penelope* (1867).[288] Nevertheless the *Warrior* has survived to this day, now preserved at Portsmouth Historic Dockyard.[289] Fairbairn's involvement in armour-plating is referred to in Chapter 7.

Some interesting ships were built at Millwall. In 1840 there was the *Iron Duke,* remarkable because she was the first *iron* steamship to cross the Atlantic, contrary to the widespread understanding that the *Great Britain* was the first in 1845.[290] Little is known about her. She was a steam barge designed to carry 40 hogsheads of sugar in her hold, and to ply the river at Demerara. The only other known details are: 110tons, 104ft 6in. long on deck, 15ft beam, draught of 3ft 4in., 24hp engine, 10ft diameter paddle wheels and speed of 7mph. She crossed the Atlantic safely, 'being fitted with lee-boards like the Yorkshire billy buoys' [*sic*].[291]

In 1841 there was a floating fire-engine for the Thames. It was driven by manpower, requiring 48 men to operate the cranked handles which drove the paddle wheels and presumably operated the pumps.[292]

The yard was obtaining orders for major ships, including the penultimate ship built by Fairbairn, the *Sir Henry Pottinger,* ordered by P&O in 1844. She was a paddle-wheeled barque to carry passengers and mail.[293] With the exception of Brunel's *Great Britain,* she was the largest steamship built at that date.[294] Her 450hp engines were by Miller & Ravenhill but it is unclear why P&O separated the contract for the ship from that for its engines.[295] The final ship built by Fairbairn

Fig.3.34: P&O Liner ***Sir Henry Pottinger.***

was HMS *Megaera,* a frigate with engines by Rennie.[296] She embodied the other great development in iron shipbuilding of these years, the replacement of paddle-wheels by propeller.[297]

What then was Fairbairn's position in the new iron ship-building industry? On the basis of tables published at the time it is clear that his firm was the major iron shipbuilder during the formative decade, 1835–44. Of 314 ships listed in an 1845 'Table of British and Foreign Steam Vessels', 94 (30%) were of iron.[298] Another 1845 list included '166 Steam Vessels' – in fact there were 168 – described as 'a considerable proportion of the most remarkable of the steam marine of this country', of which seventy (41.7%) were of iron construction, but only three did not appear on the other longer list.[299] These two lists have been combined, with duplicates eliminated, leaving 97 leading iron steamships. Of these, the only builders of three or more ships, (together accounting for 68%) are listed in *Table 3.3.*

From this table it is apparent that, in terms of numbers, Fairbairn was the leading iron ship-builder of the decade and that, unlike Ditchburn and Lairds, he had the capability to build both iron ships *and* their engines. However Woodcroft did not consider him the most influential early iron shipbuilder. He wrote that, 'It is to ... the skilful and daring experiment of Macgregor Laird in 1831, of constructing the *Alburkah* ... which was sent to the coast of Africa, and ascended the Niger with his celebrated expedition of discovery, that the real rise of iron shipbuilding should be traced'.[300] This view is followed, at least by implication, by D. Headrick, although apparently unaware of Woodcroft. Headrick points to the

Table 3.3: Iron Steamships built up to 1845 (from *Civil Engineer and Architect's Journal* and *The Practical Mechanic and Engineer's Magazine*).

Builder	[1] Iron ships and engines	[2] Iron ships with engines by others	[3] Total iron ships [1]+[2]	% of total of 97 iron ships
Fairbairn	21	5	26	26.8%
Ditchburn	-	19	19	19.6%
Lairds	-	13	13	13.4 %
R. Napier	4	-	4	4.1%
Tod & MacGregor	4	-	4	4.1%

Fig 3.35: HMS *Megaera*.

scepticism of both the public and purchasers of ships towards *iron* steamships in 1830. It would take a spectacular example to dispel their doubts. *The* spectacular example was the *Alburkah*, entering history as the first *ocean-going* iron steamship.[301] Undoubtedly this was an epic voyage,[302] but the statistics in the tables above, the technical developments and experimental work, the excitement engendered by launches at Millwall, the building of two Royal Yachts, and the export of ships to Europe, the Black Sea, the Baltic, the East India Company, South America together with Australia, all point to Fairbairn's wider influence.[303]

Fairbairn continued to be a visionary. In 1841, over half-a-century before Manchester 'accomplished a feat without precedent in modern history by transforming itself … into an inland port, fifty-five miles from the open sea',[304] Fairbairn had envisioned it.

> Any improvement which would enable vessels of 400 to 500 tons burthen to discharge their cargoes in a commodious wet dock in Hulme, would form an epoch of such magnitude in the history of Manchester, as would quadruple her population, and render her the first as well as the most enterprising city in Europe.[305]

And, speaking at the Manchester Mechanics' Institution in 1852, he concluded:

> the iron ship, of British origin, [will] yet ride triumphant on every sea, as the harbinger of peace, the supporter of commerce, and the great and only security of our national defence.[306]

Research, Reports, Papers

FAIRBAIRN'S research, reports and papers contributed to his success, in that they helped to make him widely known and, in some cases, had important commercial consequences. Some related to specific areas of work and are referred to in the sections relating to that work, notably shipbuilding and bridges. This section addresses his more general work in the sphere of materials science, together with consultancy reports, forensic reports and the importance to him of the British Association, all with particular reference to his years as sole proprietor.

The strength of material under tension has long been regarded as one of the most important characteristics required for design, quality control and life prediction of structures and plant. It is of fundamental importance to the world economy today.[307] By the mid-1830s Fairbairn and Hodgkinson had investigated the comparative strength of 57 samples of cast iron from 26 manufacturers. In 1837 the results were included, together with subsequent experiments on the

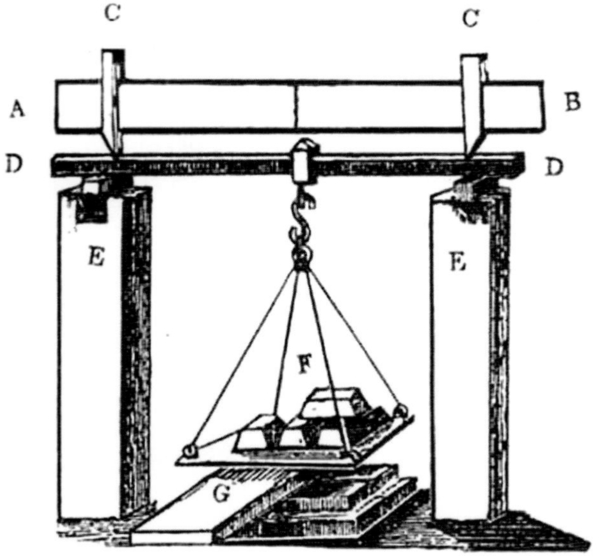

Fig.3.36: Sustained-load or creep test apparatus (1837).

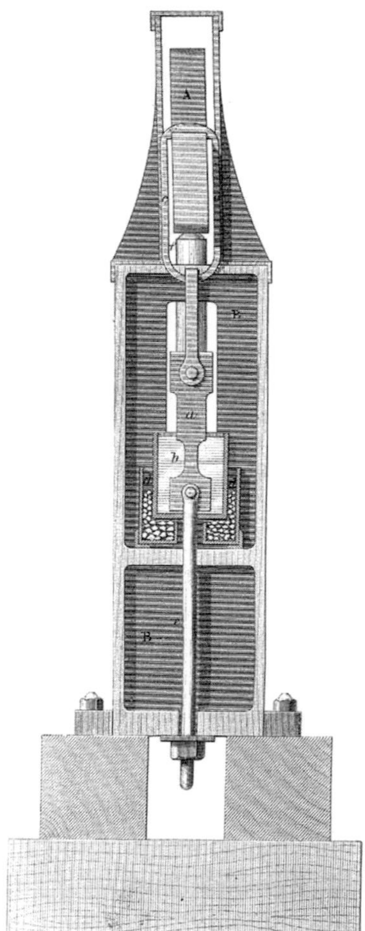

Fig.3.37: Apparatus for determining strength of wrought-iron at various temperatures (1856).

effects of time, temperature, and repeated re-melting, in Fairbairn's first paper to the Manchester Literary & Philosophical Society.[308] The tables, the most comprehensive at that time, extended to over a hundred pages. Their importance in providing practical engineers with the strengths of different irons ensured they were widely publicised. They were cited in the engineering literature on the Continent, including by Morin in France,[309] and Weisbach in Germany – the latter translated into English, Swedish, Russian and Polish.[310]

The earliest sustained-load or creep tests in this country were commenced by Fairbairn in 1837 when deflections of cast-iron bars under various loads were measured for up to seven years. Results showed that long-continued strain lessened and ultimately, however remotely, destroyed the iron's power of resistance.[311] Prior to Fairbairn the only experiments on creep were by the French engineer, Vicat, who tested four iron wires under tensile load from 1830 to 1833.[312] Creep testing is generally considered a 20th-century subject and Fairbairn's work remained without parallel for over seventy years, until the works of Chevenard, Dickenson and Lea between 1919 and 1924, and the first book on creep in metals by H.J. Tapsell in 1931.[313]

Other tests were undertaken to determine the effect on cast iron of repeated re-meltings – 'almost a modern concept'. These tests found re-meltings advantageous to strength and tenacity.[314] Fairbairn also addressed the effects of temperature on the strength of *cast iron* – one of the first recorded investigations of the influence of temperature on mechanical properties. He used the apparatus he had used for sustained-load tests, suitably modified, to obtain temperatures from below freezing to 600°F, using snow, boiling water and molten lead.[315] He found that cast iron loses strength when heated above 120°F and becomes insecure below freezing point.[316] Twenty years later he would undertake similar work with *wrought-iron* plates, having in mind structures in the cold of winter and heat of summer, and pans or boilers with boiling liquids on one side and a furnace on the other. For this he used his lever and a more sophisticated heating apparatus in what are believed to be the first tests to demonstrate the substantial increase in tensile strength in irons heated to 100–200°C, a feature demonstrated by subsequent investigators.[317] At that time he also measured the elongation of test-bars under different loads at different temperatures, realising the importance of ductility in construction materials – 'ductility is, for practical purposes, the true measure of [wrought-iron's] strength and practical utility'.[318]

Consultancy reports covered various topics including one for mill-owners on the river Bann in Northern Ireland who faced intermittent water power. This resulted in the construction of the Lough Island Reavy and Corbet Lough reservoirs, for which J.F. Bateman was engineer, and William Dargan – a future friend and client of Fairbairn – was the contractor.[319] It also led to work for Fairbairn, building at least two waterwheels on the Bann, at Hazelbank and Seapatrick.[320]

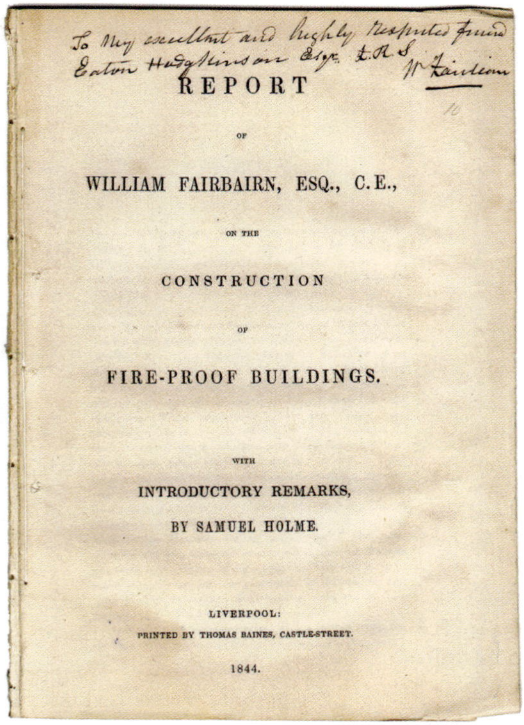
REPORT

OF

WILLIAM FAIRBAIRN, ESQ., C.E.,

ON THE

CONSTRUCTION

OF

FIRE-PROOF BUILDINGS.

WITH

INTRODUCTORY REMARKS,

BY SAMUEL HOLME.

LIVERPOOL:

PRINTED BY THOMAS BAINES, CASTLE-STREET.

1844.

Fig.3.38: Fairbairn's Report on the Construction of Fireproof Buildings. Presented to Eaton Hodgkinson.

In Liverpool huge fire losses provoked a call for 'fire-proof' warehouses and a report, subsequently published, was commissioned from Fairbairn by building contractors, S. & J. Holme.[321] His recommendations referring to non-combustible materials, compartment walls, and staircases within isolated shafts are echoed in today's Building Regulations. Two other reports on buildings were for the Anti-Corn-Law League, when Fairbairn was called to advise on the safety of the League's temporary 'pavilion' on Peter Street in 1841, and on a more permanent structure three years later.[322] Fairbairn did not take an active part in the League, but that he was presented with a gold snuff-box suggests that he undertook this work on a *pro-bono* basis.[323]

From 1841 Fairbairn undertook many investigative reports for coroners' courts, most arising from boiler explosions.[324] An early one, with Richard Roberts, followed an explosion at Jersey Mill, Ancoats.[325] In 1844 he reported, with David Bellhouse, on the dramatic collapse, with twenty fatalities, of Radcliffe's five-storey mill in Oldham which was nearing completion – a progressive collapse resulting from badly-designed cast-iron beams and ill-positioned tie-rods.[326] The range of Fairbairn's advisory and forensic reports was wide, encompassing what are now the disciplines of mechanical engineering, structural engineering and some civil engineering.

Fig.3.39: Fall of Radcliffe's new mill at Oldham, on which William Fairbairn and David Bellhouse reported at the inquest.

Fairbairn read various papers to the British Association from the mid-1830s. His participation in the Association was important in his ascent to a position of prestige and influence. From 1837 to the Manchester meeting in 1842 he delivered eight papers on subjects including the strength and other properties of cast-iron, the riveting machine, 'creep', iron as a substitute for wood in shipbuilding, and smoke prevention.[327] There was also a paper on a particularly ingenious steam-powered scoop, designed for drainage work at Haarlem, but apparently never built.[328]

In the Mechanical Science Section at the Manchester Meeting of the Association Fairbairn was amongst friends and peers – a remarkable group of Manchester engineers and national figures. President of the Section was Professor Willis from Cambridge, who was hosted by John Kennedy. Vice-Presidents were Fairbairn and Hodgkinson, together with Sir Marc Brunel and Sir John Robison, both of whom were hosted by the Fairbairns at their new home at the Polygon.[329] The Secretaries of the Section were James Thomson, Professor of Mathematics at Glasgow, whose son, James, would become Fairbairn's pupil the following year; Fairbairn's new son-in-law J.F. Bateman; J. Scott Russell and Charles Vignoles, whose son, Hutton was one of Fairbairn's pupils. The Sectional Committee included Fairbairn's old friend George Stephenson, and the Manchester engineers Roberts, Whitworth, Nasmyth and Kennedy.[330] Fairbairn's contributions to the British Association were not his only papers at this time. He spoke to the Manchester Geological Society about pumping water from coal mines,[331] and there were two papers to the Institution of Civil Engineers on his Turkish work.[332]

Conclusion

DURING the decade of sole proprietorship Fairbairn continued to build mills with their prime movers – steam engines or waterwheels – and power transmission, all of which, through optimisation, embodied most of the latest technology. This technology was diffused to many parts of Europe through publications, through visitors to the mills he had built, and by Fairbairn's personal contacts with clients in Germany, Russia and Turkey. The joint working of engineers and architects on major textile mill buildings was new. His work in the field of mill-building supports Fitzgerald's claim that in the 1830s he was 'undoubtedly the best known mill builder in the country'.[333] Indeed his mills, with their prime-movers and power transmission, mark him out as the leading mill-builder of his day.

Fairbairn's major contribution to shipbuilding was his research into the strength of wrought-iron plates and of riveted joins, coupled with emphasis on longitudinal framing and cellular construction. His experiments on the strength of riveting are believed to have been the first ever made and for the next forty years 'probably had more influence in determining the proportions adopted in practice than any others'.[334] Fairbairn was the major iron shipbuilder during the critical decade of its development, building the first iron steamship to be registered at Lloyds, the first iron steamships to reach Australia, the first iron steam yacht for royalty, and the first iron steamship to cross the Atlantic. He was also responsible, with his former pupil Albert Escher, for a classic example of the diffusion of technology, that of iron steamships to the Central European Lakes. Fairbairn was one of the founders of the British iron-shipbuilding industry, which became the greatest in the world, such that, a mere fifty years on, five out of every six ships to be found on the sea lanes of the world were British-built;[335] and naval-power capabilities enabled Britain to protect the world's trade routes.[336] The transfer of marine engine technology to textile mill engines illustrates the transfer of technology from one branch of engineering to another as does the transfer of knowledge of prefabricated ship construction, such as for the accommodation boats for the Ganges, to prefabricated buildings.

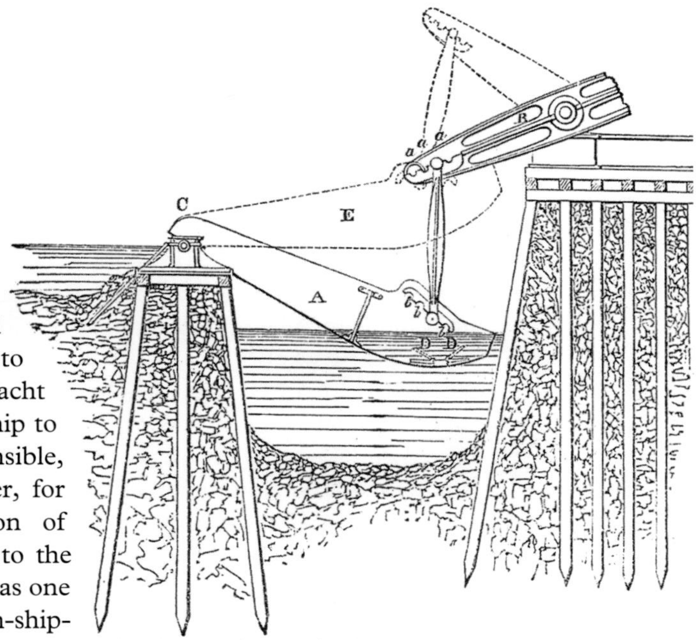

Fig.3.40: Drainage Scoop for Haarlem, Netherlands.

Experimental work undertaken by Fairbairn was far-sighted. As Loveday has pointed out, his time-dependant deformation and fracture experiments and the effects of temperature on strength have developed to become important parts of materials science, the former particularly in power engineering and the latter in aero-engines and nuclear power.[337] The many consultancy and forensic reports entrusted to Fairbairn indicate the respect in which his engineering abilities were held. That these reports covered such a wide range of expertise reflects how encompassing Fairbairn's knowledge and experience of engineering was.

Fairbairn's activities endorse Redlich's argument that entrepreneurs and their work are unpredictable; and the 'central position' of the concept of uncertainty in the study of entrepreneurship as recorded by Marino, Kreiser and Robinson.[338] This is illustrated by such as receiving and accepting unanticipated invitations to visit Russia and Turkey, with considerable commercial consequences; and experiencing the effects of matters outside his control as when economic conditions in Britain were adversely affected by the American financial panic of 1837. The riveting machine and the Lancashire boiler illustrate entrepreneurial 'innovation' in the Shumpeterian understanding, where something invented or 'hinted at' by another, was developed and marketed.[339] The riveting machine, as the keyway machine of twelve years earlier, was developed as a direct result of an industrial dispute and transformed boiler-making. The Lancashire boiler replaced the Cornish boiler to became the most widely used boiler during the remainder of the steam age. In other areas, Fairbairn contributed to on-going incremental and cumulative progression, for example the prefabricated three-storey Turkish corn mill, a forerunner of the steel-framed buildings of the 20th century, but this notable achievement was not followed up by Fairbairn. Scholars differ as to whether technological progress takes place by way of invention and the main entrepreneurial function of innovation – the Schumpeter model – or by numerous incremental and cumulative improvements. That both can be found in Fairbairn's work, and sometimes the distinction between them is blurred, provides support for the argument that simplistic models and generalisations about technological innovation are unhelpful as it takes place in a wide range of circumstances and in widely diverse ways.[340]

The decade of Fairbairn's sole proprietorship began and ended with stress, but he weathered the storms.[341] Indeed, during this decade, William Fairbairn, building on the reputation of Fairbairn & Lillie as the leading millwrights of the 1820s, established himself as an outstanding builder of mills, steam engines and boilers, as a leading iron shipbuilder and as an experimental engineer, with a growing national and international reputation.

Notes for Chapter 3

1. E Corlett, *The Iron Ship: the history and significance of Brunel's Great Britain,* (1975), p.8.
2. 'Report of a Committee, consisting of Mr J Scott Russell, Mr T Hawksley, Mr J R Napier, Mr William Fairbairn, and Professor W J M Rankine, appointed to analyse and condense the information contained in the Reports of the "Steam-ship Performance" Committee and other sources of information on the same subject', *BAAS1867,* pp.58-107.
3. S Smiles, *Industrial Biography: Iron Workers and Tool Makers,* (1863), p.328.
4. W E A Axon, *The Annals of Manchester: A Chronological Record from the Earliest Times to the end of 1885,* (1886), pp.185-6.
5. R C O Matthews, *A Study in Trade-Cycle History: Economic Fluctuations in Great Britain 1833-1842,* (1954), p.2.
6. Axon, *Annals of Manchester,* pp.199, 216.
7. J B Jefferys, *The Story of the Engineers 1800-1945,* (1945), pp.22-3.
8. *The Observer,* 14 August 1842; W Fairbairn, 'Notebook', Museum of Science and Industry, Manchester, (on display 2013); M Jenkins, *The General Strike of 1842,* (1980), pp.72-90.
9. Matthews, *Trade-Cycle History,* p.2.
10. *Life,* p.148.
11. Hayward, 'Fairbairns', pp.1.5, 2.2, A6; *Manchester Herald,* 10 August 1831.
12. See below and Chapter 4.
13. Hayward, 'Fairbairns', pp. 2.2-5, 3.4, A6.
14. *Life,* p.129.
15. Hayward, 'Fairbairns', p.1.42.
16. *Life,* p. 341, equivalent to around £3.5million at 2015 figures; A Slaven, 'The Shipbuilding Industry', in R Church (ed.), *The Dynamics of Victorian Business: Problems and Perspectives to the 1870s,* (1980), p.123.
17. [Woodcroft] VIII.
18. *Life,* p.337.
19. *Life,* p.340.
20. N Kirkham, 'Steam Engines in Derbyshire Lead Mines', *TNS,* 38, 1965-6. 72-3; Hayward, 'Fairbairns', pp.2.35-7.
21. [B Love], *Manchester As It Is,* (1839), p.213; E Butterworth, *Historical Sketches of Oldham,* (1856 ed.) p.187.
22. £2,500 in January 1837, £1,000 in December 1839, £6,000 in July 1841, £7,000 in July 1842, plus several small sums from relatives and friends (Hayward, 'Fairbairns', pp.2.16-17, based on an examination of the deeds of the property, not currently available).
23. *Life,* pp.155, 341.
24. *Life,* p.340.
25. *Life,* p.466.
26. Hayward, 'Fairbairns', pp.2.16-17, 3.4, A6.
27. *Life,* pp.148-9.
28. See Table 3.1. The figure of 2,000 men *at Millwall,* given in some secondary sources is incorrect. The 1836 and 1839 Fairbairn/Pole figures of 2,000+ look high, but Love's 1839 figure may be low. Butterworth's figure from the first half of the 1840s needs to be treated with caution as both the date and what it includes are unclear – Butterworth used it, and figures for several other firms, for purposes of comparison with Platt Hibbert. The 1842 figures are probably Fairbairn's, but their wide range makes them of minimal value. It is unfortunate that so far no published figures have come to light for the 1850s or 1860s. Hayward's figure of c.2,500 for 1851 is calculated on the basis of the figure of 704 'skilled' men in January 1852 (*The Times,* 14 January 1852), and Pole's statement that the 'turn-out' at that time 'saw the gates closed against 10,000 skilled workmen, involving the forced idleness of at least 40,000 people' (*Life,* p.323). Thus the figure is unreliable as Pole's figures are difficult to substantiate, and there is no firm evidence that the ratio of 1:3, skilled to improvers + labourers + apprentices, was applicable to Fairbairns. (Sources: 'Fifth Report from Select

Committee respecting Artisans ...', *Parliamentary Papers*, 1824, V, p.566; *MG*, 4 November 1826 - includes 12 apprentices (8.5%); *Life*, p.129; *Life*, p.157 - the autobiographical section in Pole was probably written in 1851 (*Life*, p.v); *Life*, p.339 - probably based on the previous reference; [Love], *Manchester As It Is*, pp.210-3; *CE&AJ*, 3, 1840, 397; E Butterworth, *Historical Sketches of Oldham*, (1856), p.187; E Chadwick, *Report on the Sanitary Condition of the Labouring Population of Great Britain and on the Means of its Improvement*, (1842), p.142; Hayward, 'Fairbairns', p.2.15; *Manchester City News*, 17 June 1871 - 'Shortly to be increased to over 500'.)
29. [Love], *Manchester As It Is*, pp.210-3.
30. Fairbairn, *M&MWII*, p.114.
31. B H Bradley, *The Works: The Industrial Architecture of the United States*, (1999), pp.209-11.
32. J Tann, *The Development of the Factory*, (1970), p.159. The detail is known to have been used by Fairbairn at Brereton Corn Mill (detail shown on drawings but not actually built); Travis Brook, Stockport; Brunswick Mill, Manchester; Victoria Mill, Wigan; Whittaker's Mill, Ashton-u-Lyne; and Hunslet Mill. Abroad it was used in his mills at Rouen and Gefle, and in the boiler house at the Oriental Mill in Bombay. J Longworth, *The Cotton Mills of Bolton 1780-1985: A Historical Directory*, (1987), includes illustrations of twenty-three mills with this detail; and D Gurr and J Hunt, *The Cotton Mills of Oldham*, (3rd ed. 1998), include illustrations of twelve.
33. M Williams with D A Farnie, *Cotton Mills in Greater Manchester*, (1992), pp.77-9, 98. On David Bellhouse see D R Bellhouse, *David Bellhouse and Sons, Manchester*, (1992), pp.28-36.
34. Details of a hybrid arrangement are usefully detailed by Sigsworth in respect of Black Dyke Mills in 1835 (E M Sigsworth, *Black Dyke Mills: A History*, (1958), pp.168-175). The development of procurement methods is traced by M H Port but he makes no reference to factories (M H Port, 'The Office of Works and Building Contracts in Early Nineteenth-Century England', *Economic History Review*, 20.1, 1967, 94-110).
35. *Liverpool Standard*, 27 August 1833.
36. Ure, *Philosophy*, p.33. On Ure see W S C Copeman, 'Andrew Ure, MD, FRS, (1778-1857)', *Proceedings of the Royal Society of Medicine*, 44.8, 1951, 655-662; V W Farrar, 'Andrew Ure, FRS, and the Philosophy of Manufacturers', *Notes and Records of the Royal Society of London*, 27.2, 1973, 299-324. Farrar says 'the civil engineer Fairbairn' was amongst Ure's pupils in Glasgow, but this is highly unlikely; possibly it was Peter Fairbairn.
37. 'Specification of Sundry Works intended to be done in the erection of a seven storey Fireproof Mill for Messrs. Kelly and Gilmour in Bradford Road, February 1839', (Copy held by Greater Manchester Archaeological Unit). Brunswick Mill has a lot in common with Orrell's Mill, suggesting that Fairbairn might have designed it; the builder was David Bellhouse (M Williams with D A Farnie, *Cotton Mills in Greater Manchester*, (1992), p.156).
38. *CE&AJ*, 17, 1854, 88 and Plates 12 and 13.
39. *The Builder*, 11, 1853, 773.
40. *Manchester Times*, 17 December 1853.
41. *Manchester Examiner and Times*, 12 July 1854.
42. *Manchester Times*, 17 December 1853.
43. *Manchester Examiner and Times*, 12 July 1854.
44. J Schmiechen and K Carls, *The British Market Hall: A Social and Architectural History*, (1999), pp.117-8.
45. Ure, *Philosophy*, pp.33-4 and frontispiece. The perspective drawing takes some artists licence, adding an extra storey. For a more realistic illustration as built see H Heginbotham, *Stockport Ancient and Modern*, (1892), Vol.2, p.322.
46. A Ure, *The Cotton Manufacture of Great Britain*, (1836), Vol.1, pp.296-304, 311-2.
47. A Ure, *Philosophie des Manufactures ou Économie Industrielle de la Fabraction du Coton*, (1836); A Ure, *Das Fabrikwesen in wissenschaftlicher, moralischer und commercieller Ginsicht*, (1835); A Ure, *Praktisches Handbuch des Baumwollen-Manufakturwesens. Nach der Cotton Manufacture of Great Britain*, (1837).
48. G S White, *Memoir of Samuel Slater*, (1836), pp.326-7.
49. V A C Gatrell, 'Labour, Power, and the Size of Firms in Lancashire Cotton in the Second Quarter of the Nineteenth Century', *The Economic History Review*, NS30.1, 1977. 112, quoting S D Chapman, 'Fixed Capital Formation in the British Cotton Manufacturing Industry', in J P P Higgins and S Pollard (eds), *Aspects of Capital Investment in Great Britain, 1750-1850*, (1971), pp.77-81.
50. Longworth, *Cotton Mills of Bolton*; Gurr and Hunt, *Cotton Mills of Oldham*; M Williams with D A Farnie, *Cotton Mills in Greater Manchester*, (1992).
51. Ure, *Cotton Manufacture*, Vol.1, p.304.
52. Ure, *Philosophy*, pp.33-4; Ure, *Cotton Manufacture*, Vol.1, pp.296-304, 311-2 and Plates 1 and 2; R Fitzgerald, 'The Development of the Cast Iron Frame in Textile Mills to 1850', *IAR*, 10.2, 1988, 139-41.
53. M Knierim (ed.), 'Introduction' to *Aus den Tagebüchern des fabrikanten Wilhelm Ehrenfest Jung (1780-1867) in Wuppertal-Hammerstein aus den Jahren 1844-1846*, (1984); A Oehike, 'Das englische Vorbild: Die Einführung moderner Spinnereibauten und Textiltechnik aus Lancashire', in H Bönnighausen et al, *Cotton mills for the continent: Sidney Stott und der englische Spinnereibau in Münsterland und Twente*, (2005), p.34n5. The mill was driven by a waterwheel on the Wupper, supplemented in the summer by a 50hp steam engine. It is not known if the wheel or engine were by Fairbairn, but it is possible they were. Banfield said that the mill work was English, 'we believe from Liverpool'. (T C Banfield, *Industry of the Rhine embracing a view of the social condition of the rural and manufacturing population of that district, 2, Manufactures*, (1848), pp.145-6). For illustration of this mill see H Nußbaum and H J Mahlberg, *Das Zooviertel in Wuppertal, Thiergarten, Stadion und malerisches Wohnen rund um den Märchenbrunnen*, (2004), reproduced in 'Villa Hammerstein um 1850.jpg' – Wikimedia Commons' (accessed 26 September 2016).
54. Oehike, 'Das englische Vorbild', p.34n5; Banfield, *Industry of the Rhine*, p.145.
55. A Oehlke, 'Spinnmühlen in Sachsen – Technologie-transfer und Architektinische Umsetzung einer neuen Bauaufgbe', *Mitteilungen des Chemnitzer Geschtsvereins*, Neue Folge VIII.69, 1999, 130-1. I am grateful to Roger Holden for this reference.
56. J G Kohl, *Journeys through England and Wales*, (English Translation 1844), pp.121-2.
57. E Jones, *Industrial Architecture in Britain 1750-1939*, (1985), pp.35-6; A W Skempton, 'Samuel Wyatt and the Albion Mill', *Architectural History*, 14, 1971, 53-152;
58. Tann, *Development of the Factory*, p.169.
59. Jones, *Industrial Architecture*, pp.70-4, 58-9; W Papworth (ed.), *The Architectural Publications Society Dictionary of Architecture*, (1849), Vol.8, p.11; H M Colvin, *Biographical*

Dictionary of English Architects, (1997). Fairbairn may have been introduced to Dixon by Tattersall who had recently won the competition for the Cumberland Infirmary at Carlisle. At Scharfenstein, the following year, the German architect, C F Uhlig, was involved (Oehlke, 'Spinnmühlen in Sachsen', 130-1).

60. Carlisle City Council Listing Notice 671-1/9/184 (13/11/72).

61. *Halifax Guardian,* 2 November 1844.

62. Tann, *Development of the Factory,* p.147.

63. R K Morris, 'Proof of Evidence on Behalf of the Council for British Archaeology' at 'Local Inquiry into an Application by Wood Mac Ltd for the Demolition of Robinwood Mill, Burnley Road, Todmorden (28 October 1986)', p.3.

64. Fitzgerald, 'Development of the Cast Iron Frame', 142.

65. M Binney, F Machin and K Powell, *Bright Future: The Re-use of Industrial Buildings,* (1990), pp.42-51.

66. http://www.victoriansociety.org.uk/news/leeds-mills-ontop-ten-most-endangered-buildings (dated 15 September 2015), (accessed 16. August 2016).

67. *MPICE,* 30, 1870, 461-5. On Wilson see M Chrimes, 'Wilson, Alexander' in A W Skempton (ed.), *A Biographical Dictionary of Civil Engineers in Great Britain and Ireland, Vol. 1, 1500-1830,* (2002), pp.785-6.

68. *MPICE,* 39, 1875, 259.

69. *MPICE,* 30, 1870, 465; *CE&AJ,* 4, 1841,147.

70. Fairbairn, *M&MWII,* pp.198-202; *MPICE,* 30, 1870, 463.

71. *Journal of the Society of Arts,* 25 November 1881, 50.

72. *CE&AJ,* 5, 1842, 138-9; 6, 1843, 143, 181, 191-2 + Plate VII; S B Moody, 'Description of a Water-wheel constructed by Mr. W. Fairbairn, M.Inst.C.E., and erected in Lombardy', *MPICE,* 3, 1844, 66-7; *Archeologia Industriale. Notiziario della Società Italiana per l'Archeologia Industriale, Sezione Lombadia,* 3, March 1979, 5, and 7, March 1980, 6.

73. 'Austrian Imperial Railways – Exhibition Catalogue', (2009), pp.4-5; www.biographiea.ac.at/oebl_2/206.pdf (accessed 6 October 2010).

74. See Chapter 6.

75. *Life,* pp.168-72.

76. [Woodcroft] VIII.

77. *Life,* p.172. For background see E C Clark, 'The Ottoman Industrial Revolution', *International Journal of Middle East Studies,* 5.1, 1974, 65-76.

78. W Fairbairn, 'Experimental Researches into the properties of Iron Ores of Samakoff in Turkey, and of the Hæmatite Ores of Cumberland, with a view to determine the best means for reducing them into cast and malleable states. And on the relative Strength and other Properties of cast-iron from the Turkish and other Hæmatite Ores', *MPICE,* 3, 1844, 225-45.

79. *Life,* p.173.

80. British Library, Add. MSS, 37461,ff.378-87, 37462, f.61.

81. On Howard see I Haynes, *Hyde Cotton Mills,* (2nd ed. 2005), p.40.

82. Plans held by Mr & Mrs Wood, Brereton Hall. Reproduced in T Bonson, *Driven by the Dane: Nine Centuries of Waterpower in South Cheshire and North Staffordshire,* (2003), pp.176-7.

83. There is a lingering doubt in that at least the building deviates slightly from Fairbairn's very typical drawings and it could be that the work was undertaken by a less expensive firm. Bonson believes the millwork is by Fairbairn (Bonson, *Driven by the Dane,* pp.176-7).

84. See Chapter 4.

85. Fairbairn, *M&MWII,* pp.118-26; [Scott], *Engineer and Machinist's Assistant,* Description of Plates, pp.91-8 and Plates 98-101.

86. T C Bannister, 'Bogardus Re-visited, Part 1: The Iron Fronts', *Journal of the Society of Architectural Historians* (American), 15.4, 1956, 15.

87. These included an iron house in St Petersburg in 1765 (G Casanova, *History of My Life,* (English Translation, W R Task, 1966-71), Vol.10, p.132),iron lock-keepers cottages from 1790 (*Birmingham Gazette,* 6 December 1924), iron roofs from the late eighteenth-century (S Giedion, *Space, Time and Architecture: the growth of a new tradition,* (3rd ed. 1954), pp.172-3), iron conservatories from 1812 (H-R Hitchcock, *Architecture: Nineteenth and Twentieth Centuries,* (1958), p.117), a cast-iron-framed two-storey house in Bermuda from the mid-1820s ('Restoration of the Commissioner's House' at http://www.bermudamall.com/marmuse/ commissioner/restorat.htm (accessed 11 July 2013)), cast iron shop fronts in Paris from 1830 (Bannister, 'Bogardus Re-visited', 15), open-sided buildings such as Hungerford fish market in1835, and prefabricated single-storey buildings sent to Australia in 1837 (G Herbert, *Pioneers of Prefabrication: The British Contribution in the Nineteenth Century,* (1978), p.40).

88. Herbert, *Pioneers of Prefabrication,* p.31.

89. *CE&AJ,* 4, 1841, 147.

90. H T Bernstein, *Steamboats on the Ganges. An Exploration in the History of India's Modernization through Science and Technology.* (1988), Frontispiece.

91. In March 1843 Fairbairn said the building 'was finished in 1840, and erected in Constantinople the following year'. (*MPICE,* 3, 1843, 126). In 1861 he wrote that the building was constructed in 1841 and 'when completed was exhibited at the works, Millwall'. (Fairbairn, *M&MWII,* p.116).

92. M Gayle and C Gayle, *Cast-Iron Architecture in America. The Significance of James Bogardus,* (1998), p.66.

93. Bannister, 'Bogardus Re-visited', 15, 21n34.

94. Hitchcock, *Architecture: Nineteenth and Twentieth Centuries,* p.526.

95. Fairbairn, *M&MWII,* p.190 and Plate 17; W Fairbairn, 'Description of a Woollen Mill Erected in Turkey', *MPICE,* 1843, 125-6; *MG,* 22 March 1843; *Life,* p.174;96.

96. See Chapter 4.

97. Bannister, 'Bogardus Revisited'; T F Peters, *Building the Nineteenth Century,* (1996), p.44; Herbert, *Pioneers of Prefabrication,* pp.41-2. However Professor Skempton's view should be noted, the 'mill is hardly an iron-framed building in the strictest sense and, for stability, it depended partially on a transverse brick wall supporting the main shaft of the machinery' (A W Skempton, 'The Boat Store, Sheerness (1858-60) and its Place in Structural History', *TNS,* 32, 1959, 67). In fact the *stone* wall separated the engine and boilers from the rest of the mill and served as a foundation *for the driving gear* (Fairbairn, *M&MW2,* p.121.

98. J Bogardus [and J W Thomson], *Cast-Iron Buildings: Their Construction and Advantages,* (1856), p.3.

99. M Gayle and E Gillon Jr, *Cast-Iron Architecture in New York: A Photographic Survey,* (1974), p.22.

100 . The first record of one is a 6hp engine for Birley's Chorlton Mills in 1830 (S Clark,'Chorlton Mills and their Neighbours', *IAR,* 2.3, 1978, 220).

101. [Woodcroft] VIII.

102. Fairbairn, *M&MWI I,* pp.246-7 and Plate 8. Collier states that Fairbairn's first side-lever mill engine was at Kingston Mill, Stockport, but gives no source. (D A Collier, 'A Comparative History of the Development of the Leading

Stationary Steam Engine Manufacturers of Lancashire, c.1800-1939', (Manchester University PhD Thesis, 1985), Vol.1, p.40). Hayward states that the first was at Orrell's Travis Brook Mill (R A Hayward,'The Watergrove Pumping Engine of 1838', *Bulletin of the Peak District Mines Historical Society*, 5.4, 1973, 201), but this appears to be by inference from Ure, *Philosophy*, (see Hayward, 'Fairbairns', p.2.29) whereas Ure, *Cotton Manufacture*, Plate 2, indicates a beam engine at Orrell's Mill. Haynes, suggests that Bailey Street Mill originally had beam engines, which proved inadequate and were replaced with the side-lever engines 'about 1840' (I Haynes, *Stalybridge Cotton Mills*, (1990), p.33). In this he appears to rely on what may be an incorrect date in G Watkins, *The Textile Mill Engine, Volume I*, (1970), p.42).
103. Watkins, *Textile Mill Engine*, p.42.
104. Ure, *Cotton Manufacture*, Vol.I, pp.313-4.
105. Fairbairn, *M&MWI*, p.246.
106. Fairbairn, *UIfE2*, p.235. John Bourne illustrated a 60hp Fairbairn beam engine in his *Treatise on the Steam Engine and its application to mines, mills, steam navigation and railways*, (1846), Plates 10 and 11, drawing attention to 'the peculiarity of the toothed flywheel' (p.175). Toothed flywheels were used for the 100hp engines at Saltaire in 1854 (Fairbairn, *M&MWI*, Plate V).
107. J S Russell, *A Treatise on the Steam Engine from the Seventh Edition of the Encyclopaedia Britannica*, (1841), Description of Plates, pp.vii-viii and Plates 9-11.
108. Fairbairn, *M&MWI*, p.247.
109. Collier, 'Steam Engine Manufacturers', p.40. Fairbairn installed a side-lever pumping engine at the Watergrove lead mine in Derbyshire, but there were teething troubles with it, (Kirkham, 'Steam Engines in Derbyshire Lead Mines', 71-3; Hayward, 'The Watergrove Pumping Engine', 200-14), which may explain why he built a beam engine 'on the Cornish principle' to power the pumps at the coal mine of Biolly et Fils at Verviers, Belgium, around 1840. (W Fairbairn, 'On the Economy of raising Water from Coal Mines on the Cornish Principle', *Transactions of the Manchester Geological Society*, 1, 1841, 236 and Plate 8; *Mining Journal*, 9, 1839, 90).
110. Watkins, *Textile Mill Engine 1*, p.42.
111. Watkins, *Textile Mill Engine 1*, p.48.
112. [Love], *Manchester As It Is*, p.211.
113. [Woodcroft] VIII.
114. *Life*, p.467.
115. *Life*, p.342.
116. [Love], *Manchester As It Is*, p.213, [Love's italics].
117. G N von Tunzelmann, *Steam Power and British Industrialization to 1860*, (1978), pp.79-92, 218-21; A Nuvolari and B Verspagen, 'Technical Choice, Innovation and British Steam Engineering, 1800-50', *Economic History Review*, 62.3, 2009, 685-6.
118. Fairbairn, 'On the Economy of Raising Water from Coal Mines', 179-94.
119. *CE&AJ*, 12, 1849, 316.
120. W Fairbairn, 'On the Expansive Action of Steam and a New Construction of Expansion Valves for Condensing steam Engines', *MPIME*, 1 July 1849, 22; [D Scott], *The Engineer and Machinist's Assistant*, (1850), Vol.1, Description of Plates, p.127.
121. *CE&AJ*, 12, 1849, 316.
122. Fairbairn, 'On the Expansive Action of Steam', 14-20; *MM*, 51, July-Dec 1849, 255-9; *CE&AJ*, 12, 1849, 315-7; *The Practical Mechanic's Journal*, 2, 1849-50, 162-4; *Artizan*, 7, 1849, 251-4; *Dingler's Polytechniche Journal*, 115, 1850, 1-6; [Scott], *The Engineer and Machinist's Assistant*, Vol.1, pp.126-8, Vol.2, Plate CXXXIII.
123. R L Hills, *Power from Steam: A history of the stationary steam engine*, (1989), pp.177-8.
124. An example survives in Sydney, manufactured by Thomas Perry & Son, Bilston, Staffordshire. ('Steam engine cylinder with Fairbairn drop valves, 1856-7', at http://www.powerhousemuseum.com/collection/database/?irn=212334 (accessed 17 September 2015). The reference to a patent in this article is incorrect).
125. 'Fairbairn's Valve-gear', *The Engineer*, 27, Jan-June 1869, 318.
126. Watkins, *Textile Mill Engine 1*, p.80; *Volume 2*, (1971), p.38. Cf. the tubular bridge, Chapter 5.
127. Fairbairn, *M&MWI*, p.247; Hills, *Power from Steam*, pp.157-61.
128. Appendix 3.1
129. Fairbairn, *M&MWI*, pp.235-42 , Plates 5 and 6; Hills, pp.177-8.
130. *MG*, 22 August 1857; *The Engineer*, 9, 1860, p.286.
131. See illustration.
132. See Chapter 4.
133. P C da S Telles, *A History of Brazilian Railways. Part 1 – The First Railways*, (ET by P E Waters, 1987), pp.43-5; *Engineering*, 2 February 1866; P Catchpole, *A Very British Railway*, (2003), p.182; T T dos S Cruz, *Paranapiacaba : A arquitetura e o urbanismo de uma Vila Ferroviária*, (MA Thesis, University of Sao Paulo, 2007), p.59.
134. Collier, *Stationary Steam Engine Manufacturers*, Vol.1, p.43; *The Practical Engineer*, 1890, 677; C Caine, *Cleator and Cleator Moor, Past and Present*, (1916), p.378; Walker, *Unwin*, p.60.
135. W Fairbairn, 'On Combustion of Coal, with a view to obtaining the greater Effect, and Preventing the Generation of Smoke', *BAAS1843*, pp.107-8; Patent No. 10,166 (1844); *CE&AJ*, 8, 1845, p.56 and Plate IV.
136. *Mining Journal*, 19, 1849, 415.
137. *MG*, 6 November 1862.
138. *BAAS1844*, pp.115-6, figs.1-4; *CE&AJ*, 8, 1845, 56, Plate IV; *MM*, 42, 1845, 121-2.
139. *MPICE*, 1859, 225.
140. *The Argus*, (Melbourne), 15 September 1858.
141. H H P Powles, *Steam Boilers, Their History and Development*, (1905), p.117.
142. See Chapter 7.
143. *M&MWI*, p.280
144. L E Fletcher, 'On the Lancashire Boiler, Its Construction, Equipment, and Setting', *MPICE*, 27, 1876, 59; E Leigh, *The Science of Modern Cotton Spinning*, (5th ed. 1882), Vol.2, p.264.
145. *MPICE*, 27, 1876, 91; D K Clark, *The Steam Engine: A Treatise on Steam Engines and Boilers*, (1891), Vol.I, p.687.
146. Powles, *Steam Boilers*, p.114; H W Dickinson, *A Short History of the Steam Engine*, (1938), p.159.
147. *The Engineer*, 16, 1863, 377.
148. R J Law, 'A Survey of Tank Boilers down to 1850', *TNS*, 48, 1976, 328.
149. R L Hills, *Development of Power in the Textile Industry from 1700 to 1930*, (2008), p.89.
150. J Bourne, *A Treatise on the Steam Engine in its Application to Mines, Mills, Steam Navigation, and Railways*, (1846), p.55.
151. *MG*, 6 November 1852.
152. P Quinn, 'Charles Wye Williams, Boilers and Fuel', *The Mariner's Mirror*, 93.4, 2007, 463.
153. Quinn, 'Williams', 451-2.

154. Email from glen@marina.ie 12 May 2011, referring to conversation with Dan Minchin.
155. C W Williams, 'Lecture on the Evaporative Power of Steam Boilers', in W Sturgeon (ed.), *The Annals of Electricity, Magnetism, and Chemistry; and Guardian of Experimental Science*, 8, 1842, 58-65, 161-7.
156. Quinn, 'Williams', 463.
157. Patent No. 10,166 (1844).
158. H Dircks, 'On the Smokeless Argand Furnace of Charles Wye Williams Esq', *Transactions of the Society for the Encouragement of Arts, Manufactures, and Commerce*, 55, 1845, 74-5.
159. Dircks, 'Smokeless Arand Furnace', 81.
160. C W Williams, *The Combustion of Coals and the Prevention of Smoke Chemically Considered*, (1840); C W Williams, 'On the Combustion of Coal, and the Prevention of the generation of Smoke in Furnaces', *BAAS1840*, pp.199-201.
161. *BAAS1842*, p.xxv; *BAAS1843*, p.21.
162. *The Investors' Advocate and Journal of Industry*, 122, 17 November 1841, 345.
163. M W Flinn (ed.), *Report on the Sanitary Condition of the Labouring Population and on the Means of its Improvement, (1842)*, (1968), p.223.
164. S Mosley, *The Chimney of the World: A History of Smoke Pollution in Victorian and Edwardian Manchester*, (2001), pp.119-20; R H Kargon, *Science in Victorian Manchester: Enterprise and Expertise*, (1977), pp.109-16.
165. W Fairbairn, 'On Combustion of Coal, with a view to obtaining the greatest Effect, and preventing the Generation of Smoke', *BAAS1842*, pp.107-8.
166. 'The Combustion of Coal and the Prevention of Smoke', in Sturgeon, *Annals of Electricity*, 9, 1842, 133-8.
167. Mosley, *Chimney of the World*, p.120.
168. C Flick, 'The Movement for Smoke Abatement in 19th-Century Britain', *Technology and Culture*, 21.1, 1980, 31.
169. W Fairbairn, 'On the Consumption of Fuel and Prevention of Smoke', *BAAS1844*, pp.100-20; Fairbairn, *UIfE*, pp.vii, 48-92.
170. Hills, *Power from Steam*, p.134; Williams, *Combustion of Coal*, (2nd ed. 1841), p.154.
171. *Mining Journal*, 19, 1849, 291.
172. Hayward, 'Fairbairns', p.2.46.
173. Hills, *Power from Steam*, pp.133-4.
174. H W Dickinson, *A Short History of the Steam Engine*, (1938), pp.121-2; A McEwen, *Historic Steam Boiler Explosions*, (2009), p.x.
175. Patent No, 13552, 11 March 1851. Powles, *Steam Boilers*, p.117.
176. W & J Galloway to Wm Fairbairn & Sons, 31 January 1854, quoted in W H Chaloner, 'John Galloway, 1804-1894, engineer, of Manchester and his reminiscences', in D A Farnie and W O Henderson (eds), *Industry and Innovation. Selected Essays: W H Chaloner*, (1990), p.112
177. Chaloner, 'Galloway', p.112. Between 1848 and 1891 Galloways built nearly 9,000 boilers.
178. See Chapter 7.
179. *CE&AJ*, 4, 1841,147.
180. W O Henderson, *Industrial Britain under the Regency 1814-18*, pp.5-7. See Chapter 2.
181. *MG*, 24 September 1845.
182. H Sitterding (ed.), *Escher Wyss 1805-1955: 150 Years of Development*, (1955), p.5.
183. 'schweizerschiffahrt.ch' at http://www.vssu.ch/index.php?lg=de&c=04&sc=00 (accessed 30.06.11).
184. J Dürst (ed.), *150 Jahre Dampfschiffe auf dem Zürichsee*, (1985), p.29.
185. *CE&AJ*, 1841, p.147.
186. Dürst, *Dampfschiffe*, pp.29-30.
187. 'Viel Jubel und etwas Angst. Die Jungfernfahrt des ersten Zürichsee-Dampfers, <Minerva>, am 19.Juli 1835' in *Neue Zürcher Zeitung* 19 July 2010; 'Minerva (1834)' at http://de.wikipedia.org/wiki/Minerva_(1834) (accessed 30.06.11).
188. Sitterding, *Escher Wyss*, p.17; R Stutz, 'Die Dampfschiffahrt auf dem Walensee ad 1837. Transitpost 1837 bis 1859' (2008) at www.post-www.post-und-geschichte.ch/site/.../po_2008_1_walensee _ab_1837.pdf
189. J Barrow, *Tour of Austrian Lombardy, the North Tyrol and Bavaria in 1840*, (1841), p.50; Henderson, *Regency*, p.7.
190. Sitterding, *Escher Wyss*, p.17. This figure excludes their smaller vessels. In 1969 Escher Wyss merged with Sulzer Brothers to become Sulzer Escher Wyss, a worldwide engineering group employing over 30,000 people. Fairbairn sent out other iron steamships to Central Europe including the first on the Bodensee / Lake Constance (*CE&AJ*, 1841, p.147); one for Lake Maggiore (*MG*, 26 August 1848); and others, or possibly just their engines, for Lakes Lucerne and Geneva (*Life*, p.129).
191. W Fairbairn, *Treatise on Iron Ship Building: Its History and Progress*, (1865), p.3.
192. M Laird and R A K Oldfield, *Narrative of an Expedition into the Interior of Africa by the River Niger in the Steam-Vessels Quorra and Alburkah, in 1832, 1833, and 1834.* (1837).
193. W Fairbairn to J Laird, 23 April 1834, Wirral Archives, Laird Papers, Ref.17.
194. W Fairbairn to J Laird, 29 October 1834, Wirral Archives, Laird Papers, Ref.18. The *Vulcan* was renamed *Minerva*, - see illustration.
195. *CE&AJ*, 1841, 147-8.
196. E N Kendall, 'A Tabular Statement of the Dimensions and Proportions of Forty Iron Vessels', *MPICE*, 1, 1841, 146.
197. See below.
198. A G Jones, *Birkenhead Shipbuilding*, (2011), pp.47-51. This may not include ships built by McGregor Laird.
199. *Life*, pp.336-7.
200. J Thomson, *Collected Papers in Physics and Engineering*, (1912), p.xxii; R A Hayward, *The Story and Scandal of HMS Megaera*, (1978), p.19; A J Arnold, *Iron Shipbuilding on the Thames 1832-1915: An economic and business history*, (2000), p.22.
201. [Woodcroft] VIII.
202. London Metropolitan Archives, Records of Sun Fire Office, MS 11936/555/1252067-8 and /562/1273798; H Hobhouse (ed.), *Survey of London: Volumes 34 and 44: Poplar, Blackwall and the Isle of Dogs*, (1994) pp.466-80, available at 'British History Online', at http://www.british-history.acort.aspx?compid=46519 - this contains a 'Diagram of possible layout of Fairbairn's yard in 1840' based on a deed in the Greater London Record Office; *ILN*, 13, 1848, 309 with illustration of the yard around that date. Details of machine tools sold when the yard closed in 1844 can be found in *Liverpool Mercury*, 12 February 1844.
203. W Fairbairn, *Three Lectures on the Rise and Progress of Civil and Mechanical Engineering, and on Popular Education*, (1859), p.30. Pole has 'upwards of a hundred' built at Millwall' (*Life*, pp.337-8). Smiles refers to 'upwards of a hundred and twenty' (Smiles, *Industrial Biography*, p.328). C F T Young refers to 'upwards of eighty vessels' built at Millwall (C F T Young, *The Fouling and Corrosion of Iron*

Ships: Their Causes and Means of Prevention, with Mode of Application to the Existing Iron-Clads, (1867), p.29).
204. The 75 include a small number where Fairbairn built the ship but not the engine or where he built the engine but not the ship.
205. *CE&AJ,* 4, 1841, 147.
206. Not to be confused with the wooden steamer of the same name, which made the first Atlantic crossing under steam in 1838.
207. *CE&AJ,* 4, 1841, 147; *Railway Magazine,* 3, 1837, 319; *MG,* 21 October 1837; *Transactions of the Institution of Naval Architects* 12, 1871, 306, and 14, 1873, 232; E C Smith, *A Short History of Naval and Marine Engineering,* (1938), p.97.
208. *CE&AJ,* 1, 1838, 233.
209. http://milhist.dk/vabnet/1922/ (accessed 6 August 2016); *CE&AJ,* 4, 1841, 146; *CE&AJ,* 8, 1845, 352.
210. *CE&AJ,* 3, 1840, 397.
211. *CE&AJ,* 4, 1841, 147.
212. *CE&AJ,* 3, 1840, 397, and 4, 1841, 147-8; *MM,* 38, 1840, 400; *The Sydney Herald,* 14 August 1840, 22 March 1841; W Lawson, *Steam in the Southern Pacific. The Story of Merchant Steam Navigation in the Australian Costal and Intercontinental Trades, and on the Ocean Liners of the Southern Pacific,* (1909), p.21; D Gregory, *Australian Steamships Past and Present,* (1928), pp.36-7.
213. Corlett, *The Iron Ship,* p.8.
214. *The Practical Mechanic's Journal,* 6, 1853-4, 216.
215. Young, *Fouling and Corrosion,* p.27; E C Smith, 'Thames Steam Pioneers: The Iron Shipbuilders', *Lloyds List,* 26 March 1924; W Fairbairn, *Remarks on Canal Navigation, illustrative of the Advantages of the Use of Steam, as a Moving Power on Canals …,* (1831), pp.34-40.
216. Patent No. 9,072, Sept. 1841; *Life,* p.338;
217. *The Practical Mechanic and Engineer's Magazine,* 4, 1845, 243ff. In fact it listed 93, but with duplicates there were 101.
218. Only those of Brunel and Guppy for the *Great Britain* (300hp), Seaward for the *Penelope* (320hp) and Maudslay for the *Retribution* and the *Terrible* (400hp) were more powerful.
219. T Tredgold, *The Steam Engine,* (1844), Appendix A.
220. *CE&AJ,* 6, 1843, 71 and Plates II,III; *The Practical Mechanic,* 3, 1844, 344 and Plates I,II.
221. [Scott], *Engineer and Machinist's Assistant,* Description of Plates, pp.69-71 and Plates 74-7.
222. Bourne, *A Treatise on the Steam Engine,* (1851 ed.), pp.179-80 Plate I.
223. See Chapter 4; W Fairbairn, 'On a New Description of Winding Engine', *MPIME,* 4, 1853, 137-42 and Plates 32 and 33.
224. Dickinson, *History of the Steam Engine, Appendix A,* (1844), p.123.
225. Young, *Fouling and Corrosion,* p.27; Corlett, *The Iron Ship,* p.25.
226. I am grateful to Dr Michael Bailey for this information. He advises that this boiler may have been subcontracted to the Bedlington Iron Works. (e-mail 2 March 2015).
227. *CE&AJ,* 4, 1841, 147; *Railway Magazine,* 3, 1837, 319; *Mémoires et Compte Rendu des Travaux de la Société des Ingénieurs Civils de France,* 1908, 368. A drawing, dated October 1837, signed by Wm. Fairbairn, is held in the Portefeuille of the Conservatoire des Arts et Métiers, Paris.
228. [Scott], *Engineer and Machinist's Assistant,* p.70.
229. W Fairbairn, 'An Experimental Inquiry into the Strength of Wrought-Iron Plates and their Riveted Joints as applied to Ship-building and Vessels exposed to severe strains', *Philosophical Transactions of the Royal Society,* Part 2, 1850, 677.
230. *BAAS1840,* pp.201-2; J Grantham, *Iron, as a Material for Ship-Building; being a communication to the Polytechnic Society of Liverpool,* (1842), p.16; Fairbairn, 'An Experimental Inquiry', 677-725.
231. Fairbairn, 'An Experimental Inquiry', pp.680-3, Plates LV, LVII.
232. Fairbairn, 'An Experimental Inquiry', pp.684-6.
233. Fairbairn, 'An Experimental Inquiry', p.686.
234. P Quinn, 'Wrought Iron's Suitability for Shipbuilding', *The Mariner's Mirror,* 89.4, 2003, 488.
235. Quinn, 'Wrought Iron's Suitability', 443, 455.
236. J R Crellin, 'Calvert, Frederick Crace – (1819-1873)', *Oxford DNB on-line,* (2005), (accessed 16 January 2016).
237. F C Calvert and W Fairbairn, 'On the Increased Strength of Cast Iron produced by the use of Improved Coke; with a series of Experiments on the Strength of Cast Iron smelted with Purified Coke', *MPICE,* 12, 1853, 352-74.
238. Discussion following paper by F C Calvert and W Fairbairn, *MPICE,* 12, 1853, 378-9.
239. Fairbairn, *UIfE2,* pp.318-25.
240. Fairbairn, 'An Experimental Inquiry', 688.
241. Fairbairn, 'An Experimental Inquiry', 700.
242. W Fairbairn, 'Experimental Inquiry into the Strength of Iron with respect to its Application as a substitute for Wood in Ship-building', *BAAS1840,* pp.s201-2; *CE&AJ,* 3, 1840, 388.
243. P Quinn, 'I K Brunel's Ships – First Among Equals?', *The International Journal for the History of Engineering and Technology* (formerly *TNS*), 80.1, 2010, 91,93-4.
244. *The Practical Mechanic and Engineer's Magazine,* 2nd Series, Vol.1, 1845-6, 62-4.
245. *CE&AJ,* 4, 1841, 148.
246. G D Dempsey, *Tubular and other Iron Girder Bridges,* (1850), p,14-5; R A Hayward, *The Story and Scandal of HMS Megaera,* (1978), pp.43-6.
247. [W Unwin, reporter], *First Report to the Council of the Institution of Mechanical Engineers on the Form of Riveted Joints,* (1881), quoted in Smith, *Contribution,* p.3.
248. Q Collette, I Wouters and L Lauriks, 'Evolution of historical riveted connections: joining typologies, installation techniques and calculation methods', in C A Brebbia and L Binda (eds), *Structural Studies, Repairs and Maintenance of Heritage Architecture XII,* 2011, 295-306.
249. Fairbairn, 'An Experimental Inquiry', 702-5, Plate LVIII.
250. Quinn, 'Wrought Iron's Suitability', 450.
251. Fairbairn, 'An Experimental Inquiry', 705.
252. Fairbairn, *UIfE,* p.134.
253. Quinn, 'I K Brunel's Ships', 89; Corlett, *The Iron Ship,* p.34.
254. Quinn, 'I K Brunel's Ships', pp.92-3.
255. Fairbairn, *UIfE,* pp.130-1.
256. *Mining Journal,* 18, 1848, 127; J Rapley, *The Britannia & other tubular bridges and the men who built them,* (2003), pp.157 and Plate 1.
257. Quoted in P Banbury, *Shipbuilders of the Thames and Medway,* (1971), p.247. Source not given.
258. J S Russell, 'On the Progress of Naval Architecture and Steam Navigation, including a Notice of the Large Ship of the Eastern Steam Navigation Company', *BAAS1854,* pp.160-1.
259. *The Athenæum,* 30 September 1854, 1182.
260. W Fairbairn, 'The Strength of Iron Ships', *Transactions of the Institution of Naval Architects,* 1, 1860, 11-81 +2 Plates; Fairbairn, *UIfE2,* pp.244-267; *The Engineer,* 9, Jan-June

1860, 215-6; *MM*, 1860, 151, 158-61,169-70. Similar lectures were also delivered to the Manchester Literary and Philosophical Society (*Proceedings of the Manchester Literary and Philosophical society*, 1859-60, 200-3; *MG* 18 February 1860) and to The Liverpool Polytechnic Society (*MM*, 17 February 1860, 103).
261. W Fairbairn, *Treatise on Iron Ship Building: Its History and Progress*, (1865), pp.214-8, 273-6; Fairbairn, *UIfE2*, p.xii.
262. Quinn, 'Wrought Iron's Suitability', 450; Quinn, 'I K Brunel's Ships', 93.
263. E Lehmann, 'The Historical Development of the Strength of Ships' in E Stein (ed.), *The History of Theoretical, Material and Computational Mechanics – Mathematics Meets Mechanics and Engineering*, (2014), p.270.
264. Quinn, 'I K Brunel's Ships', 93.
265. O Ozgue, P R Das, and N Barltrop, 'A comparative study of the structural integrity of single and double side skin bulk carriers under collision damage', *Marine Structures*, 18, 2005, 511-47. In the decade 1985-95, 150 bulk carriers were lost with a loss of more than 1,200 lives. Double skins are now mandatory for all new tankers, but not to keep the sea out of the ships, but rather, on environmental grounds, to keep the oil out of the sea (International Maritime Organisation, 'Construction Requirements for Oil Tankers – Double Hulls', at http://www.imo.org/en/OurWork/Environment/PollutionPrevention/OilPollution/ (accessed 5 November 2015).
266. Fairbairn, 'An Experimental Inquiry', 689; Plate LVI. This was an improved version. There is an illustration of the earlier model in *BAAS1838*, p.s161.
267. *Life*, pp.163-4, 341; Fairbairn, 'An Experimental Inquiry', 688; *Manchester Times & Gazette*, 27 October 1838.
268. R Armstrong, *The Boilers of Steam Engines: Their Calculation, Construction, and Management with a view to the Saving of Fuel..*, (new edition 1839), pp.204-5.
269. *Manchester Times & Gazette*, 27 October 1838.
270. Armstrong, *Boilers*, p.204.
271. W Fairbairn, 'On the Application of Machinery to the Manufacture of Steam-Engine Boilers, and other Vessels of Wrought Iron or Copper, subject to Pressure', *BAAS1838*, pp.160-2; *MG*, 22 June 1842.
272. *1851 Great Exhibition: Official Catalogue*, Class VI, Item 200.
273. *Engineering*, 3, 1867, 122.
274. Patent No. 7126; Hayward, p.2.50.
275. *Life*, p.164. Two forms were considered, one with a screw mechanism favoured by Smith and one with a lever mechanism, favoured by Fairbairn, who says, 'the screw would be too slow; and before any further steps were taken, I insisted on making a trial with the punching-machines which were in daily use. This was done on the following day, and Mr Smith produced as fine a specimen of riveted work as I have seen either before or since'. (*Life*, p.163).
276. *CE&AJ*, 6, 1843, 115-6.
277. A Garanger, 'Industrial Mechanisation' in M Daumas (ed.), *A History of Technology and Invention: Progress through the Ages. Volume III The Expansion of Mechanisation 1725-1860*, (English Translation by E B Hennessy 1970), p.154.
278. *CE&AJ*, 7, 1844, 356-7.
279. Dempsey, *Tubular and other Iron Girder Bridges*, pp.19-20.
280. *The Argus* (Melbourne), 2 June 1860.
281. Patent No. 9,409, July 1842; *MM*, 38, 1843, 142.
282. [Woodcroft] VII; Dempsey, *Tubular and other Iron Girder Bridges*, p.16; *The Practical Mechanic and Engineer's Magazine*, 2nd Series, I, 1845-6, 8.
283. *The Practical Mechanic and Engineer's Magazine* 2nd Series, I, 1845-6, 8. Hayward's diagrams appear to be incorrect in that they show short plates, apparently fixed horizontally (Hayward, *Megaera*, pp.35-6). The thickening of the plate edges may be another innovation where Fairbairn was indebted to a 'hint' from Bodmer who, some years earlier, had patented a system where thickened edges of the sheathing plates interlocked with a shaped back-plate (*The Practical Mechanic and Engineer's Magazine*, 2nd Series Vol.I, 1845-6, 9; Dempsey, *Tubular and other Iron Girder Bridges*, p.15).
284. Hayward, *Megaera*, pp.36-40.
285. Hayward, *Megaera*, p.40.
286. Fairbairn, *Iron Ship Building*, pp.218-20.
287. Fairbairn, *Iron Ship Building*, pp.213-7.
288. Fairbairn, *Iron Ship Building*, pp.211-3.
289. A D Lambert, *HMS Warrior*, (2011).
290. H P Spratt, *Outline History of Transatlantic Steam Navigation*, (1950), pp.27-8.
291. *CE&AJ*, 4, 1841, 148; *CE&AJ*, 8, 1845, 35; *The Practical Mechanic and Engineer's Magazine*, 4, 1845, 243ff. 'Billyboy' – 'a Humber or East Coast boat, of river-barge build' (*Shorter Oxford English Dictionary*).
292. *CE&AJ*, 4, 1841, 148.
293. *Shipping Gazette and Sydney General Trade List*, 12 September 1846, 262; at http://www.pbenyon.plus.com/Gazette/Shipping_Steam/pottinger.html (accessed 19.10.09).
294. *The Times*, 30 March 1846.
295. 'P&O Heritage. Ship Fact Sheet - Pottinger (1846)'; at http://www.poheritage.com (accessed 18.04.11)
296. Hayward, *Megaera*, pp.43-7.
297. In 1846 Fairbairn took out a patent for 'an improvement in the mode of driving the screw propeller, by the application of a large wheel with internal teeth. It is believed that a pair of engines were constructed on this plan, but it never came into general use' (*Life*, pp.338-9).
298. *CE&AJ*, 8, 1845, 352-4.
299. *The Practical Mechanic and Engineers' Magazine*, 4, 1845, 243ff.
300. [Woodcroft] VI.
301. D R Headrick, *The Tools of Empire: Technology and European Imperialism in the Nineteenth Century*, (1981), pp.26-8. Headrick makes no reference to Fairbairn. Nor does Bernstein, *Steamboats on the Ganges*, which Headrick describes as 'a model monograph in the social history of technology'. Yet Fairbairn built nine ships for the Ganges and Hoogley Rivers for the East India Company.
302. M Laird, and R A K Oldfield, *Narrative of an Expedition into the Interior of Africa by the River Niger in the Steam-Vessels Quorra and Alburkah, in 1832, 1833, and 1834.* (1837). Medicine was not as advanced as shipbuilding and when the *Alburkah* returned to Liverpool only nine of the original crew of forty-eight were alive.
303. Appendix 4.2. Robb describes Fairbairn as 'the most ardent and probably the most effective of those who supported the use of iron' for shipbuilding (A M Robb, 'Shipbuilding' in C Singer, E J Holmyard, A R Hall and T I Williams (eds), *A History of Technology: Volume V, the late nineteenth century c1850-c1900*, (1958), p. 351).
304. D A Farnie, *The Manchester Ship Canal and the rise of the Port of Manchester 1894-1975*, (1980), p.1.
305. W Sturgeon (ed.), *The Annals of Electricity, Magnetism and Chemistry*, 6, 1841, 363.
306. Fairbairn, *UIfE*, p.138.

307. M S Loveday, T Gray and J Aegerter, 'Tensile Testing of Metallic Materials: A Review', (TENSTAND –Work Package 1- Final Report, 2004), pp.2, 5.
308. W Fairbairn. 'An Experimental Inquiry into the Strength and Other Properties of Cast Iron from Various Parts of the United Kingdom', *Memoirs of the Manchester Literary & Philosophical Society,* 1842, 171-273. See also Fairbairn, *Application,* Table opposite p.74; *BAAS1837*, pp.377-415.
309. A Morin, *Résistance des Matériaux,* (3rd ed. 1862), pp.65, 71.
310. J Weisbach, *Lehrbuch der ingenieur- und maschinen-mechanik,* (3 vols. 1845-62) refered to in S P Timoshenko, *History of Strength of Materials ...,* (1953), p.123.
311. W Fairbairn, 'On the Strength and Other Properties of Cast Iron obtained from the Hot and Cold Blast', *BAAS1837*, pp.377-415; W Fairbairn, 'Appendix to a Report On the Strength and Other Properties of Cast Iron obtained from the Hot and Cold Blast', *BAAS1842,* pp.88-92; *Application,* pp.50-7; A I Smith, 'William Fairbairn and Mechanical Properties – 50,000 hours sustained-load tests on iron bars', *The Engineer,* 216, July-Dec.1963, 543-5.
312. L J Vicat, 'Note sur l'allongement progressif du fil de fer soumis à diverses tensions', *Annales des Ponts et Chausses,* 7.1, 1834, 40-4.
313. L Haworth, 'The Importance of Properties of Materials to the Designer', in T F Roylance (ed.), *Engineering Design: Papers given at the University of Nottingham Summer School of Engineering Design, September 1964,* (1966), p.114; P Chevenard, 'Sur la viscosité des aciers aux temperatures élevées', *Comptes Rendus de l'Académie des Sciences,* 169, 1919, 712-6; J H S Dickenson, 'Flow of Steels at a Low Red Heat and Scaling of Heated Steels', *Journal of the Iron and Steel Institute,* 106, 1922, 103-54; F C Lea, 'Effect of Low and High Temperatures on Materials', *MPIME,* 107, 1924, 1053-96; H J Tapsell, *Creep of Metals,* (1931).
314. Fairbairn, *Application,* pp. 59-65; Smith, *Contribution,* p.3.
315. Fairbairn, *Application,* pp.57-9; Fairbairn, 'Strength and Other Properties of Cast Iron' pp.406-15; A I Smith, 'William Fairbairn and Mechanical Properties of Metals: Effect of Temperatures on Strength', *The Engineer,* 216, July-Dec. 1963, 1015.
316. Fairbairn, *Application,* p.59.
317. Fairbairn, 'On the Tensile Strength of Wrought iron at Various Temperatures', pp.405-22 and Plates 4 and 5; Smith, 'Effect of Temperatures on Strength', 1016-7; M S Loveday, 'High temperature and time dependant mechanical testing: an historical introduction', in M S Loveday, M F Day and B F Dyson (eds), *Measurement of high temperature properties of materials,* (1982), pp.1-5.
318. Fairbairn, *Iron Ship Building,* p.22
319. W Fairbairn and J F Bateman, *Reservoirs on the River Bann, in the County of Down, Ireland, for more effectually supplying the mills with water,* (1836); J F Bateman, 'Description of the Bann Reservoirs, County Down, Ireland', *MPICE,* 1, 1841, 168-70; F Mulligan, *William Dargan: An Honourable Life 1799-1867,* (2013), p.57.
320. E R R Green, *The Industrial Archaeology of County Down,* (1963), pp.7-13
321. W Fairbairn, '*Report on the Construction of Fire-Proof Warehouses, with introductory remarks by Samuel Holme,* (1844); *MG,* 27 November and 4 December 1844; *The Practical Mechanic and Engineer's Magazine,* 4, 1845, 105-8.
322. *Manchester Times & Gazette,* 14 January 1840; *MG,* 28 January 1843; T Wyke, *A Hall for All Seasons: A History of the Free Trade Hall,* (1996), p.10.
323. *Life,* pp.466-7.
324. See Chapter 7.
325. *MG,* 13 November 1841; *CE&AJ,* 5, 1842, 51-2.
326. Fairbairn, *Application,* pp.274-9; *MG,* 6, 9 and 20 November 1844 and 9 July 1845; *MM,* 41, 1844, 348-51 and 43, 1845, 220-1; *CE&AJ,* 7, 1844, 429-32; *ILN,* 9 November 1844, 301.
327. W Fairbairn, 'On the Strength and Other Properties of Cast Iron obtained from the Hot and Cold Blast', *BAAS1837,* pp.377-415; *MM,* 27, April-Sept.1837, 453-5; W Fairbairn, 'On the Application of Machinery to the Manufacture of Steam-Engine boilers, and other vessels of Wrought-Iron or Copper, subject to Pressure', *BAAS1838,* p.160; *MM,* 29, April-Sept.1838, 473-4; W Fairbairn, 'On the effect of weights acting for an indefinite time on bars of iron', *BAAS1839,* p.s126; W Fairbairn, 'On the fan-blast as applied to furnaces', *BAAS1840,* p.s199; W Fairbairn, 'On the Strength of Iron, and its application as a substitute for wood in ship-building', *BAAS1840,* p.s201; Fairbairn, 'On Combustion of Coal', pp.s107-8; *MM,* 37, July-Dec.1842, 31-2; *MG,* 29 June 1842.
328. W Fairbairn, 'On raising water from low lands', *BAAS1840,* p.s210; *CE&AJ,* 3, 1840, 412-3; G D Dempsey, *Rudimentary Treatise on the Drainage of Districts and Lands,* (2nd ed.1854), pp.80-1; *Life,* p.180.
329. *MG,* 22 June 1842.
330. *BAAS1842,* p.ix; *Manchester Times and Gazette,* 25 June 1842.
331. Fairbairn, 'On the Economy of raising Water from Coal Mines', 179-94 and Plate 8.
332. Fairbairn, 'Experimental Researches into the properties of Iron Ores', 225-45; Fairbairn, 'Description of a Woollen Mill', 125-6.
333. Fitzgerald, 'Development of the Cast Iron Frame', 142.
334. [Unwin], *Report on the Form of Riveted Joints,* quoted in Smith, *Contribution,* p.3.
335. L T C Rolt, *Victorian Engineering,* (1970), p.78.
336. J Singleton, 'The Lancashire Cotton Industry, the Royal Navy, and the British Empire, c.1700 - c.1960' in D A Farnie and D J Jeremy, *The Fibre that Changed the World: The Cotton Industry in International Perspective, 1600-1990s,* (2004), p.66.
337. Loveday, Day and Dyson, *Measurement of high temperature properties of materials,* pp.xiii, 1-9.
338. F Redlich, 'Innovation in Business: A Systematic Presentation', *American Journal of Economics and Sociology,* 10.3, 1951, 290-1; L Marino, P Kreiser and A Robinson, 'Environmental Uncertainty and firm-level entrepreneurship' in H Landström and F Lohrke, *Historical Foundations of Entrepreneurial Research,* (2010), pp.81-97.
339. V W Ruttan, 'Usher and Schumpeter on Invention, Innovation, and Technological Change', *The Quarterly Journal of Economics,* 73.4, 1959, 597.
340. T J Pinch and W E Bijker, 'The Social Construction of Facts and Artifacts: Or How the Sociology of Science and the Sociology of Technology Might Benefit Each Other', in W E Bijker, T P Hughes and T J Pinch, *Social Construction of Technological Systems: New Directions in the Sociology and History of Technology,* (2nd ed. 2012), p.14.
341. As Bolton and Thompson argue of entrepreneurs, he had 'the ability to bounce back', (W K Bolton and J L Thompson, *Entrepreneurs: Talent, Temperament, Technique,* (2000), p.20).

Sources, Credits and Acknowledgements for Illustrations in Chapter 3

Fig.3.1 The painting was bequeathed to the Royal Society in Fairbairn's will (*MG,* 2 October 1874). Image used with the permission of the Royal Society.
Fig.3.2 R.J. Byrom, Photograph, 1961. Now demolished.
Fig.3.3 R.J. Byrom, Photograph, 1961. Now demolished.
Fig.3.4 G. Measom, *The Official Illustrated Guide to the Great Northern Railway,* (1861), reproduced in C.E. Makepeace, *Science and Technology in Manchester – Two Hundred years of the Lit. and Phil.,* (1984), p.63. (detail). Now demolished.
Fig.3.5 R.A. Hayward, 'The Fairbairn Engineering Company Limited – An Unsuccessful Victorian Limited Liability Venture', *Manchester Literary and Philosophical Society Memoirs and Proceedings,* 115, 1972–3, 64. I am grateful to Ron Hayward for permission to reproduce this map.
Fig.3.6 Fairbairn, *M&MWII,* p.114.
Fig.3.7 Fairbairn, *Application,* pp.158–60. Now demolished. Photograph R.J. Byrom, 1961.
Fig.3.8 A. Ure, *The Cotton Manufacture of Great Britain,* (1836), Vol.1, Plate 2 (part).
Fig.3.9 Ure, *Philosophy,* frontispiece. There is some artistic licence here as an extra storey has been added.
Fig.3.10 A. Oehlke, 'Spinnmühlen in Sachsen – Technologietransfer und Architektinische Umsetzung einer neuen Bauaufgbe', *Mitteilungen des Chemnitzer Geschtsvereins,* Neue Folge VIII.69, 1999, 131. I am grateful to Roger Holden for this reference.
Fig.3.11 Photograph, R.J. Byrom, (2011).
Fig.3.12 Photograph, R.J. Byrom, (1982).
Fig.3.13 Photograph, R.K. Morris, c.1985. I am grateful to Professor Morris for permission to reproduce this image.
Fig.3.14 T. Bonson, *Driven by the Dane: Nine Centuries of Waterpower in South Cheshire and North Staffordshire,* (2003), p.177. I am grateful to Tony Bonson for permission to reproduce this plan, and, through him, to Mr and Mrs Woods.
Fig.3.15 Fairbairn, *M&MWII,* p.119.
Fig.3.16 J.S. Russell, *A Treatise on the Steam Engine from the Seventh Edition of the Encyclopaedia Britannica,* (1841), Plate. Widely reproduced.
Fig.3.17 W. Fairbairn, 'On the Economy of raising Water from Coal Mines on the Cornish Principle', *Transactions of the Manchester Geological Society,* 1, 1841, Plate 8.
Fig.3.18 [D. Scott], *The Engineer and Machinist's Assistant,* (1850), Vol.2, Plate CXXXIII, Fig.3 (details).
Fig.3.19 'At the Paris Exhibition' at http://collections.vam.ac.uk/item/O145806/annexe-west-south-side-steam (accessed 6 January 2014) © Victoria and Albert Museum, London.
Fig.3.20 Walker, *Unwin,* opp. p.48.
Fig.3.21 Fairbairn, *UIfE,* pp.80–1.
Fig.3.22 J. Dürst (ed.), *150 Jahre Dampfschiffe auf dem Zürichsee,* (1985), p.31. I am grateful to Th. Gut for permission to reproduce this illustration.
Fig.3.23 A.P. Bogolyubov, *Review of the Baltic Fleet in 1848,* (Central Naval Museum, St Petersburg). For a side view of the ship see A.P. Bogolyubov, *Nevka 1838* (Central Naval Museum, St Petersburg).
Fig.3.24 D Gregory, *Australian Steamships Past and Present* (1928). For other illustrations of the *Rose* and the *Thistle,* by the marine artist J.C. Allcot, see J.H.M. Abbott, *The Newcastle Packets and the Hunter Valley,* (1942), facing pp.19 and 84.
Fig.3.25 T. Tredgold, *The Steam Engine,* (1844), Appendix A.
Fig.3.26 *The Practical Mechanic,* 3, 1844, Plate II
Fig.3.27 W. Fairbairn, 'An Experimental Inquiry into the Strength of Wrought-Iron Plates and their Riveted Joints as applied to Ship-building and Vessels exposed to severe strains', *Philosophical Transactions of the Royal Society,* Part 2, 1850, Plate LIV.
Fig.3.28 Fairbairn, 'An Experimental Inquiry', Plate LV.
Fig.3.29 Fairbairn, 'An Experimental Inquiry', Plate LVII.
Fig.3.30 Fairbairn, *UIfE,* p.135.
Fig.3.31 Fairbairn, *UIfE2,* p.249.
Fig.3.32 Fairbairn, *UIfE2,* p.258.
Fig.3.33 *1851 Great Exhibition: Official Catalogue,* Class VI, Item 200.
Fig.3.34 R.A. Hayward, *The Story and Scandal of HMS Megaera,* (1978), p.27.
Fig.3.35 Hayward, *Megaera,* p.55. Photograph c.1855 on the Black Sea during the Crimean War.
Fig.3.36 W. Fairbairn. 'An Experimental Inquiry into the Strength and Other Properties of Cast Iron from Various Parts of the United Kingdom', *Memoirs of the Manchester Literary & Philosophical Society,* 1842, p.175.
Fig.3.37 W. Fairbairn, 'On the Tensile Strength of Wrought Iron at Various Temperatures', *BAAS1856,* Plate 4.
Fig.3.38 The inscription, dating from before Hodgkinson and Fairbairn disagreed, reads, 'To my excellent and highly respected friend Eaton Hodgkinson Esq. FRS. W. Fairbairn'.
Fig.3.39 *ILN,* 9 November, 1844, 301.
Fig.3.40 *CE&AJ,* 3, 1840, 412.

CHAPTER 4

William Fairbairn & Sons

Family Business, 1842–1854

Fig.4.1

P. Westcott, *William Fairbairn*, (c.1850).

Introduction

THE third phase of Fairbairn's business career, following the initial partnership and solo practice, was one of a family business. This lasted from when some of Fairbairn's sons joined him as partners in the early 1840s until his 'retirement' in 1854.[1] This chapter and the next are about that family business – a business in which its founder remained the key player, and in which his succession planning failed. At its commencement, whilst steam-engine building and millwork at Ancoats was flourishing, shipbuilding at Millwall was in financial turmoil. Why did this occur? From 1844 no further orders for ships were accepted but the serious losses at Millwall were made good from the profits at Ancoats. Thereafter the family firm became increasingly profitable. The core business of building mills, with their prime movers and power-distribution, produced some of the most significant industrial structures of the mid-century, again endorsing Fairbairn's role as the leading mill builder of his time. He was an advocate, where appropriate, of single-storey mills, of which he built several, the forerunners of the typical 20th-century factory. He was at the forefront of elevator design, building on its beginnings in British mills and mines, an area with which his name has not previously been linked in secondary literature. Locomotive building became well-established, with the 'tank' engine – a Fairbairn 'innovation' – exhibited at the 1851 Great Exhibition. It was however the new and innovative use of wrought iron for railway bridges, the subject of the next chapter, for which Fairbairn became most famous as the Britannia and Conway Bridges excited the interest of engineers throughout Europe, and the derivative tubular-girder bridges and cranes filled gaps in the market, significantly increasing the family firm's prosperity.

Fairbairn's commitment to testing and experimentation is illustrated in this phase of his business life by work in disparate areas, with distinct differences in the reasons why each piece of work was undertaken and the approaches adopted. The testing of beams to ensure their safety continued. Forensic investigations sought to determine causes of explosions. In other cases there was systematic measuring and recording of data in order to provide useful knowledge on which engineers relied. The 'Fairbairn lever' led to him hosting experiments on the effects of pressure on solidification on which he worked with Joule, Hopkins and William Thomson. Most notable of Fairbairn's experiments were those leading to the Britannia Bridge.

A Family Business

THE family business phase of Fairbairn's career and the subsequent failure of the firm provide insights relevant to current debates among historians, and within business schools. It was attractive to have family members as partners in times of unlimited liability.[2] Investigation of thirty-three leading north-west firms of mechanical engineers founded between 1817 and 1875, showed that in nineteen (58 per cent) the succession, following the retirement or death of the founder, was by one or more family members.[3] Kinship networks often created close ties, as the fortunes of family and firm became inextricably linked – but not always so, as will be seen.[4] A dynastic business was also attractive to the 19th-century mind to perpetuate fame and family.[5] Indeed both his eulogistic biographer and the more critical Woodcroft, refer to Fairbairn's love of fame, and one might expect that it spurred the hope of a dynastic business.[6]

Fairbairn took at least four of his sons on business trips abroad but it is surprising, with their father's commitment to education, that none of his sons served a premium apprenticeship with a leading engineer, and none of the four who joined the firm went to university. There is no record of what training they had within the firm and it appears they may merely have assimilated the business by working in the firm's offices. In Thomas's case he was denied a university career because of the financial problems at the Millwall shipyard, but no reasons are recorded in respect of the other sons.[7] Around 1842 Fairbairn's two eldest sons, John (then 21) and Thomas (18), joined him as partners. The third and fifth sons, William Andrew and George followed. The fourth son, Peter, was not active in the firm but a notice of Dissolution of Partnership 'so far as regards P Fairbairn', suggests he may have had a modest share in the business.[8] He became a solicitor, was a Churchwarden at Manchester Cathedral, but died aged thirty.[9]

In the early 1840s, when John and Thomas joined their father as partners, the Ancoats works was busy and profitable in spite of the poor state of the economy, but Millwall was in trouble. John remains enigmatic: two years after becoming a partner, he left, returning to the family home terminally ill, from India, twenty-two years later.[10] Both Pole and Woodcroft avoid the question of what happened, and it remains a mystery. In 1846 the company name was changed to William Fairbairn & Sons, and, putting Millwall behind it, the family enjoyed great prosperity.[11] In 1854 Fairbairn retired from the day-to-day management of the company, taking an active consultancy role both to the company and elsewhere.

Succession is critical for the continuity and prosperity of a family business. Creative planning for succession was rare in the 19th-century family firm, its omission sometimes fatal to a business. Common weaknesses were that successful businessmen, reluctant to relinquish control, failed to delegate, and delayed retirement.[12] Fairbairn did not fall into these traps. He brought his sons in as partners at or before their coming-of-age and he retired at sixty-five.[13] Based on ability, Thomas was the obvious leader for the future, but his focus was elsewhere.[14] It has been well argued that succession based entirely on family ties can threaten the future and that bringing in able outsiders can supplement or substitute for absent talent – although Thomas was certainly not lacking in talent, it was just that it was focussed elsewhere.[15] Fairbairn tried to do this in 1866 with William Unwin, but by then it was too late and he failed.[16] He must have been bitterly disappointed that his sons did not share his love of engineering, in contrast, for example, to the sons of George Stephenson, Marc Brunel, John Rennie and John Penn. Fairbairn did not enjoy a close rapport with his sons and his relationship with Thomas, who had sole charge of the firm from 1859, was strained. The sons were not active in the Literary & Philosophical Society, the British Association, or the engineering Institutions. None is remembered for any engineering achievement. Fairbairn was far closer to William Unwin, his assistant from 1856 to 1862, and to his son-in-law, J.F. Bateman, both of whom shared his love of engineering and left engineering legacies.[17]

After the difficult start, the family business was extremely successful, especially in the fields of mills, steam engines, and tubular structures. It was said of Fairbairn by the 'friend who knew him well in business' that

> it was not till after 1845, when his sons became associated with him as partners, and their influence began to predominate in the management of the concerns, that he began to accumulate wealth, and make safe those profits of business which he had constantly earned, but had allowed to melt away again.[18]

The sons may have been undistinguished as engineers but it appears they were successful in ensuring that their father avoided unprofitable ventures.

Demise of the Millwall Shipyard

Between 1837 and 1844 events occurred which, perhaps deliberately, are blurred by Pole, and are barely covered elsewhere. These were Fairbairn's financial difficulties, his visits to Turkey, and the demise of the Millwall shipyard. Fairbairn's visits to Turkey were made hesitantly, fearing the business might suffer in his absence. His fears were fully realised. Returning from his second visit, he found both Ancoats and Millwall in 'great confusion'. At Millwall, believing an Admiralty order for a frigate had been confirmed, iron plates were bought. The order had not been confirmed, dimensions were changed, and the iron plates for which the yard had no use had to be paid for.[19] Proper management procedures appear not to have been in place.

To run the Millwall shipyard, Fairbairn had appointed his former premium apprentice, the 23-year-old Andrew Murray, with a small share in the business. Before Ancoats, Murray had attended Edinburgh Academy and Edinburgh University.[20] The appointment of this excellent but inexperienced man was a mistake, which Fairbairn acknowledged, or possibly used as an excuse:

> My young friend Murray, who was without experience, and had everything to learn, could not do much, and although he exerted himself to the utmost, it could hardly be expected that a young man could exercise all the judgment and precaution of a person whose training had attained greater maturity.[21]

John Hetherington also had a small share,[22] and Thomas Fairbairn, sent there straight from school in 1840, became a partner in1841. The 'great builders and shipwrights of the capital' opposed the new shipyard.[23] Losses were made. It was never a paying concern.[24] There were cash flow problems. Murray moved to Woolwich Dockyard in 1843 and Fairbairn made the decision to cease ship-building, and dispose of the Millwall yard.[25] At the end of that year the partnership between Wm Fairbairn, John Hetherington, Andrew Murray and Thomas Fairbairn was dissolved.[26] Thomas was based at Millwall for eight years, giving, as he wrote, his time and 'devoted attention' to bringing 'the disastrous ... concern to a close'.[27] However, that the inexperienced Thomas was given this responsibility, may suggest some continuing management naivety. The machinery was auctioned.[28] Part of the yard was disposed of, the remainder retained to complete two on-going shipbuilding contracts, and it was also used for the Britannia Bridge experiments. The yard had cost in excess of £50,000. It was sold for £12,000, a loss of over £38,000. However, the total loss at Millwall was over £100,000, indicating a loss of around £62,000 (equivalent to over four million pounds in 2015) on the shipbuilding contracts. This suggests serious mismanagement.[29]

Was it indeed mismanagement that caused the failure, or were there generic reasons beyond Fairbairn's control which affected all the shipyards on the Thames? Woodcroft acknowledged there may have been management failures, but blamed the choice of site, 'in which ambition had unwisely influenced judgement'. He highlighted the 'enormous' costs of foundations in the Thames mud, and the impossibility of Fairbairn being able to control two great works so far apart.[30] Pole suggests that his 'desire to make himself more known in the world' influenced the choice of a metropolitan site.[31] David Napier (1790–1869), who ran the yard next to Fairbairn's from 1837 to 1852, believed there was a generic reason for the failure of London shipyards – everything connected with shipbuilding in London was more expensive than in the North.[32] However, this was only partly true – John Glover showed, in 1869, that there was little difference in the cost of materials in London, although wages were higher.[33]

Arnold attributes Fairbairn's failure to 'business over-extension'.[34] Hayward, accepting that there were management weaknesses and the business was over-extended, nevertheless concludes that the cause of failure was competition from elsewhere.[35] He accepts that other Thames shipbuilders continued after Fairbairn, but cites Scott Russell and Mare as being in financial trouble in the mid-1850s.[36] These are not valid comparisons. Scott Russell had orders, but the problems with the *Great Eastern* ended his career as a shipbuilder. Mare was driven into bankruptcy by inflation at the time of the Crimean War, inadequate book-keeping, lack of cost control on the firm's civil engineering work, and his extravagant lifestyle.[37] Shipbuilding did decline on the Thames, but scholars differ as to when. Pollard dates it from 1850, Checkland and Arnold from 1866, whilst Banbury shows fine ships continuing to be built on the Thames into the 20th century.[38] Fairbairn's yard was in difficulties well before the 1850s. The operative years and causes of demise of the major iron and steel shipbuilders of the Thames are shown in *Table 4.1*. They endorse mismanagement as the cause of the failure of Fairbairn's shipyard, rather than competition from elsewhere.

Table 4.1: Demise of Thames Iron Shipbuilders.

Firm	Start	End	Cause of demise or of leaving the Thames
William Fairbairn	1835	1844/9	
J & F Napier	1837	1852	Yard sold to J Scott Russell
Ditchburn & Mare / C J Mare	1837	1856	Mismanagement? Became TI&SC – see below
J & G Rennie	1838*	1915	Moved to Wivenhoe
R & H Green	1843	1901	Changed to ship repairing
Samuda Brothers	1843	1893	Death of Joseph Samuda – no succession.
J Scott Russell & Co.	1844	1861	Retired. Financial problems
Thames I'works & S'building Co.	1856	1912	'Competition' from north and Clyde
J & W Dudgeon	1859	1875	Death/illness of partners – no succession
Millwall I'works & S'building Co.	1861	1866	Brought down by Overend Gurney's failure
John I Thorneycroft & Co.	1864	1907	Moved to Woolston, near Southampton
Yarrow & Co.	1866	1907	Moved to Scotstoun on the Clyde

*This is the date of their first marine engine work: the firm was already established as engineers in 1821 on the death of the elder John Rennie.

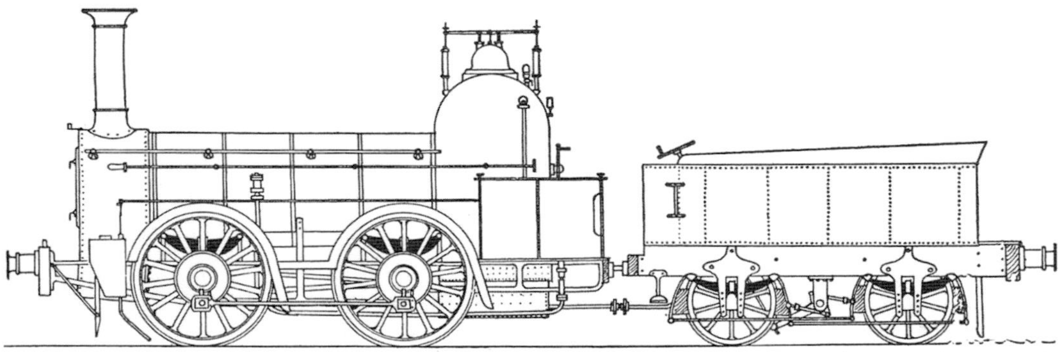

Fig.4.2: The first of the locomotives built by Fairbairns (1838).

There was no generic reason, common to all Thames iron shipyards, for the failure at Millwall. It was impossible for a man of even Fairbairn's abilities to run major manufacturing centres so far apart with inexperienced second-tier management at one of them. The cause was inadequate management and financial control. With remarkable frankness, Fairbairn admitted, 'We made many blunders as to prices ... in a business we had yet to learn'.[39] Engineering challenges and thirst for prestige had taken priority over prudent business management.

The Locomotive Department

BY mid-century Manchester was a major centre of locomotive building,[40] and Fairbairns was 'one of the most conspicuous locomotive building firms in England'.[41] As an apprentice at Percy Main in 1807, Fairbairn had made patterns for some of the first cast-iron fish-bellied rails to replace wooden rails. He attended, but did not compete in, the Rainhill trials, about which he later wrote, 'neither Mr Stephenson nor anyone else had any adequate conception of the benefits which the extension of railways has now secured'.[42] In 1831 Fairbairn & Lillie tendered, unsuccessfully, for two locomotives for the Liverpool and Manchester Railway.[43] They were tendering against Sharp Roberts who built one and Edward Bury who built two. However Fairbairn & Lillie did supply the winding mechanism for the Crown Street and Wapping tunnels at the Liverpool end of the line,[44] and the ironwork for Water Street Bridge.[45] Seven years later Fairbairn built his first locomotives – four Bury-type 0-4-0s for the Manchester & Bolton Railway.[46]

It is impossible in a book such as this to do more than to refer to some of the more interesting and important of Fairbairns' locomotives. Fuller technical details are listed elsewhere.[47] By the mid-1840s twenty-five engines had been built including four 2-2-2s for the London & Brighton Railway (1841–2), to the handsome design of Rothwell of Bolton, with 5ft 6in. driving wheels and boilers lagged with mahogany, polished and bound with brass bands;[48] and six 0-4-2s for the Manchester & Leeds Railway (1842–3), to a specification by its engineer, Thomas Gooch – a contract won against nineteen competitors.[49]

These were followed by another twenty-three 0-4-0 Bury-types for the Manchester & Leeds (1845–9),[50] twenty-one 2-2-2 'Hawkshaw singles' for the Lancashire and Yorkshire (1846–9),[51] ten 2-2-2s to the design of J.V. Gooch for the London & South Western Railway (1846–7),[52] and Fairbairns' first Irish order, six 2-2-2s for

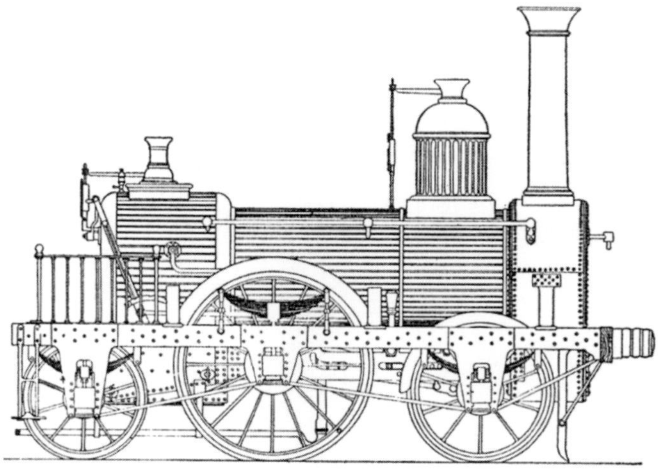

Fig.4.3: 2-2-2 for the London & Brighton Railway (1841–2).

the Midland Great Western Railway of Ireland (1846–7) to the specification of the company's engineer G.E. Hemans.[53] The significance of these locomotives is that Fairbairns were largely building to the designs and specifications of others, which would often mean competitive tendering, leading to meagre profit margins. Where they did design their own engines they were 'distinctly copyists', following the designs of Bury, Sharp, and others, though with modified details. Ahrons suggests that this was possibly because they were employing draughtsmen who had trained with these firms.[54] Fairbairn was perhaps too busy with mills, ships and steam engines to give much time to locomotives. The firm lacked a top designer such as John Haswell who, having set up the Wien-Raaber-Eisenbahn locomotive works in Vienna for Fairbairn, did not return;[55] or C.F. Beyer, who was at Sharps for twenty years before the setting up of Beyer Peacock.[56] In contrast, when Thomas moved from Millwall to Manchester, he appears to have been responsible for the locomotive department, but he was not a locomotive engineer. Nevertheless Fairbairn locomotives were solidly and well built, if perhaps lacking the flair found in some other Fairbairn products.

The first locomotives actually designed by Fairbairns are believed to have been four 2-2-2 passenger engines for the South Staffordshire Railway,[57] and one for the Shrewsbury & Birmingham Railway (1848–9).[58] The latter was described and illustrated in considerable detail by D.K. Clark in 1855.[59] It was 'a leading example of workmanship' and

> is closely similar to that made by Sharp, Stewart, and Co … and is in many respects identical with it. The example before us may therefore be accepted as an embodiment of the latest practice of two of the most conspicuous locomotive-building firms in England. It is in fact the ultimate result of the gradual development of the early six-wheel type of engine produced by Stephenson, in which, while the general design has been substantially adhered to, the details in many important respects have been materially modified and improved; and is probably the best that can be made of the species of single wheel locomotive which it represents, with respect to its details.[60]

Fig.4.4: Fairbairns' 2-2-2 *Vulcan* for the Shrewsbury and Birmingham Railway (1849).

These engines had inside cylinders and outside sandwich frames. The crank axle had four bearings (of which Clark was critical) and the carrying springs above the running plate. Driving wheels were 5ft 9in. in diameter. The firebox had a raised top and there was a plain polished dome on the middle of the three-ring boiler.

Not all early railway locomotives were well-built. A report in April 1840 complained of poor workmanship and unsatisfactory performance of five 2-2-2 locomotives which had been built by G. & J. Rennie for the London & South Western Railway in 1838. In 1841 they were despatched by rail to Fairbairns for complete reconstruction.[61] In the following year the same company sent five of its twelve locomotives built by Charles Tayleur & Co of the Vulcan Foundry, Warrington, to Fairbairns for extensive rebuilding.[62]

In the early 1840s there was considerable euphoria about 'atmospheric railways'. These were railways using differential air pressure to provide the power to move the railway vehicles. The air pressure, or partial vacuum, was conveyed to the vehicle in a continuous tube, which had a re-sealable slot to enable a piston, attached to the vehicle, to run in the tube. Much of the development of these railways was done by Joseph and Jacob Samuda of London. In 1840 the decision was taken to build an atmospheric railway from Kingstown, near Dublin, to Dalkey. Charles Vignoles was appointed engineer. The contract for the tubes was placed with the Samudas, and William Dargan was contractor for the earthworks and masonry.[63] The steam engine to power the air-pump which drove the railway was built by Fairbairn. Situated at Dalkey it had a huge 36ft diameter flywheel and was served by three Cornish boilers.[64]

At the British Association meeting in 1845 Fairbairn read a paper 'On Railway Gradients'. He referred to ongoing experiments on Hunts Bank Incline in Manchester, concluding from them that

> a great saving may be effected in the first outlay and construction of railways; and instead of spending large sums in tunnels, cuttings and embankments (in order to attain easy gradients), it will ensure much greater economy ... to increase the powers of the locomotive engine, and work lines with gradients varying from 1 in 100 to 1 in 30.[65]

The context of these experiments was that John Craven, the locomotive foreman of the Manchester & Leeds Railway, had undertaken trials of the tractive power of locomotives on steep gradients in 1844. The following year, the newly appointed Engineer to the Manchester & Leeds, John Hawkshaw, called in Fairbairn

> and the experiments he made confirmed Mr Craven's representations, which led to the abandonment of rope-haulage for trains up Hunt's Bank from Victoria Station to Miles Platting.[66]

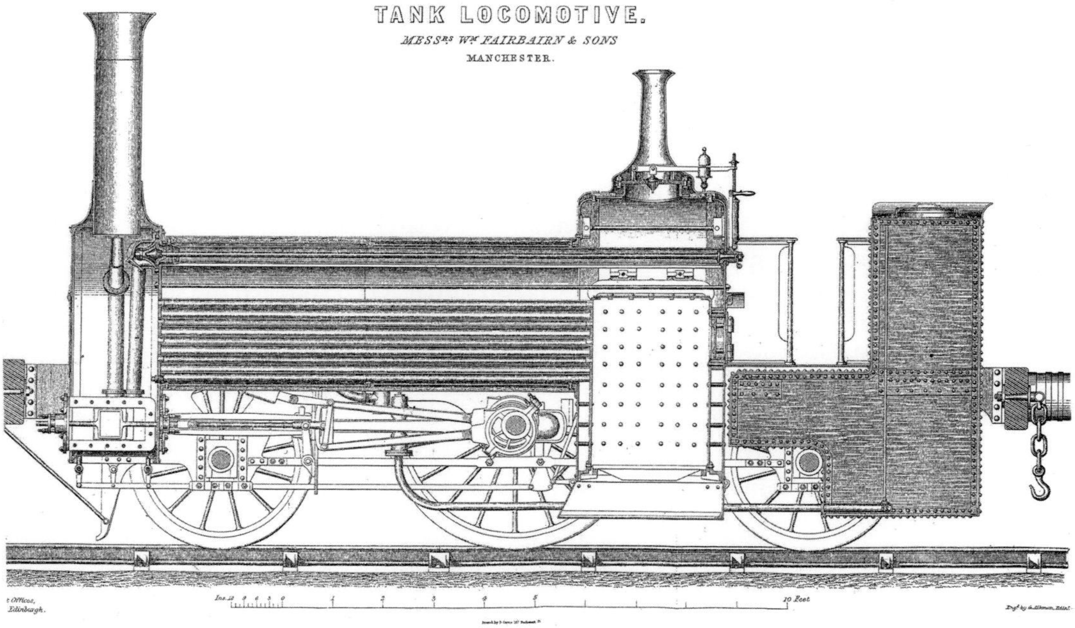

Fig.4.5: 'Fairbairn' Tank Locomotive as exhibited at the Great Exhibition 1851.

Fairbairns sought to capitalise on the mid-1840s railway boom. In 1846 Manchester engineers had enough locomotive orders in hand for two years.[67] Richard Clarke was appointed manager of Fairbairns' locomotive department – a position he held for fourteen years – and the firm built a new locomotive works.[68] It was not adjacent to a railway, which resulted in new engines, and those sent for repairs, having to be transported on horse-drawn trucks. It seems that proximity to the Fairbairn offices, foundry and pattern store was considered more important than a rail link, but this was short-sighted, and may quickly have been regretted as an axle of a truck conveying a locomotive from the Ancoats works to Bank Top station broke, causing the engine to fall on its side:

> The horses being in quick motion at the time, the engine did not fall immediately, thus allowing the escape of several men who were walking by the side of the truck, and who otherwise must have been severely injured, if not killed.[69]

The new works was spacious and well-lit, with rails and travelling cranes along the length of each of the two bays.[70] Orders were needed to fill it. Writing to Robert Stephenson, the best-connected railway engineer in the country, in July 1847 about the Britannia Bridge, Fairbairn added

> When you were last in Manchester I was speaking to you about some locomotives. I am now open for twenty or thirty to be delivered in about two years from this time. Next month we shall turn out one a week.[71]

Thomas followed up with, 'We now possess ... every kind of machinery of the first excellence'.[72] Unsurprisingly, given what followed at Conway, no orders came from Stephenson.[73] But orders flowed in from elsewhere.

Fairbairn tank engines, in which fuel and water were carried on the engines, without separate tenders, provide an example of entrepreneurial 'innovation', where the potential of an already-existing product was appreciated, developed and marketed. An indication of the standing and scale of the Fairbairn locomotive works can be gained from the fact it was visited in 1850 by the Prime Minister, John Russell, at which time fourteen or fifteen locomotives were under construction. Some were small engines, described as 'of a perfectly new construction', called tank engines. They were expected to show large savings in fuel costs.[74] The tank engine's invention has been attributed to Fairbairn, but this is incorrect.[75] There were earlier examples by others.[76] As Fairbairn had supplied the stationary steam engine for the vacuum pump for the atmospheric railway at Dalkey in 1843, it would be surprising if he was unaware of the Forrester tank engines on the adjoining Dublin & Kingstown Railway.[77] More pertinently, the engineer for the Dublin & Kingston was C.B. Vignoles, subsequently engineer to the 'Little' North-Western Railway, to which Fairbairn's first tank engines were supplied, and for which Vignoles is said to have 'drafted the specifications'.[78] Thus this may be another case of Fairbairn developing and marketing a product which existed, but was largely unknown or unexploited. If so, he did so very successfully. A Fairbairn tank engine was exhibited at the 1851 Exhibition, beneath a caption in the Catalogue describing W. Fairbairn & Sons as 'Inventors and Manufacturers'.[79] The exhibit was widely described and illustrated.[80]

Two Fairbairn tank engines were ordered for the Waterford & Tramore Railway, built by Dargan, in 1854. No.1 arrived in April 1855. The boiler was re-tubed in 1859 and a new boiler was fitted c.1865. The engine was rebuilt in 1895 with larger boiler and cab. A backsheet was added c.1907. It received a new Avonside boiler in 1924.[81] On 23 August 1935 this engine (now No.483) was hauling the 12.15pm from Waterford when it derailed, dragging the coaches down an embankment. There were injuries but no fatalities, other than the engine which was extensively damaged, and sadly scrapped.[82] It was the last Fairbairn engine and probably the last 'single' to run on a scheduled service – an 80-year working life!

From the opening of the Furness Railway in the mid 1840s, the goods traffic was well served by Bury 0-4-0 tender locomotives (of which one is preserved in the National Railway Museum, York). When further goods locomotives were required in 1854, Bury, Curtis & Kennedy had ceased trading and four very similar locomotives were ordered from Fairbairns, with tenders of Fairbairns' own design. Four more followed in 1858. All eight locomotives put in over forty years' work.[83]

Four 2-2-2 tank engines for the Liverpool, Crosby & Southport Railway, built in 1854–5, were unusual in that they had India-rubber packings in the coupling rods and axle boxes. The background to this was Patent No.2273, filed in October 1855 by Wm Andrew Fairbairn and G. Haslam. Elastic India-rubber cushions were placed between the ends of the brasses and the iron body of the coupling rod ends, and the axle boxes had elastic India-rubber packings to restore the box to its central position when leaving a curve. Wm Andrew read a paper about them to the Institution of Mechanical Engineers in 1858 – believed to be his only paper to a professional or learned society.[84] Neither arrangement appears to have been used other than in these engines and some few other experimental applications.

James McConnell was Locomotive Superintendent of the Southern Division

Fig.4.6: Bury-type 0-4-0 goods engine for the Furness Railway.

of the London & North Western Railway from 1847 until his resignation in 1862.[85] He was twenty-five years younger than William Fairbairn but they knew each other through McConnell having been an instigator of the Institution of Mechanical Engineers in 1846, of which Fairbairn was President in 1854–5.[86] It is clear that McConnell had a good rapport with Thomas Fairbairn as, in 1864 McConnell became a director of the Fairbairn company, although he does not appear to have held shares in it.[87] By then Fairbairns had ceased building locomotives and McConnell had resigned from the L&NWR.[88] In the early 1850s thirty 0-6-0 goods engines were supplied to the L&NWR by Fairbairn's, from Beyer's design, modified by McConnell. There were delays as McConnell continued to make further modifications, particularly to the fireboxes.[89] The goods engines were followed by ten McConnell 2-2-2 express locomotives – the 'McConnell Patents'. Fairbairns' tender was £2,950 for each but they also agreed to take ten old engines at £450 each in part payment. The express engines were to have 18in. cylinders, 7ft 6in. diameter driving wheels, and a low centre of gravity. There was an outside frame with bearings for the three axles. The firebox was unusually large and the 300+ tubes were shortened to 7ft. With various other innovations, Harry Jack finds D.K. Clark's description of them as 'a bundle of novelties' is justified.[90] Powerful and fast, they created much media interest and were pictured in *The Illustrated London News*.[91] G.A. Sekon, first editor of *The Railway Magazine*, wrote that, having been 'introduced with a vast amount of publicity, they became a nine days' wonder, and sank into quiescent mediocrity'.[92] Yet seventy years on Ahrons described them 'like many of McConnell's engines' as 'years in advance of the period'.[93]

Fig.4.7: 'McConnell's Patent' 2-2-2 express passenger engine No.300 under construction.

Although further applications of the Patent boiler were blocked by the General Locomotive Committee of the L&NWR, McConnell persisted. He produced drawings of a different and smaller version, a 2-2-2, which were published in 1855, but it was never built.[94] However a similar McConnell engine was built by Fairbairns for the Paris Exhibition of 1855, named after the Empress of France, *Eugénie.*[95]

William Fairbairn was much involved with the Paris Exhibition. He was an active member of its Manchester Committee,[96] a juror in the mechanical section,[97] an exhibitor of a columnar engine,[98] the subject of one of six exhibited marble busts by the sculptor, Patric Park;[99] and, commissioned by the Board of Trade, he produced a report on the exhibition for government.[100] Of the 17 locomotives in the Exhibition only two were English.[101] In published extracts from his report in *Useful Information for Engineers,* Fairbairn makes only passing reference to that by Stephensons but devotes three pages to *Eugénie,* omitting to say that Fairbairns built it.[102] The engine was purchased by Chemin de Fer du Nord. Jack states that complications with the firebox led to frequent repairs and the engine was withdrawn in 1866, after only 11 years use,[103] and Davies that lack of adhesion led to withdrawal by 1870 and scrapping in 1872.[104] No further orders followed for *Euégnie*-type locomotives. McConnell and Fairbairns must have been bitterly disappointed.

There were failings in the locomotive department. In 1850 Archibald Sturrock, who had worked for Fairbairn as a journeyman, became Locomotive Engineer of the Great Northern Railway. He immediately placed orders for forty 0-6-0 goods locomotives, twenty from E.B. Wilson at £1,790 each and twenty from Fairbairns at £2,000 each, for delivery in 1851. There were delays. Fairbairns were summoned to a board meeting, at which they promised to start deliveries in October. This date was missed and the order got caught up in the engineers'

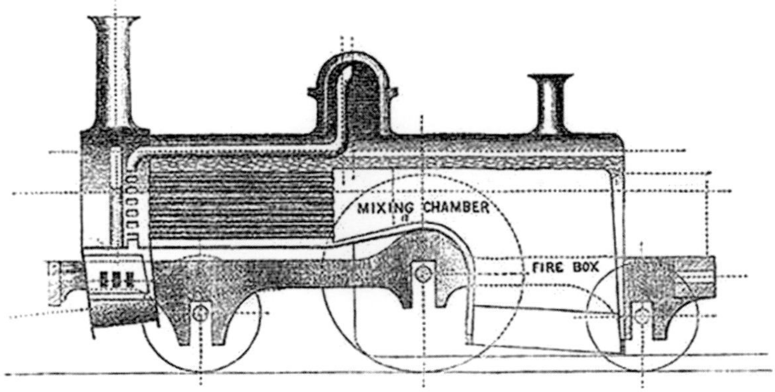

Fig.4.8: Longitudinal section through *Eugénie*.

dispute. Deliveries did not start until May 1852, by which time half the order had been cancelled and placed elsewhere.[105] The Great Northern placed no more business with Fairbairns. The reasons for the initial delays are unclear. The department was not over committed, as average annual production in the three years 1850–2 was below that for the previous and subsequent periods of three years.[106] It may be that the disruption caused by the engineers' dispute was wider than has been appreciated. On a more positive note, Fairbairns completed their first foreign order in 1853 – three 2-2-2 tank engines for the Estrada de Ferro Mauá, the first railway in Brazil. Of these, the *Baroneza,* is the only Fairbairn locomotive which survives in a largely unaltered state.[107]

Unlike Nasmyths, there is no 'works list' of Fairbairn locomotives. Locomotives were not given 'works numbers'. This is a major reason for the comparative obscurity surrounding Fairbairns as locomotive builders and the number of locomotives which the firm built remains unclear.[108] Appendix 4.1 lists 488 locomotives (or 477 if rebuilds are omitted). Of the 488, 89 (18%) went to foreign railways and 72 (15%) went to Ireland. Future discoveries may well take the total above 500. Up to 1863, when they ceased locomotive building, Fairbairns was the second largest locomotive builder in Manchester.[109] Sharps, Manchester's leading locomotive builder at that time, had built many more, but Nasmyths, who also started locomotive building in 1838, had only built 113 by 1863.[110] Beyer Peacock, who went on to build over 8,000 locomotives, was not established until 1854.[111] In spite of the limited in-house design capability, with a locomotive being produced, on average, every nine days in 1853 and 1854, Fairbairns had a significant locomotive department at the time of William's 'retirement'.[112]

Fig. 4.9: *Baroneza*, the first railway engine in Brazil, 1853, on display at the Engenho de Dentro Railway Museum, Rio de Janiro.

William Fairbairn & Sons – Two National Events 1851–2

The Great Exhibition

William Fairbairn & Sons were involved in two national events of 1851–2 which impacted on engineers, and both of which had commercial consequences for the firm – the Great Exhibition and the engineering dispute. For the Exhibition, William served on the Manchester Committee and its Machinery Sub-committee – interrupted by three months in Russia in the summer of 1850.[113] He was amongst friends – the Sub-committee included Whitworth and Nasmyth.[114] Fairbairns exhibited a steam engine, a tubular crane, a tank locomotive, a riveting machine, a flour mill and 'specimens of corn-mill work'.[115] This was a major commitment in terms of time and money. While it is difficult to identify orders arising directly from the Exhibition or quantify the benefits, the indications are that Fairbairns received substantial commissions for their exhibited products in the following years. A specific example was the tubular cranes at Portsmouth Dockyard where, since 1846, Andrew Murray had been Chief Engineer. When visiting the Exhibition, he made specific notes of several pieces of equipment, including the tubular crane exhibited by his previous firm.[116] There is no doubt that, with this major display at the Exhibition immediately following the completion of the Britannia Bridge, the firm was very well known and its reputation rode high. William had the prestige of being a juror in the machinery section of the Exhibition.[117]

The Engineers' Dispute

Within months of the euphoria of the Great Exhibition the Fairbairn business was seriously interrupted by the engineers' dispute of 1851–2. The importance of this dispute in the development of mechanical engineering in Britain has not been adequately appreciated. As *The Times* said at the start of it, the cry of the engineers is 'the cry of the handloom weavers – a cry against inevitable and irresistible laws'.[118] It was a dispute in which Fairbairns was an important player although William Fairbairn took no active part, leaving it to the most able of his sons, Thomas, to represent the firm.[119] In fact Thomas did far more than represent the firm – he acted as unofficial spokesman for the employers and in this role was undoubtedly effective, if perhaps scurrilous. For three decades there had been strikes about the number of apprentices who might be employed and about the operating of machines by non-millwrights, who were producing work quicker and more accurately than the traditional millwrights did by hand.[120] These strikes were symptomatic of the immense changes in the engineering industry, which took place from 1820, occasioned by the introduction and growth of machine tools.[121] Workers beyond the apprentice stage can be expected to resist new techniques in so far as innovation makes their skills obsolete and thus reduces their expected life-time earnings.[122]

The 1851–2 dispute involved the newly formed Amalgamated Society of Engineers, Machinists, Millwrights, Smiths, and Pattern-makers. Spurred on by its official, William Newton, 'a powerful and ebullient orator if less able as an administrator'[123] and an advocate of co-operative engineering workshops,[124] the Society had two major aims, the abolition of systematic overtime and the abolition of piece-work. There was also a third issue, the dismissal of non-union machine operators, so-called 'illegal men', later claimed by Newton to be 'a local issue'.[125] The dispute initially centred on Hibbert and Platt, textile machine makers of Oldham, one of the largest engineering firms in the world and a pioneer in capital-intensive machinery.[126] In November the ASE advised that no overtime or piece-work would take place after 31 December 1851. In mid-December thirty-four North-west employers, employing over 10,000 men, agreed to close their factories if employees at any one of them withdrew their services.[127] Thomas, as a spokesman for the employers, wrote to *The Times* under the *nom-de-plume*, *Amicus*.[128] His letter of 20 December set out the demands of the ASE, including the third issue, the dismissal of non-ASE machine operators, some of whom had been employed for years:

Fig.4.10: Fairbairn corn mill, and jib of Fairbairn crane: Machinery Hall, Great Exhibition 1851.

> That the masters at once, and without reserve, discharge the class of persons engaged in and long trained to the working of self-acting machines, and employ in their stead mechanics, members of the Union.
>
> These are the formal demands; but it is understood that the council are prepared to advocate an equalisation of the rate of wages; to lend themselves, in fact, to an agitation for a trial of the ingenious doctrines of M. Louis Blanc.[129]

This letter aimed to set non-ASE men against the union, and, with reference to the French socialist, Louis Blanc, traded on the nervousness of the middle classes following the 1848 French Revolution. There was a prompt, long and angry denial from William Newton.[130] In reply *Amicus* quoted documents signed by Newton requiring Platts to dismiss non-union men.[131] Newton asserted that he had never signed any such documents, and *Amicus* called on Hibbert & Platt to confirm the position.[132] They obliged by a letter to *The Times*: Newton had attended meetings at Oldham on 5 and 17 May when the demands to dismiss non-ASE members were made, and copies of documents signed by him were quoted.[133] At a meeting of the ASE on 30 December, Newton was at pains to point out that the dismissal of men by Hibbert & Platt was a local matter between them and their workmen. He had attended at Oldham 'in his individual capacity'.[134] A further blistering letter from *Amicus* in *The Times* of 5 January accused Mr Newton of having 'an utter disregard of consistency, honesty of purpose or truth'.[135] And so the correspondence continued. Employees declined to work overtime on 10 January 1852, and the employers closed their doors.[136] In early February there were reports of ill-feeling 'between the labourers and society men'.[137] Scenting success, employers advertised 'immediate employment'[138] to men who would sign a Declaration, which included:

> I will not whilst in your employment, call in question the right of any man to follow any honest calling in which he may desire to engage, or of any employer to make what arrangements or engage what workmen he pleases, on whatever terms they choose mutually to agree.[139]

By 8 March 6,056 men had signed the Declaration.[140] The right to employ non-ASE members to operate new machines was a major issue for employers. Around 21 March, ten weeks into the strike, the ASE reduced weekly payments to its members.[141] On 22 March *Amicus* struck again:

> Men are being rapidly trained to the use of lathes and machines ... the unskilled men ... have shown an extreme anxiety to learn and better their positions ... Thus, when the followers of Mr. Newton are prepared at his bidding to resume work they will find their places filled.[142]

Towards the end of March there was heavy picketing of the Fairbairn works because men had been taken on from Scotland in an attempt to scotch the strike.

> On Wednesday night, a crowd of about 3,000 persons ... collected in Canal-street, about six o'clock, and hooted the men as they left Messrs. Fairbairn's establishment; some of the men were pursued down Store-

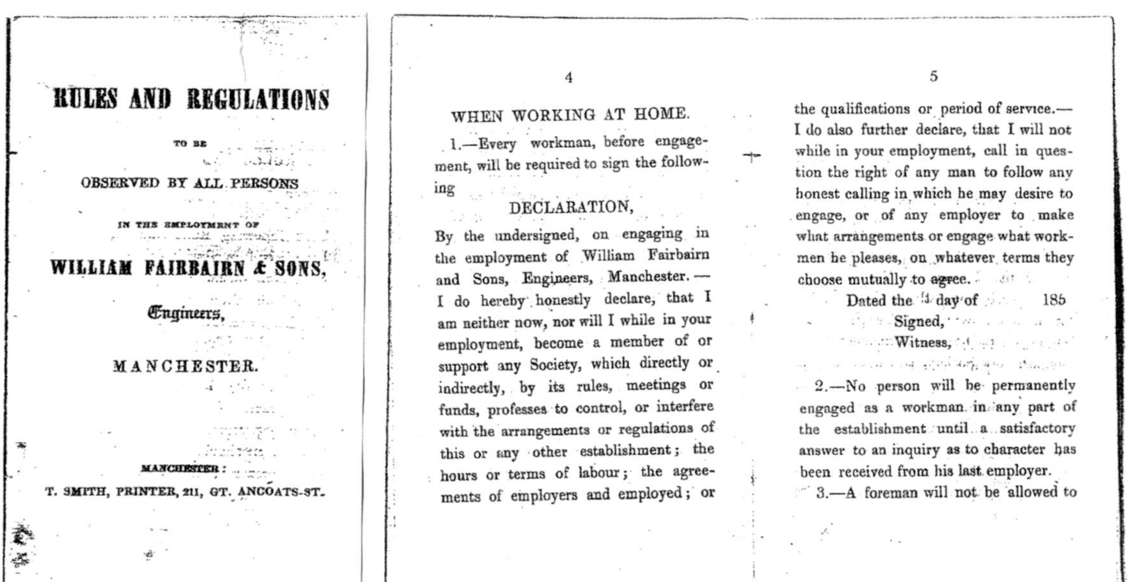

RULES AND REGULATIONS

TO BE

OBSERVED BY ALL PERSONS

IN THE EMPLOYMENT OF

WILLIAM FAIRBAIRN & SONS,

Engineers,

MANCHESTER.

MANCHESTER:
T. SMITH, PRINTER, 211, GT. ANCOATS-ST.

4

WHEN WORKING AT HOME.

1.—Every workman, before engagement, will be required to sign the following

DECLARATION,

By the undersigned, on engaging in the employment of William Fairbairn and Sons, Engineers, Manchester.—I do hereby honestly declare, that I am neither now, nor will I while in your employment, become a member of or support any Society, which directly or indirectly, by its rules, meetings or funds, professes to control, or interfere with the arrangements or regulations of this or any other establishment; the hours or terms of labour; the agreements of employers and employed; or

5

the qualifications or period of service.—I do also further declare, that I will not while in your employment, call in question the right of any man to follow any honest calling in which he may desire to engage, or of any employer to make what arrangements or engage what workmen he pleases, on whatever terms they choose mutually to agree.

Dated the day of 185

Signed,

Witness,

2.—No person will be permanently engaged as a workman in any part of the establishment until a satisfactory answer to an inquiry as to character has been received from his last employer.

3.—A foreman will not be allowed to

Fig.4.11: Title page and extracts from Fairbairns' *Rules and Regulations* 1852.

> street by the crowd and were pelted with stones. An inspector of police and 30 constables were on the spot ... and exerted themselves for the protection of the obnoxious men.[143]

Fleeming Jenkin, one of Fairbairn's pupils at the time

> had the full benefit of every possible groan and bad language. But the police cleared a lane through the crowd, the pupils were suffered to escape unhurt, and only the Knobsticks followed home and kicked with clogs.[144]

Sixteen years later Jenkin wrote a scholarly defence of trades unions, reprinted by the London School of Economics in1931, concluding, 'Let us neither seek to destroy trade-unions, strong as they are for good and evil, nor yet fear with a firm hand to set a legal limit to their power'.[145] Another of Fairbairn's pupils at the time, Guilford Molesworth, recorded that many of the men resented the interference of the union as they valued the opportunity of overtime at increased rates.[146]

On 5 April the ASE offered to withdraw their demands if the employers would withdraw their Declaration.[147] Sidney Smith, the Secretary of the Central Association of Employers replied at once: there was no way.[148] By now 10,939 men had signed the Declaration.[149] By 24 April the ASE was unable to pay its members.[150] It was all over. The united action of the employers, and the large numbers of semi-skilled men keen to operate self-acting machines, had put the ASE in an impossible position. Thomas Hughes QC, Christian Socialist and author of *Tom Brown's Schooldays*, wrote about Thomas Fairbairn's reference to Louis Blanc:

> Anything more untrue could scarcely have been invented. This body of the highest- paid workmen in England were as notoriously opposed to equalisation, in the sense here used, as they were to the reduction of wages. However the shaft went home and the wound rankled, and much of the bitter feeling of the men ... 'that they had been misrepresented', must be laid to the account of this reckless anonymous writer.[151]

Reckless they may have been, but Thomas's letters were politically astute. In deliberately widening the Union's stance by emphasising the local threat to the livelihoods of semi-skilled machine operators and by introducing revolutionary undertones, *Amicus* inflicted fatal wounds before the strike even began.

Industrial disputes usually bring losses. Employers lost money as a result of this dispute, as Fairbairns did. The cancelled order for locomotives for the Great Northern Railway was worth £1.4 million at 2015 figures, and there must have been other losses in a dispute lasting over three months. The operatives and their families suffered, 'the savings of years and household treasures were parted with'.[152] Some of the unskilled labourers, trapped between the union members and the employers, were in a state of destitution as the strike dragged on.[153] The ASE was a loser – it had paid out around £40,000.[154] The engineering industry suffered as factories in Belgium and France took 'every advantage',[155] and as skilled men emigrated to Belgium, Australia and Canada.[156] The employers recovered – it took Fairbairns until the end of the year.[157] Most of the employees recovered. The employers did not achieve the destruction of the ASE. In spite of the Declarations, ASE branches were gradually re-established.[158] It is clear that many Declarations were 'signed by men who did not leave, and never meant to leave, the Amalgamated Society' which, wrote Hughes, 'was inexcusable'.[159] One of the consequences of the emigration was that ASE branches were formed in Sydney, Montreal, Hamilton and Toronto.[160] The 1852 dispute was the last time that engineering employers did not allow their employees to negotiate collectively.[161] The real losers were the old aristocrats of the workshops, the millwrights, who never recovered.

The employers' success in the 1852 dispute had profound effects on the British engineering industry. The employers were free to employ whomsoever they wished – skilled, semi-skilled or unskilled men or boys – to operate their machines. The backward-looking restrictive practices were swept away, opening the door to automated production-line factories. It was a fundamental change, without which the British mechanical engineering industry would have fallen even further behind the American industry. A late starter, America had overtaken Britain in mechanical engineering. Whitworth, visiting there in 1852, was impressed with the results of the application of machinery, 'rendered still more remarkable by the fact that combinations to resist its introduction there are unheard of', and that 'the intelligent and educated artisan is ... free to earn all that he can, by making the best use of his hands, without let or hindrance by his fellows'.[162] One outcome of the dispute was a period of relative stability in the industry. It also led to increased social mobility – restrictions which had held back the able and hard-working were now a thing of the past in this industry.[163] Another outcome of the 1852 strike was that it helped to open the way for the flagship engineering factory of the 1850s, the Royal Small Arms Factory at Enfield – single storey, top-lit, steam-driven, automated with production lines, and, significantly, with a large majority of semi-skilled or unskilled machine operators.[164]

Given the pivotal importance of the 1851–2 engineering dispute, it is remarkable that it has received so little – and inaccurate – attention. In 1969 and 1972 Keith Burgess wrote two papers about the dispute,[165] but with no mention of two of the main protagonists, William Newton and Thomas Fairbairn. Nor did he refer to the important letter of Hibbert & Platt of 1 January 1852. In a biting critique of these papers P.J. Murphy refers to 'the way in which political inclinations have come to hamper and distort critical faculties ... Only that evidence conducive to their maintenance is perceived, while any material which threatens the ideal apparently goes unseen.'[166]

William, supportive of the activities of trades unions as friendly societies in the relief of the old and sick, was opposed to collective bargaining and restrictive practices, sincerely believing that they harmed the working man.[167] This was in the tradition of the Liberalism of the Manchester School – free trade, minimal state, individual liberty. 'What has given us our Armstrongs, Whitworths, Fairbairns,' Cobden asked the House of Commons, 'but the free industry of this country?'[168]

It should not be overlooked that at a meeting with Shaftsbury, Edwin Chadwick and representatives of cotton operatives in 1851, the seconder of the vote of thanks had 'long known that Mr Fairbairn was devoted to the interests of the working class'.[169] An inkling of why William chose to take a back seat in the 1852 dispute may perhaps be gleaned from a passage in his letter to Elizabeth Gaskell, following the publication of her *North and South* three years later:

> Poor old Higgins, with his weak consumptive daughter, is a true picture of a Manchester man. There are many like him in this town, and a better sample of independent industry you could not have hit upon. Higgins is an excellent representative of a Lancashire operative – strictly independent – and is one of the best characters in the piece.[170]

Nicholas Higgins was a trade union organiser and strike leader.[171] William had an affinity with, and respect for, the craftsman millwright, with whom he had worked 'on the tools', in a way that Thomas had not.

The Leading Mill Builder

DURING the decade of the family business, Fairbairn confirmed his position as the best-known mill builder in Britain by the construction of some of the largest and most advanced factories of the time. The construction of an industrial building is very different from constructing a batch of locomotives on a production line within the controlled space of a workshop. Each industrial building is site-specific with opportunities and constraints such as levels and ground conditions, access to water-power or coal, access to railway or canal, or the cost of urban land encouraging multi-storey construction. It is specific to an industry, to house particular machinery or warehouse particular goods. It is specific to the client's brief in terms of scale, budget and aesthetics. It is also specific to time, as machinery and processes change, as structural systems allow greater spans, as health and safety become of more concern, and as architectural tastes change.

Fairbairn first made his name with industrial buildings and his practice in them remained important throughout his career and into 'retirement'. In this, industrial buildings differed from other areas of his work – shipbuilding, locomotives, bridges and cranes for example, where construction was undertaken for limited periods of opportunity. His mill-building work exemplifies best practice in industrial building over the half-century 1820–70, from the plain multi-storey textile mills of the 1820s, to the integrated mills with weaving sheds of the 1830–50s, culminating in Saltaire; and the development of single-storey mills looking forward to the 20th century. In contrast there were corn mills, reaching the zenith of milling with millstones in the automated mill at Taganrog in Russia, and including the unique floating corn-mill for Balaclava.

Fairbairn's buildings served diverse industries – cotton, wool, silk, flax, corn-milling, armaments, locomotive works, ironworks, gunpowder and sugar production – and were situated in diverse locations – Lancashire, Yorkshire, London, the Midlands, Ireland, Continental Europe, Turkey, Russia, India, Egypt – such that any patterns can only be very general. Nevertheless, over five decades of rapid change, there are some constants in his approach linking the disparate phases of his business life and the varied industries for which he worked.

First the development of industrial buildings with their prime-movers and power transmission was incremental and cumulative. There were Fairbairn 'innovations' such as the increased efficiency of the power transmission at Murrays' mill, the Hodgkinson beams, fan-driven ventilation, the linear layout of corn-mills, the Lancashire boiler and the wrought-iron frame for the Dublin warehouse; but there were also many innovations by others, which Fairbairn incorporated into his industrial buildings. These included the suspension water-wheel, the north-light roof, the passenger lift and the single-storey integrated textile mill. Fairbairn was essentially an optimiser, seeking the optimal solution by selecting the best elements from the available alternatives, and improving wherever he could see a way of doing so. Secondly, because he optimised, his mills tended to be at the leading edge for their time, and for their industry. Usually of 'fireproof' construction,[172] with functional planning, and usually with efficient generation and transmission of power, they were widely visited and praised, continually enhancing Fairbairn's reputation. There was another consequence of this approach, that the comparative cost of Fairbairn's mills and millwork was high. This drew clients who wanted the best available and were prepared and able to pay for it. Thus his leading-edge work, of high quality, was self-perpetuating.

Saltaire – Heroic Yorkshire Worsted Mill

AN outstanding example of a manufacturer who wanted the best, and for whom cost was not a limiting factor, was Titus Salt. His factory at Saltaire, Yorkshire, was at the summit of Victorian mill-building and best known of the mills on which Fairbairn worked. Salt had developed a new branch of textiles, spinning and weaving alpaca fibre from South America. Having decided to move from congested Bradford to the Aire valley, he appointed Lockwood & Mawson, architects of Bradford.[173] An engineer was also needed and Salt visited the works of Benjamin Hick and of William Fairbairn, selecting the latter.[174] If Salt had built his mill twenty-five years earlier, he would probably have gone to a firm of mill-wrights such as Fairbairn & Lillie. Now architect and engineer worked together, producing the six-storey T-shaped, 'fire-proof' building with extensive north-light combing and weaving sheds. 'Vast in scale and architecturally magnificent, it drew on all Fairbairn's previous experience'.[175] He was responsible for the iron structure of Salt's factory, most of which was cast locally,[176] for the lay-out of the machinery, the design and manufacture of the boilers, steam engines and line-shafting, and for a tubular-girder bridge over the Aire. The factory floors, although arched, were of hollow brick construction to reduce weight.[177] Steam was generated by ten multi-tube boilers, driving two independent condensing engines with power transmitted by nearly two miles of line-shafting.[178] Enough wool was spun to serve 1,200 power looms. Notable at Saltaire was the increasing emphasis on building services engineering. The largest mill of its kind in the world, it was spacious and well-lit, and incorporated every known improvement 'to secure an agreeable temperature and healthy ventilation'. For its 5,000 lights the mill had its own gas works and gasometer. Soft water was collected from the roofs to be used for processes in the mill. To supply drinking water to the whole of Saltaire, a well was sunk and a pump, powered by one of the steam engines, raised water to a reservoir above the town.[179] Writing of the Saltaire factory three years after its completion, John James recognised it as 'supremely at the head of those in the worsted department'. He drew attention to the emphasis on the 'health and comfort' of the workpeople, indicative of its relative novelty. In the weaving shed all the shafting was located under the floor to prevent accidents and free the shed from 'the giddy whirl of the shafting and gearing'.[180] This was a logical development from the one main sub-floor shaft at the Sedgwick Mill in 1818, but it was not novel, having been used at

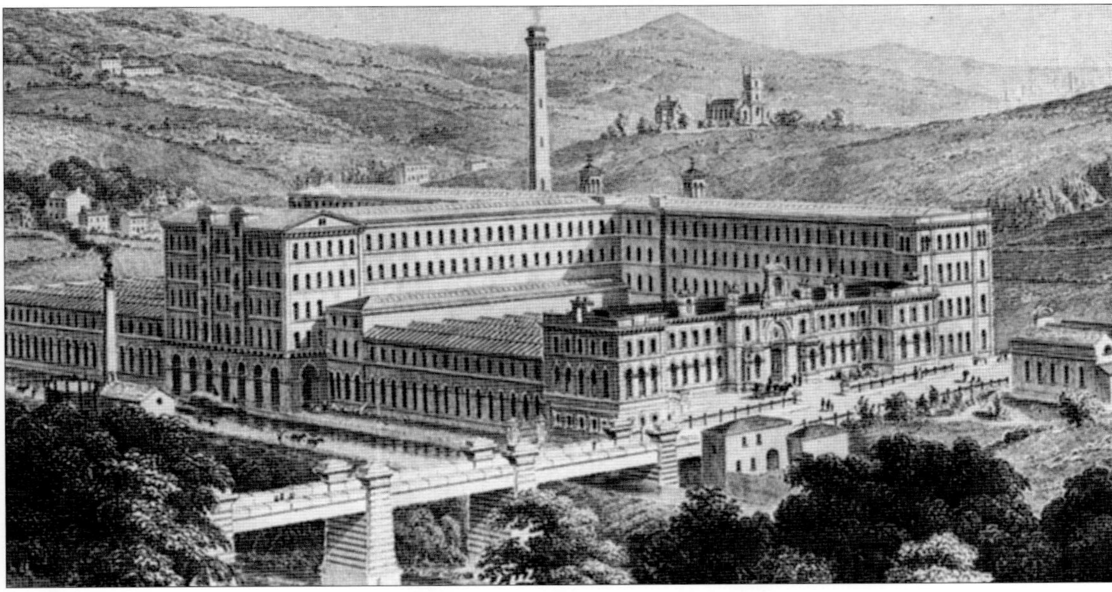

Fig.4.12: Saltaire showing the main office entrance and tubular-girder bridge.

Marshall's Temple Mill in 1840.[181] Again Fairbairn was optimising, incorporating whatever had been shown to be advantageous.

Always aware of risks of flaws in castings, Fairbairn tested the beams, as he had first done at Leeds in 1824. His approach, testing a beam to destruction and then proof-loading the subsequent beams to one-third that which broke the first beam, was described by A.I. Smith in the 1970s as 'quite modern'.[182] Four hundred Saltaire beams were tested in this way.[183]

Fairbairn made the most of this exceptionally high profile commission. In his reply to Salt's toast to architect and engineer at the opening he referred to the nearly two miles of shafting and the 5,638 miles of cloth per annum which 'would almost reach to the native mountains of the alpaca'. The event was widely reported, including a large illustration of the factory in *The Illustrated London News*.[184] Fairbairn immediately published plans in *On the Application of Cast and Wrought Iron to Building Purposes* which ran to four editions and was translated into French and Spanish.[185] Saltaire attracted many visitors, 'not only men of business, but of science and philanthropy', so much so that it became necessary to restrict access because of industrial espionage.[186] Fairbairn admired Salt, and was justly proud of Saltaire, his largest work of mill construction.[187]

Single-storey Textile Mills

FAIRBAIRN was an early advocate of single-storey integrated textile mills which represented a radical re-thinking of textile mill design. He did not initiate single-storey industrial buildings, but typically developed and publicised them. Saw-tooth or north-light roofs were introduced in parallel with the mass production of power looms around 1827. Having the looms on solid ground was considered beneficial. On upper floors their reciprocating motion tended to set up vibrations which were detrimental to the building and to the weaving process. With a solid ground floor there was slightly more moisture, which was beneficial to the weaving, as was the good, even daylight on the work surfaces.[188] The first reference to a north-light roof by Fairbairn & Lillie was at Gotts in 1829 – the year in which Fieldens built a weaving shed of an acre in Todmorden.[189] A step towards the integration of spinning and weaving in a single-storey factory was taken by Fairbairn at Bailey Brothers' mill in Stalybridge in 1836, which, instead of the typical multi-storey building, had only two stories. It had 23,000 mule spindles in one room on the first floor and 1,000 power looms on the ground floor, with space for more.[190] Thus vertical movement of goods and personnel was reduced, supervision was easier, and the space more flexible. No illustrations of this mill are known to exist, but parts of Fairbairn's two-storey Victoria Mill in Wigan from around the same time have survived.[191] The first significant integrated single-storey textile mill was Marshall's Temple Mill in Leeds, 1838–41, by James Marshall and the architect Joseph Bonomi, with square vaulted bays and conical roof-lights, following the weaving shed erected by James Smith at the same time as the Fairbairn & Lillie waterwheels at Deanston in 1830.[192] Marshall's Temple Mill was the influence for Fairbairn's single-storey integrated woollen mill in Turkey.[193] The Royal Small Arms Factory at Enfield, and his single-storey integrated cotton mills in India followed. These buildings were signposts to the standard single-storey factories of today.

Fairbairn's water-powered integrated woollen mill for Izmet in Turkey was designed in 1843. Modelled on Marshall's Temple Mill, it was lit and ventilated by 'circular apertures in the centre of each compartment of the roof', but the construction of the roof was described thus:

> On the pillars are placed long cast-iron beams, from which spring … segmental arches forming the roof, and consisting of ribs between the opposite pillars, with intermediate spaces filled in with wrought-iron plates in two thicknesses, between which is a non-conducting material.[194]

This sandwich roof construction appears to have been without precedent – the forerunner of today's typical insulated sandwich construction, used worldwide.[195] The original intention was to have prefabricated iron-panelled walls, but in the event stone was used. There was a partial undercroft but it was used as a dye-house and did not accommodate the shafting. This mill was publicised by a paper to the Institution of Civil Engineers, and a description and plates in *Mills and Millwork*.[196]

Single-storey integrated mills were further advocated by Fairbairn's Indian cotton mills. The second, third and several other early cotton mills in India were at Bombay, by Fairbairns, who were responsible for the planning, structure, steam-engines and shafting.[197] One of these mills, the Oriental Spinning and Weaving Company, was a large single-storey building, rectangular in plan with north-light roof.[198] Commercially successful, it was the mill chosen by Fairbairn to illustrate 'Cotton Mills' in *Mills and Millwork*.[199] The move to single-storey factories was slow. Land costs in urban areas were a factor. Evan Leigh, author of the *Science of Modern Cotton Spinning* (1871), which ran to at least five editions, had reservations, yet his speculative chapter on the future of cotton mills was illustrated by a plan based on the Oriental.[200] Further integrated single-storey woollen mills were built in Yorkshire after 1870,[201] but single-storey cotton mills in Britain tended to be one-function as at Cromer Ring (spinning) Mill, Middleton (1905), and Harp (ring doubling) Mill, Castleton, Rochdale (1908) both designed by

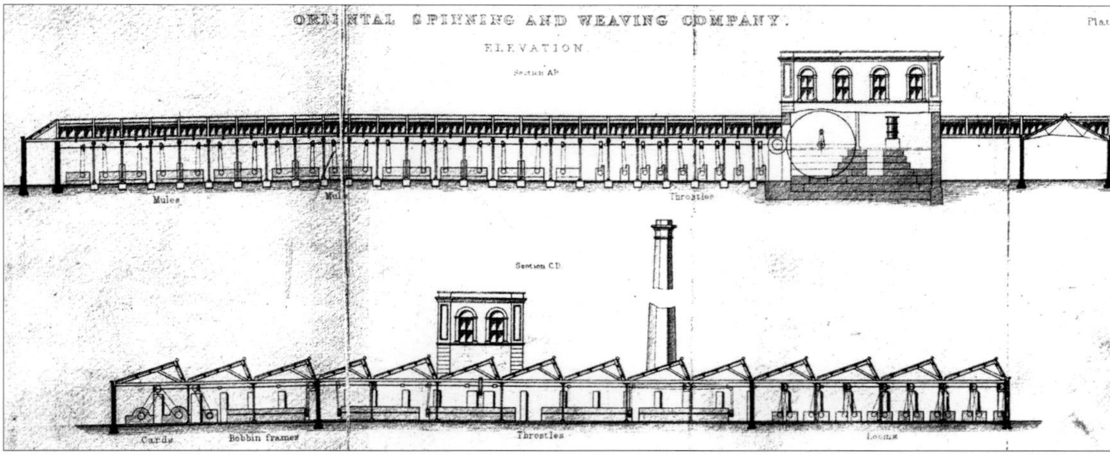

Fig.4.13: Oriental Spinning and Weaving Company, Bombay, 1855.

P. Sidney Stott of Oldham, and closely resembling Fairbairn's Oriental mill in appearance and construction.[202] One mill which did not conform to this trend was the innovative single-storey Premier Mill at Stalybridge, built in 1906, also to the design of P. Sidney Stott. It housed both spinning and weaving machinery, electrically driven from a public supply, and is described by Roger Holden as 'perhaps the most forward looking mill in Lancashire at the time'.[203]

Enfield – Prototype of Today's Factories

IMMEDIATELY following his work at Saltaire, Fairbairn was involved in the Royal Small Arms factory at Enfield. This was the first major factory outside America to employ 'the American System'.[204] That said, it is important to recognise the precursors in Bramah and Maudslay's locks at the end of the 18th century and Marc Brunel and Maudslay's mechanised manufacture of ships' blocks at Portsmouth in the early 19th century.[205] Enfield's single-storey layout contained production lines with the latest automated machinery capable of producing 100,000 rifles a year with interchangeable parts.[206] Five circumstances provided the impetus for this factory – the prospect of the Crimean war, the failure of British musket-makers to meet the army's needs, the American rifles at the 1851 London Exhibition, the example of Colt's revolver factory in Pimlico,[207] and the outcome of the 1851–2 engineers' dispute which led to semi-skilled men and boys operating machines.[208] The Government sent a Commission, including Fairbairn's friend Joseph Whitworth, ostensibly to visit the 1852 New York Exhibition, but in fact to report on American small arms manufacture.[209] They found British mechanisation overtaken, with two American factories each capable of producing 30,000 muskets a year.[210] In 1853 with war declared, the War Office appointed Nasmyth, also well-known to Fairbairn, to chair a Committee to establish rifle manufacture at Enfield.[211] John Anderson, who had proved himself in charge of the Royal Gunpowder Factory and who as a journeyman had worked for Fairbairn, was given executive responsibility.[212] He proposed a factory of self-acting machines, with

> everything ... passing consecutively on from one stage to another, never passing over the same ground twice, so that the raw materials which go in at one side shall come out a finished musket at the other.[213]

Fig.4.14: Royal Small Arms Factory, Enfield. Built 1854.

A delegation was sent to America to purchase machinery.[214] Fairbairn was engaged to work with Anderson on the plan and layout of machinery, and to construct the north-light roof, steam engines, power transmission by way of shafting and belts, and the heating.[215] On this occasion the shafting was overhead. The main room was 300ft square with ten lines of shafting driving 800 machines.[216]

In one sense there was nothing new at Enfield. Single-storey weaving sheds had led to single-storey factories. Automated machinery was already operational in America. What Enfield did was to bring these two features together for the first time in Europe on this scale. Fairbairn was involved here in a project which developed the marriage-value of existing discrete structures and technologies. Ames and Rosenberg cite this factory as 'the beginning of the movement of mass-production techniques from the United States to Europe'.[217] Enfield also illustrates the growing necessity for teamwork in a major project. The partnering of Anderson in a project management role, with Fairbairn providing an engineering 'design & build' input, James Burton, a Master Armourer brought from America to oversee the installation of machinery,[218] and architects Lockwood & Mawson, with whom Fairbairn had worked at Saltaire,[219] foreshadows much of today's construction industry teamwork. As so often, Fairbairn's appointment at Enfield, came from peer engineers – and the great Manchester triumvirate of Whitworth, Nasmyth and Fairbairn all had a hand in this innovative and important building.

Fig.4.15: RSAF Barrel Room in 1897.

Fig.4.16:
The great
Manchester
engineering
triumvirate.

William Fairbairn
(1789–1874).

James Nasmyth
(1808–1890).

Joseph Whitworth
(1803–1887).

The factory quickly became a national showpiece.[220] Early visitors included Baron Smola of the Austrian Artillery service, General Barriero from Portugal and Prince Oscar of Sweden. The French military regarded the factory with interest, and there were suspicions that it led 'to the commencement of a similar enterprise in Russia'.[221] The Royal Small Arms Factory at Enfield was, when it was built, the most advanced factory in Europe, and a forerunner of many single-storey production-line factories of the 20th century.

European Flax and Cotton Mills

PERSONAL contacts continued to be important in Fairbairns' obtaining and progressing work, particularly in Europe. Manchester had close links with Rouen,[222] where Fairbairn was appointed engineer for the first 'fireproof' factory in France – the flax mill, La Foudre. It was one of the largest industrial buildings in France, with elevations by a French architect – such that it was visited by Napoléon III and Eugénie.[223]

Swedish entrepreneurs at Gefle (or Gävle) appointed William Owens, a Manchester cotton man resident in

Fig.4.17:
La Foudre, Flax
Mill, Rouen,
1845.

Fig.4.18: Gefle Manufaktur Aktiebolag, Strömsbro (1851).

Sweden, to manage the building and running of a cotton mill. Naturally turning to Manchester, he had plans prepared by Fairbairn and negotiated the contracts.[224] Fairbairns supplied structural ironwork, the waterwheel and power transmission.[225] Owens and a director went to Manchester to sign the contracts.[226] Fairbairn visited Gefle in June 1850 with his son Peter,[227] and again in June 1851 with an engineer 'to prepare for the installation of the water wheel'.[228]

Around the same time as this work in Sweden, Fairbairn was involved with a water-powered flax mill at Narva in Russia for the financier and philanthropist Alexander von Stieglitz.[229]

The Norwegian entrepreneur Halvor

Figs 4.19 & 4.20: Waterwheel at Gefle.

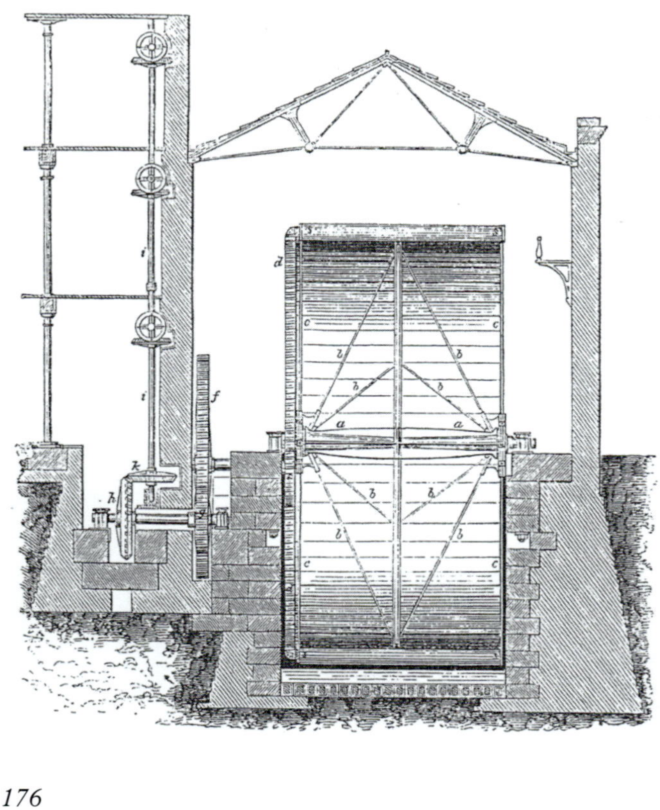

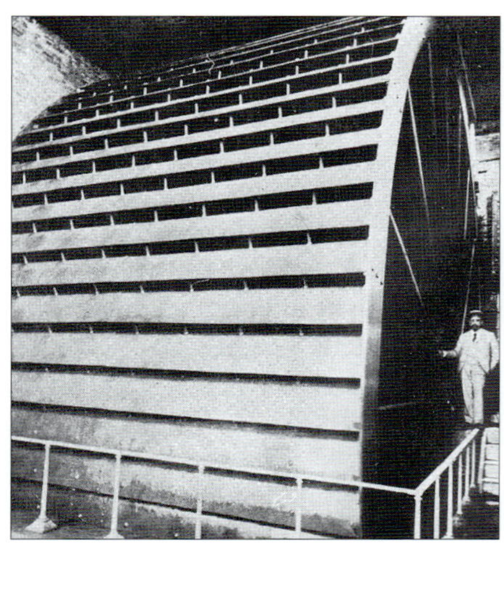

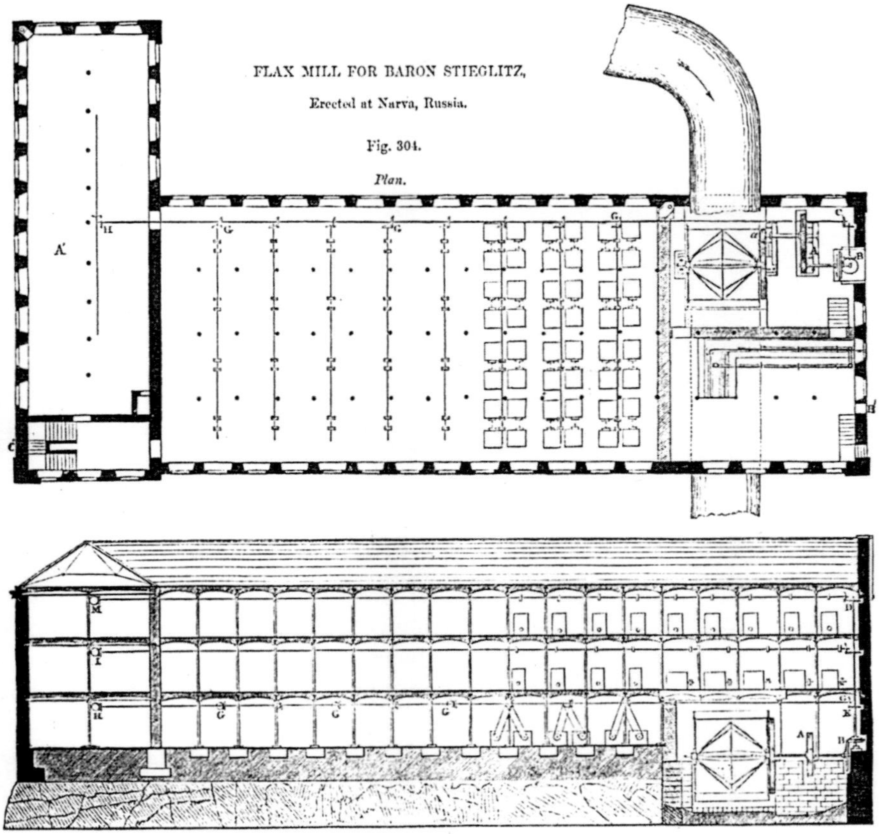

Fig.4.21: Flax Mill at Narva, Russia (c.1851).

Schou regularly visited Britain.[230] Around 1849 Fairbairn supplied the engine for Schou's Brenneriveien Weavery.[231] A few years later Schou moved his business to a water-powered site on the Aker River in Oslo, changing the company name to Hjula Weavery, which became the largest textile company in Norway. Fairbairn appears to have acted as a consultant. Schou relied heavily on him for guidance on machinery and costs. In March 1854 Fairbairn wrote to Schou giving an estimate of £2,600 for a 60hp waterwheel, gearing, shafting and heating apparatus, including boiler.[232] In November he wrote offering to help 'in our line ourselves, or to give you the full benefit of our advice in procuring elsewhere. We will see Hetherington's designs and that they are upon the best system and arrangement'.[233] Fairbairn was in contact with Schou from 1850 to 1861, and recruited five of the British workers for Hjula.[234] It is unclear whether or not he supplied any machinery. There are references to payments of £361 in 1854, made through the British agents, Sewell & Neck, who maintained an office in Christiania.[235]

Fairbairn's established reputation, supplemented by personal contacts, provided a sequence of commissions for prestigious industrial buildings for the family business, throughout Europe and beyond, as well as mills nearer home such as that of John Whittaker and Brothers at Ashton-under-Lyne, completed in the early 1850s when Fairbairn wrote, 'it is the last mill, upon a large scale, which has been built in these districts, and embodies all the new improvements'.[236]

Fig. 4.22: Whittakers' Mill, Ashton-under-Lyne (c.1850).

In 2003 The International Committee for the Conservation of the Industrial Heritage prepared a draft list of internationally important *existing textile* sites.[237] A measure of Fairbairn's standing in the field of industrial building may be gauged from the fact that he worked at nine of the sites on this list, as listed in *Table 4.2*, way ahead of any other engineer, architect, contractor or entrepreneur. Supplementing this list by Fairbairn's textile mills that are known to have been demolished, his corn mills and buildings for engineering and other industries, his waterwheels, steam engines and boilers, establishes him as the leading engineer in the field of industrial buildings during the half-century 1820–70.

Table 4.2: Existing Textile Mill Sites on the draft list of TICCIH, at which Fairbairn worked.

Country	Mill	Date of Fairbairn Involvement	Work done by Fairbairn (For main references see Appendix 2.1)
England	Murray Mill, Ancoats, Manchester (cotton)	1818	Renewal of millwork (shafting etc)
England	McConnel & Kennedy, Sedgwick Mill, Ancoats, Manchester (cotton)	1818, 1866	Layout, millwork. In 1866 moving of columns to give wider bays
Ireland	Malcolmsons, Portlaw (cotton)	1826	Design and other work, but full extent not known
Scotland	James Finlay & Co, Deanston (cotton)	1830	Waterwheels and lade
Ireland	Herdman's, Sion Mills (flax)	1835–9	Waterwheels and leat
France	La Foudre, Rouen (flax)	1845	Design and other work, but extent not known
Finland	Finlaysons, Tampere (cotton)	1849	Waterwheel and millwork, and probably extension
Russia	Steiglitz Flax Mill, Ivangorod (flax)	1850	Design, millwork, waterwheel
England	Saltaire (alpaca)	1851–3	Design, structural ironwork, millwork, engines

* For main references see Appendix 2.1.

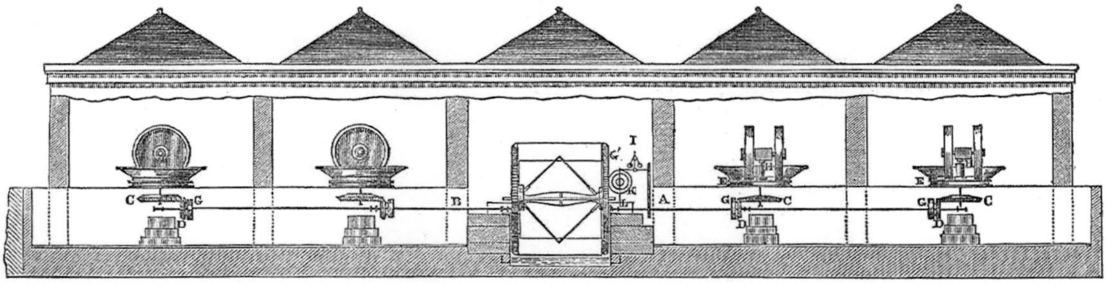

Gunpowder Mills

Fig.4.23: Gunpowder Mill, Waltham Abbey, mid 1850s.

WHEN steam engines were first introduced to drive powder mills they were sited about 200 yards from the mills in order to minimise the risk of explosion, with power transmission by compressed air. However it was found that over half the power of the engine was lost by this arrangement and it was largely abandoned. Where mills needed to be near to each other, they were separated by butts or mounds of earth. In the mid 1850s a new water-driven powder mill was built at Waltham Abbey, with which Fairbairn was involved. Again John Anderson was the client. Here the mills were separated by solid stone walls, each covered by a lightweight roof, intended in the case of explosion to save the adjoining mill. The mills were driven directly from the axle of the waterwheel, two on each side, with the shaft running beneath the floor, and a governor to control the speed. The engaging and disengaging gear was worked by a wheel on the outside of the building. All the gear wheels had wood and iron teeth as, in powder mills, 'iron working into iron is always attended by danger'. Fairbairn adds, 'How far this arrangement will answer the purpose has not been determined, as no explosion has taken place since the mills were finished'![238]

Corn Milling

IN 1850 *The Engineer and Machinist's Assistant* saw Fairbairn occupying 'a most distinguished position' in the development of corn-mills and consequently 'extensively employed' in their erection throughout Great Britain and the Continent.[239] Further afield, and an indication of his standing, an Australian newspaper advertisement for the sale of a corn-mill advised that, 'The works of Wm. Fairbairn, of Manchester, and C. and O. Evans of Philadelphia, were constantly consulted in its erection'.[240] At the Great Exhibition, Fairbairn's flour-mill exhibit – 'Improvements in the manner of driving, in the means employed for adjusting and regulating the grinding stones, and in the means of feeding' – won a Council Medal.[241] At the base of Fairbairn's flour mill there was a substantial cast iron framework which supported the main horizontal driving shaft and the vertical spindle. The framework was surmounted by a funnel-shaped casting which supported the lower fixed millstone. The upper millstone revolved with the spindle. Both millstones were enclosed in a sheet-iron casing. Within the base framework there was a screw mechanism which enabled the particular pair of stones to be disengaged.[242] Above the stones, and held by a tripod, was 'the silent feeding apparatus' which Fairbairn claimed to have introduced.[243]

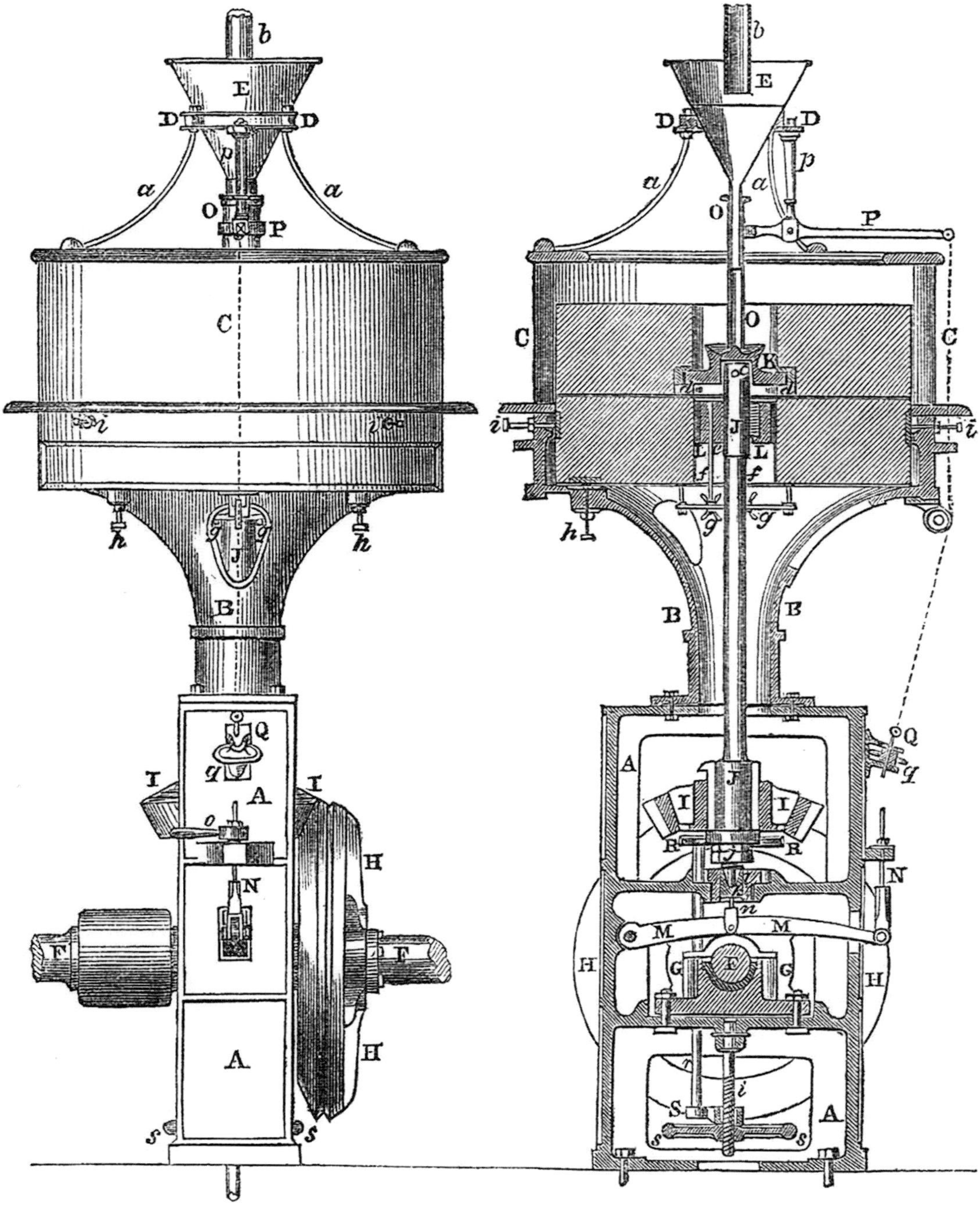

Fig.4.24:
Fairbairn's Flour Mill.

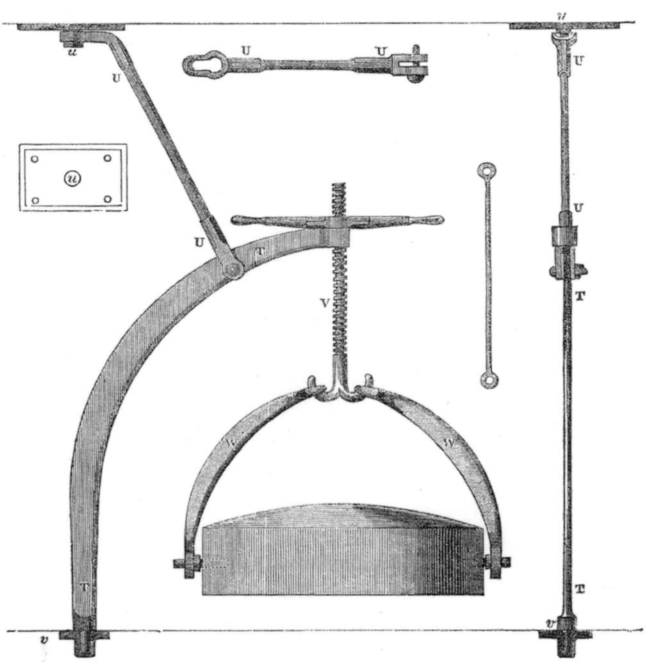

Fig.4.25: **Fairbairn's Stone-lifting Apparatus.**

Ancillary equipment included elevators, consisting of long endless chains of small buckets formed of tin plate, mounted on a leather or canvas band, passing over pulleys in cast-iron housings at top and bottom and enclosed with wooden boxing;[244] and 'creepers', made by Fairbairns, consisting of a long Archemedian screws enclosed in wooden boxes or troughs.[245] Fairbairn is credited with having improved the earlier 'rude, cumbrous and expensive devices'.[246] The portable stone lifting apparatus provided by Fairbairns consisted of a near quadrant shaped arm and a rod. The former turned in a step in a cast-iron floor plate and the rod was fixed to a stud on the beam above. These were located so that one position served two mills. At the end of the arm there was a screwed rod which enabled the running stones to be raised. They could then be swung round so that they were above the floor, inverted, and lowered to the floor, so that the stone-dressers could work on them.[247]

In 1851 a fire destroyed the corn mill of J. & J. Norton in Wolverhampton. Fairbairns were engaged to rebuild the mill as a state-of-the-art five-storey 'fire-

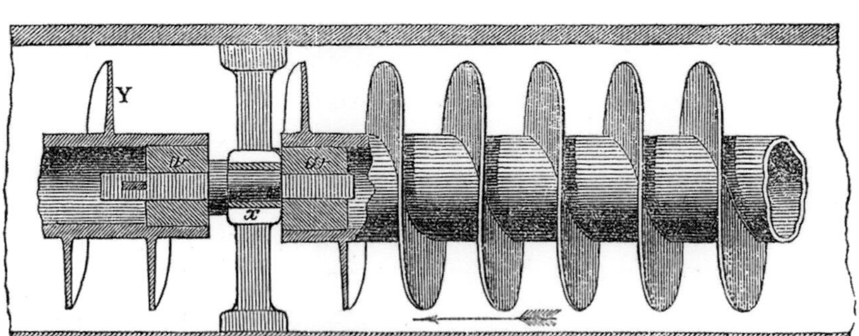

Fig.4.26: **Fairbairn's 'Creeper'.**

Right and below:
Figs 4.27 & 4.28: Wolverhampton Corn Mill in 1961.

Right:
Fig.4.29: Wolverhampton Corn Mill in 1981.

proof' mill with 18 pairs of millstones, albeit on a constricted site. The engines had been installed by Fairbairns in 1846,[248] and were 'comparatively uninjured'. [249] The new building was increased in size by building over the canal basin so that unloading of corn and loading of flour could take place within the mill and coal barges could pass under the mill to the side of the boiler house. Fairbairn's preferred arrangement was millstones in a straight line but the18 pairs of stones would not fit in one line, and examination of the structure and plans suggests there were two parallel rows of millstones.[250] The roof is of cast- and wrought-iron trusses, supported on the main brick walls, and is partially glazed. Sadly this fine building was demolished in 2015 despite efforts to save it by the Victorian Society and others.[251] This followed fire damage to the eastern end of the building in or around 2009 – significantly to the bays where the brick arch floors had been removed around 1910 to install silos.[252]

The point was made in Chapter 2 that the quality of Fairbairn's work led to large and prestigious projects, but it is incorrect that he was concerned *only* with such projects. Hipwell & Sons were the proprietors of the modest water-powered corn mill at Sharnbrook on the Ouse. In 1853 they wanted a new waterwheel and

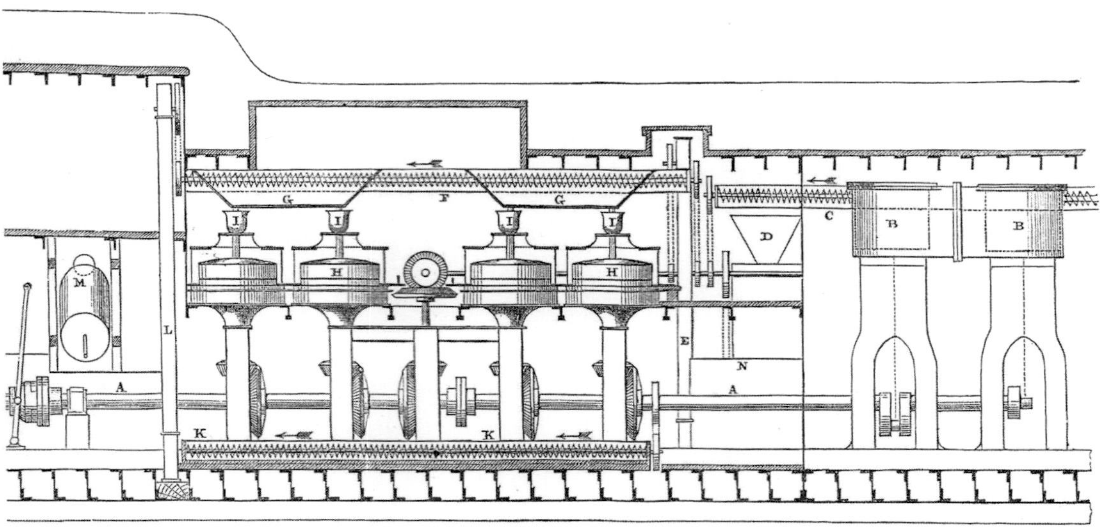

Fig.4.30: HMS *Bruiser* – part Long Section.

machinery. They contacted Fairbairns. William, now aged 64, was as busy as usual. Saltaire was nearing completion, there was work in various overseas countries, including mills, bridges and cranes. And yet on 18 April he wrote to Hipwells:

> Gent^{m}.
> My presence in London will not be needed until Thursday next and if perfectly convenient I would visit Stoke Mills on the following Saturday. Will you inform me if that day will suit you in a note which will find me at 5 Duke Street, Westminster.[253]

Here is a Fellow of the Royal Society, a member of l'Academie des Sciences and of the Athenaeum Club,[254] who is not short of money, visiting a modest corn mill in rural Bedfordshire on a Saturday – four years before the railway got there![255]

Fairbairn flour-mills were commissioned by John Anderson for a unique project for the Crimea. Once again a peer engineer, facing a novel challenge, turned to Fairbairn. Two ships were fitted out by Fairbairns in less than twelve weeks, one as an automated corn-mill with four pairs of stones, and the other a bakery. With 80hp engines by Robert Stephenson & Co, they steamed to Balaclava, produced 24,000lb. of flour per day from poor local wheat, and baked 6,000 3lb.-loaves per day. Wheat could be ground even when the vessel was at sea in a heavy swell.[256] They were an eminent success.

Fairbairn brought traditional milling with millstones to the height of its development with a steam-driven mill with thirty-six pairs of millstones at Taganrog in Russia – designed before the Crimean War and probably built after it. This was one of the largest and most mechanised corn-mills of the mid-19th century and represents the culmination of fifty years of Fairbairn's corn-mill work. The mill was four stories high, plus an attic. Its upper floors were, untypically in a Fairbairn mill, of timber. The grain entered the mill from the nearby granary by a chute below ground floor level and was carried to the top floor by a continuous bucket elevator. In a typically Fairbairn arrangement the millstones were in a single row on the first

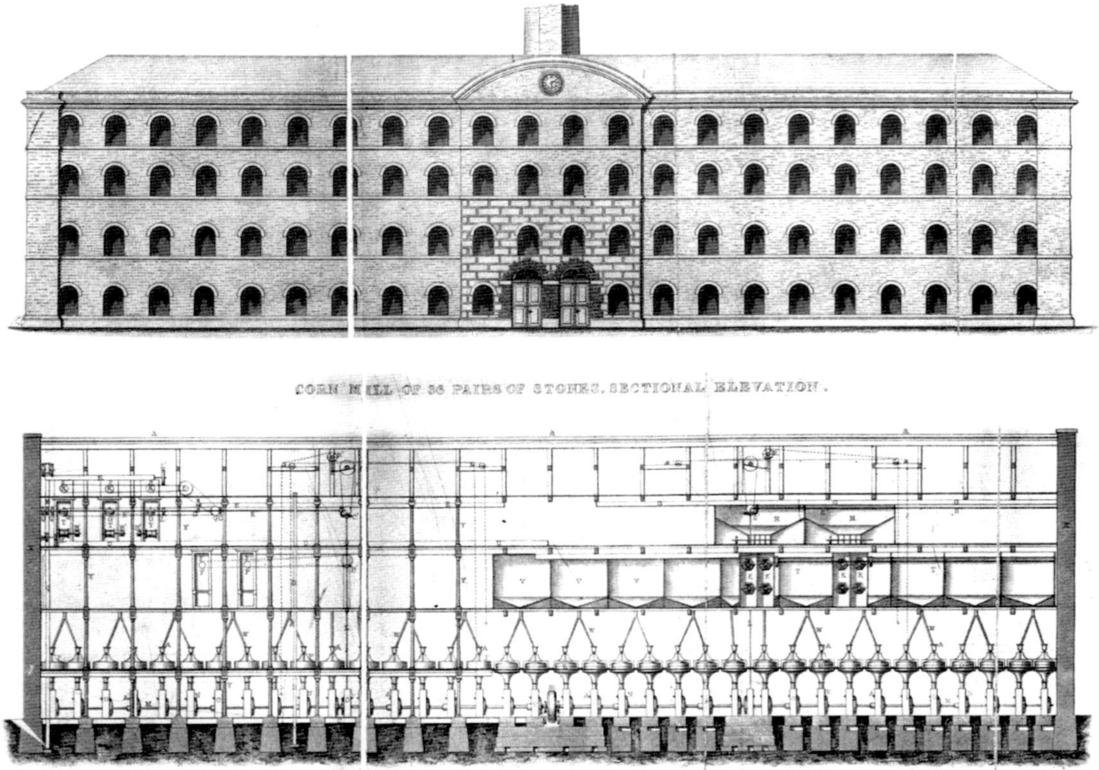

Fig.4.31: Corn-mill with 36 pairs of millstones, Taganrog, Russia. Designed c.1852. Built c.1860.

floor, driven by shafting and gearing on the ground floor. The whole mill was a mechanised process on a grand scale.[257]

Fairbairn did not introduce mechanisation or automation to corn-milling. These went back to the previous century, to Boulton & Watt and Rennie at the Albion Mill in London, and to Oliver Evans in America.[258] What Fairbairn did for corn-milling, as in several other areas of his work, was to optimise by bringing together and improving all that was best in the practice of the day. For two thousand years man's staple diet had been milled with stones, but by 1860 millstones were near their end. The future lay with the radically different roller milling, which was faster, used less power and provided a better product, and in the development of which Fairbairn played no part.[259] His fame, in this sphere, rests on having been the engineer who brought traditional corn-milling to its zenith.

Coal Mining – Engines, Elevators and Experiments

FAIRBAIRN'S apprenticeship was with a colliery engineer and over forty years later he was still involved with colliery engineering. Reference was made in the previous chapter to his pumping engine at Verviers in Belgium, 1839–40. A substantial pumping engine was installed at Duke Pit in Oldham in 1846.[260] Fairbairn's best known work in this field was at Astley Pit, Dukinfield, begun in 1847, and significant in a number of respects. Firstly, and typically, he was working at what was the deepest pit of its day, on what the *Manchester Guardian* called 'one of the most stupendous works undertaken in the annals of mining'.[261] His work included innovative pumping and winding engines, thought to be the most powerful of their time, and encompassing the transfer of technology from marine engines to mine engines. It also involved aspects of the development of the elevator which have been largely ignored. Finally this work provides an illustration of how Fairbairn's engineering work meshed with his experimental work – in this case working with leading scientists and with the British Association.

The pumping engine at Astley was on the Cornish principle, whereby the engine raised the plungers and pump-rods, and then, as these descended, they forced the water up to the next level. However, instead of the almost invariable working beam above the cylinder, Fairbairn substituted two beams – one counterweighted – below the cylinder, resting on a platform level with the ground, thus saving the expense of a high building and massive masonry. This was similar to the marine engines developed by Fairbairn for textile mill use, but without the flywheel.[262] There were sets of plunger pumps, each raising the water 200ft. Each stroke of the engine raised 30,000 gallons. Fairbairn described the engine to the Institution of Mechanical Engineers.[263] His paper and illustrations were reproduced in *The Engineer*, and in *The Mechanics' Magazine.*[264] The latter drew a response from an engineer in Edinburgh, 'The arrangement is not novel', and details were given of five similar earlier engines.[265] Whether Fairbairn's pumping engine was a development of his marine engines of the 1830s, or the development of a more recent engine by someone else is unclear. In any event the gifted engineer Charles Beyer, having viewed Fairbairn's engine at work, thought it 'one of the finest pumping engines he had seen'.[266]

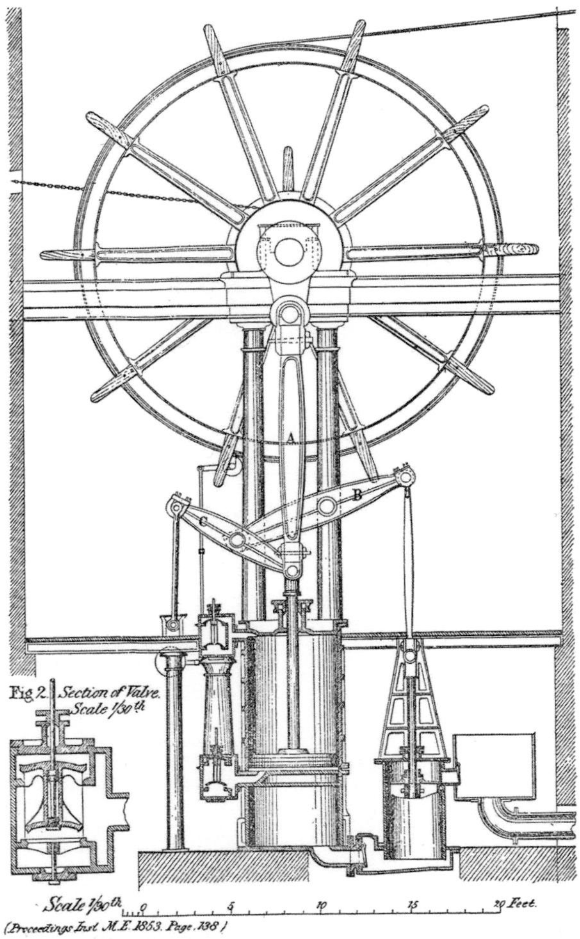

Fig.4.32: Winding Engine, Astley.

The winding engine was within a 50ft-high engine-house. Four massive cast-iron columns rose to an entablature which supported the crankshaft and two fly-wheels, on the peripheries of which wound the 'flat wire ropes'. The engine was of the 'direct action' type with 'grasshopper' motion, similar to Fairbairn's widely acclaimed marine engines for the frigates *Vulture, Odin* and *Dragon.* Again there was transference of technology from marine engines to a mine engine, although the pumping and winding engines at Astley were very different from each other, as illustrated.[267] The winding engine was also the subject of a paper to the Institution of Mechanical Engineers.[268]

Fairbairn had hands-on involvement with elevators in textile mills and coal

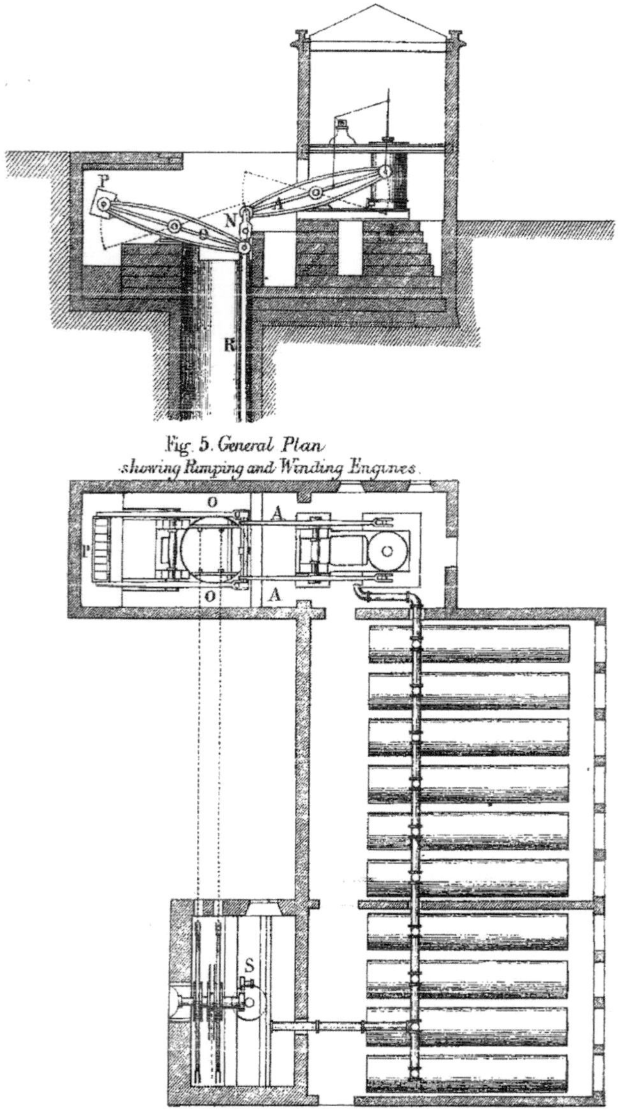

Fig.4.33: Section through Pumping Engine, Astley, and plan showing pumping engine, winding engine *(bottom left)* and range of 10 boilers.

mines but little acknowledgement has been given to any British engineers in the development of the elevator, which was included in a recent list of the *100 Greatest Science Inventions of All Time,* because it has enabled modern cities to be built by making tall buildings accessible.[269] Its invention has been attributed to James Bogardus and to Elisha Otis,[270] but its initial development – incremental and cumulative – took place in two parallel but discrete areas of British industry in the first half of the 19th century, textile mills and coal mining. In both of these industries Fairbairn had hands-on involvement with elevators. Strutt's early 19th-century North Mill at Belper had a steam-driven 'crane' with a counter-weighted basket to convey workers to the upper floors. It was made by Frost of Derby.[271] In 1835 Fairbairn included two developed examples of this lift at Orrell's Mill, illustrated by Ure in 1835 with a sectional drawing showing passengers in the car.[272] This was immediately reproduced in *The Penny Magazine,* including its American edition published in Boston and New York, nearly two decades before Bogardus and Otis.[273]

At Astley, Fairbairn's winding engine was larger than any previously built. The two 'cages' counterweighted each other, lifting coal to the surface and conveying personnel to and from the workings. Each cage, carrying four 8cwt coal boxes, was raised in one minute, at 20mph over the equivalent of a 200-storey building – and this three years before Bogardus and Otis.[274]

Astley Pit also illustrates how there was sometimes a link between Fairbairn's engineering work, his experimental work and the British Association. At the Association's 1847 meeting the mathematical geologist, William Hopkins, presented a paper about the fluidity, solidification, form and thickness of the earth's crust.[275] Wishing to experiment on the influence of pressure on solidification, he asked if Fairbairn's lever was still available, which it was, and a grant was obtained for the experiment, subject to Joule being involved.[276] Using the lever, pressure was put on various substances in a brass cylinder, which was then heated. William Thomson, a former pupil of Hopkins (and later to become

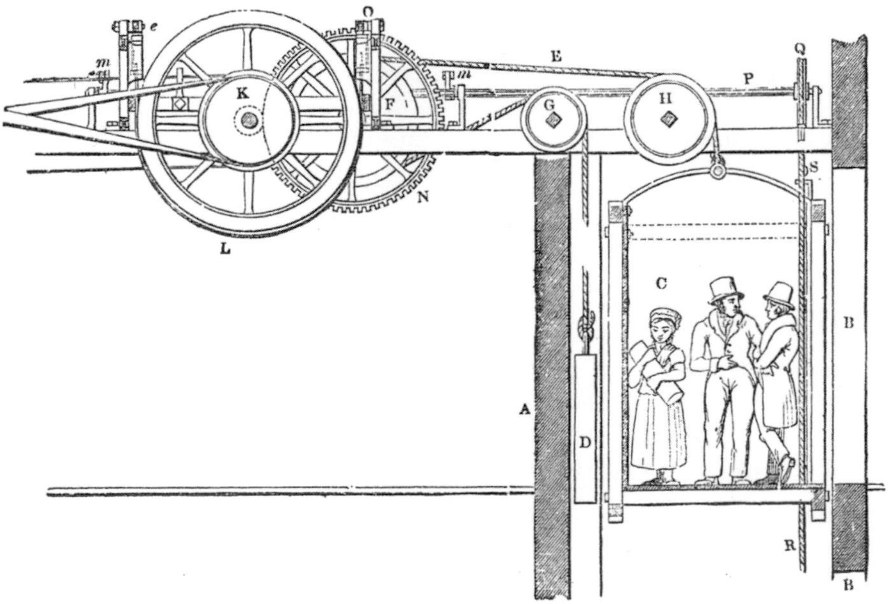

Fig.4.34: Ure's Illustration of an Elevator at Orrell's Mill, Stockport, 1835.

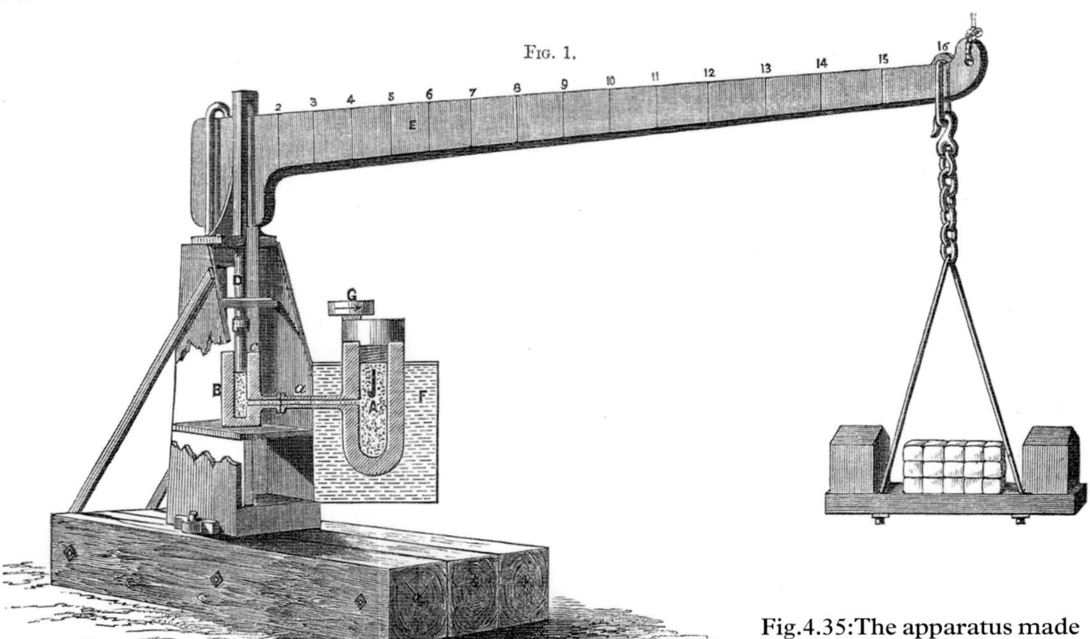

Fig.4.35: The apparatus made by Fairbairn and used by Hodgkinson, as adapted for the experiments of Hopkins, Joule, Fairbairn and Thomson.

Lord Kelvin), designed a device for the experiments to show when fusion took place.[277] The results were encouraging. Joule wrote to Fairbairn on 12 August 1853:

> Our experiment this morning was satisfactory, the needle having fallen at 373, the exact temperature expected by Mr Hopkins, and which shows that the temperature of fusion rises with the pressure in arithmetic progression.[278]

In his Presidential Address to the British Association in 1853 Hopkins referred to the experiments in which 'Mr Fairbairn, Mr Joule and myself ... have been engaged at Manchester'.[279] Subsequently both Hopkins and Fairbairn read papers on the experiments to the British Association.[280] In a paper to the Royal Society in1857, Hopkins

> felt bound to express in the strongest terms my obligations to my friends Mr Fairbairn and Mr Joule. Without the aid of the former of these gentlemen I should have been unable even to commence the series of experiments which I have now nearly concluded.[281]

An important aspect of this research was that it accorded with the thermodynamics research of William and James Thomson, and provided a vital part of the context for William Thomson's studies of the cooling of the earth.[282]

Fairbairn also measured the temperatures of the strata at various depths during the sinking of the Astley shaft, finding a rise of 1°F per 76.8ft.[283] He read papers on this work,[284] and his temperature measurements were reproduced in Hull's standard work, *The Coal-Fields of Great Britain.*[285] These various experiments enhanced Fairbairn's prestige, not least through his association with men of the standing of Hopkins, Joule and Thomson. His engineering work at Astley was drawn to the notice of a wider audience when Astley was one of the offered excursions at the British Association meeting in Manchester in 1861 under Fairbairn's presidency, and members, accompanied by several ladies, descended the shaft and 'drank the Queen's health deep beneath the earth's surface'.[286]

Conclusion

FROM the mid-1840s Fairbairns was a family business as at least four of the sons, without the benefits of premium apprenticeships or university education, joined their father. Here would seem to be the dynastic spur identified by Schumpeter and Perkin.[287] Here too was succession without recruitment of people of ability from outside the family, providing support for Mary Rose's advocacy of the need to plan creatively for leadership succession.[288] There is little evidence of the roles of the sons in the business other than that Thomas, the most able of them, was at Millwall for nearly eight years and there are strong indications that on his return to Manchester he ran the locomotive department. He also played a significant part in the engineers' dispute of 1851–2. At Ancoats the firm designed mills and manufactured steam-engines, boilers, millwork and locomotives, with tubular-girder bridges from 1847 and cranes from 1850. There are no records of profit margins or the scale of profits being made. There is, however, ample evidence that the work at Ancoats was sufficiently profitable to make good the substantial losses at Millwall; that when the sons' influence began to predominate, profits began to accumulate; and that the tubular-girder bridges were highly profitable at the end of the 1840s.[289] It is likely that the locomotive department was less profitable, as tendering for locomotives designed by railway company engineers was very competitive, although margins for overseas locomotives are likely to have been higher.

The family firm's construction of mills and millwork continued to enhance William Fairbairn's standing in this field. They ranged from the vast scale and quality of Salt's Mill to the advocacy of single-storey factories which set the pattern for the next century. The mills, together with their prime movers and power transmission, were exercises in optimisation, bringing together the best from every source. With so many examples of Fairbairn's mill work there is need for a re-appraisal of Tann's somewhat dismissive assessment.[290] Fairbairn's colliery engines provide a further illustration of the transfer of technology highlighted by Rosenberg and Vincenti, in this case from marine engines to colliery engines.[291] Attention has been drawn to colliery winding machinery in the history of the elevator and, for the first time, to Fairbairn's role in constructing the elevator with the longest and fastest travel at that time, a lacuna in Gray's work.[292] The firm's display at the Great Exhibition has been highlighted and there is little doubt that it was beneficial to the business.

Throughout this decade Fairbairn's experimental testing and measuring continued to establish his reputation in this field – beams at Saltaire, the causes of boiler explosions, temperatures at the Astley Pit, the effect of pressure on solidification, and, in the next chapter, his most famous experimental work – that which led to the Britannia Bridge. Active participation in the Institution of Civil Engineers and the British Association kept his name to the fore amongst peers – and a large amount of work came from peer engineers.

It may not have been until his sons' influence predominated that Fairbairn accumulated wealth, but it was their father's vision and impetus that generated that wealth. He did so through the mills he planned, the technology he transferred, his drive to ensure that his commissions incorporated the best available, and his development of derivatives from the tubular bridge experiments. It was their father who networked among potential clients and who publicised the company's achievements. The sons' training in engineering was meagre, but this was of small consequence whilst their father was in full spate. The situation would change dramatically within a few years of his retirement. The family business phase of Fairbairn's career ended on the highest of notes. Three sons inherited one of the largest, most respected, innovative, versatile and profitable engineering companies in Britain.

Notes for Chapter 4

1. The definition of a family firm adopted here is that referred to inthe Introduction - a firm where a family owns enough of the equity to be able to exert control over strategy and is involved in top management positions.
2. M B Rose, 'Beyond Buddenbrooks: the family firm and the management of succession in nineteenth-century Britain', in J Brown and M B Rose (eds), *Entrepreneurship, networks and modern business,* (1993), p.130.
3. Unpublished research by the author.
4. S Pollard, *The Genesis of Modern Management,* (1965), p.173.
5. J A Schumpeter, *The Theory of Economic Development,* (1934), p.93.
6. *Life,* p.472; [Woodcroft] X.
7. *Life,* pp.168, 223, 342, 363, 371.
8. *MG,* 22 July 1859. Or this may, just possibly, have been Fairbairn's brother, Peter.
9. Hayward, *Fairbairns,* p.6.9; *MG,* 26 March 1856; *The Gentleman's Magazine,* October 1859, 432.
10. *London Gazette,* 3 January 1845; *Life,* p.426; W Fairbairn to T R Robinson, 23 February 1867, Cambridge University Library, Stokes Papers, Add 7342, TR38.
11. *Life,* pp.317-9.
12. Rose, 'Beyond Buddenbrooks', pp.135-6.
13. *Life,* pp.317, 342, 329.
14. See Chapter 8.
15. Rose, 'Beyond Buddenbrooks', p.137.
16. See Chapter 7.
17. See Chapter 7; *Life,* pp.191-2; P Russell, 'John Frederic La Trobe-Bateman (1810-1889) Water Engineer', *TNS,* 52, 1880-81, 119-380.
18. *Life,* pp.465-6.
19. *Life,* p.339. See Chapter 3. The problems must have occurred during the second visit as the only frigate built by the yard was the *Megaera,* which originated from drawings submitted to the Admiralty by Fairbairns in 1843. (R A Hayward, *The Story and Scandal of HMS Megaera,* (1978), p.43).
20. *MPICE,* 36, 1872-3, 270; Hayward, *Megaera,* p.20.
21. *Life,* p.155.
22. *MPICE,* 101 Part 3, 1890, 295.
23. W Fairbairn, *Treatise on Iron Ship Building: Its History and Progress,* (1865), p.4; J H Clapham, *An Economic History of Modern Britain: the Early Railway Age, 1820-1850,* (2nd ed.1939), p.440.
24. *Life,* p.335.
25. *Life,* p.341. Murray went on to a successful career, becoming Chief Engineer at Portsmouth Dockyard (*MPICE,* 36, 1872-3, 271-2).
26. *London Gazette,* 20278, 10 November 1843, p.3669.
27. *Life,* p.342.
28. *Liverpool Mercury,* 12 February 1844.
29. *Life,* pp.341-2. Arnold sees the £100,000 as the operating loss, in addition to the capital loss, but this would appear to be a misreading of his source. (A J Arnold, *Iron Shipbuilding on the Thames, 1832-1915,* (2000), p.27).
30. [Woodcroft] VIII.
31. *Life,* p.336.
32. D Bell (ed.), *David Napier, Engineer, 1790-1869: An Autobiographical Sketch with Notes,* (1912), p.47.
33. J Glover, 'On the Decline of Shipbuilding on the Thames', *Journal of the Statistical Society of London,* 32.3, 1869, 289.
34. Arnold, *Iron Shipbuilding,* pp.26, 152.
35. Hayward, 'Fairbairns', p.5.11.
36. *The Engineer,* 1, 1856, 69, 196.
37. P Banbury, *Shipbuilders of the Thames and Medway,* (1971), pp.246-7; Arnold, *Iron Shipbuilding,* pp.53, 59-61.
38. S Pollard, 'The Decline of Shipbuilding on the Thames', *The Economic History Review,* NS 3.1, 1950, 72, 76; Arnold, *Iron Shipbuilding,* pp.153-4; S C Checkland, *The Rise of Industrial Society in England 1815-1885,* (1864), pp.45, 147; Banbury, *Shipbuilders,* p.70.
39. *Life,* pp.154-5.
40. During the steam age over 32,000 steam railway locomotives were built in the north-west of England, 15,000 of them in Manchester.
41. D K Clark, *Railway Machinery: A Treatise on the Mechanical Engineering of Railways: embracing the Principles of Construction of Rolling and Fixed Plant,* (1855), p.201.
42. Fairbairn, *UIfE2,* p.241.
43. R H G Thomas, *The Liverpool and Manchester Railway,* (1980), p.157.
44. Thomas, *Liverpool and Manchester Railway,* pp.109-13.
45. See Chapter 2.
46. E Craven, 'The Early Locomotives of the Lancashire and Yorkshire Railway', *Journal of the Stephenson Locomotive Society,* 32, 1956, 187; J Marshall, *The Lancashire and Yorkshire Railway,* Vol.3, (1972), p.29.
47. R J Byrom, 'Fairbairn Railway Engines', (unpublished).
48. [G F Burtt], *The Locomotives of the London, Brighton & South Coast Railway, 1839-1903,* (1903), p.28; C H Ellis, *The London, Brighton & South Coast Railway,* (1960), p.34.
49. Marshall, *Lancashire and Yorkshire,* p.23.
50. Marshall, *Lancashire and Yorkshire,* pp.59-60, 217-8; Rush, *Lancashire and Yorkshire,* pp.26-8.
51. Marshall, *Lancashire and Yorkshire,* pp.60-3; Rush, *Lancashire and Yorkshire,* p.40.
52. [C] H Ellis, *The South Western Railway: Its Mechanical History and Background 1838-1922,* (1956), pp.51-2.
53. E L Ahrons, *Locomotive and Train Working in the Latter Part of the Nineteenth Century, Volume 6,* (1954), p.50; W E Shepherd, *The Midland Great Western Railway of Ireland: An Illustrated History,* (1994), p.17.
54. Ahrons, 'Famous Firms', p.184.
55. P Ransome-Wallis, *Illustrated Encyclopaedia of World Railway Locomotives,* (2001 ed.), p.499; See Chapter 6.
56. R L Hills, *Life and Inventions of Richard Roberts 1789-1864,* (2002), pp.192-3; R L Hills and D Patrick, *Beyer Peacock: Locomotive Builders to the World,* (1982).
57. H Jack, *Locomotives of the LNWR Southern Division: London & Birmingham Railway, London & North Western Railway and Wolverton Locomotive Works,* (2001), pp.212-6.
58. Ahrons, 'Famous Firms', p.184; Jack, *Locomotives of the LNWR,* p.217.
59. Clark, *Railway Machinery,* pp.198-202 and Plates 1-3.
60. Clark, *Railway Machinery,* p.201.
61. D L Bradley, *Locomotives of the London and South Western Railway, Part 1,* (1965), pp.35-8.
62. Bradley, *Locomotives of the London and South Western Railway,* pp.29-30, 36. There is however some confusion surrounding these engines. Kidner and Marshall say that

four of the five were *sold* to Fairbairns in 1842, and that five, apparently new, engines were delivered in 1843 (C F D Marshall and R W Kidner, *A History of the Southern Railway, Volume 1,* (2nd ed. 1963), pp.159-60).
63. C Hadfield, *Atmospheric Railways: A Victorian Venture in Silent Speed,* (1967), p.41.
64. Hadfield, *Atmospheric Railways,* p.108.
65. W Fairbairn. 'On Railway Gradients', *BAAS1845,* p.s94.
66. *MPICE,* 90.4, 1886-7, 421. See also M Beaumont, *Sir John Hawkshaw 1811-1891: The Life and Work of an Eminent Victorian Engineer,* (2015), p.45.
67. *Mining Journal,* 16, 1846, 159.
68. *The Engineer,* 9, 1860, 52; Marshall, *Lancashire and Yorkshire,* p.60; Hayward, 'Fairbairns', p.2.15. The new works was in Mill Street, Ancoats, a short walk from their other premises. The work caused delays to an order for twenty-three 0-4-0s for the Manchester & Leeds Railway.
69. *MG,* 24 December 1847.
70 . *The Engineer,* 1, 1856, 324.
71. W Fairbairn to R Stephenson 26 July1847, No.291, Stephenson Papers, Institution of Civil Engineers.
72. T Fairbairn to R Stephenson 18 August 1847, Nos.312-4, Stephenson Papers, Institution of Civil Engineers.
73. See Chapter 5.
74. *MG,* 6 April 1850.
75. *Life,* p.317. There is sometimes confusion between the 'Fairbairn Tank' and the LMS 'Fairburn Tank', designed by Charles Fairburn and built 1945-51.
76. By Forrester on the Dublin & Kingston Railway and the London & Greenwich Railway, by Sharp Brothers on the Manchester & Birmingham Railway, and by Jones & Potts on the London & Blackwall Railway (E L Ahrons, *The British Steam Railway Locomotive from 1825 to 1925,* (1927), pp.43, 83; Lowe, *British Steam Locomotive Builders,* p.176).
77. Hadfield, *Atmospheric Railways,* p.108; Lowe, *British Steam Locomotive Builders,* p.176 – they were built in 1834 and converted to tank locomotives 1837-40. Forrester ceased locomotive building c.1847.
78. K H Vignoles, *Charles Blacker Vignoles: Romantic Engineer,* (1952), pp.48ff, 146-7.
79. *1851 Great Exhibition: Official Catalogue,* Class V, No.732. Fairbairn was either boosting his claim as the inventor of the tank engine, or seeking a public persona as an inventor, or both.
80. *ILN,* 9 Aug. 1851, 195; *The Engineer and Machinist,* 3, 1851, 197; *The Practical Mechanic's Journal,* 4, 1851-2, 271-2 and Plates 89-91.
81. H Fayle and A T Newham, *The Waterford & Tramore Railway,* (2nd ed. 1972), pp.22-6.
82. Fayle and Newham, *Waterford & Tramore Railway,* pp.44-5.
83. R W Rush, *The Furness Railway 1843-1923,* (1973), pp.64-5, 108. The withdrawal dates given by Rush for at least Nos 9 and 15 – 1901 and 1899 respectively – are incorrect as both these engines were photographed by Harold Hopwood in 1902 (B Arman, 'The H L Hopwood Collection 1901-1926. Part 2: The Furness Railway at Barrow in 1902', *Railway Archive,* 19, 2008, 44-7).
84. *MPIME,* 1858, 166-70; *The Engineer,* 6, July-Dec.1858, 171-2; Ahrons, *British Steam Railway Locomotive,* pp.126-7.
85. On McConnell see Jack, *Locomotives of the LNWR,* pp.49-70.
86. R H Parsons, *A History of the Institution of Mechanical Engineers 1847-1947,* (1947), pp.10-1, 287.
87. R A Hayward, 'The Fairbairn Engineering Company Limited – An Unsuccessful Victorian Limited Liability Venture', *Manchester Literary and Philosophical Society Memoirs and Proceedings,* 115, 1972-3, 62, 71.
88. See Chapter 8; Jack, *Locomotives of the LNWR,* p.68.
89. Jack, *Locomotives of the LNWR,* pp.177-9.
90. Jack, *Locomotives of the LNWR,* pp.189-93.
91. *ILN,* 18 December 1852, 552.
92. quoted in Jack, *Locomotives of the LNWR,* p.193.
93. Ahrons, 'Famous Firms', p.185.
94. In W Johnston, *Imperial Cyclopaedia of Machinery,* (1855) – referred to in Jack, *Locomotives of the LNWR,* p.195.
95. Jack, *Locomotives of the LNWR,* p.195.
96. *MG,* 26 July1854.
97. *MM,* 63, Jul.-Dec. 1855, 35.
98. *MG,* 1 October 1855.
99. *MG,* 3 March 1855. The others were the Emperor Napoleon III, Mark Philips, John Platt, Joseph Whitworth, and 'a lady'.
100. W Fairbairn, *On the Comparative Merits of the Machinery of the Paris Universal Exhibition, 1855,* (1855). Extracts in Fairbairn, *UIfE3,* pp.125-58.
101. *MM,* 63, Jul.-Dec. 1855, 483.
102. Fairbairn, *UIfE3,* pp.132-4.
103. Jack, *Locomotives of the LNWR,* p.195.
104. J Davies, 'List' of Fairbairn locomotives (unpublished).
105. T Vernon, *Archibald Sturrock Pioneer Locomotive Engineer,* (2007), pp.17, 60-1, 67, 72-3.
106. See Appendix 4.1.
107. P C da S Telles, *A History of Brazilian Railways, Part 1 – The First Railways,* (ET by P E Walters 1987), pp.16-7, 48. There is also a Fairbairn locomotive existing at Braga in Portugal but it appears to have been much altered (The Stephenson Locomotive Society Photographic Collection, List 16, (2009); http://geoff-plumb.fotopic.net (accessed 21 December 2010)).
108. Musson suggested that Pole's 'more than 600' was far too high. In this he relied on Ahrons, and Dewhurst (A E Musson, 'Introduction' to the 1970 facsimile edition' of *Life,* p.xiv). Ahrons had listed 392 and estimated a total of 'rather more than 400' with only 10 per cent for foreign railways. Dewhurst surmised that the missing ones could not have been more than twenty-five or so (Dewhurst, 'An Early Brazilian Locomotive', 15 March 1930, 80-1). Terence King, Mike Page and John Davies have all done subsequent research in this area but none of it has been published. King and Page, working independently of each other, listed 413 and 443 respectively (T King, WL 10308, Terence King Collection, Stephenson Locomotive Society Archive; M Page, WL8733/18, Stephenson Locomotive Society Archive). The most complete list is that by Dr J Davies of Queensland, who lists 451 including major rebuilds, and has graciously shared his research.
109. See Chapter 8.
110. E L Ahrons, 'Short Histories of Famous Firms No.XVII: Sharp, Stewart & Co Limited, originally of Manchester, later of Glasgow', *The Engineer,* 136, July-Dec.1923, 194-7; J Cantrell, *Nasmyth, Wilson & Co. Patricroft Locomotive Builders,* (2005), p.138.
111. Hills and Patrick, *Beyer Peacock.*
112. Appendix 4.1.
113. W Fairbairn to T Worthington, 20 October 1850, Manchester Central Library Archives M6/3 11/73. Worthington was at that time a young architect from a Cross Street family who was secretary to the Manchester Committee

(A J Pass, *Thomas Worthington: Victorian Architecture and Social Purpose*, (1988), pp.26-7).
114. *MG*, 22 June 1850, 4 January 1851, 18 January 1851.
115. *Official Catalogue of the Great Exhibition of the Works of all Nations*, (1851), Class V, Nos. 26, 417, 522, 732; Class VI, Nos. 200, 403, 421. No. 522 refers to a tank engine which appears to be duplicated at 732.
116. D Evans, *Building the Steam Navy: Dockyards, Technology and the Creation of the Victorian Battle Fleet 1830-1906*, (2004), p.89; Fairbairn, *UIfE2*, p.296.
117. *Manchester Examiner & Times*, 30 April 1851.
118. *The Times*, 12 January 1852.
119. *Life*, pp.323-4.
120. For example, in the North of England there were strikes at Fairbairn & Lillie 1825 (*MG*, 19 and 26 February 1825; *Manchester Gazette*, 19 February 1825); Dobson & Barlow 1831 (J B Jefferys, *The Story of the Engineers 1800-1945*, (1945), p.21; J A Cantrell, *James Nasmyth and the Bridgewater Foundry: A study of entrepreneurship in the early engineering industry*, (1984), p.227); Peter Fairbairn 1833 (*Fortunes Made in Business*, pp.272-3; E Gauldie (ed.), *The Dundee Textile Industry 1790-1885, from the papers of Peter Carmichael of Arthurstone*, (1969), p.47); Lillie 1834 (*MG*, 12 July 1834); Nasmyth 1836 (S Smiles, *Industrial Biography: Iron Workers and Tool Makers*, (1863), pp.293-4; Cantrell, *Nasmyth*, pp.240-2; William Fairbairn 1837 (*Life*, pp.163-4, 420); Jones & Potts, Newton-le-Willows 1846 (Jefferys, *The Engineers*, pp.26-7); Harrison, Blackburn 1851 (Cantrell, *Nasmyth*, p.228): Carr, Bingley, Milner & Ellis, Wakefield and Kitsons at Leeds (K Burgess, 'Trade Union Policy and the 1852 Lock-out in the British Engineering Industry', *International Review of Social History*, 17, 1972, 648-50).
121. Cantrell, *Nasmyth*, pp.224-8; A E Musson and E Robinson, *Science and Technology in the Industrial Revolution*, (1969), pp.489-509.
122. J Mokyr, 'The Political Economy of Technological Change: Resistance and Innovation in Economic History' in M Berg and K Bruland (eds) *Technological Revolutions in Europe: Historical Perspectives* (1998), p.55.
123. H Pelling, *A History of British Trade Unionism*, (2nd ed. 1971), p.50.
124. E Frow, 'The Engineers' Strike of 1852', *Marxism Today*, 7, 1963, 119; Jefferys, *The Engineers*, p.43.
125. *The Times*, 29 December 1851.
126. Jefferys, *The Engineers*, pp.35-6; J Burnett, (ed.), *Useful Toil: Autobiographies of working people from the 1820s to the 1920s*, (1974), p.310.
127. *The Times*, 17 December 1851.
128. *Life*, pp.323-4.
129. *The Times*, 22 December 1851. Louis Blanc (1811-82) - French politician and historian, who advocated co-operatives.
130. *The Times*, 25 December, 1851.
131. *The Times*, 27 December 1851.
132. *The Times*, 31 December 1851.
133. *The Times*, 01 January 1852.
134. *The Observer*, 04 January 1852.
135. *The Times*, 05 January 1852.
136. *The Observer*, 11 January 1852.
137. *The Observer*, 2 and 9 February 1852.
138. *Manchester Examiner & Times*, 7 February 1852.
139. *The Observer*, 9 February 1852; William Fairbairn & Sons, *Rules and Regulations to be Observed by All Persons in the Employment of William Fairbairn & Sons, Engineers, Manchester*, (1852), pp.4-5. The Declaration became known in trade union circles as the 'odious document' – see for example J Mitchell, *Organised Labor: Its Problems, Purposes and Ideals and the Present and Future of American Wage Earners*, (1903), p.43.
140. *MG*, 10 March 1852.
141. *The Observer*, 21 March 1852.
142. *The Times*, 22 March 1852.
143. *MG*, 27 March 1852.
144. R L Stevenson, 'Memoir of Fleeming Jenkin', in S Colvin and J A Ewing (eds), *Papers Literary, Scientific. &c. by the Late Fleeming Jenkin, FRS, LLD*, (1887), p.l.
145. [Jenkin, H CF,] 'Trade-Unions: How Far Legitimate', *North British Review*, 9(NS), 1868, 1-62. Reprinted in *The Graphic Representation of the Laws of Supply and Demand, and other Essays on Political Economy*, (1931), pp.1-75.
146. E J Molesworth (ed.), *Life of Sir Guilford L Molesworth KCIE*, (1922), pp.20-2 - the dates here are incorrect as the ASE was not formed until 1851. Towards the end of his very long life Moleworth wrote a polemic against socialism, G [L] Molesworth, *The Curse of Socialism*, (1918).
147. *The Observer*, 11 April 1852.
148. *The Times*, 09 April 1852.
149. *The Times*, 09 April 1852.
150. *The Observer*, 25 April 1852.
151. T Hughes, *Account of the Lock-out of Engineers, &c. 1851-2*, (1860), p.14. This was prepared for the National Society for the Promotion of Social Science, at the request of the Committee of Trades Unions.
152. *Life*, p.325.
153. Smiles, *Nasmyth*, op. cit., p.298.
154. Hughes, *Account*, p.22.
155. *The Observer*, 23 February 1852.
156. *The Observer*, 15 February 1852; Jefferys, *The Engineers*, p.44; Hughes, *Account*, p.22.
157. W Fairbairn to A Morin, quoted in *Life*, pp.326-7. At the time Morin was Director of the Conservatoire National des Arts et Métiers, where in 1852 he opened the first engineering teaching laboratory. http://www.gap-system.og?~history/Biographies/Morin.html (accessed 17.11.11).
158. Jeffreys, *The Engineers*, p.44.
159. Hughes, *Account*, p.22.
160. Hughes, *Account*, p.22.
161. S and B Webb, *A History of Trade Unionism*, (1894), p.183.
162. N Rosenberg, *The American System of Manufactures. The Report of the Committee on the Machinery of the United States 1855 and the Special Reports of George Wallis and Joseph Whitworth 1854. Edited with an Introduction by Nathan Rosenberg ...* , (1969), pp.387-9.
163. Fairbairn, *UIfE2*, pp.213-4.
164. See below.
165. K Burgess, 'Technological Change and the 1852 Lock-out in the British Engineering Industry', *International Review of Social History'*, 14, 1969, 215-36; K Burgess, 'Trade Union Policy and the 1852 Lock-out in the British Engineering Industry', *International Review of Social History*, 17, 1972, 645-60.
166. P J Murphy, 'The Origins of the 1852 Lock-out in the British Engineering Industry Reconsidered', *International Review of Social History*, 23, 1978, 265-6.
167. Fairbairn, *M&MWI*, p.ix. See also *Amicus* in *The Times*, 11 February 1852.
168. House of Commons Debate, 7 July 1862, quoted in S Smiles, *Industrial Biography*, (1863), pp.331-2.
169. *MG*, 22 November 1851.

170. Quoted in A W Ward, 'Introduction' to E Gaskell, *North and South,* (Vol.IV of *The Works of Mrs Gaskell,* 'The Knutsford Edition', 1906).
171. Having inevitably lost his job when the strike failed, Higgins pleaded (successfully due to Margaret Hale's intervention) for re-employment with Thornton, to feed the orphaned children of Boucher whose intemperate violence, in the face of Thornton importing Irish labour to break the strike, had been one of the causes of the strike's failure.
172. Robinwood Mill at Todmorden (1842-3) and the Taganrog corn mill (see below) were exceptions.
173. E Jones, *Industrial Architecture in Britain 1750-1939,* (1985), pp.95-7. They had just won the competition for Bradford's St George's Hall, but as yet were relatively unknown.
174. R Balgarnie, *Sir Titus Salt, Baronet: His Life and its Lessons,* (1877), pp.117-20.
175. R S Fitzgerald, 'The Development of the Cast Iron Frame in Textile Mills to 1856', *IAR,* 10.2, 1988, 143. Giles, C and Goodall, I H, *Yorkshire Textile Mills: The Buildings of the Yorkshire Textile Industry 1770-1930,* (1992), many references to Saltaire.
176. Fitzgerald, 'Development of the Cast Iron Frame', 143.
177. Fairbairn, *Application,* Plate II. There were precedents for this at Strutt's 1792 mill and in the 1830s by Telford at St Katherine's dock (Fitzgerald, 'Development of Cast Iron Frame', 143).
178. The boilers consumed fifty tons of coal a day. There must have been problems with them as they were replaced after only ten years by eighteen Galloway boilers (*Engineering,* 11, 1866, 248).
179. Fairbairn, *Application,* pp.173-6.
180. J James, *History of Worsted Manufacture in England,* (1857), pp.467-9.
181. W G Rimmer, *Marshall's of Leeds, Flax-spinners 1788-1886,* (1960), p.206; C Tomlinson, *The Useful Arts and Manufactures of Great Britain – The Manufacture of Linen Flax,* (no date, but mid nineteenth-century), p.2.
182. Smith, *Contribution,* pp.2-3.
183. Fairbairn, *Application,* pp.169-71.
184. *ILN,* 1 October 1853, 287-9.
185. Fairbairn, *Application,* pp.165-76 and Plates 1 and 2 (the 3rd edition (1864) also contained a fold-out plate as frontispiece); C Condit, 'Buildings and Construction', in M Kranzberg and C W Pursell Jr (eds), *Technology in Western Civilisation,* (1967), Vol.1. p.376; W Fairbairn, *De l'Application de la fonte, du fer et de la tôle dans les constructions,* (translated by L Perret-Porta 1857); W Fairbairn, *Allicacion del hierro a las construcciones,* (translated by D E Saavedra 1857), with *Suplemento,* (1859).
186. Balgarnie, *Salt,* pp.149-50.
187. Fairbairn, *Application,* p.166.
188. R N Holden, *Manufacturing the Cloth of the World: Weaving Mills in Lancashire,* (to be published), p.53; R S Fitzgerald, 'Structural Development', in M Binney, R S Fitzgerald, K Langenbach and K Powell, *Satanic Mills – Industrial Architecture of the Pennines,* (1979), p.66; *M&MWII,* p.115.
189. Fairbairn to Gott, 19 February 1829, MS193, Gott Papers, Brotherton Library, Leeds; *Fortunes Made in Business,* Vol.1, p.420.
190. I Haynes, *Stalybridge Cotton Mills,* (1990), p.33; A Ure, *The Cotton Manufacture of Great Britain,* Vol.1, pp.313-4 - here Ure refers to 40,000 mule spindles and 1,250 power looms.
191. R Pollard and N Pevsner, *Buildings of England: Lancashire: Liverpool and the South West,* (2006), p.675. This mill features on Eyre Crowe's oil painting, '*The Dinner Hour, Wigan*' (1874) in Manchester City Art Gallery.
192. J Combe, 'Description of a Flax Mill recently erected by Messrs. Marshall & Co at Leeds', *MPICE,* 2,1842, 142; *The Penny Magazine,* Supplement June 1843, 247-8; Jones, *Industrial Architecture,* pp.102-3. An alternative multi-storey proposal 'by Fairbairn' was rejected. Rimmer assumes this was Peter Fairbairn of Leeds, who regularly worked with Marshalls (Rimmer, *Marshall's of Leeds,* p.202) but Jenkins refers to the plans as submitted by William Fairbairn, (D T Jenkins, *The West Riding Wool Textile Industry 1770-1835,* (1975), p.63), which seems likely, given Fairbairn's statement, 'Messrs. Marshall, with whose works I have been professionally connected for some years' (Fairbairn, *M&MW2,* p.193.
193. Fairbairn, *M&MWII,* pp.188-92 and Plates 17 and 18; W Fairbairn, 'Description of a Woollen Mill erected in Turkey', *MPICE,* 3, 1843, 125-6; *MG,* 22 March 1843, 20 December 1843.
194. *MG,* 22 March 1843.
195. Herbert writes that, 'Many manufacturers from Laycock (1843) onwards attempted to cope with the heat ... problem' (G Herbert, *Pioneers of Prefabrication: The British Contribution in the Nineteenth Century,* (1978), p.197n99). However a doubt lingers as to whether the sandwich form was actually built at Izmet as other sources indicate there was a plaster lining (Fairbairn, 'Description of a Woollen Mill', 125-6; *MM,* 39, July-Dec.1843, 175; *Life,* p.174).
196. Fairbairn, 'Description of Woollen Mill', 125-6; Fairbairn, *M&MWII,* pp.188-95 and Plates 17, 18. The weaving section contained both hand- and power-looms.
197. It is not clear which were the other mills that Fairbairns built in Bombay, but by autumn 1861 there were already six cotton mills there with several others under construction, and yet others in the planning (A W Silver, *Manchester Men and Indian Cotton 1847-72,* (1966), p.308). Fairbairn's first Bombay mill, for the Bombay Spinning and Weaving Company at Tardeo, was built 1854-5. Its machinery - it had 17,000 spindles - was from Hetheringtons in Manchester (S D Mehta, *TheCotton Mills of India 1854-1954,* (1954), pp.17-21).
198. Fairbairn, *M&MWII,* pp.179-87 and Plates 15 and 16. The mill was 379ft 6in. by 315ft 6in. and contained 23,040 mule and 8,604 throstle spindles, and 480 power looms (E Leigh, *The Science of Modern Cotton Spinning* (1871), p.49). The machinery was obtained from three different Lancashire firms, Platt Brothers, Hetheringtons and Sharp Roberts, in order to compare the relative efficiencies (D A Farnie, 'The Role of Cotton Textiles in the Economic Development of India, 1600-1990' in D A Farnie and D Jeremy (eds), *The Fibre that Changed the World: The Cotton Industry in International Perspective, 1600-1990s* (2004), p.402).
199. S M Rutnagur, *Bombay Industries: The Cotton Mills,* (Bombay 1927), pp.9-10; Fairbairn, *M&MWII,* pp.179-87 and Plates 15 and 16.
200. Leigh, *Cotton Spinning,* pp.49-51, 289-90, and Plates 14, 39. Leigh attributes the design to Hetherington, who presumably supplied the machinery, and misleadingly titles the plate 'Bombay Spinning and Weaving Company'.
201. At Alverthorpe (1870-2) and Yeadon (1888), (Giles and Goodall, *Yorkshire Textile Mills,* pp.107-8).
202. M Williams with D A Farnie, *Cotton Mills in Greater Manchester,* (1992), p.41; The Lancashire Cotton

Corporation Limited, *The mills and organisation of The Lancashire Cotton Corporation Limited,* (1951), p.21. Single-storey cotton-weaving mills included Martin's Mill, Rochdale (1850s) and Carrs Mill, Ashton-under-Lyne (1884), (Williams and Farnie, *Cotton Mills,* pp.77, 97).
203. R N Holden, *Stott and Sons: Architects of the Lancashire Cotton Mill,* 1998), p.119.
204. J Sawyer, 'The Social Basis of the American System of Engineering', *The Journal of Economic History,* 14, 1954, 377. There were modest precedents in late eighteenth century France (J Mokyr, 'Technological Inertia in Economic History', *Journal of Economic History,* 52.2, 1992, 332). There was also Colt's small factory of 1851-2 in Pimlico (N Rosenberg (ed.), *The American System of Manufactures: The Report of the Committee on the Machinery of the United States 1855 and the Special Reports of George Wallis and Joseph Whitworth 1854. Edited with an Introduction by Nathan Rosenberg* ... , (1969), pp.17, 45).
205. L T C Rolt, *Tools for the Job: A Short History of Machine Tools,* (1965), pp.83-91.
206. *MM,* 23 August 1861, 110.
207. Rosenberg, *The American System of Manufactures,* pp.17, 45.
208. By 1848 there were 1,250 men and boys employed – 250 skilled and 1,000 semi-skilled or unskilled, a ratio of 1:5 (M E Bowbelski, *The Royal Small Arms Factory,* (Edmonton Hundred Historical Society Occasional Paper New Series No. 35, 1977), p.7).
209. Rosenberg, *The American System of Manufactures;* N Atkinson, *Sir Joseph Whitworth, 'The World's Best Mechanician',* (1996), pp.190-9. George Wallis and John Wilson were the other two members. Wallis was at the Royal Manchester Institution 1832-7, when he became friendly with Whitworth. Later he was head of the Manchester School of Design, 1843-5 (J Sharman, 'George Wallis, a pioneer of industrial art', at http://www.historywebsite.co.uk/genealogy/wallis/wallis)!.htm (accessed 11 December 2013).
210. D Pam, *The Royal Small Arms Factory Enfield & Its Workers,* (1998), p.47.
211. In earlier years Nasmyth was a regular visitor to Fairbairn's home, and Fairbairn had purchased machine tools from him (*Life,* pp.155-6; Cantrell, *Nasmyth,* p.31n57, 1836 two planing machines and a nut-cutting machine, p.114, 1853-6 lathes; S Smiles, (ed.), *James Nasmyth Engineer; An Autobiography* (1895 ed.), p.348).
212. *MPICE,* 86, 1886, 347. On Anderson see Rosenberg, *American System of Manufactures,* pp.80-6.
213. 'Select Committee on Small Arms', *Parliamentary Papers,* XVIII-1854, p.27.
214. Pam, *Royal Small Arms Factory,* p.9.
215. *Life,* p.327.
216. *MM,* 23 August 1861, 110.
217. E Ames and N Rosenberg, 'The Enfield Arsenal in Theory and History', *The Economic Journal,* 78.312, 1968, p.827.
218. J H Lewis, 'The Development of the Royal Small Arms Factory (Enfield Lock) and its Influence upon Mass Production Technology and Product Design c.1820-1880', (PhD Thesis Middlesex University 1996), p.23.
219. Jones, *Industrial Architecture,* p.102.
220. Rosenberg, *American System of Manufactures,* p.53.
221. J Tuff, *Historical, Topographical and Statistical Notes of Enfield,* (1858), p.195.
222. In the late eighteenth century the Normandy Chamber of Commerce referred to Manchester as 'the Rouen of England' (quoted in A Young, *Travels During the Years 1787, 1788, & 1789 undertaken more particularly with a view of ascertaining the Cultivation, Wealth and Resources, and National Prosperity of the Kingdom of France,* (2nd ed.1794), pp.523-4). But see J-P Chaline, 'The Cotton Manufacturer in Normandy and England during the Nineteenth Century', *Textile History,* 17-1, 1983, pp.19, 25
223. http://thiriez.org/filaturesJN/LaFoudre.htm (accessed 16 September 2016); http://www.petit-quevilly.fr/histoire/la-foudre/ (accessed 16 September 2016); Chaline, 'Cotton Manufacturer in Normandy', p.24.
224. S Söderberg, *Förstaden vid Testeboån: arkivanteckningar och traditioner om Strömsbros uppkomst och utveckling,* (1966), pp.59-63; *Manchester Guardian,* 31 July 1850; Nilson, 'Invisible Hand', p.14, reprinted in *A Case of Identities: Festschrift to Martin Peterson,* (Göteborg University 2006).
225. P Carlberg, 'Personal Contacts Between the Manchester Area and Gefle in Sweden a Hundred Years Ago: A Communication', *Transactions of the Lancashire and Cheshire Antiquarian Society,* 70, 1960, 58; *UIfE2,* pp.229-31.
226. Nilson, 'Invisible Hand', p.12.
227. *Life,* p.365.
228. Nilson, 'Invisible Hand', p.18.
229. Fairbairn, *M&MWII,* pp.198-211.
230. K Bruland, *British Technology and European Industrialisation: The Norwegian Textile Industry in the mid-nineteenth century,* (1989), p.62.
231. Bruland, *Norwegian Textile Industry,* pp.42, 114.
232. Bruland, *Norwegian Textile Industry,* pp.76-7. Letter Fairbairn to Schou, 31 March 1854, Copy obtained from Norsk Teknisk Museum, Oslo, where the original is held. The machinery appears to have been supplied by Hetherington.
233. Bruland, *Norwegian Textile Industry,* pp.76-7. Letter Fairbairn to Schou, 8 November 1854.
234. Bruland, *Norwegian Textile Industry,* pp.87, 114-5, 170.
235. Bruland, *Norwegian Textile Industry,* pp.93, 103.
236. Fairbairn, *Application,* pp.158-60. See also I Haynes, *Cotton in Ashton,* (1987), p.39.
237. The International Committee for the Conservation of the Industrial Heritage, *The International Context for Textile Sites,* (2003). Note that these are *existing textile* sites – the list would obviously be very much longer if demolished mills and non-textile sites were included.
238. *M&MW2,* pp.255-61.
239. [D Scott], *The Engineer and Machinist's Assistant,* (1850), Vol.1 Description of Plates, p.91, Vol.2 Plates 98-106.
240. *The Maitland Mercury and Hunter River General Advertiser,* 2 April 1859. The only writing on corn mills by Fairbairn prior to this date was a section of his report 'On the Comparative Merits of the Machinery of the Paris Universal Exhibition, 1855' which was not widely available until reprinted in 1866 (Fairbairn, *UIfE3,* pp.141-4) and the reference is probably to the descriptions and illustrations of Fairbairn's work in *The Engineer and Machinist's Assistant,* Plates 98-106. The Evans' work would probably be Oliver Evans, *The Young Mill-Wright and Miller's Guide,* which had reached its fourteenth edition by 1859.
241. *Exhibition 1851: Reports of the Juries,* (1852), pp.203-4.
242. [Scott], *Engineer and Machinist's Assistant,* Vol 1, pp.94-7; Fairbairn, *M&MWII,* pp.147-52.
243. Fairbairn, *M&MWII,* p.157; Fairbairn, *UIfE3,* p.143.
244. Fairbairn, *M&MWII,* pp.139-40.
245. Fairbairn, *M&MWII,* pp.140-1.

246. [Scott], *Engineer and Machinist's Assistant,* Vol.I, p.97.
247. Fairbairn, *M&MWII,* pp.161-3.
248. *Wolverhampton Chronicle,* 14 Jan 1846.
249. R A Hayward, 'Old steam Mill, Wolverhampton – A Milestone in Fireproof Construction, *Staffordshire Industrial Archaeology Society Journal,* 3, 1972, 40.
250. For relevant plan see R K Morriss, *The Old Steam Mill Wolverhampton: An Outline Archaeological & Architectural Analysis,* (1999), p.20, although the implications of the plan are not discussed.
251. *The Victorian,* 51, 2015, 24-5.
252. Morriss, *Old Steam Mill,* pp.16, 35 and Figs.7-10.
253. Bedford and Luton Archives and Records Service, Doc. Ref. X106/205.
254. *Life,* pp.241-254.
255. 'Sharnbrook Timeline' at http://209.85.229.132/search?q=cache:zewn-_108vkJ:www.galaxy.bedfordshire.gov.uk (accessed 10.01.09). Estimates followed: £380 for a waterwheel to drive three pairs of stones and £360 for the machinery, with alternatives for a larger wheel (Bedford and Luton Archives and Records Service, Doc. Ref. X106/123). The indications are that a water wheel was supplied by Fairbairns, but the present evidence is inconclusive.
256. Fairbairn, *M&MWII,* pp.132-8; W Fairbairn, 'On a Floating Corn Mill for the Navy', *MPICE,* 17, 1858, 159.
257. Fairbairn, *M&MWII,* pp.127-32 and Plates 11-14.
258. A W Skempton, 'Samuel Wyatt and the Albion Mill', *Architectural History,* 14, 1971, 53-152; S Smiles, *Lives of the Engineers,* (Vol.2, 1862), pp.136-42; O Evans, *The Young Millwright and Miller's Guide,* (1795); S Giedion, *Mechanization Takes Command: a contribution to anonymous history,* (1948), pp.84-5.
259. See Chapter 8.
260. G Fanning, *Oldham Coal,* (2001), pp.135-8.
261. *MG,* 16 July 1858.
262. See Chapter 3; Fairbairn, *M&MWI,* pp.246-7 and Plate 8. Also illustrated in J Scott Russell, 'Steam-Engine' in *Encyclopaedia Britannica 7th Edition*; and in G Watkins, *The Textile Mill Engine,* (1970), Vol.1, p.43.
263. W Fairbairn, 'Description of a New Construction of Pumping Engine', *MPIME,* 6, 1855, 177-82 and Plates 33-36. The cylinder was 70in. in diameter and the stroke 8ft.
264. *The Engineer,* 1, Jan.-June 1856, 183-4; *MM,* 64, 1856, 337-40.
265. *MM,* 64, 1856, 370.
266. *MPIME,* 6, 1855, 182.
267. For illustrations of the engine of the *Vulture* see *The Practical Mechanic.* 3, 1844, Plate 2; for that of the Odin see J Bourne, *A Treatise on the Steam Engine in its application to Mines, Mills, Steam Navigation, and Railways,* (1851), Plate I; and for that of the *Dragon,* see [Scott], *Engineer and Machinist's Assistant,* Plates 74-8.
268. W Fairbairn, 'On a New Description of Winding Engine', *MPIME,* 4, 1853, 137-42 and Plates 32 and 33. The cylinder was 60 inches in diameter and the stroke 8ft.
269. K F Haven, *100 Greatest Science Inventions of All Time,* (2007), p.120.
270. S Giedion, *Space, Time and Architecture,* (3rd ed. 1954), pp.196-7, 206-8.
271. A Rees, *Cyclopaedia or Universal Dictionary of Arts, Sciences and Literature,* (Vol.1, 1812-3), referred to in L E Gray, *From Ascending Rooms to Express Elevators: A History of the Passenger Elevator in the 19th Century,* (2002), pp.1-8.
272. Ure, *Philosophy,* pp.45-54.
273. Gray, *Passenger Elevator,* p.8.
274. Fairbairn, 'Winding Engine', 137-2. Reporting to the Coroner's court on the failure of a platform hoist at Miles Platting Locomotive Works in 1867, Fairbairn pointed out that in 'factories, mills and warehouses' hoists had 'catches which acted with great precision' stopping the hoist if the chain or rope broke, as had happened at Miles Platting (*MG,* 30 January 1867, 2 March 1867). This was fourteen years after the dramatic demonstration in New York when Otis cut the rope which was raising the platform on which he stood – and it did not crash to the ground. (Giedion, *Space, Time and Architecture,* p.207). It is unclear if safety catches were introduced into Britain before or after the spectacular Otis demonstration.
275. W Hopkins, 'On the Geological Theories of Elevation and Earthquakes', *BASS1847,* pp.33ff.
276. *Life,* pp.289-91. On Joule see D S L Cardwell, *James Joule: A Biography,* (1989).
277. Fairbairn, *UIfE3,* pp.90-5.
278. *Life,* p.295.
279. W Hopkins, 'Presidential Address', *BAAS1853,* p.li.
280. W Hopkins, 'An Account of some Experiments on the Effect of Pressure on the Temperature of Fusion of Different Substances', *BAAS1854,* pp.s57-8; W Fairbairn, 'On the Density of Various Bodies when subjected to enormous compressing Forces', *BAAS1854,* p.s56.
281. W Hopkins, 'Experimental Researches on the Conductive Powers of various Substances, with the Application of the Results to the Problem of Terrestrial Temperature', *Philosophical Transactions of the Royal Society,* 1857, 805.
282. C Smith and M N Wise, *Energy and Empire: A Biographical Study of Lord Kelvin,* (1989), pp.553-9.
283. Fairbairn, *UIfE3,* p.89.
284. W Fairbairn, 'On the Temperature of the Earth's Crust as exhibited by Thermometrical Observations obtained during sinking of the deep mine at Dukinfield', *Proceedings of the Manchester Literary and Philosophical Society,* 1861, 64-6; W Fairbairn, 'Report on the Temperature of the Earth's Crust, as exhibited in the Dukinfield Deep Mine', *BAAS1861,* pp.s53-6.
285. E Hull, *The Coal-Fields of Great Britain* (1861), pp.223-9,
286. *The Engineer,* 12, July-Sept.1861, 182.
287. Schumpeter, *Theory of Economic Development,* pp.93-4; H Perkin, *The Origins of Modern English Society 1780-1880,* (1969), pp.83-5.
288. Rose, 'Beyond Buddenbrooks, pp.133-5.
289. *Life,* pp.318-9, 342, 466.
290. J Tann, *The Development of the Factory,* (1970), p.105.
291. Rosenberg and Vincenti, *Britannia,* pp.49-52.
292. Gray, *Passenger Elevator.*

Sources, Credits and Acknowledgements for Illustrations in Chapter 4

Fig.4.1 Westcott's original portrait is displayed at the Institution of Civil Engineers, London. T.O. Barlow's subsequent engraving was published by Thomas Agnew & Sons in 1852.

Fig.4.2 R.W. Rush, *The Lancashire and Yorkshire Railway and its Locomotives 1846–1923* (1949), p.25.

Fig.4.3 [G.F. Burtt], *The Locomotives of the London, Brighton & South Coast Railway, 1839–1903* (1903), p.34.

Fig.4.4 J .W. Lowe, *British Steam Locomotive Builders* (1975), p.168.

Fig.4.5 *The Practical Mechanic's Journal,* 4, 1851–2, Plate 89.

Fig.4.6 W.M. Gradon, *The Furness Railway 1846–1923* (1946), p.47.

Fig.4.7 *ILN*, 18 December 1852, 552.

Fig.4.8 Fairbairn, *UIfE3,* p.133.

Fig.4.9 P C Dewhurst, 'An Early Brazilian Locomotive by Fairbairn's', *The Locomotive,* 15 March 1930, pp.80–1.

Fig.4.10 L. Haghe, *The Great Exhibition: Moving Machinery* (1851) (Detail). Also reproduced in A. Briggs, *The Power of Steam: An Illustrated History of the World's Steam Age* (1982), pp.26–7.

Fig.4.11 William Fairbairn & Sons, *Rules and Regulations to be observed by all persons in the employment of William Fairbairn & Sons, Engineers, Manchester* (1852), Title page, pp.4–5.

Fig.4.12 J. Reynolds, *Saltaire. An introduction to the village of Sir Titus Salt* (1985), Front cover.

Fig.4.13 Fairbairn, *M&MWII,* Plate 15.

Fig.4.14 V. Carter (ed.), *Treasures of Enfield: discovering the buildings of a London Borough,* (2000), pp.40–1. Reproduced by courtesy of The Enfield Society.

Fig.4.15 http://www.gracesguide.co.uk/File:Im1897SYQ4-Enfield.jpg (accessed 18 September 2013).

Fig.4.16 *Life,* Frontispiece; Engraving from a photograph by Elliott & Fry, 1882; S. Smiles (ed.), *James Nasmyth, Engineer, An Autobiography* (1895 ed.), Frontispiece.

Fig.4.17 Painting by Auguste Bertrand, engraved by Edward Ethrington and published in J. Turgan, *Les Grandes Usines* (1860s). Also reproduced at http://thiriez.org/filaturesJN/LaFoudre.htm

Fig.4.18 Image 10418526, Science Museum/Science & Society Picture Library.

Fig.4.19 P. Carlberg, 'Personal Contacts Between the Manchester Area and Gefle in Sweden a Hundred Years Ago: A Communication', *Transactions of the Lancashire and Cheshire Antiquarian Society,* 70, 1960, fig.VI.

Fig.4.20 Fairbairn, *UIfE2,* p.230.

Fig.4.21 Fairbairn, *M&MWII,* pp.200–1.

Fig.4.22 Image EPW028908, © Historic England.

Fig.4.23 Fairbairn, *M&MW,* p.256.

Fig.4.24 Fairbairn, *M&MWII,* p.148.

Fig.4.25 Fairbairn, *M&MWII,* p.162.

Fig.4.26 Fairbairn, *M&MWII,* p.140.

Fig.4.27 R.J. Byrom, Photograph 1961.

Fig.4.28 R.J. Byrom, Photograph 1961.

Fig.4.29 R.J. Byrom, Photograph 1981.

Fig.4.30 Fairbairn, *M&MWII,* p.134.

Fig.4.31 Fairbairn, *M&MWII,* Plates 11 and 13.

Fig.4.32 W. Fairbairn, 'On a New Description of Winding Engine', *MPIME,* 4, 1853, Plate 32.

Fig.4.33 W. Fairbairn, 'Description of a New Construction of Pumping Engine', *MPIME,* 6, 1855, Plate 35.

Fig.4.34 Ure, *Philosophy,* p.48.

Fig.4.35 Fairbairn, *UIfE3,* p.91.

CHAPTER 5

Family Business

Part 2: Tubular Structures

Introduction

A major component of the work of the family business was an innovative use of wrought iron, by way of tubular structures, which put the seal on Fairbairn's fame. This sphere of work had its roots in commercial pressure for a railway to Holyhead, for the passage to Ireland, coupled with onerous Admiralty requirements for bridging the Menai Straits. Fairbairn undertook experiments which resulted in the Britannia Bridge, increasing the longest wrought-iron girder spans of its day from 31ft 6in. to 460ft, far longer than any before it. It was a leap into the unknown, achieved from a base where relevant knowledge, theoretical or practical of an iron structure of this magnitude, did not exist. Engineers came from all over Europe to see it. On the Continent it promoted the adoption of girder bridges for long spans in place of suspension bridges, it made an important contribution to the knowledge of buckling, it spawned various derivatives, and it anticipated the box-girder bridges of the late 20th century. Yet few other tubular bridges were built, and soon after Britannia was constructed it was assumed to be a cul-de-sac in bridge-building. Why was this? And what was Fairbairn's role in the development of tubular structures?

There was rapid growth in the building of the derivative tubular-girder bridges but their construction was largely limited to a single decade of opportunity. Why was this, what took their place, and was this a matter of theoretical calculations superseding rule-of-thumb and trial-and-error? The timing of the collapse of Stephenson's Dee Bridge is shown to be crucial to these questions.

From the tubular-girder there were further derivatives, most spectacularly the swan-neck tubular crane – the 'Fairbairn crane'. For twenty-five years it was the dockside heavy crane of choice, and continued to be built into the early years of the 20th century. How did it diffuse to other countries and to Continental manufacturers, and what were its derivatives?

Bringing together Fairbairn's work on these various tubular structures emphasises their inter-relationship, which occurred within the context of two of the most formative decades in the history of bridge-building, 1840–60, when the increasing availability of wrought iron, the growing demands of the railways and the development of the theory of structures came together, bringing rapid change, in which Fairbairn played a major role.

Tubular Bridges

No sooner had Fairbairn decided to cease shipbuilding than, in 1845, another unexpected opportunity arose through a link from earlier days, George Stephenson, who 'arranged' for his son Robert, now Engineer to the Chester & Holyhead Railway, to meet with Fairbairn.[1] The challenge was to bridge the Menai Straits, where the Admiralty required 105ft headroom for a 460ft span. This ruled out an arch. A bowstring design appears to have been considered but rejected – maintaining the required headroom during construction would have been difficult.[2] A suspension bridge was ruled out by feared instability.[3] Recalling the launch of the *Prince of Wales* when she stuck supported only at bow and stern, without breaking her back, Robert Stephenson suggested a tube.[4] The Board of Trade adviser, General Palsey, suggested a suspended deck, formed into a lattice but with the chains retained.[5] Stephenson amalgamated the ideas, proposing decks on which the tubes would be constructed, after which the chains would support the tubes, and Fairbairn was commissioned to pursue the proposal. In May Robert Stephenson gave evidence to a House of Commons Select Committee:

> I am instituting a series of experiments in conjunction with Mr Fairbairn of Manchester, who is already in possession of experiments on iron ships, which place the thing beyond all doubt. He has ascertained that a vessel of 250ft in length supported at the ends will not yield with all the machinery in the middle.
>
> Q. Have your calculations been submitted to any other engineers?
>
> A. I have made them, in conjunction with Mr. Fairbairn of Manchester, whose experience is greater than any other man's in England.[6]

The Bill having been passed, Fairbairn began experiments in July, at Millwall where his last two ships were nearing completion.[7] He suggested that Hodgkinson, with his mathematical knowledge, be engaged to assist. Initial experiments used 'Fairbairn's Lever' on various tubes of aerodynamic circular and elliptical shapes. These were considered first because of concern about wind-loading, but the rectangular section was soon found superior because of the tendency of the others to buckle – the flange in a rectangular section is larger and more efficient than in a circular or elliptical cross-section.[8] He also found that the top flange in compression needed to be strengthened with closed cells to minimise the risk of buckling. He was surprised: results 'were not always in accordance with established theories … On the contrary, weakness was found where strength was expected'.[9]

These experiments were the earliest shell-buckling tests on thin-walled tube structures under axial compression and bending, with failure through elastic instability.[10] They have some affinity to Fairbairn's later experiments with boiler flue

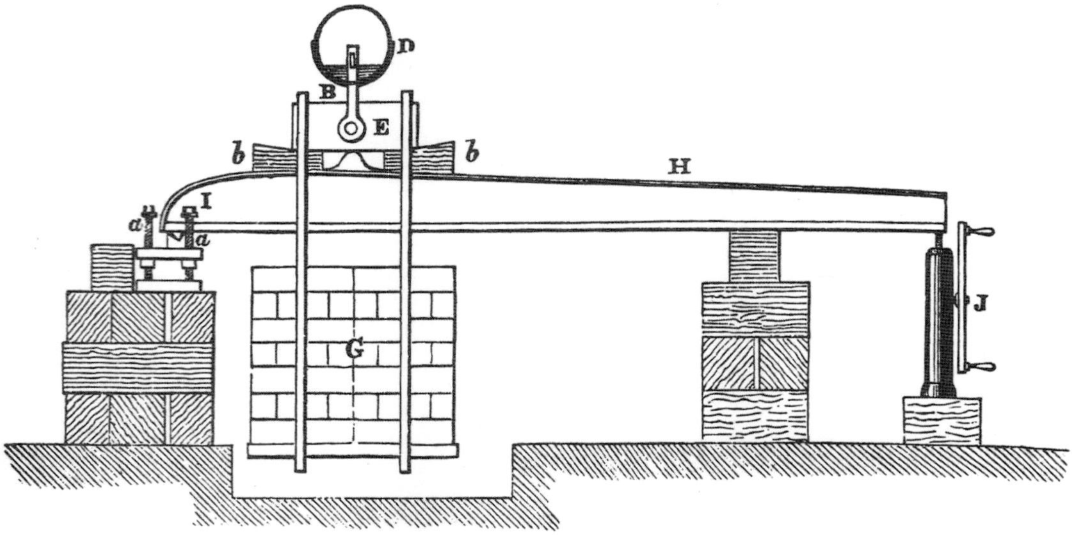

Fig. 5.1: Fairbairn's Lever as used for testing tubes at the start of the Britannia experiments.

tubes under external pressure.[11] For Fairbairn and Hodgkinson the problem was limited to lateral buckling of a compressed member, but elastic stability has since developed into a major area of engineering science, with thin shells of prime importance in modern aircraft construction.[12] Well over a century later these classical experiments continue to hold 'great interest for engineers working with thin walled structures'.[13]

In February 1846, Stephenson, Fairbairn and Hodgkinson delivered individual reports to the Railway Company's Directors, all positive but differing about the chains – Hodgkinson recommended retention, Stephenson appeared undecided and Fairbairn was firmly against their retention.[15] By May, masonry was underway at Conway and Menai, anticipating satisfactory test results on a ⅙th scale model, 75ft long with six cells at the top, based on earlier experiments.[16] Because of the cost, Fairbairn was limited to one model and proceeded by parameter variation – keeping all the components, save one, constant and noting the effects of changing that one component. This approach had some affinity to that of Smeaton's experiments on cement for the Eddystone lighthouse, and Joule's work on electromagnets.[17] The bottom flange failed, was strengthened and re-tested, until in the sixth test, the top buckled. Fairbairn had increased the failure load from 35.5 tons to 86 tons with only 20 per cent more material.[18] These were the first structural model tests on a British railway bridge:[19] 'Nothing', said Stephenson's trusted assistant, Edwin Clark, 'could be more satisfactory than the result'.[20]

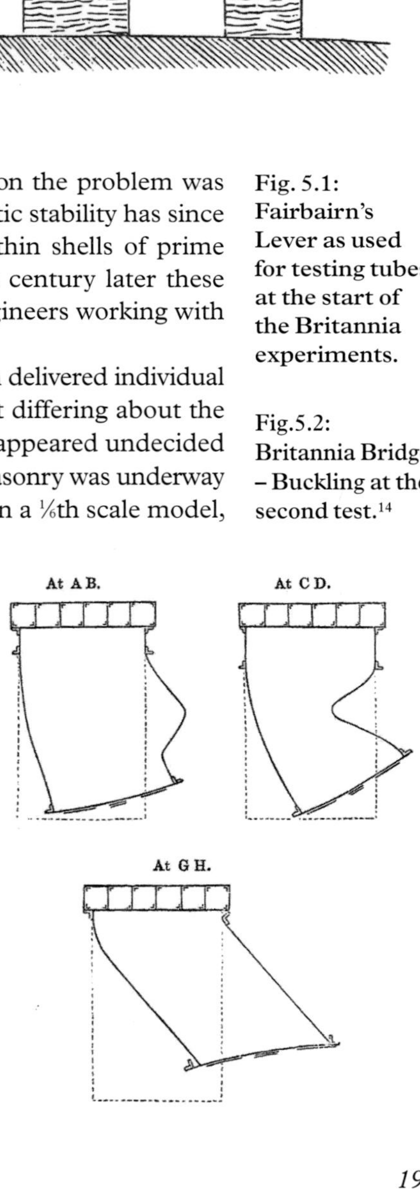

Fig.5.2: Britannia Bridge – Buckling at the second test.[14]

Following the reports of February 1846, Fairbairn's long-standing friendship with Hodgkinson soured. Fairbairn had provided facilities, at his expense, for

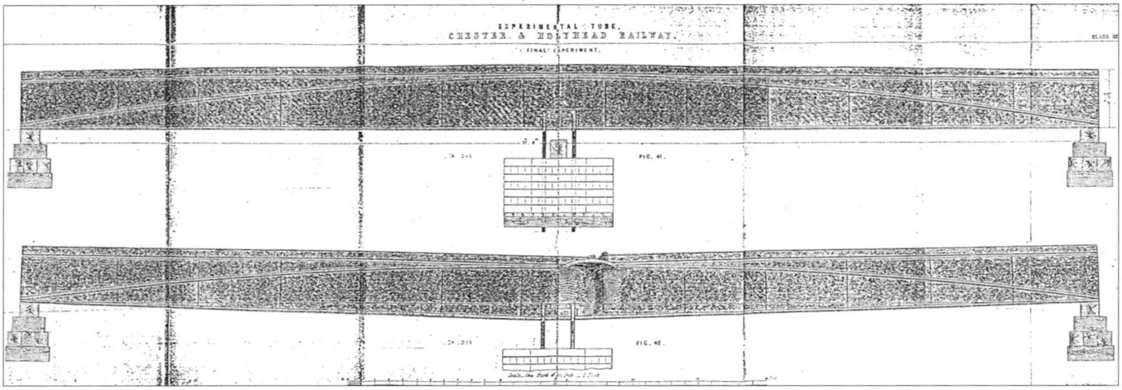

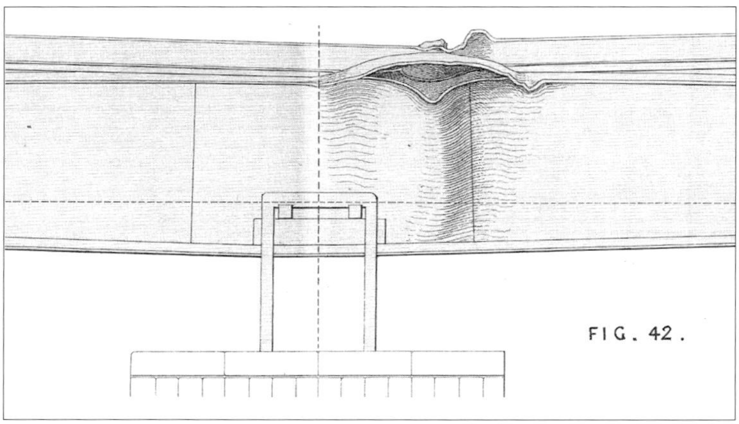

Figs 5.3 & 5.4: Final testing of the Britannia model.

Hodgkinson's experiments over two decades, for which Hodgkinson expressed his thanks in 1840 in glowing terms,[21] and Hodgkinson had been a trusted tutor of Fairbairn's sons.[22] In 1843 there had been a suggestion that all was not well, when Fairbairn wrote:

> I think, but I may be wrong, that I have observed some faint appearance of an ugly monster which they call ... [*sic*] which sometimes darkens the amiable and bright side of your character ...You must never look upon me as your competitor, but as your friend and fellow labourer in a totally different department of science.[23]

In March 1846, Hodgkinson suggested to Stephenson that Fairbairn's tests might be proved wrong, adding, 'I do not wish to exclude Mr Fairbairn, except from taking the lead in those matters which his want of education and experience render him utterly incapable of'.[24] Fairbairn stood by his results, and offered to withdraw.[25] His expertise was crucial to the project at this point and it was probably the threat of withdrawal that led to the board appointing Fairbairn 'to superintend the construction of the Conway and Britannia bridges, in conjunction with Mr Stephenson'.[26] Hodgkinson continued his experiments slowly at Ancoats, reporting to Stephenson. Time was an issue and it became apparent that the design would have to proceed without formulae from Hodgkinson. His results, whilst providing useful knowledge played little part in the design of the bridges and his involvement soon ceased.[27] At the British Association meeting in September Hodgkinson claimed the credit for suggesting the cellular tops to the tubes.[28] Stephenson deplored the jealousy that the investigation had given rise to and on 9 October Fairbairn assured him that 'the misunderstanding between Mr Hodgkinson and myself ... is now all settled', but on 28 October Clark wrote to Stephenson, 'They continue to hate each other most enthusiastically'.[29] However, following Hodgkinson's decease Fairbairn spoke of him to the British Association, 'For a long series of years he and I worked together ... I look back to the days of our joint labour with unalloyed pleasure and satisfaction'.[30]

By May 1846 masonry was underway at both Conway and Menai on a 'fast track'.[31] By July, Fairbairns had completed basic drawings, and tenders were invited for the supply of iron plates.[32] Fairbairn, continuing to oppose the chains, proposed floating the tubes into position and jacking them up. Stephenson eventually accepted this and chains were abandoned.[33] In August Fairbairn invited tenders for constructing the tubes. In October the board instructed that further tenders be obtained. However the masonry contractor at Conway, William Evans, offered a price for constructing the Conway tubes on site, floating them into position and raising them. The work was placed with him.[34]

Fairbairns had submitted one of the initial tenders which the Board rejected. Before further tenders were obtained Thomas Fairbairn, now managing the Millwall works, and another unsuccessful tenderer, Ditchburn & Mare, submitted an unsolicited joint bid for half the work. This, Thomas claimed, was done *'without the knowledge, in any way whatever, of Mr Fairbairn'* and accepted by the Directors *'without any communication with him as their acting Engineer'*.[35] Thomas then transferred Fairbairns' share of the contract to Ditchburn & Mare – presumably netting a profit. The Directors were not pleased that this was done without them being consulted. They were also uneasy about the costs of the experiments. As a consequence they rescinded William's authority to certify payments.[36] In his book William Fairbairn maintained that when he became aware of the Directors' disquiet, it was he who urged the transfer of the contract to Ditchburn & Mare, in order to 'maintain a perfectly independent position'.[37] However, given that he again tendered his resignation, this may indicate a damage-limitation exercise. At Stephenson's behest the resignation was withdrawn.[38]

Fabrication of the first Conway tube started in April 1847, on timber staging on the shore. Stone piers were built under each end of the tube and, when it was complete, the timber staging was cut away and the beam load-tested.[39] On 6 March 1848, with both Stephensons, I.K. Brunel, J.M. Rendel and William Fairbairn atop, the 1,300-ton tube was floated down the turbulent tidal Conway on six pontoons,[40] with Fairbairn having 'the direction on the Conway side'. There was a glitch. The tube caught on a rock, only about twelve feet from its destination.[41] It took five days to free it, after which two hydraulic presses lifted it to 18ft above high water.[42]

Conway was followed by Britannia, with its immense, self-supporting, wrought-iron tubes and its tall towers, built for possible chains, adding to its grandeur. The mode of construction, without chains, was due to Fairbairn who has not been given the appropriate credit for this. Without the towers Britannia would have been a visually undistinguished structure, but of much greater importance is that

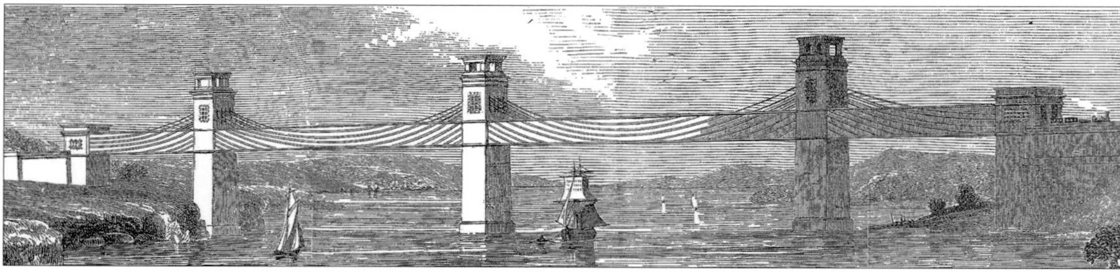

Fig.5.5: Britannia Bridge as Stephenson originally envisaged it.

the omission of suspension chains paved the way for major girder bridges in Continental Europe, in lieu of suspension bridges, as discussed later in this chapter.

Britannia was the most famous bridge of the 1850s with the longest box-girder span in the world until after World War 2.[43] Fairbairn's drawing office prepared the

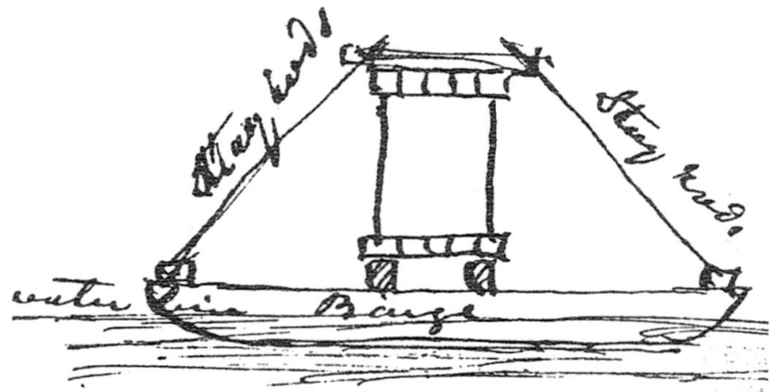

Fig.5.6: Fairbairn's initial diagram of the method of floating the tubes.

detailed drawings. He organised the site layout with the tubes to be built on timber staging between high and low water, which would enable pontoons to be inserted to lift the tubes on a rising tide.[44] He designed the arrangement for raising the tubes, by chains from hydraulic presses supported on massive cast-iron beams near the tops of the towers.[45] However Edwin Clark has claimed that the original idea of floating and raising the tubes was his.[46] But before the tubes were lifted Fairbairn was no longer involved.

He was not involved because of circumstances at Conway which changed his career path. When John Lucas painted Robert Stephenson with Britannia's

Fig.5.7: Floating the first Conway tube

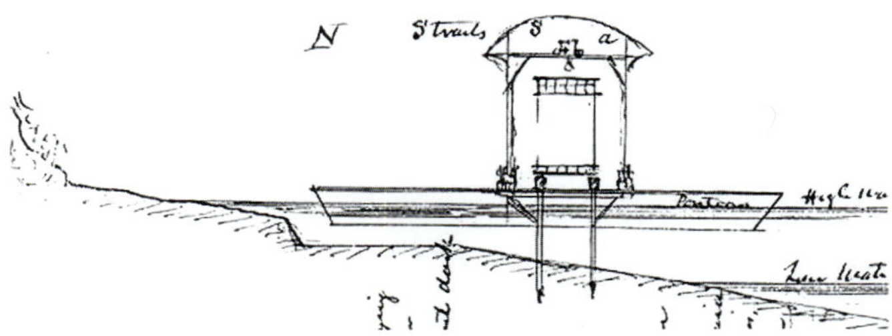

Fig.5.8: Fairbairn's early proposals for building the Britannia tubes on the Carnarvon shore.

engineers, Fairbairn was not included.[47] When Queen Victoria opened the bridge he was not present. He never worked with Stephenson again. The successful testing of the first Conway tube, on its temporary stone piers, was the defining moment. After three years of 'prognostications of failure', a euphoric Fairbairn wrote to several eminent engineers and professors describing what he saw as a personal triumph. Recipients included Professor Moseley, Charles Babbage, Professor Willis and George Rennie.[48] These letters apparently 'gave dissatisfaction to Mr Stephenson' although Fairbairn subsequently disclaimed any intention of appropriating to himself 'the merit of the undertaking'.[49] The letter to Babbage has survived and its wording contains nothing which should have caused concern to Stephenson.[50] On completion of the first tube, the Conway gentry hosted a dinner to which Fairbairn was invited. Some time before that dinner, a discussion at the Society of Arts had given the credit for the Britannia Bridge to Fairbairn. Stephenson told him that he intended to settle the matter at the dinner. However Fairbairn declined to attend and wrote suggesting he withdraw but was unanswered. Stephenson's Conway speech was self-adulatory, claiming that his wrought-iron bridge of a few years earlier at Ware was the inspiration for Britannia. On 22 May 1848 Fairbairn resigned.[51] He then wrote his book about the bridges, pouring scorn on Stephenson's explanation of their genesis from the Ware Bridge:

Fig.5.9: The one remaining section of a Britannia tube.

> [the Ware Bridge] is constructed upon the principle of the common cast-iron girder bridge, each separate beam being formed of wrought-iron plates connected together by angle-irons ... This

Fig.5.10: The Britannia Bridge as it was in 1962.

> form of wrought-iron girder had been long in use before the erection of the Ware Bridge; and it is defective, as well in principle as in construction: a great body of the material is not in the top flanches, as it ought to be, in order to attain the section of greatest strength. Besides, it is impossible to trace any analogy between a combination of this form of beam and a tubular girder with a cellular top. The beams in the Ware Bridge do not offer a united resistance to strain, in the manner in which beams with a cellular top do; on the contrary, each beam has its own distinct part of the load to carry, and that very imperfectly for want of a due proportion in the top and bottom flanches. An immense tube, upwards of a thousand tons in weight, suspended in the air, and through which railway trains should pass, was a magnificent conception; surely Mr. Stephenson does injustice to his original idea as well as to the spirit with which it was realised, when he couples it with so imperfect a structure. While I freely award to Mr. Stephenson the honour of this conception, I claim for myself the credit of having rendered it practicable.[52]

A riposte was inevitable. It came by way of Edwin Clark's substantial book, on Stephenson's behalf.[53] Profession and press were divided about attributing credit for the bridges.[54] In 1877 Woodcroft wrote: Fairbairn 'showed but little political sagacity, and a strong tendency to grasp a larger share, both of profit and fame, than it could be at all expected that Stephenson would … accord to him'.[55] Today many would agree with Dempsey's 1849 assessment that 'these great

Fig.5.11: Robert Stephenson (1803–1859).

works owe their design and construction to [their] *joint* labours'.[56] Britannia would not have been built without either man. Yet, these disagreements were 'not without advantage to [Fairbairn], inasmuch as they brought his name more prominently before the world'.[57]

Today mental stress amongst those at the cutting-edge of construction is a subject of international study. In 1848 it was not. The stresses faced by both Stephenson and Fairbairn must have been immense.

Karl Culmann, the German railway engineer who became professor of structural engineering at Zurich Polytechnic, visited England and America in 1849. On his return he published an extensive study of English and American bridges.[58] He was impressed by the tubular bridges; but in his later writings he considered it was mistaken to have entrusted the study for Britannia to a man with insufficient theoretical knowledge, who experimented to discover what could have been learnt from existing books. However, no less an authority than Stephen Timoshenko, writes that:

> Culmann's disparaging remarks are ill-considered because little was known at that time about thin-

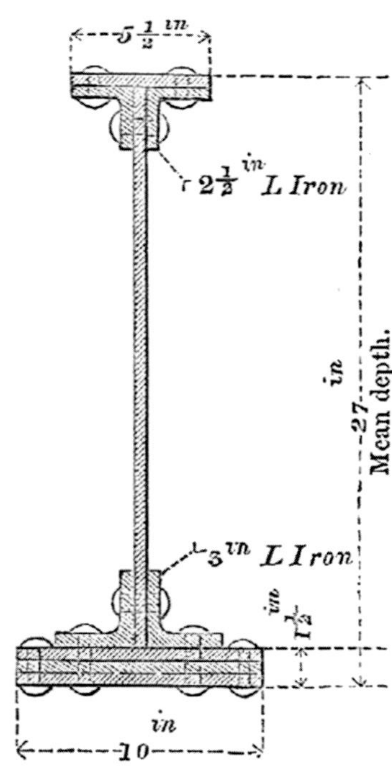

Fig.5.12. The beams which Stephenson claimed were his inspiration for the Britannia Bridge.

> walled structures. In fact, Fairbairn's experiments first drew the attention of engineers to the importance of the stability questions in designing compressed iron plates and shells.[59]

Donald Cardwell states that the size of the bridge made an application of the theories of the French engineers hazardous, not least because they had only been tested on *cast* iron;[60] and Josef Singer points out that no theoretical methods were available to assess buckling strength.[61] These all accord with statements made at the time by Hodgkinson, 'The laws of resistance of plates to wrinkling were utterly unknown',[62] and Clark, 'No one knew, *à priori,* the resistance of plates to buckling, which was a new fact altogether, and one not involved in any of the formulæ hitherto employed'.[63] The suggestion that lattices should have been used failed to take into account that Macneill's recently completed lattices in Ireland were unsuccessful because the joints were inadequate to deal with the loads, the structures became dangerously springy, and the top chords tended to buckle.[64] Nor should it be overlooked that the major Prussian lattice bridges, over the Vistula and the Rhine, discussed below, were opened seven and nine years respectively after Britannia.

The French engineer B.P.E. Clapeyron (1790–1864), describing Britannia as 'this magnificent structure' nevertheless considered the plates 'too thin at the points of support'.[65] But Timoshenko points out that Clapeyron appears not to have taken the mode of assembling the beams into account.[66] Todhunter and Pearson in *The Theory of Elasticity* (1886) wrote of the influence of the Britannia and Conway bridges in France:

> We have given sufficient evidence to show the important contributions to physical knowledge made by these great engineering works ... They are frequently referred to by Saint-Venant in his edition of Navier's *Leçons*.[67]

The Russian railway engineer D.J. Jourawski (or Zhuravski) (1821–91), analysed the tubes showing that the number of rivets could have been considerably reduced without detriment.[68] He was critical of the design of Britannia and of Fairbairn's model testing.[69] That Continental engineers of the standing of these men were discussing Fairbairn's design work for the Britannia Bridge bears testimony to the importance of that work.

Few tubular bridges were built. Nevertheless Britannia was influential in bridge development through its derivative tubular-girder bridges and tubular cranes and, more particularly, in the demise of the suspension bridge in Continental Europe. To cross the Rhine at Cologne a suspension bridge was proposed, involving splitting trains into sections, raising them to bridge level, and towing them across with horses. The Prussian Ambassador, Christian von Bunsen, visiting Manchester, met Fairbairn and suggested he should visit Cologne and Berlin and put alternative proposals to the Prussian government. Within days Fairbairn was off to Germany, taking Dorothy and son George with him, but before going he set out his terms of engagement in a letter to Bunsen: there would be no repeat of the Britannia situation.

> On the construction of this great work, should it be executed, it is not my intention to become the contractor; the government or the authorities of

> Cologne shall make their own selection as to those who shall do the work, but I shall give all the designs and working drawings, superintend, and take responsibility of the execution and security of the work, and that upon some scale of remuneration which may hereafter be agreed upon.[70]

He was looked after in Germany by the elderly Alexander von Humboldt – the two had considerable mutual respect for each other – who arranged for Fairbairn to dine with King Frederick William IV.[71] Fairbairn proposed a bridge that would carry complete trains, with inclines to get them to the correct level. A Prussian delegation visited Fairbairn in 1849 and he took them to see recently finished tubular-girder bridges, and Britannia under construction. After an abortive competition, which Fairbairn refused to enter, and the visit of a second delegation to Britain, Prussian engineers were appointed to design a lattice-girder bridge.[72] This was a blow to Fairbairn who had devoted much effort to designs for what would have been the longest girder spans in the world.[73] Yet Fairbairn did influence the outcome of the Vistula and Rhine bridges. Karl Lentze was responsible for the proposed suspension bridge over the Vistula (Weichsel) at Tczew (Dirschau), and also for procuring the Cologne Bridge. A suspension bridge was chosen for the Vistula because 'it was the sole proven means of achieving a large clear span'.[74] Lentze was part of the first delegation to visit Fairbairn, and was aware of Macneill's iron lattice-girder bridge in Dublin, probably from his visit there in the winter of 1844–5.[75] At Tczew work stopped soon after it started, due to the unrest of 1848. When it started again, Lentze having visited Fairbairn, had abandoned the suspension bridge in favour of rigid girders, albeit lattice girders.[76] Tczew eclipsed all previous lattice bridges and provided the model for Cologne.[77] Now in Poland it still exists in a poor state. Since 2004 it bears a plaque:

> This historic bridge is the first example of a long span lattice truss bridge on the European mainland. By combining the original American idea of wooden trusses with the tubular concept of the Britannia Bridge, Wales, a new type of dense lattice truss structure made of iron was developed. This pattern of engineering was then disseminated throughout the Continent.[78]

As Cywinski maintains, the Britannia Bridge together with the American *Town*-system of bridge constructions became the model for the historic bridges in Tczew (Wisla) and Malbork (Nogat).[79] As such they influenced bridge-engineering throughout Europe. K-E Kurrer sees the structure of Britannia as the type of innovation that occurs only once in a century, and so complex that it cannot be calculated, but only safeguarded by tests. A practical effect of Britannia was that for the Tczew Bridge, Lentze abandoned the large-scale tests he had proposed and deemed it sufficient to rely upon calculations, because Fairbairn had done tests for Britannia which had been built successfully.[80] In addition Fairbairn's proposal for the Cologne Bridge 'over which trains could pass unimpeded', was adopted.[81]

The tubular bridge, stunning and influential as Britannia was, could not compete with trussed girder bridges. Fairbairn built no more tubular bridges, and Stephenson only two, Brotherton, and Victoria,[82] plus three in Egypt where the track was on top of the tubes.[83] In addition at least two were built in Europe.[84]

The importance of the Britannia Bridge and Fairbairn's work in connection

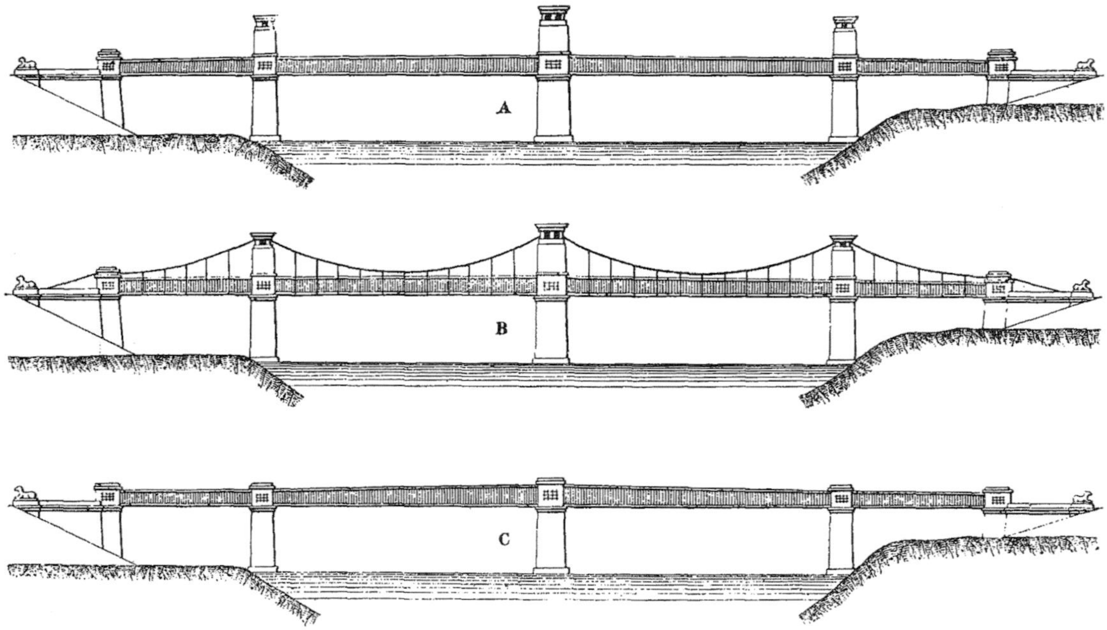

Fig.5.13:
The aesthetics of the Britannia Bridge.

with it – a completely new concept to meet a seemingly impossible challenge and so well documented – has made it an on-going focus of debate. It has been used to illustrate the theory that innovation in technology does not involve the application of knowledge derived from science;[85] to illustrate a theory of the inventive process,[86] and a theory of design;[87] and as a paradigm of 'failure-driven design'.[88] A claim has been made that Britannia owes its existence to American timber lattice bridges – that 'a source of vaunted "British" invention was "American"'.[89] The timber lattice bridge was a forerunner of the wrought-iron lattice but emphatically not of the Britannia Bridge. Throughout the 20th century it has been the subject of aesthetic debate.[90] At a meeting in 1901, five years after Louis Sullivan's 'form ... follows function',[91] Sir Benjamin Baker 'found architects and engineers holding up as a model a bridge designed for suspension chains which had never been erected!'[92] In 1910 J.A. Lux hailed the engineer as 'the true architect of the modern age'.[93] Eighty years on, Baker's aesthetic still carried weight,[94] but it is not justified. To get the railway open, speed was of the essence. On 15 July 1846 Fairbairn wrote to Stephenson suggesting that the tubes be raised by hydraulic lifting gear as this would *save time and cost* – possibly the first explicit statement relating technological method to saving construction time, which has become so important today.[95] It is also probably the case that Britannia was the first recorded example of 'fast-track' parallel planning, in which the normal linear construction sequence is decoupled and different parts – in this case the experimentation, the construction of the masonry piers and the decision on the superstructure – run in parallel in order to gain time. As Peters has argued aesthetic judgements in engineering need to take account of 'process' and not just the end result.[96] But even if this was not the case, Britannia illustrates the 'power and harmony' which

Fairbairn saw resulting from the collaboration between engineer and architect – in this case Francis Thompson (1808–95). In this it endorses both Fairbairn's disparagement of the functional tradition in early industrial buildings, and his teamwork with architects for the mills at Carlisle, Saltaire and Enfield.[97] Similarly Fairbairn's proposals for Cologne included 'monumental' piers.[98]

Britannia has not been without its critics, notably Henry Petroski. Both he and Rolt compare Britannia unfavourably with Brunel's Saltash Bridge which was seen to be 'altogether lighter, more scientific and economical'.[99] However, whilst a great bridge, Saltash had little influence, being opened nine years after Britannia and never repeated. Specific criticisms of Britannia included its cost – twice the cost of Saltash on a per foot of single track basis and 'what might today be termed an environmental failure' because of the smoke-filled air within the tubes.[100] Remarkably Petroski does not make any reference to what many would consider the major criticism of this bridge, the working conditions of those constructing the cells:

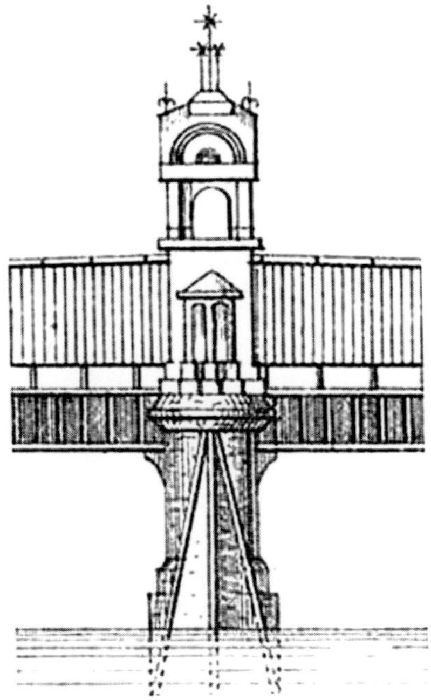

Fig.5.14: Proposed central pier, Cologne.

> A tiny rivet-boy – we observed one little mite only ten years of age – in clothes professionally worn into holes at the knees and elbows – crawling heels foremost for a considerable distance into one of these flues [the structural cells] ... slowly followed by his huge lord and master the holder-up, who exactly fits the flue, for the plain and excellent reason that ... the flue was purposely predestined to be exactly big enough to fit him; and as, buried alive in this receptacle, he can move but very slowly, he requires some time, advancing head foremost , to reach the point at which he is to commence his work ... unless we had witnessed the operation, we could scarcely have believed that any set of human beings, or rather of fellow-creatures, could professionally work from morning till night, stuffed horizontally into a flue of such small dimensions – that they could endure the confinement which only allows them, by changing from one side to another, to throw sometimes the right leg and sometimes the left leg over the elastic handle of the hammer – and above all that they could bear the deafening noises created close to and immediately thundering into their very ears![101]

Woodcroft saw tubular bridges as 'little more than a curious and startling episode in engineering history' never likely to be repeated.[102] Over a century later Petroski concurred: 'the tunnel vision created by so narrow a focus on the dominant structural problem of bridging the Menai Strait resulted in a monument to engineering short-sightedness'.[103] They were both wrong. Gustav Lindenthal

1850–1935, famed as a bridge engineer in America, said, 'The Britannia Bridge … marks a milestone in bridge engineering', and, 'Too much credit cannot be given to that galaxy of early English bridge engineers … Stephenson, Fairbairn, Telford, Tierney Clark – for the originality and daring of their plans and constructions. They did their own thinking; they did not wait for precedents, but created them'.[104] Rosenberg and Vincenti conclude, 'The most broadly influential aspect of the tubular-bridge experience, economically as well as technologically, was its role in the shift from cast iron to wrought iron as the material of choice for engineering structures'.[105] Sean Dooley notes that knowledge created by Fairbairn and Hodgkinson about thin-walled tubes certainly aided Fowler and Baker build the Forth bridge in 1890.[106]

Fairbairn and Stephenson had been ahead of their time. Their design concept was followed up over a century later as box girder bridges in steel or concrete became a popular choice during the road-building expansion of the 1960s.[107] Björn Åkesson lists nine characteristics of Britannia that 'made it identical to a modern box-girder bridge'.[108] But in the early 1970s there was a set-back when five steel box girder bridges collapsed, in Australia, Wales, Germany and Austria,[109] revealing an on-going lack of understanding of the slender compressive elements present in such structures.[110] Ironically one was the West Gate Bridge over the Yarra-Yarra near Melbourne, a few miles from where Fairbairn had successfully bridged the same river with a tubular-girder bridge in the 1850s.[111] If the West Gate Bridge had been lifted as a complete unit, as Britannia was, buckling and deflection problems would have been avoided.[112]

Years later Fairbairn wrote of his thoughts on his resignation from Britannia:

> the question was, shall I continue to build bridges. I chose to do so, and I believe I did right, as the principle was quite new, and no one understood the construction so well. I therefore gave designs, and received orders for more than one hundred bridges in the course of a very few years. Up to the present time, 1870, I have built and designed, with the assistance of the Fairbairn Engineering Company, nearly one thousand bridges.[113]

Tubular-Girder Bridges

FAIRBAIRN developed the tubular-girder bridge – comprising wrought-iron tubular girders on each side, with the deck spanning between them – during the Britannia experiments. It was a direct derivative of the tubular bridge, as Fairbairn instinctively applied the principle to a solution for medium spans. There was a primitive form of tubular girder, with inclined sides, from c.1840 on the Pollok & Govan Railway, and at Govan Iron Works,[114] but it is doubtful if Fairbairn was aware of it. Thrust into prominence by a spectacular – and, for Fairbairn, opportune – failure of a trussed cast-iron girder bridge, tubular-girder bridges provided an available and reliable alternative, at a time when lattice and Warren girder bridges were still in their embryo stages. In doing so they brought substantial rewards to the family firm during the short window of opportunity of only a decade or so before more theoretical trussed solutions rendered tubular-girder bridges obsolete.

Fairbairn's first reference to a tubular-girder bridge was in a letter to Robert Stephenson in February 1846, proposing one for the Dee crossing, but Stephenson preferred a cast-iron beam trussed with wrought-iron rods.[115] The tubular-girder bridge was the subject of a patent, filed October 1846 and granted April 1847, in Fairbairn's name.[116] Stephenson shared the cost of the Patent but then declined to be involved.[117] Fairbairn proposed to charge parties using the patent a licence fee of £1 per foot of span.[118] On 20 April 1847, fourteen months after his proposal to Stephenson, Fairbairn presented what purported to be a short paper to the Institution of Civil Engineers about a relatively minor accident at a small provincial mill. A cast-iron beam, trussed with wrought-iron rods, had failed:

> The three large girders, although not of the best form, were ... a near approximation to it, when acting without trusses; but with those auxiliaries they were decidedly disproportionate, and more particularly defective in the top ribs, which rendered them exceedingly precarious, and decidedly unfit for supporting the load placed on them.[119]

The atmosphere in the meeting must have been electric, for Fairbairn was taking on the engineering establishment with a condemnation of the principle of the trussed beam. In the previous four years at least thirty-four trussed cast-iron girder bridges had been built, eight of them by Stephenson.[120] Robert Stephenson

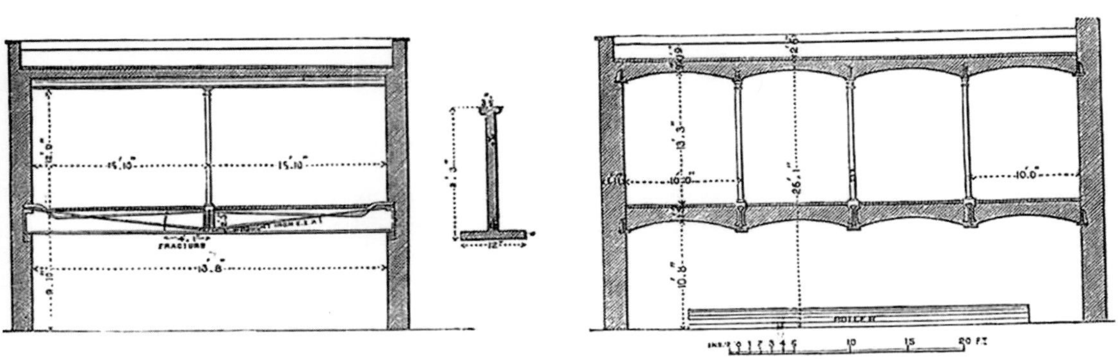

Fig.5.15: Gray's Mill, Manchester, prior to collapse.

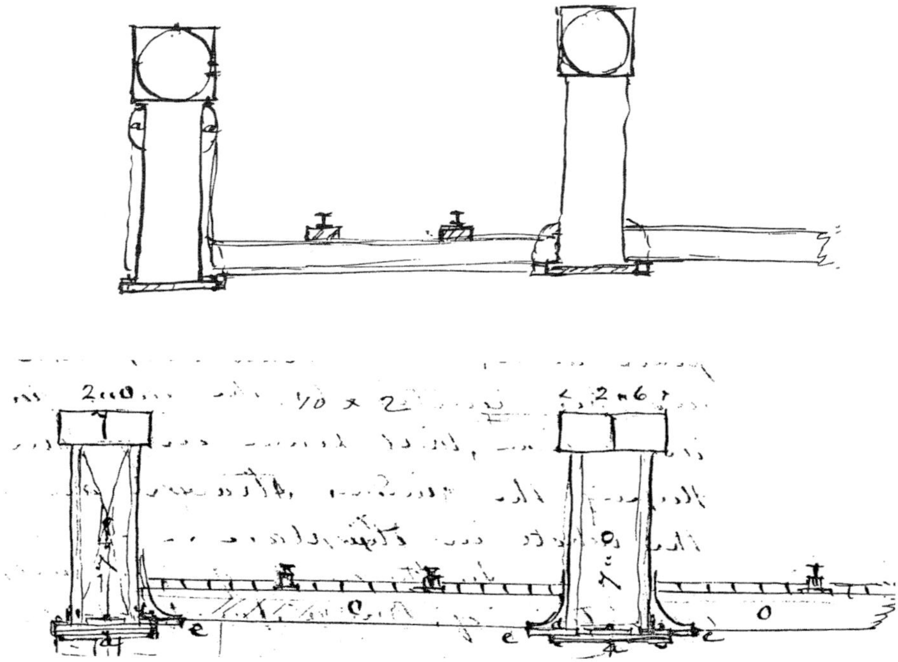

Figs 5.16 & 5.17: Fairbairn's first and second sketches for the Dee Bridge.

was present, as were Bidder and Vignoles (who had built the first trussed-girder bridge in 1831). They and others sought to justify this form of construction. Only George Rennie was clearly with Fairbairn.[121] A month later, on 24 May 1847, Stephenson's Dee Bridge collapsed, vindicating Fairbairn's paper. Within three weeks Fairbairn wrote again to Stephenson, again suggesting a tubular-girder bridge, but again it was rejected.[122]

In spite of Stephenson's claim, supported by Locke and others, that the accident was caused by a derailment, the inquest jury criticised the design.[123] The outcome of the collapse and inquest was that the Board of Trade sought details of iron bridges in use or under construction from 226 railway companies, obtaining replies from 174,[124] and a Royal Commission was set up to inquire into the Application of Iron to Railway Structures. It produced its substantial Report in 1849, with evidence from many engineers. Stephenson considered girders formed of separate castings with a tension-rod along the bottom to be 'as good a form as any'. Fairbairn and Brunel were critical of this form of construction. The Commissioners were equivocal, advising caution, but not condemning trussed girders.[125] Fairbairn, entirely typically, put in hand a comprehensive and systematic series of experiments on various trussed beams:

> On the safety of these tension-rods I have always had serious apprehensions; but as many other persons of highly distinguished attainments hold a different opinion, it may not be considered irrelevant if I adduce my reasons for the view which I take, and the experiments upon which those reasons are founded.

An additional motive may have been to boost the tubular-girder. In any event

MR. FAIRBAIRN'S WROUGHT-IRON GIRDER-BRIDGE OVER THE LEEDS AND LIVERPOOL CANAL.

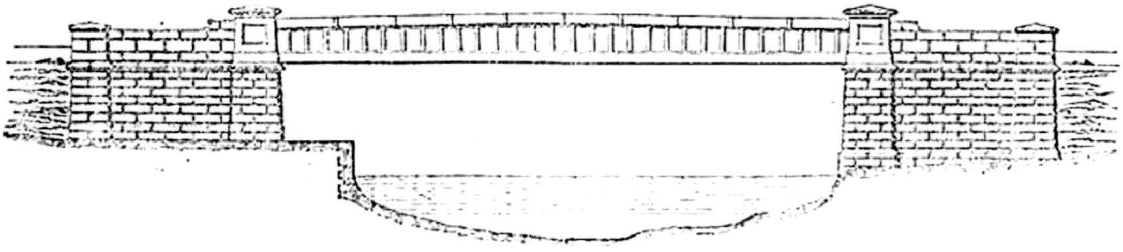

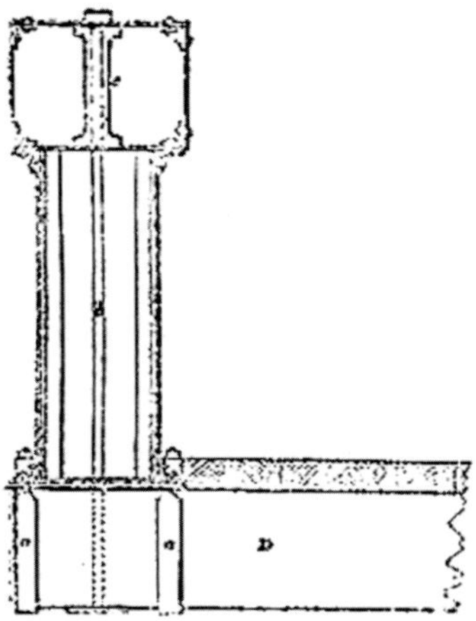

Fig. 3.—Transverse Section of Girder enlarged.

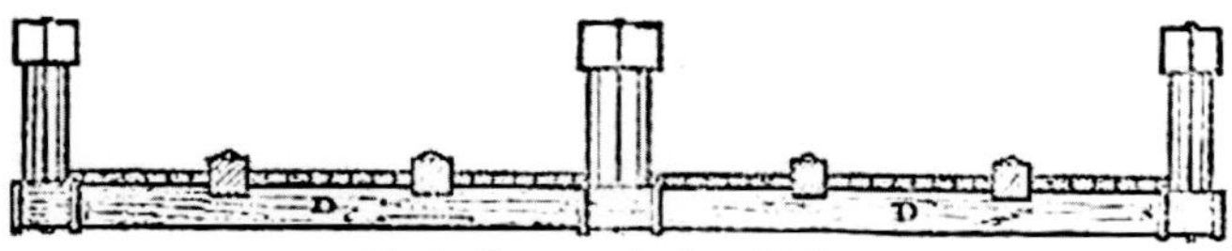

Fig. 2.—Transverse Section of Bridge.

Fig.5.18: **The First Tubular-Girder Bridge, at Blackburn,** 1847.

his results were clear: 'What is … infinitely preferable, is a well-constructed malleable beam'. He emphasised the differences between the two metals, notably the ductility of wrought iron causing elongation when acted on by a tensile force, and the difference in 'set' when the two metals are released from a tensile force. If the two metals could be brought to act in perfect concert, there could be advantages, but this was impractical. He concluded that 'within comparatively small limits of load, a truss-beam may pass from a condition of perfect security and safety to one of uncertainty and danger'.[126]

One outcome of the Dee disaster and inquest was that many engineers immediately stopped building trussed-girder bridges. For example E.T. Bellhouse was building nine of these bridges on the Manchester South Junction Railway at the time. Work was stopped and cast-iron segmental arch bridges were substituted.[127] For medium and long spans cast-iron arches were expensive, and where headroom was an issue they were unsuitable. The Commission had dismissed lattice bridges in a single sentence following the views of Stephenson and Brunel.[128] It was unable to express an opinion on tubular-girders because they were 'of such recent introduction'; but noted that engineers were 'for the most part exceedingly favourable towards them', thus encouraging their use. Examples from Blackburn and Ardwick were illustrated.[129] The Dee disaster could not have come at a more opportune time for Fairbairn, a month after the patent was granted, as the first tubular-girder bridges at Blackburn and Bolton were almost complete, and before there was confidence in lattice or other trussed bridges.[130] The tubular-girders filled a void that the collapse created. By the end of 1847 there were around twenty-four tubular-girder bridges completed or under construction,[131] not all by Fairbairn as some had cast-iron tops.[132]

With the patent in place and inundated with orders from railway engineers with deadlines to meet, Fairbairn was in an exceptionally strong position. The bridges commanded 'high and remunerative prices', such that Pole recorded, 'the manufacture of these alone realised a fortune'.[133] It is unlikely that the initial flood of orders could all have been built at Ancoats, and some were probably sub-let or built under licence.

The first two Fairbairn tubular-girder bridges were built for C.B. Vignoles, Engineer to the Blackburn, Darwen & Bolton Railway, to carry the railway over a road and a canal at Blackburn.[134] They were widely illustrated.[135] In June 1847 Fairbairn was on site at Blackburn and, ever alert to the value of experimental testing, ran three coupled engines, 60 tons in total, over the road bridge at speeds up to 25mph. This established that deflection was the same at all velocities. He then affixed wedges to the rails, causing the locomotives to rise up and then drop down onto the rails. Deflection increased but the girders 'restored themselves to their original position'. This is believed to be the first impact test on a wrought-iron structure, and was a harbinger of subsequent fatigue testing.[136]

Around the same time Fairbairn was involved in two tubular-girder bridges in Bolton, for James Thomson, Engineer to the Liverpool and Bury Railway, which appear to have been fabricated by Garforths of Dukinfield.[137]

One of the engineers who abandoned intended trussed cast-iron girders was John Fowler (1817–98), of the Manchester, Sheffield & Lincolnshire Railway, who would go on to design the Forth Bridge with Benjamin Baker.[138] Faced with the need to act quickly, Fowler opted for Fairbairn's tubular-girder bridges. By the end of 1847 he was committed to at least six of them.[139] Gainsborough, with two spans of 154ft and completed in 1849, was the longest tubular-girder bridge then erected. It was fabricated by Fairbairns. Fowler and Fairbairn supervised its erection. To keep the navigation open, the two spans were joined together on one bank and the 336ft tubes were rolled out over the river, cantilevering 130ft before receiving support from a temporary roller carried on piles.[140]

Another of Fowler's Trent bridges was at Torksey, where controversy led to a deeper understanding of the continuity of beams over their supports, and Fairbairn's mathematical weaknesses became clear. The girders, Fowler claimed, followed Fairbairn's guidance for an earlier bridge, but it appeared that Fairbairn had changed his guidance.[141] As at Gainsborough the two spans were assembled together on a bank and rolled out over the river using intermediate stagings. No thought was given to the effects of continuity over the supports, which occurred, unplanned, as a sequel to the mode of erection.[142] In December 1849 Captain Simmons tested and declined to accept the bridge, despite leading engineers declaring their convictions about its sufficiency.[143] He noted the beams 'were not built in a very line, nor were they very accurate in form; and several rivets appeared to be not very perfect. The deflection was more than he anticipated'.[144] Concern about Government interference in engineering was mounting in the profession and Fairbairn was asked to read a paper about Torksey at the Institution of Civil Engineers in March 1850. Many leading engineers were present. Apart from the concerns about interference, there were two primary issues. First, it appeared that Fairbairn had changed his design recommendations. Fowler found it extraordinary that proportions which Fairbairn had endorsed for another bridge

Fig.5.19: Gainsborough Bridge, built 1849.

which had performed efficiently for two years, he now deemed insufficient – a point to which Fairbairn gave no explanation. This suggests that the Torksey girders may have been built under licence by Fowler, and not at Ancoats.[145] Fairbairn took the view that whilst the girders did not attain the strength he recommended, he considered them 'sufficiently strong to render the bridge perfectly secure'. The second issue was that Fairbairn had not taken into account the increase in strength due to continuity over the central support.[146] It was a highly charged meeting from which Fairbairn apparently walked out.[147] Knowledge of the effects of continuity over intermediate supports was available in Henry Moseley's celebrated book on the mechanical principles of engineering

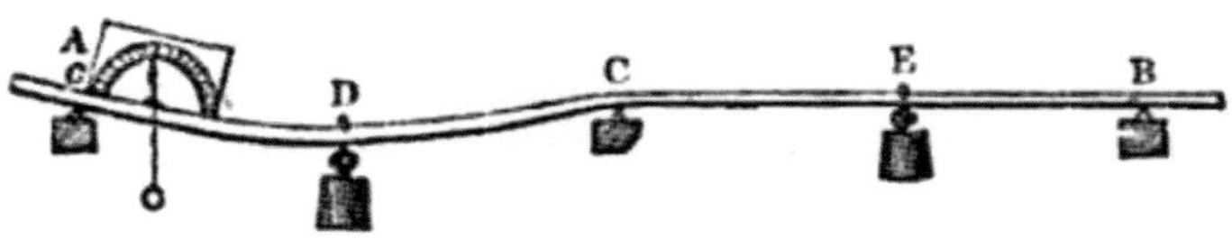

Fig.5.20: Moseley's diagram of the effects of continuity over a support (1843).

(1843).[148] C.H.Wild (1819–57) produced calculations showing that Simmons' requirements were met and trenchantly criticised Fairbairn's dismissive approach to the continuous beam issue – Fairbairn was 'not only unphilosophical, but positively dangerous'.[149] He produced a diagram of a model he had tested. William Pole (1814–1900) exhibited a very similar diagram and presented an exhaustive mathematical analysis, using calculus, of the behaviour of the Torksey bridge relating to deflection and stresses due to traffic.[150]

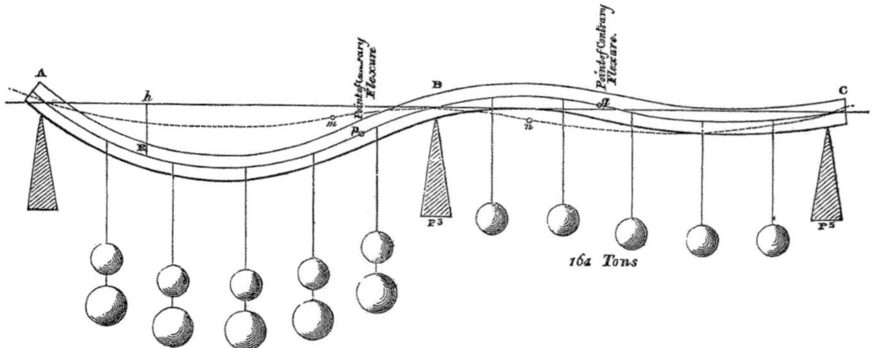

The dotted line represents the position of the girder as deflected by its own weight only, viz. 154 tons distributed over each span.

Fig.5.21: C.H.Wild and W. Pole's Torksey diagram of the effects of continuity over a support.

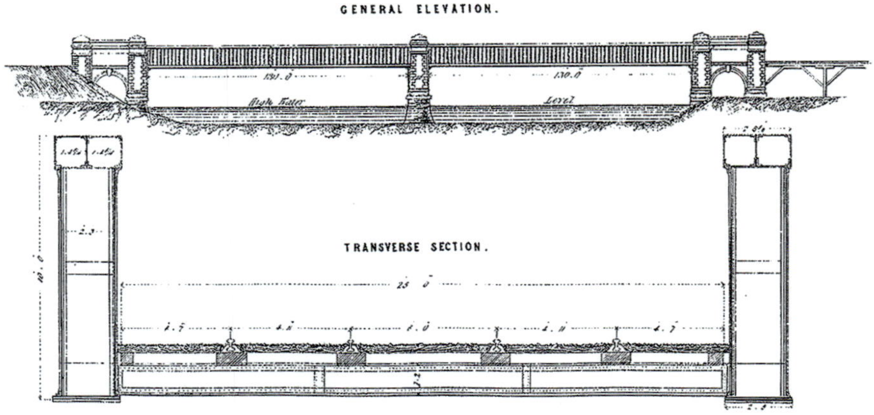

Fig.5.22: Torksey Tubular Girder Bridge, 1849–50.

On 28 March, without any strengthening of the bridge, but following further extensive testing and the theoretical arguments about the effects of continuity over the intermediate support, Simmons allowed the line to open.[151] Whilst this was a vindication of Fairbairn's claim of the sufficiency of the bridge structure, it also pointed to his lack of appreciation of the effects of continuity, highlighting his mathematical weakness. The importance of Torksey was the development of continuous beam theory, of which it was the catalyst. The knowledge diffused quickly.[152] It was particularly valuable for the Britannia Bridge, on which Pole did the theoretical work,[153] and which Lindenthral saw as 'laying the foundations for the development of the modern cantilever bridge'.[154] Fairbairn, perhaps somewhat grudgingly, included Pole's results in his *Application of Cast and Wrought Iron to Building Purposes* with the qualification that he would 'recommend the exercise of caution in trusting to theoretical formulae',[155] a view echoed after the collapses of the 1970s by the Merrison Committee's surprise 'at the paucity of research … put into examining the behaviour of steel box girders'.[156] Yet Pole's work on

Fig.5.23: Torksey Bridge, 2008.

Fig.5.24: **Barwon Road Bridge, Victoria** (1855–9).

the strength and deflection of the Britannia Bridge, including the method of equalising bending moments due to self-weight at mid-span and support points, was of great importance. In the 18th century the Bernoullis and Euler had studied geometrical forms of elastic curves.[157] A century later Todhunter and Pearson wrote that the Britannia and Conway Bridges comprised 'probably the most important problems to which the Bernoulli-Eulerian theory was ever or ever will be applied',[158] and today Pole's work is considered to be 'especially significant in the history of theory of structures'.[159] The Torksey episode highlighted the rapid transition from rule-of-thumb and empirical formulae to structural calculations, a transition with which Fairbairn was not entirely at ease.

By 1851, Fairbairn had erected more than a hundred tubular-girder bridges and more followed.[160] Known bridges with which Fairbairn was involved are listed in Appendix 5.1 where details of the main sources are given. Examples in Ireland at Cahir, Ballinasloe, and Galway, were associated with the engineer G.W. Hemans,[161] and the contractor, William Dargan.[162] In Scotland there were bridges, associated with Joseph Mitchell,[163] over the Findhorn and the Spey. Fairbairns exported tubular-girder bridges to Canada, Spain, Portugal, and Australia. The Barwon Road Bridge, Victoria (1855–9), had the longest bridge spans (210ft) in Australia until the Fitzroy River suspension bridge in 1881.[164]

The influence of the tubular-girder bridge during its short window, as seen in its diffusion to other manufacturers, was limited. In France there was a brief vogue for the bridges in the period 1851–6.[165] In 1857 Charles Marcellis erected a tubular-girder swing bridge at Antwerp harbour, 'an almost exact replica of a type [of tubular-girder] Fairbairn had patented more than 10 years earlier'.[166] In Australia there were box girders by John Fowler over the Nepean at Menangle and Penrith, and over the Maribyrnong at Keilor, all fabricated at Peto & Brassey's Birkenhead works.[167] The five-span Taradale Viaduct was also made in England.[168] Its Australian contractors imported a riveting machine from Fairbairns.[169] In Canada there were a number of tubular-girder bridges on the Grand Trunk Railway, supplied by Jackson, Peto, Brassey & Betts,[170] and one with three 100ft-spans over the Saint John River in New Brunswick, designed by A.L. Light.[171]

Within seventeen years of having built their first tubular-girder bridge, Fairbairns built their last. The experience at Cologne was followed by similar for bridges at Drogheda, and Calcutta. James Barton was one of a new generation of Irish engineers influenced by engineers on the Continent. He was appointed engineer, under John Macneill, for a bridge to span the Boyne at Drogheda. Macneill, under whose direction the first wrought-iron lattice girder bridges had been built at Raheny and Dublin, proposed a three-span lattice bridge but problems with his

Fig.5.25: Fairbairn Lattice Bridge over the Tay at Dalguise (1862–3).

earlier bridges were causing second thoughts. The matter was resolved when they met C.H. Wild who had a plan of 'Warren's patent girder', marked up with the 'strains and compressions' that the different parts of the girder were subjected to by a given load.[172] With Wild's help Barton modified the design for Drogheda. He then constructed and tested 60ft lattice and tubular girders, finding the lattice advantageous.[173] At the British Association meeting in Belfast in 1852, Barton read a paper referring to his tests.[174] Fairbairn was present and came out of the discussion rather badly.[175] Barton's bridge was built and is believed to have been the first iron bridge designed in accordance with stress calculations.[176]

George Turnbull left England in March 1850 to become Chief Engineer to the East India Railway Co. whose consultant engineer in London was J.M. Rendel, President of the ICE (who had been one of those present atop the first Conway tube when it was floated in 1848).[177] In August 1850 Turnbull sought Rendel's approval for a tubular-girder bridge to cross the Ballee Khal, near Calcutta. Approval was not forthcoming. Since Turnbull had left England, advised Rendel, the Warren truss had attracted attention. Composed of comparatively small pieces, it could be erected by unskilled mechanics whereas tubular-girders were sent out in lengths which needed to be riveted together on site. Rendel obtained estimates of £4,170 for the tubular-girders, and £3,090 for the Warren trusses, had tests undertaken, and on 7 July 1851 wrote to the East India Company, 'I am now in a position to recommend this description of bridge [i.e. Warren trusses] without qualification'. The recommendation was followed and there do not appear to be any tubular girder bridges in India.[178]

In April 1855, Barton read a paper at the Institution of Civil Engineers, and described his Drogheda Viaduct.[179] The discussion continued over three evenings, concentrating on the comparison of tubular girders, Warren girders and lattice trusses. Criticisms of tubular structures on the grounds of uneconomic use of material were rebutted unconvincingly by Stephenson and Bidder, both in their fifties. They were seen as 'the old guard', being replaced by the new, such as James Barton, W.B. Blood, W.T. Doyne and J.M. Heppel, all in their thirties, and all wanting to be considered part of the more theoretically educated engineering elite, found on the Continent.[180] Further attacks on tubular structures followed from Zerah Colburn, the American who edited *The Engineer,* drawing on his experience in the United States.[181] In 1867 Benjamin Baker's, *Long Span Railway Bridges,* identified the tubular-girder as the most unfavourable type of long-span bridge because of its uneconomical use of material.[182] But by then the contest was over. The new guard had won decisively. Theoretical calculations had supplanted empirical formulae and the rule-of-thumb. By the early 1860s the tubular-girder bridge was discarded, and Fairbairns, like everyone else, had moved on to lattice

Fig.5.26: Liverpool Landing Stage.

or other trusses.[183] One of their first was over the Tay at Dalguise, and Fairbairn, well into 'retirement', tested a quarter-size model, but he did stress the need for calculations of the horizontal strains to which the top and bottom were subject – calculations which were undertaken by William Unwin.[184]

Compared with the great lattices at Tczew and Cologne, that over the Tay was trivial. Leadership in lattice girder bridges had passed to Continental Europe and America. In 1864–6 Fairbairns built the Warehouse Point Bridge in Connecticut on the Hartford & New Haven Railway, to the design of James Laurie. It was the most extensive iron bridge in the United States, 1,525ft long with seventeen spans of up to 177ft – plate girders for the short spans and Whipple trusses for the longer spans.[185] In 1871 Fairbairns undertook its largest contract, 140 lattice and plate girder bridges for the Intercolonial Railway of Canada. The Intercolonial chose Fairbairns' double-lattices for intermediate spans, but for the longest spans it chose pin-jointed trusses from a firm in Philadelphia.[186] Neither Fairbairns nor Britain were any longer in the forefront of bridge design. That tubular-girder bridges did flourish for the decade from 1847 was primarily due to them being readily available following the demise of the trussed cast-iron girder, just prior to lattice and other wrought-iron trusses establishing their viability. It may be, as James suggests, that the dominance of tubular-girder bridges for 100–200ft spans during this decade hindered development of other truss forms, contributing towards the decline of British bridge design in the late nineteenth century'.[187] If this is correct, the decline stems from the collapse of Stephenson's Dee Bridge. If the Dee Bridge had failed five years later, it is likely the void would have been filled by lattices or other trusses, rather than by tubular girders.

Fairbairn had the insight to identify parallel uses for his 'inventions'. In doing so his influence was wide. He brought to bridge-building knowledge from his experiments on wrought-iron plates and riveted joints, and from his shipbuilding experience. He then advocated the transference of cells from his bridge experiments back to shipbuilding, seen most notably in Brunel and Scott Russell's *Great Eastern.*[188] Fairbairn's adoption of a rectangular tube with sandwich flanges was novel and advanced: today stiffened plates are basic components of bridges, buildings, motor vehicles, aircraft, offshore platforms and ships.[189]

George's Landing Stage at Liverpool, 505ft by 80ft, located 120ft from the quay wall, was designed by William Cubitt in 1847–8.[190] To reach it Fairbairn constructed a tubular-girder access ramp 150ft long, hinged at each end, to rise and fall with the tide. Further similar ramps followed at Liverpool and elsewhere.[191] Fairbairn was also involved in the early development of tubular-

Fig.5.27: Keadby railway swing bridge over the Trent, 1863–4.

girder railway swing bridges. Around the same time as the first Liverpool Landing Stage, Cubitt was the engineer for a railway swing bridge at Folkstone Harbour,[192] and it seems likely that this used tubular girders. In 1849–50 both of Fairbairn's proposals for the Cologne Bridge involved swing sections at each end.[193] There was a swing bridge at Perth,[194] and a swing section in his Lough Attalia Bridge at Galway of which Humber illustrates the technical details.[195] In 1860 Fairbairns built a swing railway bridge over the Desjardins Canal at Hamilton, Ontario,[196] in 1862 over the Caledonian Canal at Clachnaharry,[197] culminating in 1863–4 with Keadby Bridge over the Trent – providing two 60ft navigation channels, with the bridge manually operated by two men in four minutes.[198]

As early as 1846, Fairbairn advocated a wrought-iron aqueduct over the river Weaver, a 100ft tubular girder 6ft deep by 2ft wide,[199] but it is not known if it was built.

Alongside tubular-girder bridges were wrought-iron plate-girder bridges, replacing cast-iron beams. These were single-web beams built up from wrought-iron plates, which Fairbairn popularised and which were used by engineers around the world in their thousands for relatively short-span bridges.[200] For example, in the mid 1850s Fairbairns built one for John Fowler to carry the Sheffield Goods Branch railway over a canal and tramway. The distance between abutments was 158ft and there were two wrought-iron plate girders168ft long supported at intermediate points by cast-iron columns, dividing the bridge into two unequal spans of 60ft and 96ft.[201]

Whilst tubular beams had been used in buildings for some years, their scale increased. In 1847 Fairbairn received a commission for a 63ft tubular girder to support the ends of several trusses over the new Liverpool Lime-street Station.[202] This roof, designed and built by Richard Turner of Dublin,[203] was 374ft long with a clear span of 153ft 6in. – an 'unprecedented span'[204] at that time. It was considered visionary and Turner looked to Fairbairn and James Kennedy (of the Clarence Foundry, Liverpool) for support. This was forthcoming but, typically, Fairbairn advised:

> I should, however, recommend Mr Locke, the Engineer, and the Directors, not to trust to computation, but to submit your principal rafters, truss-bars, frames, &c to direct experiment.[205]

Floors with built-up wrought-iron beams were also seen, at least by their manufacturers, to have their genesis in the great tubular-bridge research. Thus a letter from Cooper and Hewitt of New York in 1854, puffing their fire proof floors, refers to:

> wrought iron girder construction on the most approved philosophical principles developed in the thorough experiments of Messrs Stephenson,

Hodgkinson, and Fairbairn, made in reference to the construction of the Great Tubular Bridge over the Menai Straits.[206]

Around 1850 major works were undertaken at Keyham dockyard largely under the direction of the very able Deputy Director of Admiralty Works, William Scamp.[207] At the time Fairbairn was supplying cranes to Keyham.[208] Scamp proposed a rectangular floating wrought-iron caisson for the sea lock, retractable into a recess at the side of the lock.[209] Caissons, or floating dams, for use in place of larger lock gates, went back to the end of the 18th century, but pumping out the water from them and hauling them out of the way, was cumbersome and slow.[210] A dock gate on wheels had been proposed by the Dutch engineer Singels in 1839.[211] Scamp's design was for a rectangular iron caisson with an air chamber and a ballast chamber containing water, such that by a minor adjustment of the water level in the ballast chamber the caisson could be raised several inches, so that it would float. The lock entrance was 80ft wide at the top and 65ft 8in. at the base, by 42ft high. Scamp consulted Fairbairn. Reference was made to 'the enormous strength of the malleable iron constructions for the Conway and Britannia bridges'. Fairbairn prepared drawings and constructed the caisson. It was 14ft 10in. between faces, with a roadway on the top, and weighed 290 tons. Yet it only required a few men on a capstan to draw it into (or out of) its recess and the whole operation took less than ten minutes. True to form, in July 1850 immediately after completion, Fairbairn measured the deflection of the caisson at four heights, with water levels from 16ft to 38ft at increments of 1ft.[212]

Rolt and Cardwell have both put forward the view that following the strength of box sections having been demonstrated at Conway and Britannia, Fairbairn's friend, Whitworth, extensively adopted the box section in his machine-tool designs, 'the effect being to exaggerate their uncompromising, rectilinear appearance'.[213] The heavy frames of machine tools were replaced by lighter, hollow ones, as 'the thin but strong metal tube became a feature of engineering structures down to the present day'.[214]

Tubular Cranes – Icons of the Victorian Waterfront

THE major derivative from the tubular-girder bridge was the tubular crane, which used a tapering, curved, inverted version of the tubular-girder as its jib. Fairbairn did not invent the curved jib,[215] but he grasped it and married it to tubular wrought-iron, to the extent that he gave his name to this genre of crane, which became icons of the Victorian waterfront in many parts of Europe. The crane emphasises Fairbairn's technological perception – visualising a level tubular bridge girder as a quadrant forming the jib of a crane. It illustrates the transfer of technology from one iron structure to another. And this was coupled with the drive and investment to realise the vision – to build the cranes and persuade a conservative Admiralty to purchase these new machines. Diffusion of the crane was rapid, to harbours, dockyards, and quays in Europe and beyond. They rode the crest of a wave in the two decades 1850–70. Thereafter, although some were still being built as late as the first decade of the 20th century, they faded – overcome by the advent of steel, by new forms of crane, new sources of power, and the need for greater speed and flexibility.

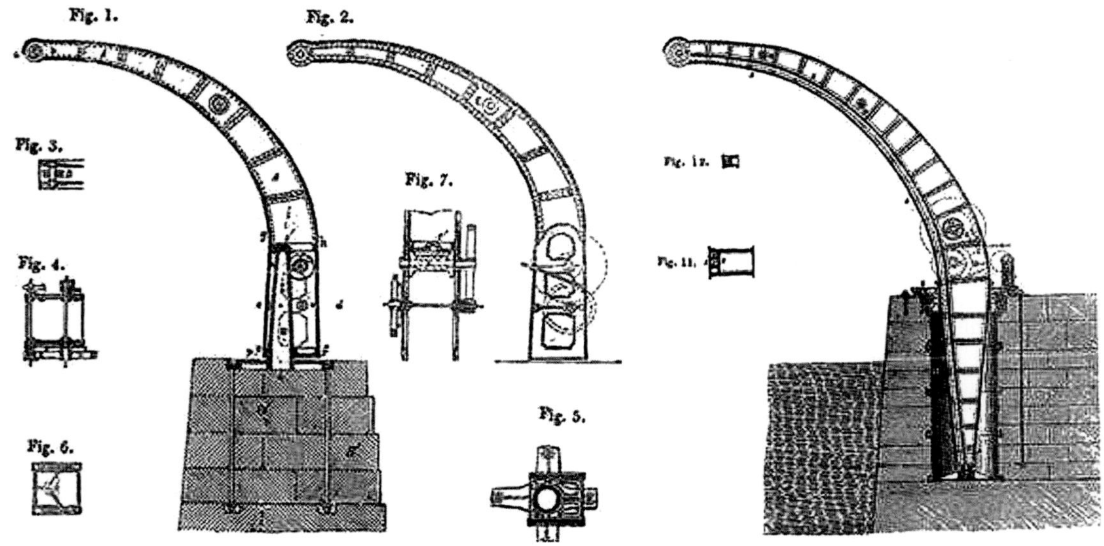

Fig.5.28: Drawings from the Fairbairn Crane Patent.

Fairbairn's understanding of wrought iron facilitated the transfer of its technology from one form of structure to another – the perception that although they had totally different functions, the hull of a ship, a bridge, and the jib of a crane had similarities, and could all be treated as tubular beams. The patent for the crane was filed in November 1850, and was for constructing the jib from metal plates, forming a series of tubular or cellular compartments. The patent illustrated two versions of the crane, the first, for small loads, was mounted on a traditional crane post. The second, for heavier loads, had the jib strengthened by three cells on the concave side of the tube, and below ground a much more substantial foundation with a tubular structure tapering as an elongated cone, within a cylindrical cast-iron casing, and restrained at ground level.[216] Curved jibs had an advantage over straight jibs which prevented bulky loads, such as ships' boilers, being raised to the full height because they would foul the jib, a point illustrated by Fairbairn.[217]

Fairbairn's crane was designed primarily for dockside use. Its tapering, curving, box section was instantly recognisable. The cover of *The Mechanics' Magazine* of 17 May 1851 carried illustrations from the Patent. A paper about the crane was read at the British Association meeting in 1850, and widely reported.[218] In 1851 Fairbairns exhibited a 5-ton crane at the Great Exhibition,[219] and in 1853 at the Dublin exhibition.[220] In that year the 4th edition of Ure's *Dictionary of Arts, Manufactures and Mines* included an illustrated entry on these cranes.[221] By this time they were generally referred to as 'Fairbairn cranes' and the designation was universally adopted, even though later examples were built by other manufacturers.[222] Nothing like them had been seen before in the United Kingdom, although a correspondent to the *Mining Journal* in 1850 claimed that a tubular crane of boiler plate had been made in Belgium in 1846.[223] They met a need and quickly became 'the cranes of choice in harbours'.[224]

The first order came from the Admiralty for cranes to lift 12 tons, but to be proved to 24 tons, for the new docks at Keyham and Devonport. Fairbairn's tender – the highest – was accepted subject to only one crane being supplied in the first instance and tested at Fairbairn's expense.[225] The tests were satisfactory and another five were supplied.[226] At the same time cranes were constructed at Southampton and Birkenhead. The first international orders were from Russia, following Fairbairn's visit to St Petersburg and Cronstadt in July 1850.[227] By 1857 there was a colossal 60-ton crane at Keyham. The point of the jib was 60ft above the quay – the height of a six-storey building – and the base of the inverted cone 23ft below ground level. The crane swept a circle 106ft in diameter. The applied power was increased by a factor of 632, by the gearing and pulleys.[228] It was the most powerful crane in the world.

The next step was to bring steam power to the cranes. John Trickett, Chief Engineer at Keyham did this for his cranes around 1858.[229] In 1859 a 60-ton crane at Portsmouth had steam power from the start.[230] A change in construction had occurred by this time in all but the largest cranes – T-iron ribs were substituted for the cells, giving adequate strength more economically.[231] Around 1859 Fairbairn proposed a crane 120ft high – equivalent to a typical twelve-storey building – as a substitute for the worn-out masting shears at Woolwich, but it was not built.[232] After Fairbairn's 'retirement', the firm continued to build cranes[233] and exports *probably* included Helsingør,[234] Newcastle NSW,[235] Seville,[236] and Venice.[237]

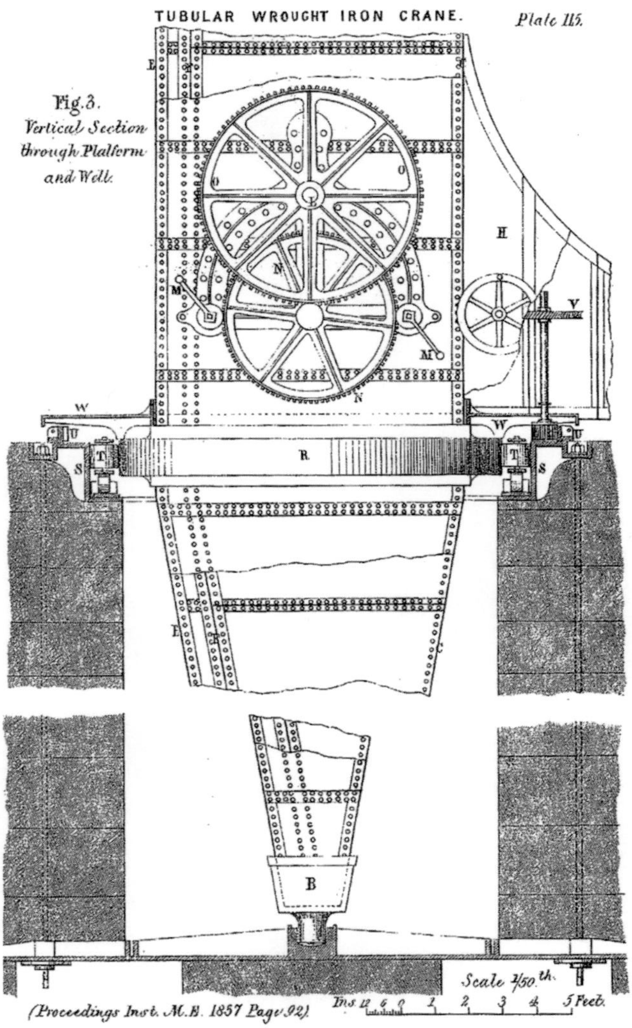

Fig.5.29: Details of 60-ton Fairbairn manually-operated crane at Keyham.

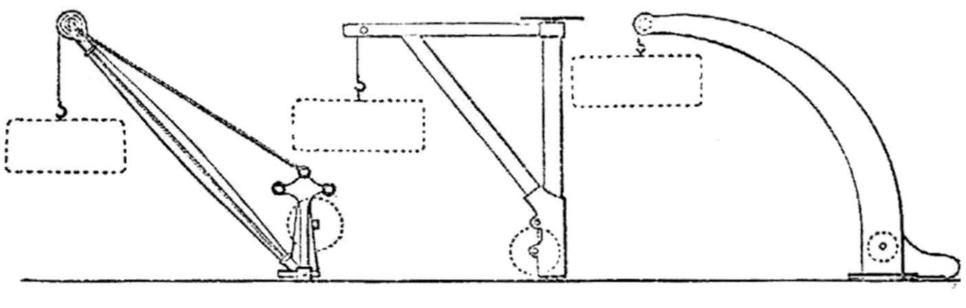

Fig.5.30: Straight and Curved Jibs.

Fig.5.31: 'Fairbairn' Crane at Yokusuka, Japan.

Diffusion of Fairbairn's cranes can be traced to Holland and thence to Japan. They were manufactured in Holland from 1861 at the Prins van Oranje factory at The Hague, managed in the 1860s by Cornelius Hotz.[238] Their design and detailing were very similar to cranes built in Ancoats, suggesting they may have been built under licence.[239] The alternative is that they were copied or the patent had expired. There were close links between the Prins van Oranje factory and British engineering firms – an advertisement of 1867 listed four British firms for which Prins van Oranje was agent, but Fairbairns was not one of them.[240] The factory's records have not survived, and it remains unclear how Prins van Oranje came to build 'Fairbairn' cranes.[241] They built at least twenty-five such cranes and probably more.[242] In the 1850s, after 200 years of isolation, Japan began to open to western nations. In 1862 a Japanese delegation visited the west, notably the International Exhibition in London. In Holland they visited the Prins van Oranje factory.[243] In 1866 the Tokugawa government built the Yokasuka military arsenal and naval base.[244] By 1872 it had a 30-ton 'Fairbairn' crane. It was unusual in that its height was increased by being built on a cast-iron and masonry plinth.[245] Whilst documentary proof is not available, all the indications are that this crane was manufactured at the Prins van Oranje factory.

Fairbairn's influence can also be followed through cranes built by other manufacturers after the patent expired or after Fairbairns ceased trading. The best known is Stothert & Pitt's Bristol crane, the only 'Fairbairn' *steam* crane to have survived.[246] Other British examples included Davis & Primrose's crane at Sunderland;[247] and what is believed to have been the largest and strongest 'Fairbairn' crane – 100-ton capacity, 80ft high with a radius of 60ft, at Hartlepool, built in 1906 by Cowans Sheldon of Carlisle to handle ships' boilers.[248] Cowans Sheldon built several 'Fairbairn' cranes including two at Malta Naval Dockyard as late as 1908.[249] Cail, Halot & Cie of Brussels built a 50-ton crane at Vlissingen in the Netherlands in 1878.[250] Bachman states that numerous German firms constructed 'Fairbairn' cranes but does not identify locations or firms.[251] 'Fairbairn' cranes remain at Frankfurt an der Oder built by Gruson Maschinen Fabrik of Buckau Magdeburg, and on Lake Constance. Elsewhere there are existing cranes

Fig.5.32: Section of large tubular crane at Fairbairn's yard. (Roof of Royal Albert Hall behind).

Fig.5.33: Erection of crane at Maassluis, Netherlands.

on the Donau at Vienna and at Visby on the Swedish island of Gotland.[252] All these cranes, their manufacturers and geographical locations, are witness to Fairbairn's influence and the spread of his technology from Ancoats. Appendix 5.2 lists known 'Fairbairn' quayside and similar cranes. Fairbairn's cranes were fabricated at the Ancoats works, assembled in their yard and then sent to site in sections.[253] Evidence of the means of site erection is meagre, but it is known that at Maassluis in 1892 sheerlegs were used.[254]

Scholars who 'cast doubt on the importance of individuals often rely on a dispensability axiom: if an invention had not been made by *X* it would have been made by *Y*'.[255] It is very hard to imagine that such an axiom would have applied to the wrought-iron, swan-neck, tubular crane, if Fairbairn had not invented it.

Fig.5.34: Fairbairn crane in the company's own boiler shop, with Lancashire boilers, c.1860.

Derivatives and Demise of the Fairbairn Crane

THERE were derivatives of the Fairbairn crane, especially in Germany – not strictly Fairbairn cranes, but where the influence is indisputable. Weisbach and Herrmann refer to ten hydraulic cranes on crane-posts at Gustemünde, c.1866, 'constructed according to *Fairbairn's system*'.[256] Otto Lueger in the 1904 edition of his technical dictionary illustrated a small 'Fairbairnkran' which was post-mounted with only a partially curved jib; and the 'Fairbairn-Ausleger', a steam-powered crane on tracks.[257] Later editions of *Meyers Konversations-Lexikon* illustrated a 'Fairbairn-Kran' with tapered conical section below ground and largely straight jib.[258] Anton Böttcher's encyclopaedic book about cranes illustrates travelling and hydraulic cranes that are clear Fairbairn derivatives, mostly relatively small and dating from the beginning of the 20th century.[259] Some Fairbairn cranes were used in industry. Fairbairn had one in his own boiler shop. Most spectacular in this field, and a clear testimony to Fairbairn's on-going influence, were four cranes built in 1877 at the famous Schneider works at Le Creusot in France to serve the most powerful steam hammer in the world. These steam-powered swan-neck cranes lifted ingots weighing up to 120 tons and with their tubular curving jibs, gearing mechanisms and tapering bases within cylinders, were in every respect 'Fairbairn' cranes.[260]

Fairbairn's influence can also be appreciated in the genesis of the railway breakdown crane. During the 1850s, he applied tubular construction to a travelling railway crane. It lifted 15 tons to a height of 18ft above the rails and swivelled in a 25ft diameter circle, but its height limited its use to rails unobstructed by bridges.[261] However Appleby Brothers subsequently modified it, by hinging the jib, leading to the typical railway breakdown crane of the 20th century.[262]

There were criticisms of the 'Fairbairn' cranes – that they were uneconomic in their use of iron, they were difficult to paint and maintain within the jib, and they could only lift and lower loads on the circle described by the point of the jib. William Anderson, after a premium apprentice with Fairbairn, moved to Courtney

Stephens in Dublin in 1855 and developed a 'Fairbairn' crane with lattice jib and base, to address the first two of these criticisms.[263] However B.B. Stoney, in his *Theory of Strains* (1869) points out that 'the lattice web is not well suited for bent cranes exceeding 10 tons'.[264] To address the third criticism some cranes were built with the upper part of the jib horizontal, supporting a wheeled bogey, but this approach was only suitable for small cranes.[265] It was clear that more adaptable cranes were needed. The area in which the Fairbairn cranes scored best was where heavy lifting was involved, particularly of bulky items, such as ships' boilers. In this field they were the preferred crane for twenty-five years, until the advent in the mid 1870s of such radical new cranes as Armstrong Mitchell's 160-ton hydraulic cranes at La Spezia and elsewhere,[266] James Taylor's 70-ton crane with hinged jib at Dundee, and his Titan at Colombo, followed in 1889 by a 100-ton, steel, hinged jib crane at Belfast.[267] 'Fairbairn' cranes, having had a twenty-five year window – considerably longer than the tubular-girder bridges from which they sprang – nevertheless lingered on, with examples being built into the 20th century, notably by Cowans Sheldon of Carlisle.

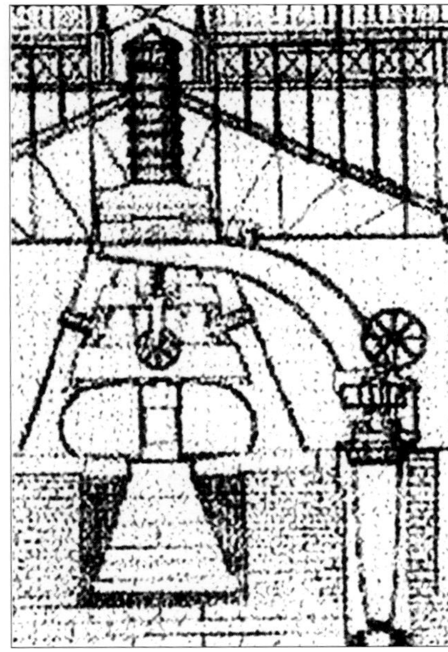

Fig.5.35: Steam hammer with 'Fairbairn' crane, Le Creusot, France.

Only around thirty cranes built by Fairbairn have been identified.[268] The total number built is unlikely ever to be known. In 1861 the *North British Review* reported that, 'All Her Majesty's dockyards have been supplied with this invaluable piece of Machinery, and great numbers have been supplied to other parts of the

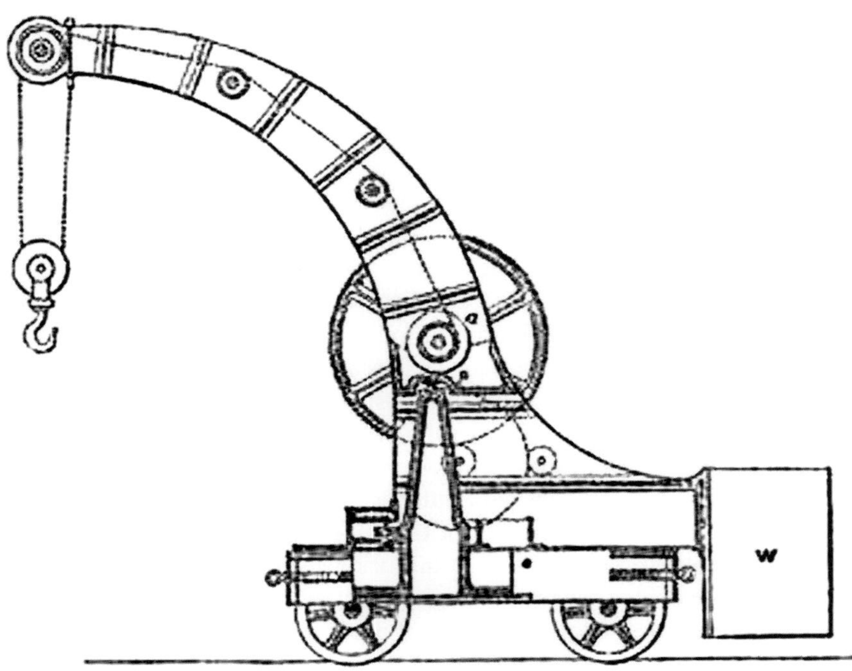

Fig.5.36: Fairbairn railway crane, pre-1860.

Continent'.[269] It is probable that the thirty represent less than ten per cent of the Ancoats-built total. Most Fairbairn cranes were scrapped, unrecorded, during the middle quarters of the 20th century. The only one remaining in the British Isles that was built by Fairbairns is at Dover.[270] In 2014 this crane was, according to the new plaque, 'lovingly restored' by the Port of Dover, but its iconic form – its primary harbourscape feature – has been seriously compromised by the introduction of a prop beneath the point of its jib.[271] It is to Middelberg in the Netherlands that one must look for the most thorough conservation of a 'Fairbairn' crane. In 2009 this Prins van Oranje crane, was lifted out of its foundation by a mobile crane, transported by road to a factory at Vlissingen where the corroded below-ground section was repaired, prior to the crane's re-erection.[272] The move from demolition to preservation of these unique cranes in the last half-century is an indication of the growing appreciation of Fairbairn's importance and influence, as well as interest in industrial archaeology. A more recent development is a 'flickr' site dedicated to Fairbairn cranes as photographers have discovered their photogenic qualities, for example the Helsingør crane, thought to be the oldest *existing* crane built in Ancoats, framing Hamlet's castle.

Conclusion

FORTUNE accrued to Fairbairn and the family business from steam engines and millwork after the closure of the loss-making Millwall shipyard. Fame and further fortune accrued from bridges and cranes after Fairbairn seized the unanticipated opportunity to undertake experiments leading to the longest railway bridge of the mid 19th century. He applied to the task knowledge from prior experiments, and experience gained in shipbuilding. Fairbairn's Britannia experiments, very different from many of his others, demonstrated his adaptability, and pragmatic approach. Many of his experiments involved the systematic measuring, recording and analysis of data leading to applicable mathematical formulae, but the Britannia experiments relied on parameter variation to achieve a solution. As such they were used by Rosenberg and Vincenti to support the thesis that the work of the civil engineer did not involve the application of theoretical knowledge.[273] In one sense Britannia, with the repeated testing of the model tube until it failed, was a fruit of failure, yet at a time when no adequate theoretical approach existed, it produced an impressive result. That result was a bridge which probably could not have been built with the available technology ten years earlier, and would not have been built ten years later because by then lattice or Warren girders had become established. It was due to Fairbairn's vision and persistence that Britannia was a true girder bridge without suspension chains, and as such influenced the adoption of girder bridges in Continental Europe. In spite of the euphoria with which it was greeted, the Britannia Bridge was soon deemed to be a cul-de-sac in bridge construction. Contrary to the norm, where seemingly promising developments often turn out to be dead ends with little place in technological history and soon forgotten, Britannia has come out of the cul-de-sac and is extensively written about today. Petroski's description of this bridge as 'a monument to engineering short-sightedness'[274] is far from the mark in the light of Åkesson's analysis of Britannia's far-sighted characteristics that made it the forerunner of the modern box-girder bridge.[275]

Fig.5.37: Fairbairn Crane, Helsingør, Denmark, c.1862, believed to be the oldest existing crane manufactured at Ancoats, photographed, 2014, by J. Hallbæck-Andersen, framing Hamlet's castle.

The tubular bridges spawned the more numerous tubular-girder bridges, and from them the technology was transferred to tubular cranes. Tubular-girder bridges were built during the short interval following the collapses of Gray's Mill and Stephenson's Dee Bridge in 1847 and the development of viable lattice and Warren trusses. Without those collapses just before the viability of the lattice was established, it is likely that very few tubular-girder bridges would have been built. Whilst the links between Gray's Mill and the Dee Bridge have been noted by others, the effect of the Dee Bridge collapse on the rapid success of tubular-girder bridges has not been appreciated. During their short window, Fairbairn's tubular and tubular-girder bridges contributed to the understanding of buckling, continuous beams and fatigue. The transition from empirical rule-of-thumb to theoretical calculations is exemplified by reference to continuous beam theory at Torksey where continuity over the supports was incorporated unwittingly as a result of the mode of erection. The Government Inspector's concerns led to exhaustive mathematical analysis, which also highlighted Fairbairn's weaknesses in this area. The tubular crane was the preferred dockside crane for heavy loads for twenty-five years, with its iconic swan-neck form gracing Victorian waterfronts, yet there is almost no secondary literature about its development, construction, diffusion or demise – a gap which this book has sought to fill. Interesting as its derivative structures are, Fairbairn's outstanding achievement as an experimental engineer and bridge-builder was his involvement in the design and construction of the Britannia Bridge. At the time it was built there was no available relevant practical or theoretical knowledge, yet

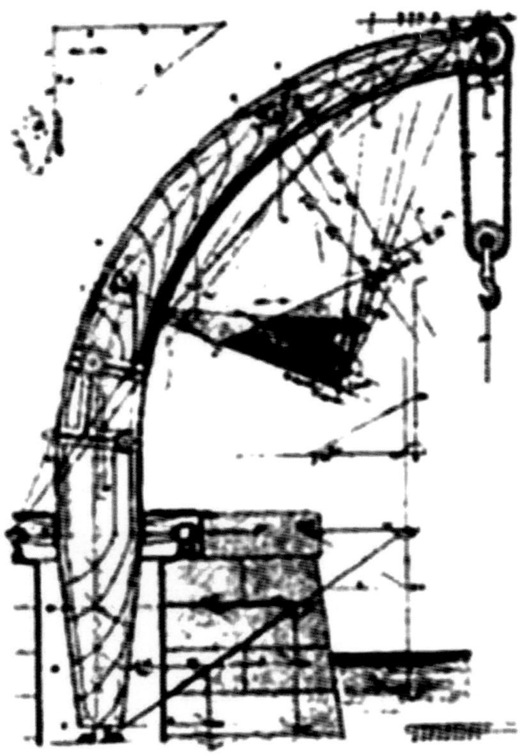

Fig.5.38: Culmann's working drawing of the stresses in Fairbairn's crane.

what was produced was a novel, chainless, visually powerful bridge, of far greater magnitude than any girder bridge that previously existed, which excited the interest of Europe's leading engineers, bringing Fairbairn to their notice.

Epilogue to Chapter 5: Fairbairn Cranes and Orthopaedics

IT is more than a little surprising to find Fairbairn's influence, through his cranes, extending into 21st-century orthopaedics. Karl Culmann had a remarkable insight into the power of graphical techniques of analysis. When he visited England in 1849 in connection with his study of bridges, he met Fairbairn around the time of the first Keyham cranes.[276] In *Die graphische Statik* (Zurich 1866) Culmann analysed, by way of graphical statics, the stress patterns in the jib of a Fairbairn crane, which he illustrates.[277] This stress drawing and a photograph of Stothert & Pitt's Bristol 'Fairbairn' crane are reproduced in recent orthopaedic texts.[278] How did this happen? In 1855 Culmann was appointed Professor of Theory of Structures at the Institute ofTechnology in Zurich.[279] At the same time Georg von Mayer (1815–92) was Professor of Anatomy at Zurich University.[280] In July 1866 Culmann attended a meeting at which Mayer read a paper 'On the significance of bones as machine parts and the structure of cancellous bones'.[281] The professors realised that there was a similarity between the stress lines in the crane and the trabeculae in the upper part of the femur.[282] Von Meyer's paper, published in 1867, referred to this similarity.[283] The two men co-operated in the field of bone biomechanics, providing a good example of the interdisciplinary transfer of technical knowledge.

Julius Wolff (1836–1902) was an orthopaedic surgeon in Berlin who, in the late 1860s, became aware of the work of von Meyer and Culmann. He took their work a step further, establishing the concept of bone adaptation occurring in response to mechanical stress, postulated in Wolff's law, and making his name part of today's orthopaedic lexicon.[284] Wolff published a paper in 1870 with an illustration showing the similarities between the stress lines in the crane and the forces acting on the interior of a human femur.[285] This became a basis for his *magnum opus*, *Das Gesetz der Transformation der Knochen* (1892, ET 1986, *The Law of Bone Remodelling*). Wolff describes von Meyer and Cullman's meeting as 'an incident extremely fortunate to science' and states that 'merit for the great discovery of the mathematic significance of this [bone] architecture belongs exclusively and singly to the Zurich mathematician, Prof. Culmann'.[286] The esteem for Wolff today is marked by the recent translation of his major work into English and

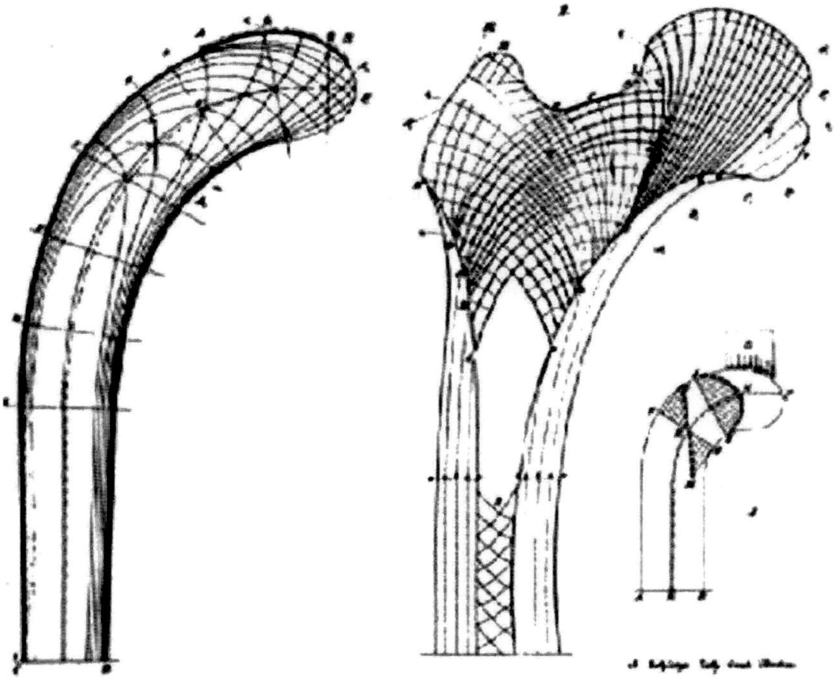

Fig.5.39: Culmann's stress patterns for a crane *(left)* and von Meyer's trabecular drawing of the proximal femur *(right)*.

over 690 citations from it between 1980 and 2005.[287] Culmann's stress trajectories for cranes are illustrated in such 21st-century textbooks as Marco Viceconti's *Multiscale Modeling of the Skeletal System* and Brian Hall's *Bones and Cartilage: Developmental and Evolutionary Skeletal Biology.*[288] That is the orthopaedic establishment's accepted history, and the effect of the apparent eureka moment between von Meyer and Culmann at Zurich.[289] The same story featured in a recent congress of the International Union of Theoretical and Applied Mechanics.[290] But there is good reason to believe that it was no eureka moment, and that Wolff's law was foreseen over nine years before 1866. In his *Application of Cast and Wrought-Iron to Building Purposes* (2nd edition 1857–8) Fairbairn likens the form of the Britannia and Conway bridges to the hollow bones of birds and the stems of grasses, grains and bamboos, the 'unerring laws', as he puts it, 'of the great Architect of nature'.[291] This approach goes back to Galileo, 'The quill of

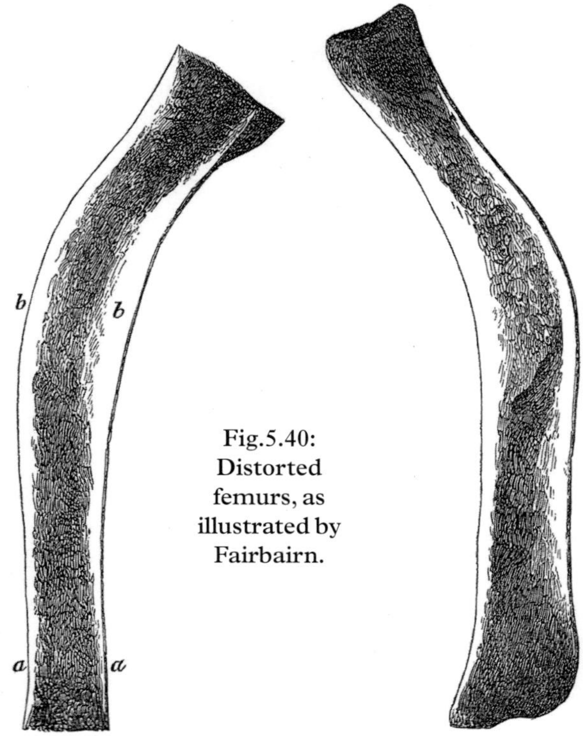

Fig.5.40: Distorted femurs, as illustrated by Fairbairn.

the bird's feather, the hollow shaft of a reed, the thin tube of the wheat-straw bearing its heavy burden in the ear, are all illustrations which Galileo used in his account of this mechanical principle'.[292] Fairbairn then turns to the cells at the top and bottom of the Britannia tubes, 'Indeed all bones when examined microscopically are found to be composed of minute cells, adding greatly to the strength and lightness of the structure'. He illustrates this by reference to two longitudinal sections of femur, 'taken from a rickety subject, where distortion had taken place', and continues:

> in order to compensate for the bent form of curvature of the bone (which in a more healthy state, to act as a pillar, would have been nearly straight), the whole of the porous or cellular interior is incased in a thin shell or tube of hard bony substance, as dense and compact as the finest ivory. Had the subject been healthy and the limb straight, the envelope of ivory would have been thin, and something like the form shown at a a ; but owing to the curvature, and in order to compensate for that defective form in its resistance to vertical pressure, Nature, in her workings supplies the deficiency by filling up the concave side with a thicker stratum of hard bone, and a proportionate thinner stratum on the convex side, to make up for the loss of strength arising from the curvature of the pillar as shown at b b.[293]

There is no doubt about Wolff's importance in the history of orthopaedics. But in Fairbairn's words above is the link between the femur and the Fairbairn crane, and the genesis of Wolff's Law. This text and Fairbairn's illustration, nine years before the famous Zurich meeting, must have been known to Culmann, as Fairbairn's books, widely read in Europe, would have been available at the Zurich Institute of Technology.[294] The situation raises many questions. How did Fairbairn come by rickety femurs, and what prompted him to cut them down their centres? Was he working with one of his medical friends from the Manchester Literary and Philosophical Society, or from the British Association? It can hardly be coincidence that Fairbairn and van Meyer both used femurs as their examples. With these questions as yet unanswered, the fact remains that Wolff's Law originated in Ancoats – a point that has not been appreciated in the current literature. It is an insight into the innate understanding of engineering which Fairbairn possessed. Nor has it been noted that this linking of the rickety femur and the tubular crane, of natural history and engineering, was in the Humboldian tradition.[295]

Notes for Chapter 5

1. [R] J [M] Sutherland, 'Iron Railway Bridges' in M R Bailey (ed.), *Robert Stephenson – The Eminent Engineer*, (2003), p.318.
2. Sutherland, 'Iron Railway Bridges', pp.334-5.
3. T F Peters, *Building in the Nineteenth Century*, (1996), p.166; S B Hamilton, 'Building and Civil Engineering Construction', in C Singer et al (eds.), *A History of Technology*, (1958), Vol.4, p.460. Stephenson had had a negative experience with one on the Stockton & Darlington Railway – the trains caused a wave in the deck 'which racked the bridge to pieces in a few years'.
4. E Clark, *The Britannia and Conway Tubular Bridges*, (1850), Vol.1, p.30. Fairbairn is also said to have witnessed the event (L T C Rolt, *Victorian Engineering*, (1970), pp.84-5. However, Rolt gives no source for this statement). Two other sources of the idea, both from 1845, have been given – a Mr Randall on the Commons Committee examining the Chester & Holyhead Bill, and John de la Haye of Liverpool (S Tyson, 'Notes on the History, Development and Use of Tubes in the Construction of Bridges, *IAR*, 2.2, 1977-8, 143).
5. Clark, *Britannia and Conway Tubular Bridges*, Vol.1, pp.65-7.
6. 'Select Committee on Railway Bills: Group 2: Chester and Holyhead Railway Bill', 5 May 1845, quoted in Clark, *Britannia and Conway Tubular Bridges*, Vol.1, pp.48-50. It may be significant that Stephenson refers to Fairbairn's 'experience' rather than his mathematical abilities.
7. R A Hayward, *The Story and Scandal of HMS Megaera*, (1978), pp.26-8, 43-7. They were the *Sir Henry Pottinger* for P&O, and the frigate, HMS *Megaera*.
8. B Åkesson, *Plate Buckling in Bridges and Other Structures*, (2007), p.26.
9. Fairbairn, *B&CTB*, p.210. The results are shown on pp.209-81.
10. J Singer, J Arbocz and T Weller, *Buckling Experiments: Shells, Volume 2: Shells, Built-up Structures, Composites and Additional Topics*, (2002), p.623.
11. See Chapter 7; J Singer, 'Experimental Studies in Shell Buckling', in N F Knight, Jr and N P Nemeth (eds.), *Stability Analysis of Plates and Shells: A Collection of Papers in Honor of Dr Manuel Stein*, (1998), p.21.
12. J Singer, 'Stiffened Cylindrical Shells' in J G Teng and J M Rotter, *Buckling of Thin Metal Shells*, (2004), p.287. In 1965 H L Cox wrote that Fairbairn's experimental work was 'very advanced, indeed modern aircraft have just got back to it' (H L Cox, *The Design of Structures of Least Weight*, (1965), p.1).
13. S P Timoshenko and J M Gere, *Theory of Elastic Stability, Second Edition*, (1961), p.432.
14. Fairbairn, *B&CTB*, Plate 19; also reproduced in T F Peters, *Building in the Nineteenth Century*, (1996), p.167.
15. Fairbairn, *B&CTB*, pp.33-47.
16. C A Gresham, 'William Fairbairn and the Conway Tubular Bridge', *Transactions of the Caernarvonshire Historical Society*, 9, 1948, 51.
17. D Cardwell, *James Joule: A Biography*, (1989), p.34.
18. Rosenberg and Vincenti, *Britannia*, pp.25-7.
19. D Smith, 'Structural model testing and the design of British railway bridges in the nineteenth century', *TNS*, 48, 1976-7, 88.
20. Clark, *Britannia and Conway Tubular Bridges*, Vol.1, p.184.
21. *Life*, p.181.
22. W Fairbairn to E Hodgkinson, 17 August 1843, *Proceedings of the Manchester Literary and Philosophical Society*, 17, 1877-8, 145-6.
23. Fairbairn to Hodgkinson, 17 August 1843, *Proceedings of the Manchester Literary and Philosophical Society*, 145.
24. Sutherland, 'Iron Railway Bridges', p.323; E Hodgkinson to R Stephenson, 18 March 1846, ECLB/96, Stephenson Papers, Institution of Civil Engineers.
25. Fairbairn, *B&CTB*, p.59.
26. Sutherland, 'Iron Railway Bridges', p.323;
27. Rosenberg and Vincenti, *Britannia*, p.29; B Warburton, 'Eaton Hodgkinson (1789-1861) and the Science of Strength of Materials', (PhD Thesis UMIST 1971), p.4.22.
28. *CE&AJ*, 9, 1846, 307-8; *Report of the Commissioners appointed to inquire into the Application of Iron to Railway Structures*, (1849), pp.118, 123-4. Hodgkinson was one of the Commissioners.
The Report included Hodgkinson's substantial paper, at Appendix AA (pp.115-180), entitled 'Experimental Inquiries to determine the Strength of Wrought-Iron Tubes, made at the instance of Robert Stephenson Esq MP, FRS, to supply data for the erection of his Tubular Bridges across the Menai Straits, and the River Conway'; and a report from the *Railway Chronicle*, 3 October 1846, of Hodgkinson's paper at the British Association. In it he said, 'I expressed a conviction that the tubes then tried, and others proposed, would not be the best for the intended purpose, though they would afford valuable introductory knowledge. I urged the tube to bear the greatest weight must be formed as a large beam or girder, having its top and bottom equally capable of resistance ... I suggested that the top should be formed of cylindrical tubes, as I felt that these tubes, or something analogous to them, would best resist the strain to which the top would be exposed.' (p.123).
29. Fairbairn, *B&CTB*, pp.60, 108; E Clark to R Stephenson, 28 October, 1846, ECLB/3, Stephenson Papers, ICE.
30. W Fairbairn, 'Presidential Address', *BAAS1861*, pp.lxiv-v.
31. Gresham, 'Fairbairn and the Conway Tubular Bridge', 51. 'Fast-track' i.e. the masonry was being built before the design of the ironwork was finalised.
32. Rosenberg and Vincenti, *Britannia*, p.30; J Rapley, *The Britannia and other Tubular Bridges*, (2003), p.68; Clark, *Britannia and Conway Tubular Bridges*, Vol.1, p.805.
33. Fairbairn, *B&CTB*, pp.90-3; Rosenberg and Vincenti, *Britannia*, p.40.
34. Clark, *Britannia and Conway Tubular Bridges*, Vol.1, p.807; J Millar, *William Heap and his Company*, (1976), p.22.
35. T Fairbairn, *Truths and Tubes on Self-supporting Principles; A few words in reply to the Author of 'High-ways and Dry-ways'*, (1849), pp.46-52. Thomas Fairbairn's italics.
36. Clark, *Britannia and Conway Tubular Bridges*, Vol.1, pp.810-11.
37. Fairbairn, *B&CTB*, p.128.
38. Clark, *Britannia and Conway Tubular Bridges*, Vol.1, p.812.
39. Gresham, 'Conway', 54-5. The tube deflected by $7\frac{7}{8}$in., against a calculated deflection of 8in. Loads of 300 tons were introduced into the tube and deflection increased to 11in., returning to $8\frac{1}{4}$in. when the load was removed.
40. *Mining Journal*, 18, 1848, 127; *Liverpool Mercury*, 7 March 1848. The tube was 412ft long and 25ft 6in high.

41. *CE&AJ*, 11, 1848, 126.
42. The Conway bridge is still in use and is of world heritage standing (E DeLony, *Context for World Heritage Bridges*, (1996), p.14).
43. Åkesson, *Plate Buckling*, p.28.
44. Fairbairn, *B&CTB*, p.99.
45. J Rapley, *The Britannia & Other Tubular Bridges*, (2003), pp.100-2.
46. E Clark, manuscript note at the bottom of p.90 of his copy of Fairbairn, *B&CTB*, in Archive of ICE. Transcript in Peters, *Building in the Nineteenth Century*, p.412n28.
47. Rapley, *Britannia*, Plate 1.
48. Henry Moseley (1801-1872), Professor of Natural and Experimental Philosophy and Astronomy, King's College, London. Charles Babbage (1797-1871), Mathematician and Computer Pioneer. Robert Willis (1800-1875), Jacksonian Professor of Natural and Experimental Philosophy, Cambridge. George Rennie (1791-1866), Engineer, eldest son of John Rennie.
49. Fairbairn, *B&CTB*, p.163.
50. Fairbairn to Babbage, 31 January 1848, Add.MSS 37194/106, Babbage Correspondence Vol.12, British Library.
51. Fairbairn, *B&CTB*, pp.172-4; MG, 20 May 1848.
52. Fairbairn, *B&CTB*, pp.113-4.
53. Clark, *Britannia and Conway Tubular Bridges*.
54. *Quarterly Review*, 85, June-Sept. 1849, 399-452, reprinted as F B Head, *High-ways and Dry-ways or, the Britannia and Conway Tubular Bridges*, ([1849] 1968); *North British Review*, 13, 1850, 399-446; T Fairbairn, *Truths and Tubes*; A Ure, *A Dictionary of Arts, Manufactures and Mines ...* (4th ed. 1853), pp.668-687.
55. [Woodcroft] VIII. See below.
56. G D Dempsey, *Tubular and other Iron Girder Bridges*, (1849), p.viii, [Dempsey's italics].
57. *Life*, p.241.
58. Timoshenko, *Strength of Materials*, pp.190-1; K Culmann, 'Der Bau der eisernen Brücken in England und Amerika', *Allgemeine Bauzeitung*, 17, 1852, 163-222.
59. It seems that initially Culmann obtained his information from Fairbairn (Timoshenko, *Strength of Materials*, p.192-3). S P Timoshenko (1876-1972) was Professor at the University of Michigan and later at Stanford. He wrote twelve textbooks which revolutionised the teaching of mechanics and were translated into thirty-five languages.
60. D Cardwell, *The Fontana history of technology*, (1994), p.257.
61. J Singer, J Arbocz and T Weller, *Buckling Experiments: Volume 1: Basic Concepts, Columns, Beams and Plates,* (1998), p.453.
62. *Report of the Commissioners appointed to inquire into the Application of Iron to Railway Structures*, p.115.
63. Clark, *Britannia and Conway Tubular Bridges*, Vol.1, p.104.
64. P Sibly, 'The Prediction of Structural Failures', (DPhil Thesis, UCL, 1977), p.22.
65. B P E Clapeyron, 'Calcul d'une pouter élastique reposant librement sur des appuis inégalment espacés', *Comptes Rendus hebdomadaires des Séances de l'Académie des Sciences*, 45, 1076; quoted in T M Charlton, *A history of theory of structures in the nineteenth century*, (1982), pp.24-5.
66. Timoshenko, *Strength of Materials*, pp.160-1.
67. I Todhunter and K Pearson, *A History of the Theory of Elasticity and of the Strength of Materials from Galilei to the Present Time*, (1886), p.795, referring to C-L Navier, *Résumé des Leçons*, (3rd ed. edited by M M de Saint-Venant, 1864).
68. Timoshenko, *Strength of Materials*, pp.142-4; D J Jourawski, 'Sur la résistance d'un corps prismatique et d'une pièce compose en bois ou en tôle de fer à une force perpendiculaire à leur longueur', *Mémoires Annales des Ponts et Chaussées,* 12, 1856, 328-51, (this is a French translation of part of the paper presented at the Russian Academy of Sciences in 1854).
69. Timoshenko, *Strength of Materials*, pp.160-2; *Mémoires Annales des Ponts et Chaussées*, 20, 1860, 113, (this is also a French translation of Jourawski's paper, originally presented in Russian).
70. *Life*, p.219.
71. *Life*, pp.221-3.
72. *Life*, pp.217-37.
73. Fairbairn, *Application*, pp.261-72, 286-9, Plates 3-5. Moorsom's prize-winning (but never built) proposal for a lattice girder was longer, with two spans of 600ft (*The Engineer*, 19 August 1859, 138).
74. K-E Kurrer, *The History of the Theory of Structures: From Arch Analysis to Computational Mechanics*, (2012 ed.), p.44; G C Mehrtens, 'Zur Baugeschichte der alten Eisenbahnbrücken bei Dirschau und Marienburg', *Zeitschrift für Bauwesen*, 43, 1893, 101.
75. K Lentze, 'Bemerkungen über die grösseren Brükenbauwerke in Frankreich England und Irland auf einer Reise im Winter 1844/45 gesammelt', *Vorhandlungen des Vereins zur Beföderung des Gewerbfleisses in Preussen,* 1846, 88-114, referred to in J G James, 'The Origins and Worldwide Spread of Warren-Truss Bridges in the mid-Nineteenth Century. Part 1. Origins and Early Examples in the UK', in N Smith (ed.), *History of Technology*, 11, 1986, 71-2, 116n19.
76. W Ramm, 'History and construction of the Old Vistula Bridges in Tezew', in S Huerta (ed.), *Proceedings of the First International Congress on Construction History, Madrid, 20th-24th January 2003,* (2003), pp.1702-3; Kurrer, *History of the Theory of Structures,* p.44.
77. J G James, 'Overseas Railways and the Spread of Iron Bridges, c.1850-70', (Typescript, ICE Library), p.48; Ramm, 'Old Vistula Bridges', p.1704; 'Dombrücke' at http://en.structurae.de/structures/data/index.cfm?ID=s00001788 (accessed 16 August 2013).
78. Z Cywinski, 'ASCE Honor for the Tczew Bridge in Poland', *ASCE Civil and Environmental Engineering Reports,* 3, 2009, 15. This was one of 240 plaques erected worldwide by the American Society of Civil Engineers.
79. Cywinski, 'ASCE Honor', 17.
80. Kurrer, *History of the Theory of Structures*, p.44.
81. Life, p.235.
82. Brotherton: Rapley, *Britannia*, pp.49-54. Victoria: Rapley, *Britannia*, pp.128-39; Millar, *William Heap*, pp.50-8.
83. Rapley, *Britannia*, pp.126-7. Henry Petroski states that 'had the tube concept given way to one in which the train would ride atop a (box) girder, Stephenson's railroad bridge across the Menai Strait might have played an entirely different role in the history of structural design'. No explanation is given. The trains did ride atop the Egyptian tubes but this appears to have had little influence in the subsequent history of structural design (H Petroski, *Design Paradigms: Case Histories of Error and Judgment in Engineering,* (1994), p.113).

84. One by the Couillet Works in Belgium was over the Sambre at Namur (*CE&AJ*, 14, 1850, 35). Another, by the Creuzot Works, was on the Berne & Lausanne line, over the Sorine near Fribourg, where foot traffic was accommodated in the interior and the railway 'outside', presumably on top of the tube (*The Engineer*, 14, July-Dec.1862, 78).
85. Rosenberg and Vincenti, *Britannia*, p.71.
86. N Rosenberg and W G Vincenti, 'The Tubular Bridges and Usher's Theory of Invention', Appendix to Rosenberg and Vincenti, *Britannia*, pp.75-9.
87. S Dasgupta, 'Testing the Hypothesis Law of Design: The Case of the Britannia Bridge', *Research in Engineering Design*, 6, 1994, 38-57.
88. Petroski, *Design Paradigms*, pp.99-120.
89. G K Dreicer, 'Building Bridges and Boundaries: The Lattice and the Tube, 1820-1860', *Technology and Culture*, 51.1, 2010, 126-163.
90. H-R Hitchcock, *Architecture Nineteenth and Twentieth Centuries*, (1958), pp.179-80; D P Billington, *The Tower and the Bridge: The New Art of Structural Engineering*, (1985), pp.47-8.
91. L Sullivan, 'The Tall Office Building Aesthetically Considered', *Lippincott's Magazine*, March 189 6, 408.
92. Discussion following J Husband, 'The Aesthetic Treatment of Bridge Structures', *MPICE*, 145, 1901, 206-7.
93. B Hub, 'Architect Versus Engineer: Monumentality Versus Dematerialization', in M Rinke and J Schwartz (eds.), *Before Steel: The Introduction of Structural Iron and its Consequences*, (2010), pp.159-60; J A Lux, *Ingenieuraesthetic*, (1910). Hub seeks to reconcile the two positions in a debate in the work of Peter Behrens turbine hall (p.170).
94. Billington, *The Tower and the Bridge*, pp.47-8,
95. W Fairbairn to R Stephenson 15 July 1846, No.88, Stephenson Papers, ICE; Fairbairn, *B&CTB*, pp.90-1; Peters, *Building in the Nineteenth Century*, p.171.
96. Peters, *Building in the Nineteenth Century*, pp.177-8.
97. Fairbairn, *M&MWII*, p.114; see Chapters 3 and 4.
98. Fairbairn, *Application*, Plates 4 and 5.
99. L T C Rolt, *George and Robert Stephenson: The Railway Revolution*, (1960), p.317; Petroski, *Design Paradigms*, p.116.
100. Petroski, *Design Paradigms*, pp.116-8.
101. Head, *High-ways and Dry-ways*, pp.181-3.
102. [Woodcroft] VIII.
103. Petroski, *Design Paradigms*, p.101.
104. G Lindenthal, 'The Continuous Truss Bridge over the Ohio River at Sciotville, Ohio, of the Chesapeake and Ohio Northern Railway', *Transactions of the American Society of Civil Engineers*, 85, 1922, 912.
105. Rosenberg and Vincenti, *Britannia*, p.60.
106. S C Dooley, 'The Development of Material-Adapted Structural Form', (École Polytechnique Fédérale de Lausanne, DSc Thesis, 2004), p.A141.
107. D Beckett, *Stephensons' Britain*, (1984), p.132.
108. B Åkesson, *Understanding Bridge Collapses*, (2008), pp.179-92.
109. Åkesson, *Understanding Bridge Collapses*, – Fourth Danube Bridge at Vienna, pp149-54; Cleddau Bridge, pp.171-8; West Gate Bridge, pp.179-92; Rhine Bridge at Koblenz, pp.193-200; Zeulenroda Bridge in former East Germany, pp.201-12. R J Bridle and F A Sims, 'The effect of bridge failures on UK technical policy and practice', *MPICE Engineering History and Heritage*, 162.1, 2009, 39-49.
110. A C Burton, 'Lessons Learned in the Design and Erection of Box Girder Bridges from the West Gate Collapse', (MEng Thesis, MIT, 2007), Abstract and p.71.
111. J G James, handwritten 'Index' at the ICE, referring to P Alsop, 'Bridging the Rivers, Part 5', *Investigator*, 4.2, 1969, pp.84-101.
112. Åkesson, *Plate buckling*, p.50.
113. *Life*, p.213.
114. G D Dempsey, *Tubular and other Iron Girder Bridges*, (1850), pp.84-6.
115. W Fairbairn to R Stephenson, 4 February 1846, No.44, Stephenson Papers, ICE; Sutherland, 'Iron Railway Bridges', pp.311-5.
116. Patent No.11,401; CE&AJ, 10, 1847, 143.
117. *Life*, pp.212-3.
118. Clark, *Britannia and Conway Tubular Bridges*, Vol.2, p.813.
119. W Fairbairn, 'On some Defects in the Principle and Construction of Fire-proof Buildings', *MPICE*, 6, 1847, 213-7 and Plate; Fairbairn, *Application*, pp.279-84; Life, p.186. See also C R Gagg and P Lewis, 'The Rise and Fall of Cast Iron in Victorian Structures', *Engineering Failure Analysis*, 18.8, 2011, 1964.
120. Sibly, 'Prediction of Structural Failures', Fig.2.10.
121. 'Discussion' following Fairbairn, 'On some Defects', *MPICE*.6, 1847, 217-24.
122. W Fairbairn to R Stephenson, 10 June 1847, No.265, Stephenson Papers, ICE.
123. P R Lewis, *Disaster on the Dee: Robert Stephenson's Nemesis of 1847*, (2007), pp.105-8, 117-8.
124. 'Returns and Plans of Iron Bridges, 1847', MT8/1, National Archive, Kew.
125. *Report of the Commissioners on the Application of Iron to Railway Structures*, pp.xvii, 267-8, 413-5.
126. Fairbairn, *Application*, pp.28-49.
127. D R Bellhouse, *David Bellhouse and Sons, Manchester*, (1992), p.48.
128. *Report of the Commissioners on the Application of Iron to Railway Structures*, pp.xvii, 340, 359.
129. *Report of the Commissioners on the Application of Iron to Railway Structures*, pp.xvii, and Plates, 6.1, 7.6, 7.6*, 7.7, 7.7*.
130. W Fairbairn to R Stephenson, 21 July 1847, Nos.284-6, Stephenson Papers, ICE.
131. Based on 'Returns and Plans of Iron Bridges, 1847', with maximum span in brackets: Blackburn Darwen & Bolton 2 (60'); East Lancashire 4 (62'); Lancashire & Yorkshire 4 (100'); London & Blackwall 2 (60'); London & North Western 10 (60'); Manchester Sheffield & Lincolnshire 2 (154'); (Silby, 'Prediction of Structural Failures', Table 2.1).
132. S Smith, 'The Development and Use of the Tubular Beam, 1830-1860', *History of Technology*, 14, 1992, 115 and132n42.
133. *Life*, pp.318-9.
134. K H Vignoles, *Charles Blacker Vignoles: romantic engineer*, (1982), p.177. Simultaneously Stephenson was responsible for a tubular-girder bridge at Camden designed by his resident engineer, R B Dockray, but with a cast-iron top (*CE&AJ*, 1847, 376; R J M Sutherland, 'The Introduction of Structural Wrought Iron' in R J M Sutherland (ed.), *Structural Iron 1750-1850*, (1997), p.284). Stephenson was also responsible for two tubular-girder bridges at Gateshead in the late 1840s, differing from Fairbairn's in that the girders had tapering sides (R W Rennison, 'The Influence of William Fairbairn on Robert Stephenson's bridge designs: four bridges in north-east England', *IAR*, 20, 1998, 40-3).

135. *CE&AJ*, 1847, 376-7; *Report of the Commissioners on the Application of Iron to Railway Structures*, (1849), Plate 6.1; *MM*, 52, 1850, 282-5 and figs.1-5; Dempsey, *Tubular and other Iron Girder Bridges*, pp.24-6; C Tomlinson, *Cyclopædia of useful arts, mechanical and chemical, manufactures, mining and engineering*, (1852), pp.237-8.
136. *Report of the Commissioners on the Application of Iron to Railway Structures*, pp.324, 410-1, Appendix 6 Plate 1; *MM*, 52, 1850, 282-5 and figs.1-5; Fairbairn, *Application*, pp.228-9. The height of the wedges was firstly 1 inch and then 1½ inches. Subsequently Fairbairn undertook far more sophisticated experiments on impact, vibration and fatigue. See Chapter 7.
137. Rapley, *Britannia*, p.120; W Fairbairn to R Stephenson, 21 July 1847, Stephenson Papers, ICE Archive, Nos.283-4. In 1848 Thomson engineered the first lattice girder bridges to be built in England, to the designs of the Irish engineer, John Macneill (*Bolton Chronicle*, 18 November 1848; M Beaumont, *Sir John Hawkshaw 1811-1891: The Life and Work of an Eminent Victorian Engineer*, (2015), p.44).
138. T Mackay, *The Life of Sir John Fowler, Engineer, Bart. …* (1900).
139. Sibly, 'Prediction of Structural Failures', Fig.2.10.
140. *MG*, 18 July 1849.
141 Discussion following W Fairbairn, 'On Tubular Girder Bridges', *MPICE*, 9, 1850, 246, 237.
142. K Keen, 'Torksey Railway Bridge', *Journal of the Railway & Canal Historical Society*, 197, 2007, 520-1.
143. *The Sheffield and Rotherham Independent*, 2 February 1850.
144. Discussion following Fairbairn, 'On Tubular Girder Bridges', 276-7.
145. Keen states that the girders were manufactured at Fairbairn's works (Keen, 'Torksey Railway Bridge', 520-1) but concedes that this is based on hearsay (letter K B Keen to R J Byrom, 22 May 2016). If Fairbairn had fabricated the girders, he would surely have followed his own latest guidance.
146. Discussion following Fairbairn, 'On Tubular Girder Bridges', 237, 246; Mackay, *Sir John Fowler*, pp.98-9.
147. Rapley, *Britannia*, p.121. No source given, but Fairbairn was not present for most of the discussion.
148. H Moseley, *The Mechanical Principles of Engineering and Architecture*, (1843), pp.xvi, 521-5.
149. Discussion following Fairbairn, 'On Tubular Girder Bridges', 254-7.
150. T M Charlton, 'Contributions to the Science of Bridge-Building in the Nineteenth Century by Henry Moseley, Hon. Ll.D., F.R.S. and William Pole, D.Mus., F.R.S.', *Notes and Records of the Royal Society of London*, 30, 1976, 176.
151. Discussion following Fairbairn, 'On Tubular Girder Bridges', 282-3, 286-7.
152. Fairbairn, 'On Tubular Girder Bridges', 233-87; *CE&AJ*, 13, 1850, 142; *The Practical Mechanic's Journal*, 3, 1850-1, 43; Clark, *Britannia and Conway Tubular Bridges*, Vol.2, pp.774-86;
153. W Pole, 'Strength and deflection of the Britannia Bridge Considered as a Continuous Beam', in Clark, *Britannia and Conway Tubular Bridges*, Vol.2, pp.774-87; Charlton, *History of theory of structures*, pp.19-24.
154. Lindenthral, 'Continuous Truss Bridge', p.212.
155. Fairbairn, *Application*, pp.237-46.
156. R J Bridle and F A Sims, 'The effect of bridge failures on UK technical policy and practice', *Proceedings of the Institution of Civil Engineers: Engineering History and Heritage*, 162, 2009, 39-49.
157. Timoshenko, *Strength of Materials*, pp.25-36.
158. Todhunter and Pearson, *Theory of Elasticity*, Vol.1, para.1489.
159. Charlton, *History of theory of structures*, pp.20-1.
160. *Life*, pp.213, 318-9; MG, 6 April 1850. Fairbairn stated in 1861 that he 'had made perhaps more than two hundred bridges upon his formula' (*The Builder*, 28 September 1861, 669). Pole says the figure was subsequently increased to a thousand (*Life*, p.319) but this is wrong because it includes plate girder and other types of bridge (p.213).
161. 'Obituary of George W Hemans', *MPICE*, 85, 1886, 394-9. Hemans, under Macneill, was responsible for the design of the first iron lattice girder bridges in the British Isles.
162. F Mulligan, *William Dargan: An Honourable Life 1799-1867*, (2013), pp.133, 202; B MacAongusa, 'William Dargan: Great Railway Builder and Patriot', *Journal of the Irish Railway Record Society*, 24.171, 2010, 4-12; R Balgarnie, *Sir Titus Salt, Baronet: His Life and Its Lessons*, (1877), pp.149-53.
163. R M Birse, rev. M Chrimes, 'Mitchell, Joseph (1803-83)', *Oxford DNB* on-line, (2010), (accessed 15 August 2013).
164. G Vines and K McInnes, *Metal Road Bridges in Victoria, Part 1*, (rev. ed. 2010), pp.18-9.
165. James, Overseas Railways, p.45. James refers to 'the brief vogue for plate and box girders following the euphoria of Britannia', noting Fox Henderson's Pont de la Quarantine, Lyon (2x220ft spans), and the others from Gouin & Cie at Clichy (75ft), Asnières (5x100ft), Barsac (100ft), Langon (210ft, 240ft, 210ft), Aiguillon (210ft) and Moissac (2x150ft + 3x220ft spans).
166. K Verswijver, I Wouters, I Bertels and E De Kooning, 'Cast-iron bridges of Belgian industrialist Charles Marcellis (1798-1864)', in C A Brebbia and L Binda (eds.), *Structural Studies, Repairs and maintenance of Heritage Architecture XII*, (2011), pp.214, 217.
167. James, 'Overseas Railways', p.20; *The Engineer*, 18 December 1863, 361; *The Sydney Morning Herald*, 27 May 1864; Rapley, *Britannia*, pp.124-5.
168. *ILN*, 13 August 1864, 161-2; *The Argus* (Melbourne), 10 January 1862; http://museumvictoria.com.au/railways/image.aspx?PID-4036 (accessed 13 December 2009). Fairbairns apparently declined to tender for Tarradale because they considered the specification to be over-designed (*The Argus*, 22 November 1860). It was manufactured by Rowland Brotherhood of Chippenham, Wiltshire.
169. *The Argus* (Melbourne), 2 June 1860.
170. James, 'Overseas Railways', pp.25-6.
171. Rapley, Britannia, p.124.
172. James, 'Origins and Worldwide Spread of Warren-Truss Bridges', 96, quoting J Macneill, *The Boyne Bridge*, (1860).
173. C O'Mahony, 'Iron Rails and Harbour Walls: James Barton of Farndreg', *Journal of the County Louth Archaeological and Historical Society*, 22.2, 1990, 134-6; J Barton, 'On the Economic Distribution of Material in the Sides or Vertical Portion of Wrought Iron Beams', *MPICE*, 14, 1855, 451-2.
174. J Barton, 'On the Calculations of Strains in Lattice Girders with practical deductions therefrom', *BAAS1852*, p.123.
175. *Belfast Newsletter*, 6 September 1852.

176. O'Mahony, 'Iron Rails and Harbour Walls', 139. D J Jourawski had done stress calculations for a timber bridge in 1844 (Timoshenko, *Strength of Materials*, pp.142-3).
177. *Liverpool Mercury*, 7 March 1848.
178. James, 'Origins and Worldwide Spread of Warren-Truss Bridges', 98-9.
179. Barton, 'Economic Distribution', 443-90.
180. Smith, 'The development and use of the tubular beam', pp.120-2. Moorsom, who was older, was also an advocate of lattices.
181. *The Engineer*, 7, 1859, 138, 389-90; 8, 1859, 138; 9, 1860, 157-8.
182. B Baker, *Long Span Railway Bridges*, (1867), p.8.
183. For example in Scotland: J Mitchell, 'On the Construction and Works of the Highland Railway', *BAAS1867*, pp.151-61; *The Engineer*, 24, 1867, 233-4; *Engineering*, 4, July-Sept, 1867, 205-7, 223-4, 264, 288; *ILN*, September 1893, 288.
184. Fairbairn, *Application*, (3rd ed. 1864), pp.123-36.
185. H M Burt, *Burt's Illustrated guide to the Connecticut Valley*, (1867) pp.60-5; G L Vose, *Manual for Railroad Engineers and Engineering Students*, (1873), pp.262-6; *Engineering*, 2, 1866, 83-5.
186. S Fleming, *The Intercolonial: A Historical Sketch of the Inception, Location, Construction and Completion of the line of Railway uniting the Inland and Atlantic Provinces of the Dominion, with maps and numerous illustrations*, (1876), pp.138, 147 and Plates; The Intercolonial Railway. *The Genesis of its bridges, with the official documents laid before Parliament*, (1875); James, 'Overseas Railways', p.27; 'Correspondence relating to a claim of Mr H B Higginson and Queen's University against the Government of Canada. Printed to afford information respecting the character and merits of the claim', (1903), p.10; See Chapter 8.
187. J G James, 'Some Steps in the Evolution of Early Iron Arched Bridge Design', *TNS*, 59, 1987-8, 153.
188. See Chapter 3. W Fairbairn, 'An Experimental Inquiry into the Strength of Wrought-Iron Plates and their Riveted Joints as applied to Ship-building and Vessels exposed to severe strains', *Philosophical Transactions of the Royal Society, Part 2, 1850*, 677-725; E C Smith, *A Short History of Naval and Marine Engineering*, (1938), pp.109-10; W Fairbairn and T Tate, 'On the Strength of Materials considered in relation to the Construction of Iron Ships', *BAAS1865*, pp.253-4.
189. Cox, *Design of Structures of Least Weight*, pp.1-2; Burton, 'Lessons Learned in the Design and Erection of Box Girder Bridges', p.24.
190. Dempsey, *Tubular and other Iron Girder Bridges*, pp.27-9; Fairbairn, 'The Rise and Progress of Manufactures and Commerce', Vol.2, pp.xcix-c; *MG*, 2 June 1847; *ILN*, 10, 1847, 373. For background see B Dempster, Liverpool's Floating Landing Stages, (MSc Dissertation, University of Liverpool, 2003), pp.50-1. I am grateful to Peter Rowlands for bringing this to my notice.
191. Fairbairn, 'The Rise and Progress of Manufactures and Commerce', pp.c-ci; *CE&AJ*, 18, 1855, *The Engineer*, 38, 1874, 132. There was a precedent for hinged ramps in bowstrings by Marc Brunel (J G James, 'The evolution of iron bridge trusses to 1850', *TNS*, 52, 1980-1, 72).
192. British Listed Buildings, 'Folkstone Harbour Viaduct and Swing Bridge', ID 1404114, 'Kent Rail – Folkstone Harbour' at http://www.kentrail.org.uk/folkstone_ harbour.htm (accessed 14 May 2016).
193. Fairbairn, *Application*, Plates V and VI.
194. Rapley, *Britannia*, p.120 – no primary source given.
195. W Humber, *Cast and Wrought Iron Bridges and Girders*, (1857), pp.43-9 and Plates 27, 28, 29.
196. *Hamilton Spectator*, 22 March 1860.
197. *Engineering*, 4, July-Dec. 1867, 205-6; *MM*, NS21, Jan.-June 1869, 84-5; R and S Byrom, 'The Railway Swing Bridge at Clachnaharry', *Towpath Action Group Newsletter*, 83, 2012, 15-6. .
198. *MG*, 16 May 1864.
199. R Rawlinson, *Report on Liverpool's Water Supply*, (1846), p.25 (Appendix B). It is believed one was erected over the Crowden Brook near Longdendale as part of the Manchester Waterworks scheme – I am grateful to Bill Slatcher for this information.
200. DeLony, *Context for World Heritage Bridges*, p.6; Fairbairn, *Application*, pp. 255-7; Fairbairn, 'On the Application of Cast and Wrought-iron to Railways', pp.464-5.
201. Humber, *Cast and Wrought Iron Bridge Construction*, pp.76-7 and Plates 39 and 40.
202. R Turner, 'Description of the Roof over the Railway Station, Lime-street, Liverpool', *MPICE*, 9, 1850, 207.
203. Turner, 'Description of the Roof', 204-14 and Plate; Fairbairn, *UIfE3*, pp.216-20; *CE&AJ*, 14, 1851, 174-5; C E Peterson, 'Inventing the I-beam: Richard Turner, Cooper & Hewitt and Others', *Bulletin of the Association for Preservation Technology*, 12.4, 1980, 11.
204. R Dixon and S Muthesius, *Victorian Architecture*, (1978), pp.100-1.
205. Turner, 'Description of the Roof', 210.
206. Peterson, 'Inventing the I-beam', 20 .
207. D Evans, *Building the Steam Navy: Dockyards, Technology and the Creation of the Victorian Battle Fleet, 1830-1906*, (2004), pp.88-106.
208. See below.
209. W Fairbairn, 'Description of the Sliding Caisson at Keyham Dockyard', *MPICE*, 13, 1854, 444-5; *MM*, 60, 1854, 468; *CE&AJ*, 17, 1854, 237, 258-63; *Life*, pp.319-20.
210. Fairbairn, 'Description of the Sliding Caisson', 444; S Clegg jn, 'Sliding Caisson at Keyham Docks', *CE&AJ*, 19, 1854, 259-60.
211. Fairbairn, 'Description of the Sliding Caisson', 458.
212. Fairbairn, 'Description of the Sliding Caisson', 444-63, Plate 4.
213. L T C Rolt, *Tools for the Job: A Short History of Machine Tools*, (1965), p.120.
214. Cardwell, *Fontana History of Technology*, p.260.
215. A quayside crane with two curved cast iron jibs was constructed by Hick of Bolton in 1818 (*Mechanic*, 1818, plate C11, reproduced in *Industrial Archaeology News*, 179, 2016, 11. Not 1834 as O Bachmann, H-H Cohrs, T Whiteman and A Wislicki, *The History of Cranes*, (1997), pp.45-6). There were two cranes with curved jibs in the Grand Junction Railway's erecting shop at Crewe in 1849 (*ILN*, 24 March 1849, 189).
216. Patent No, 13317; *MM*, 1851, 381-3.
217. Fairbairn, *UIfE2*, p.289; *Life*, p.321.
218. 'On a Wrought Iron Tubular Crane, designed by William Fairbairn, CE, FRS. Communicated by Sir David Brewster ...', *BAAS1850*, pp.s177-9; *The Engineer and Machinist*, 3, March-Dec. 1851, 174-6.
219. *Official Catalogue of the Great Exhibition of the Works of all Nations*, (1851), Class 5, pp.230-1. The description however was of the first 12ton Keyham crane, whereas that exhibited and illustrated was a 5ton crane.
220. J Sproule, (ed.), *The Irish Industrial Exhibition of 1863. A Detailed Catalogue of its Contents*, (1854), p.178.

221. A Ure, *A Dictionary of Arts, Manufactures and Mines; containing a clear exposition of their Principles and Practice*, (4th ed. 1853), pp.549-53. Also published in America.
222. For example *Appleby's Illustrated Handbook of Machinery*, (1882 ed.), pp.2-3, advises that this type of crane is 'generally known as the Fairbairn crane, from the fact that it was originally designed by Sir William Fairbairn'.
223. *Mining Journal*, 20, 1850, 244.
224. Bachmann et al, *History of Cranes*, p.47.
225. Evans, *Steam Navy*, p.74.
226. Fairbairn, *UIfE2*, pp.282-9. The cranes were to lift to a height of 30ft and sweep in a circle of 65ft diameter.
227. Fairbairn, *UIfE2*, p.290; *The Engineer and Machinist*, 3, 1851, 179 and Plate; *Life*, pp.366-7.
228. Fairbairn, *UIfE2*, pp.290-4; Fairbairn, 'Description of Large Tubular Wrought Iron Crane', 87-96 and Plates; *Newton's London Journal of Arts and Sciences*, 1858, 171-176.
229. Fairbairn, *UIfE2*, p.295; *MPICE*, 95, 1889, 392.
230. *The Engineer*, 7, Jan-June 1859, 272; Fairbairn, *UIfE2*, p.296.
231. Fairbairn, *UIfE2*, pp.290-1.
232. Fairbairn, *UIfE2*, p.295.
233. For example Folkstone: *ILN*, 34, 22 January 1859, p.81. Chatham: *The Practical Mechanics Journal*, March 1861, 329 - see also Evans, *Steam Navy*, pp.185-7. West India Docks: British History on Line at http://66.102.9.104/search?q=cache:Vs8yWKEA3IUJ: www.british-history.ac.uk/repo (accessed 15.03.07).
234. 'Cranes Kraner Krane' at www.my1287.dk/index.php?option=co... (accessed 28.01/11).
235. *The Engineer*, 15, Jan-June1863, 26; *Sydney Morning Herald*, 21 November1862.
236. http://maps.google.com/maps/ms?ie=UTF8&t= m&vpsrc= 6&source=embed&oe=UTF.. (accessed 28 July 2012); http://portal.apsevilla.com/wps/portal/puerto_es/cultura_es?WCM_GLOBAL_CONT... (accessed 28 July 2012).
237. http://www.practicalmachinist.com/vb/showthread.php?t= 114790 (accessed 02.06.08). There is no hard evidence that this crane was built in Ancoats, but on the present balance of probabilities, it was. Indeed for none of these existing cranes, apart from Newcastle, have I yet found any documentary evidence they were manufactured in Ancoats – hence the cautious '*probably*'.
238. Werkgroep Industrieel Erfgoed Zeeland (WIEZ), 'Stadskraan Middelburg teruggeplaatst' at http://www.industrieelerfgoedzeeland.nl/stadskraam-middelburg-teruggeplaatst/ (accessed 03 August 2012); W H P M van Hoof, 'De ijzergieterij en machinefabfiek "De Prins van Oranje" te 's-Gravenhage', *Industriële Archeologie*, 35,1990, 50-7.
239. On licences in the textile machine industry see C Macleod, 'Strategies for innovation: the diffusion of new technology in nineteenth-century British industry', *Economic History Review*, 45, 1992, 294-6, 303.
240. Advertisement of 1867. Photocopy in the author's possession, courtesy of Prof. Van Hooff.
241. http://www.kadekraankampen.nl (accessed 17 July 2012). This website was revised on 23 July 2012 and no longer has the information about the Prins van Oranje factory.
242. W H P M van Hooff, *In het rijk van de Nederlandse Vulcanus: de Nederlandse machinenijverheid 1825-1914*, (1990), p.213.
243. M Sano, *Alcock no Edo: Shodai Eikoko koshi ga mita Bakumatsu-Nihon*, (2003). This book is in Japanese, but there is a helpful review in English at http://yagian.blogspot.co.uk/2011/01/strange-destiny-of-interpreter-in-end.html (accessed 20 July 2012); Van Hooff, 'De ijzergieterij', 50-5. The visit resulted from an acquaintance between Hotz and the Dutch anti-colonial author Multaluli - the nom de plume of E D Dekker who wrote *Japansche gesprekken*, (1862) in connection with the Japanese mission to Western Europe.
244. T Hashimoto, 'Introducing a French Technological System: The Origin and Early History of the Yokosuka Dockyard', *East Asian Science, Technology and Medicine*, 16, 1999, 53-72.
245. K Tamagawa, 'The Historic Cranes in Japan, Part 1', *Cargo Handling JAPAN*, 49.5, 2004, 526-7. The crane's location is shown on a bird's-eye view of the Dockyard in 1883 (Hashimoto, 'Yokosuka Dockyard', 56-7).
246. A King, *Fairbairn Steam Crane 1876* (City of Bristol Museum and Art Gallery – no date, but c.2000); Listing Notice Ref. 901-1/42/1317.
247. 'Davis and Primrose' at http://www.gracesguide.co.uk/wiki/Davis_and_Primrose (accessed 16 April 2010); Advertisment in Bristol City Museums, M Shed Archive.
248. SINE Project, 'Structure Details for Son of Swan Neck', at http://sine.ncl.ac.uk/view_structure_information.asp?struct_id=1227 (accessed 10 August 2010); 'Port Cities – Hartlepool. A journey into the port half a century ago', at http://portcities,hartlepool.gov.uk/server.php?show=ConNarrative.138&chapterId=669 (accessed 8 November 2005).
249. 'Cumbria Archive Service Catalogue' at http://www.archiveweb.cumbria.gov.uk/CalmView/default.aspx (accessed 4 April 2016); *Engineering*, 87, Jan-June 1909, 240.
250. *De Ingenieur*, 19, 1904, 577. It was the strongest crane in the Netherlands.
251. Bachmann et al, *History of Cranes*, p.47.
252. http://www.flickr.com/groups/fairbairncranes/ (accessed 3 August 2012). At Rorschach on the Swiss side of Lake Constance, there is an electrically-powered 'Fairbairn' crane by Stahl (visited 2012).
253. J Bennison, 'Essay', (1869 Lancashire County Archive DDX/184); R J M Sutherland, 'The birth of stress: a historical review' in W Addis, (ed.) *Structural and Civil Engineering Design*, (1999), p.249; Hayward, 'Fairbairns', p.3.19. David Evans states that the 40ton Keyham crane was 'to be made on the spot', and that 'Fairbairn's engineer reckoned on finishing it by the end of May'. It is possible it was made on site, but unlikely. A probable explanation is that it was delivered in parts to be put together on site (Evans, *Building the Steam Navy*, p.105n10).
254. Information from Professor Van Hooff.
255. J Mokyr, *The Lever of Riches: Technological Creativity and Economic Progress*, (1990), p.13.
256. J Weisbach and G Herrmann, *The Mechanics of Hoisting Machinery ...* , (ET of the second German ed. by K P Dahlstrom, 1907), pp.258-9 – their italics.
257. O Lueger, *Lexikon der gesamenten Technik*, (2nd ed. 1904, Stuttgart and Leipzig) – Article 'Krane (1)', Figs.5 & 12.
258. *Meyers Konversations-Lexikon*, (6th ed., Leipzig, 1902-8), Article 'Krane', Plate Krane III.
259. A Böttcher, *Cranes: Their Construction, Mechanical Equipment and Working*, (Translated and Supplemented with English, American, and Continental Practice by A. Tolhausen, 1908), pp. 55, 60, 64, 68

260. American Society of Mechanical Engineers, 'The Creusot Steam Hammer', (Leaflet on the occasion of visit to Le Creusot 16 Sept.1981).
261. Fairbairn, *UIfE2*, pp.296-7.
262. J Jessop & Son, *Catalogue*, (1892), p.13; Böttcher, *Cranes*, p.375.
263. W Anderson, 'Description of a Six-ton Crane, with Curved, Diagonally braced Jib, at the Pigeon-house Fort, Dublin', *Transactions of the Institution of Civil Engineers of Ireland*, 6, 1863, 188-92 and Plate I.
264. B B Stoney, *The Theory of Strains in Girders and Similar Structures, with Observations on the Application of Theory to Practice and Tables of the Strength and other Properties of Materials*, (1869), pp.129-36.
265. For example 40ton hydraulic crane at Portsmouth Dockyard by The Hydraulic Engineering Co, Chester (*The Engineer,* 57, Jan-June 1884, 3). In 1892 Edward Marks described a 4ton forge crane with a bogy traversing 6ft 6in. (E C R Marks, *Notes on the Construction of Cranes and Lifting Machinery,* (1892), pp.109-11).
266. http://www.veniceinperil.org/projects/armstrong-mitchell-crane (accessed 15 May 2015).
267. 'James Taylor – A Great Birkenhead Engineer', at http://www.oldmerseytimes.co.uk/jamestaylor.html (accessed 22 August 2013), pp.13-14; *The Engineer*, 14 June 1889, 501.
268. Appendix 5.1.
269. *North British Review*, 1861, 172.
270. Listing Notice No. 685/0/10036 16-Dec-09, 'Wellington Dock and associated structures, including crane situated on Esplanade Quay'; Dover District Council, 'Dover District Heritage Strategy', (Draft, June 2012), p.153 – this 635-page report devotes all of three lines to the crane, referring to the Fairburn [*sic*] Engineering Co.
271. http://www.doverport.co.uk/about/news/historic-crane-restoration-gives-heritage-trail-a/12922/... (accessed 1 January 2015).
272. Industrieel Erfgoed Zeeland, 'Stadskraan Middelburg teruggeplaatst' at http://www.industrieelerfgoedzeeland.nl/stadskraan-middelburg-teruggeplaatst/ (accessed 3 August 2012); 'wiez' Photograph November 2009 at http://www.flickr.com/photos/ wiez/4738069101/in/pool-fairbairncranes (accessed 07.08.10).
273. Rosenberg and Vincenti, *Britannia*, p.71.
274. Petroski, *Design Paradigms,* p.101.
275. Åkesson, *Understanding Bridge Collapses*, pp.179-92.
276. Charlton, *History of theory of structures*, pp.169-70.
277. K Culmann, *Die Graphische Statik*, (1866), pp.264-70 and Fig.1 of Plate 11.
278. C Lehmann, and B Maurer, *Karl Culmann und die graphische Statik*, (2006), p.127; J G Skedross and S L Baucom, 'Mathematical analysis of trabecular "trajectories" in apparent trajectorial structures: The unfortunate historical emphasis on the human proximal femur', *Journal of Theoretical Biology*, 244, 2007, 21.
279. Timoshenko, *Strength of Materials*, p.194.
280. J G Skedross and R A Brand, 'Biographical Sketch: Hermann von Meyer (1815-1892)', *Clinical Orthopaedics and Related Research*, 469.11, 2011, 3072-6.
281. G H von Meyer, 'Die Architectur der Spongiosa. (Zehnter Beitrag zur Mechanik des menschlichen Knochengerüstes)', *Archiv fur Anatomie, Physiologie und wissenschaftliche Medicin*, 1867, 615-628.
282. B Rüttimann, 'A Noteworthy Meeting of the Society for Nature Research in Zurich. Two Important Precursors of Julius Wolff: Carl Culmann and Hermann von Meyer', in G Regling (ed.), *Wolff's law and connective tissue regulation: modern interdisciplinary comments on Wolff's law of connective tissue regulation and rational understanding of common clinical problems*, (1993), pp.13-5; Skedross and Baucom, 'Mathematical analysis', 38-9.
283. Skedross and Baucom, 'Mathematical analysis', 17; Rüttiman, 'A Noteworthy Meeting', p.15.
284. R A Brand, 'Biographical Sketch: Julius Wolff, (1836-1902), Clinical Orthopaedics and Related Research, 468.4, 2010, 1056-65.
285. J Wolff , 'Ueber die innere Architectur der Knochen und ihre Bedeutung für die Frage vom Knochenwachsthum', *Virchows Arch Pathol Anat Physiol,* 1870, 50, 389-450, English Translation and abridgement by M O Heller, W R Taylor, N Aslanidis and G N Duda, 'The Classic: On the Inner Architecture of Bones and its Importance for Bone Growth', *Clinical Orthopaedics and Related Research*, 468.4, 2010, 1056-65.
286. A Schreiber, H A C Jacob, and A H Huggler, 'Model Investigations on the Loading of Bone in the Area of Hip-Joint Prostheses, and New Prosthesis Designs in Consideration of Wolff's Law', in Regling (ed.), Wolff's law, p.243.
287. Skedross and Baucom, 'Mathematical analysis', 16.
288. M Viceconti, *Multiscale Modeling of the Skeletal System*, (2011), p.146; B K Hall, *Bones and Cartilage: Developmental and Evolutionary Skeletal Biology*, (2005), pp.409-10.
289. The same is related, with the Culmann and von Meyer diagrams, by the zoologist D'Arcy Wentworth Thompson, *On Growth and Form* (2nd ed. 1945), pp.976-9.
290. S C Cowin, 'Nature's Structural Engineering of Bone on a Daily Basis', in S R Bodner, J Singer, A Solan and Z Hashin (eds.), *Theoretical and Applied Mechanics 1992*, (IUTAM 1993), pp.145-7. IUTAM is the major international body in the field of mechanics. Its 1992 congress was at Haifa, Israel. See also R D Martin, D B Burr and N A Sharkey, *Skeletal Tissue Mechanics,* (1998), pp.226-7.
291. Fairbairn, *Application*, p.213.
292. D'A W Thompson, *On Growth and Form*, p.971, citing Galileo, *Dialogues Concerning Two New Sciences*, ([1638] E T Crew and Salvio, New York 1914)); Cardwell, *Fontana history of technology*, p.260.
293. Fairbairn, *Application*, pp.214-5.
294. Timoshenko, *Strength of Materials*, p.123.
295. M Dettelbach, 'Humboldtian Science' in N Jardine, J A Secord and E C Spary (eds.), *Cultures of Natural History*, (1996), p.287; K Olesko, 'Humboldt, Alexander von' and 'Humboldtian Science' in J L Heilbron (ed.), *The Oxford Companion to the History of Modern Science*, (2003), pp.383-7; W H Brock, 'Humboldt and the British: A Note on the Character of British Science', *Annals of Science*, 50.4, 1993, 369, refers to Fairbairn meeting Humboldt; *Life*, p.217-37.

Sources, Credits and Acknowledgements for Illustrations in Chapter 5

Fig.5.1 Fairbairn, *B&CTB*, p.212.
Fig.5.2 Fairbairn, *B&CTB,* Plate 19.
Fig.5.3 Fairbairn, *B&CTB,* Plate 20.
Fig.5.4 Fairbairn, *B&CTB,* Plate 20 (part).
Fig.5.5 *ILN*, 26 September 1846, 197.
Fig.5.6 W. Fairbairn to R. Stephenson, 18 July 1846, No.92, Stephenson Papers, Institution of Civil Engineers. I am grateful to the ICE for permission to reproduce this and illustrations 5.8, 5.16 and 5.17.
Fig.5.7 S. Smiles, *Lives of the Engineers with an Account of their Principal Works*, Vol.3, p.440.
Fig.5.8 W. Fairbairn to R. Stephenson, 15 August 1846, No.103, Stephenson Papers, Institution of Civil Engineers.
Fig.5.9 Photograph by Velela (2005) at: https://commons.wikimedia.org/wiki/File:Britannia_Bridge_wrought_iron_section.jpg (accessed 21 September 2016).
Fig.5.10 Photograph: R.J. Byrom, 1962.
Fig.5.11 Smiles, *Lives of the Engineers*, Vol.3, opp. p.301.
Fig.5.12 Fairbairn, *B&CTB,* p.113.
Fig.5.13 Discussion following J. Husband, 'The Aesthetic Treatment of Bridge Structures', *MPICE,* 145, 1901, 215.
Fig.5.14 Fairbairn, *Application,* Plate 5 (detail).
Fig.5.15 Fairbairn, *Application*, pp.280–1.
Fig.5.16 W. Fairbairn to R. Stephenson, 4 February 1846, No.44, Stephenson Papers, ICE.
Fig.5.17 W. Fairbairn to R. Stephenson, 10 June 1847, No.265, Stephenson Papers, ICE.
Fig.5.18 *CE&AJ,* 10, 1847, 377.
Fig.5.19 Gainsborough Bridge before it was reconstructed in 1992. I am indebted to Michael Lewis for this photograph. The curved angle-irons appear to be decorative rather than structural.
Fig.5.20 H. Moseley, *The Mechanical Principles of Engineering and Architecture*, (1843), p.525.
Fig.5.21 Fairbairn, *Application,* p.238.
Fig.5.22 W. Fairbairn, 'On the Security and Limit of Strength of Tubular Girder Bridges constructed of Wrought Iron', *Memoirs of the Manchester Literary and Philosophical Society,* 14, 1850, Plate.
Fig.5.23 Photograph, R.J. Byrom, 2008.
Fig.5.24 State Library, Victoria, reproduced in G. Vines and K. McInnes, *Metal Road Bridges in Victoria, Part 1 – History of Metal Road Bridges in Victoria,* (rev.ed. 2010), p.20.
Fig.5.25 Fairbairn, *Application*, (3rd ed.1864), p.129.
Fig.5.26 W. Fairbairn, 'The Rise and Progress of Manufactures and Commerce, and of Civil and Mechanical Engineering, in Lancashire and Cheshire', in T. Baines, *Lancashire and Cheshire, Past and Present,* (1869), Vol.2, Plate II.
Fig.5.27 Crowle Community Forum: Keadby Bridge at http://crowle.org/?p=4068 (accessed 21 October 2016).
Fig.5.28 *MM,* 1851, 381.
Fig.5.29 W. Fairbairn, 'Description of Large Tubular Wrought Iron Crane recently erected at Keyham Dockyard, Devonport', *MPIME,* 1857, Plate 115.
Fig.5.30 Fairbairn, *UIfE2,* p.289.
Fig.5.31 K. Tamagawa, 'The Historic Cranes in Japan, Part 1', *Cargo Handling JAPAN*, 49.5, 2004, 526.
Fig.5.32 R.J.M. Sutherland, 'The birth of stress: a historical review' in W. Addis, (ed.) *Structural and Civil Engineering Design*, (1999), p.249. There is a similar photograph in Hayward, 'Fairbairns', p.3.19. showing the tapering foot, and there is a sketch plan of the yard showing the crane in J. Bennison, 'Essay', (1869, Lancashire County Archive DDX/184).
Fig.5.33 I am grateful to Prof. Van Hooff for this photograph.
Fig.5.34 G. Meason, *The Official Illustrated Guide to the Great Northern Railway* (1861), p.393.
Fig.5.35 American Society of Mechanical Engineers, 'The Creusot Steam Hammer', (Leaflet on the occasion of visit to Le Creusot 16 Sept.1981), (Detail).
Fig.5.36 Fairbairn, *UIfE2,* p.296.
Fig.5.37 https://www.flickr.com/photos/janhallback/14537775620/in/photolist-wWkXN1-uJg (accessed 26 March 2016). I am grateful to Jan Hallbæk-Andersen for permission to reproduce this photograph.
Fig.5.38 K. Culmann, *Die Graphische Statik*, (1866), Fig.1 of Plate 11; also reproduced in C. Lehmann, and B. Maurer, *Karl Culmann und die graphische Statik:Zeichnen, die Sprache des Ingenieurs*, (2006), p.127 and in K-E Kurrer, *The History of the Theory of Structures: From Arch Analysis to Computational Mechanics* (2012 ed.), p.704.
Fig.5.39 http://www.engin.umich.edu/class/bme456/boneadapt/boneadapt.htm (accessed 23 July 2012), based on the widely reproduced illustration in J. Wolff, 'Ueber die innere Architectur der Knochen und ihre Bedeutung für die Frage vom Knochenwachsthum', *Virchows Arch Pathol Anat Physiol,* 1870, 50.
Fig.5.40 Fairbairn, *Application*, p.214.

CHAPTER 6

Spreading Influence

Education, Fitters and Ancoats Alumni

Introduction

FAIRBAIRN'S commitment to education was lifelong. It started with his self-education programme in apprenticeship days and progressed to the education and training of young engineers, particularly through the founding of educational institutions, and through training his own pupils. One of the many ways in which Fairbairn's influence was diffused was through those who worked for him and those who trained or gained experience with him. Over the fifty-eight years of its existence the Fairbairn company employed thousands of men but, so far, only around 115 have been identified.[1] An influential but neglected group is the fitters and erectors, often peripatetic, who put machinery or structures together, ensured they worked, trained operatives and were the human face of Fairbairns around the world. Some did not return but built careers for themselves in new countries. Fairbairn's pupils or premium apprentices were outstanding, many becoming important engineers and teachers of engineering. These areas of influence were on-going throughout Fairbairn's career but it is appropriate to place them in the chronological context of the successful years leading to Fairbairn's 'retirement'. They include the next generation of engineers.

Educational Foundations

CONSCIOUS of the limitations of his own education, Fairbairn had a lifelong concern to make the advantages of education available to the young, particularly to young engineers, believing that education led to success in engineering as in other spheres.[2] Many Manchester engineers agreed – as shown by the founding of the Mechanics' Institution – but this approach was by no means universal. It was not shared, for example, by the Stephensons or I.K. Brunel. So far as theoretical training was concerned – as distinct from knowledge gained from experience – Brunel possessed a caution amounting to suspicion which was not untypical amongst mid-century British engineers.[3] Fairbairn's motives were the benefit to the working man and the benefit to society. 'I am desirous of raising the faculties of every working-man, for the double purpose of making him more useful to society, as well as to those with whom he associates and for whom he labours', he said to the young men at Bolton Mechanics' Institute.[4] It reflected his experience. In 1842 he had told the Poor Law Commissioners:

> The best educated are invariably intrusted [*sic*] with the most important parts of the work. If there be any intricate work, or anything that requires close mental application, as a class we always select the men of the best school education first. In out-door works, when, for example, there is a steam-engine, or a water-wheel, or millwork to erect, a foreman or some responsible workman must be chosen, and the choice, in nine cases out of ten, falls on the man of the best school education. It is then found to be very useful to have a man capable of making a drawing, taking dimensions, or sending a letter. The best educated are always the best paid.[5]

A decade later, he stressed the opportunities for social mobility:

> an educated man skilled in his trade, with a knowledge of the laws on which it is founded, and of steady habits, is much more valuable to his employer and himself, than the mere workman whose actions are akin to those of a machine, whose mind is a waste, and who seldom thinks. The first ... is sure of employment, and the best and most difficult operations will be entrusted to his care; he attains the confidence of his employer, probably becomes a foreman, and ultimately he acquires a share in the business, or sets up for himself. How many deserving young men have done this! How many have I myself recommended to situations of trust; and how many from small beginnings and little promise have risen to distinction in their respective professions![6]

At the same time Fairbairn was pointing to the superior technical education available on the Continent, notably the Conservatoire des Arts et Métiers, and the Ecole Centrale des Arts et Manufactures, in France, and calling for similar institutions in Britain.[7] This view intensified towards the end of his life, driven by his growing concern that British engineering was being eclipsed by other countries. In his final paper in 1873 he pointed out that Britain was on many occasions far behind France, Switzerland, Germany and the United States, whose engineers 'are better educated than ourselves'. The response must be 'exertion in every scientific pursuit'.[8] Such was his own very modern response, 'a man's education ... *does not finish until he dies*'.[9]

This commitment to education was expressed in Fairbairn's active involvement in the establishment of the Manchester Mechanics' Institution, the Ancoats Lyceum, the Royal Victoria Gallery, the Manchester School of Design and the Chair of Engineering at Owens College. The Mechanics' Institution had its genesis at a chance meeting of Fairbairn, Richard Roberts and Thomas Hopkins. Wishing to teach young men 'the application of science to mechanical and manufacturing art', they each agreed to subscribe ten pounds, and to persuade others to do so.[10] The original impetus appears to have come from the success of George Birkbeck's course for working men at the Andersonian Institution in Glasgow.[11] A meeting was convened on 7 April 1824, chaired by Benjamin Heywood. Various motions were passed unanimously including one moved by John Kennedy and seconded by Fairbairn that 'an Institution be now formed, to be called the Manchester Mechanics' Institution'.[12] It was the first Mechanics' Institution outside London, where one had been founded under Birkbeck's influence, the previous year.[13] Free tickets attracted 1,400 to the initial lectures, but enthusiasm soon waned and after two or three years long lecture courses on science gave way to shorter courses. A purpose-built Institution in Cooper Street was opened in 1827, claimed as 'the first erected in this country to include accommodation for all the objects of a Mechanics' Institution'. It included a lecture hall to seat up to 1,000 people.[14] However, the Institution was considered to have been disappointing in its failure to attract 'the operative classes', although this is not a universal view.[15] As early as 1829 it was at a low ebb, with Rowland Detrosier leading a break-away faction. The issue was that management was imposed rather than elected.[16] Changes were made and by the mid 1830s the Institution had largely recovered.[17] In 1829 a class teacher in mechanical drawing claimed that his pupils were 'now distributed in different parts of the world having been sent by their masters in consequence of their abilities in the art of mechanical drawing'.[18] The Manchester Statistical Society, giving evidence to a Select Committee in 1835, reported that 'many individuals attribute their advancement in life to the opportunity which this institution has afforded them of improving themselves'.[19] The following year J. Wheeler noted that 'mechanics and others educated at the Institution are particularly sought after by machinists and master tradesmen of the town'.[20] In 1845 'Ladies Classes' were established.[21] Fairbairn's support for these classes is clear from his lectures on education, 'the minds of women ... for penetration, quickness of perception, and high attainments in mental culture ... are fully equal, if not superior,' to men, although he nevertheless envisaged the major influence of the mother to be within her family.[22] Certainly by mid century the Institution was flourishing and a larger building was needed. Fairbairn was amongst those who 'offered to promote the erection of a new building'.[23] It was a fine building, designed by J.E. Gregan in 1854 in the Italian Palazzo style,[24] with an unusual partially-glazed roof structure by Fairbairn.[25] Even so there were continuing concerns. Fairbairn's view was that Mechanics' Institutes 'have failed to serve the particular classes for whom they were originally intended, and they have probably been still less efficient as regards the advancement of the industrial arts'.[26] This prompted a call for the establishment of Lyceums, mechanics'-institute-like organisations whose distinctiveness lay in their lower membership fees, more democratic ethos, and provision of newsrooms'.[27] They were to be located where working people actually lived.

In 1838 William Fairbairn and Benjamin Heywood provided the impetus for Manchester's first Lyceum, at Ancoats although Heywood had established the Miles Platting Mechanics' Institution the previous year, with the ethos of a Lyceum.[28] A building was obtained at the corner of Great Ancoats Street and Union Street next to the McConnel & Kennedy and Murray mills, and near to Fairbairn's works.[29] At the first General Meeting, Fairbairn was elected President and Heywood, recently created a Baronet, Treasurer.[30] Those present and elected as Vice-presidents included Richard Cobden,[31] James Heywood MP,[32] William Langton (cashier at Heywoods' Bank and a promoter of the Manchester Statistical Society)[33] and William M'Connel (son of James M'Connel). Others present included Eaton Hodgkinson, John Dalton and J.E. Taylor (editor of the *Manchester Guardian*). Yet again it was with the movers of society that Fairbairn was associating. In recognition of the more democratic approach, those elected Directors were largely men with Ancoats addresses. The only Director who became well known was Abel Heywood, the Chartist newsagent, who twice became Mayor of Manchester, and in 1876–7 guided Manchester Town Hall to completion.[34] The Ancoats Lyceum initially attracted the young men and women whose education was Fairbairn's concern.[35] In 1844 J.A. Nicholls, from another Cross Street family, was appointed secretary. He established day schools for 'short-timers' in the factories.[36] But within a decade the annual meetings exuded gloom as numbers decreased year on year.[37] At the 1852 meeting, chaired by Fairbairn who was still President, the attendance was small.[38] It was the last annual meeting. Fairbairn had remained loyal and supportive throughout the Lyceum's 14-year life, from initial success to failure and extinction. He found it hard to understand, perhaps because of his own innate attributes, why in an age of such progress and opportunity, so many manual-working people displayed 'apathy and indifference' towards learning.[39] The Lyceum had failed to compete with the conviviality of the pub.[40]

In 1839, by which time at least half the lectures at the Mechanics' Institution were on non-scientific subjects, H.H. Birley chaired a meeting to discuss the formation of an institute of practical science, modelled on the Royal Adelaide Gallery in London. Its aims were to illustrate the progress in industry and practical science, present experimental demonstrations, stimulate research and invention, and arouse the interest of young people – as Donald Cardwell puts it, 'Manchester was to have a science museum'.[41] Fairbairn spoke strongly in support. The institution was named the 'Royal Victoria Gallery for the Encouragement of Practical Science'. Birley was chairman and directors included engineers Fairbairn, Hodgkinson, Whitworth and John Hawkshaw. The patron was no less than Queen Victoria. William Sturgeon was appointed superintendent of 'a major addition to science in Manchester'. The vigour of the gallery and Sturgeon's energies were not matched by local support and, after only two years, the Gallery failed and closed its doors.[42]

Fairbairn was one of the founders of the Manchester School of Design. In 1837–38 the painter Robert Haydon toured Britain advocating art education for artisans.[43] On 25 January 1838 he addressed a meeting at the Manchester Mechanics' Institution, chaired by James Frazer (or Fraser) and with James Heywood present.[44] On 2 February he visited Fairbairn's works and in the evening dined with Fairbairn and Frazer when they 'planned the School'.[45] On 19 February Haydon was back for another meeting, when eleven resolutions which he, Frazer and Fairbairn had framed, were passed. The second resolution proposed by Fairbairn and seconded by Leo Schuster – a friend from journeyman days – read, 'That a society to be called "The School of Design" be established in Manchester'. Its Provisional Council included engineers Fairbairn and Nasmyth.[46] The School opened in October 1838, the second such School in the country, preceded only by the Royal School of Art in London, founded in 1837.[47]

Owens College had its genesis in a lecture to the Manchester Statistical Society by H.L. Jones in 1836,[48] following which a large committee, including Fairbairn, was appointed to further the project.[49] A bequest from John Owens in 1846 enabled goals to be realised.[50] In 1866 a meeting of engineers decided it was time to establish a chair of engineering. Subscriptions were sought and a committee appointed – Whitworth, Fairbairn, Beyer and Robinson. They led from the front: Beyer donated £3,000, Whitworth and Robinson £1,000 each and Fairbairn £500.[51] In March 1868 they appointed the 25-year-old Osborne Reynolds as Manchester's first Professor of Engineering.[52] Fairbairn was disappointed – he had hoped for the appointment of his former assistant William Unwin.[53]

From the mid 1860s there had been moves to extend Owens College. Fairbairn was on the committee charged to raise the money.[54] Land was acquired on Oxford Road. The Duke of Devonshire was President and Waterhouse was appointed architect. Fairbairn became a Governor. He was at the laying of the foundation stone in 1870 and the opening of the new buildings in 1873.[55]

Ancoats Lyceum was a failure. The Mechanics' Institution developed into UMIST – the University of Manchester Institute of Science and Technology. Owens College became the Victoria University of Manchester, which joined with UMIST in 2004 to become the University of Manchester. Osborne Reynolds' engineering department has become the School of Mechanical, Aerospace and Civil Engineering, ranked fourth in the UK for engineering and technology. The School of Design has been successful and is now a Faculty in Manchester Metropolitan University. It is notable that Fairbairn, with so little formal education, should have had a part in the foundation of each of these academic institutions.

Fairbairns' Fitters and Erectors

As British engineers exported to the world, their products were accompanied by fitters, erectors, engineers and instructors, to assemble and commission the products and advise on operation and maintenance. In 1839 Love wrote of men '*located* in various parts of Europe, who are employed under the direction of Mr Fairbairn, in superintending the erection of work manufactured in these premises'.[56] This employee group of site-based fitters and erectors has received little attention generally, and none so far as Fairbairn is concerned.

The practice of sending men with their steam-engines was developed by Boulton & Watt. For their customers, having access to a skilled mechanic was a crucial factor in purchasing a steam-engine. They sent some of their best men to major customers, but this led to difficulties. The erection of a distant engine could take a mechanic away for three years. Several failed to return and by the start of the 19th century Boulton and Watt were recruiting mechanics for foreign customers upon receipt of a firm order, with the expectation that they would remain abroad.[57]

Initially Fairbairn's fitters and erectors, engineers and instructors, were to be found in various British and Continental locations.[58] Later the locations extended to Turkey, North Africa, Russia, Australia, India and the Americas. Because of the destruction of the Fairbairn business records, the number of such men is impossible to quantify, but it must have been many hundreds over the life of the company. Of these only around twenty have been identified. In addition pupils sometimes helped with work on site.[59] A 36ft-diameter Fairbairn waterwheel, 16ft wide, required 185 man-weeks to construct in the workshop and 65 man-weeks to erect on site, suggesting that, at least so far as waterwheels were concerned, approaching a third of the labour force might have been on site.[60] However it is likely that this would have involved some local labour working with Fairbairn fitters. A similar ratio might apply to millwork and possibly to larger steam engines, but the ratio was very much lower for ships and locomotives. Any overall ratio can be little more than a guess, but a figure between five and ten per cent of the total labour-force might be of the right order. Some of them were peripatetic, moving from job to job. Some put down roots overseas.

In this way Britain – and Fairbairns – lost some very able engineers. One such was the brilliant locomotive engineer John Haswell (1812–97). A Scot, he studied at the Andersonian in Glasgow and then with Claud Girdwood, possibly as a premium apprentice, moving to Fairbairns in 1835.[61] In 1838 the Wien-Raaber-Eisenbahn commissioned Fairbairn in connection with the construction of a locomotive works in Vienna. Fairbairn sent Haswell to set it up. The railway company's Engineer recognised Haswell's abilities and appointed him to manage the works.[62] Haswell remained there over forty years during which he founded Austrian locomotive building, built the first iron-foundry in Vienna and introduced industrialisation into Austria. The works became known throughout Europe and its locomotives were exhibited at the international exhibitions.[63] Haswell's expertise was lost to Fairbairns and to Britain.

Another engineer who did not return was John Fyfe (1803–89), originally from Glasgow where he had been apprenticed at Baird & Co. In 1840 Australia's first steamship company, the Hunter River Steam Navigation Co. of Newcastle, NSW, ordered an iron steamship from Fairbairn.[64] She was the *Rose*, 153ft 6in. long, a paddler with a pair of side-lever engines to supplement her sails.[65] Fyfe, a journeyman at Fairbairn's Millwall shipyard, worked on her engines and when she steamed down the Thames in November 1840 he was on board as engineer. It seems he was accompanied by his wife, so maybe they had planned to emigrate.[66] Nearly six months later the *Rose* reached Australia – the first iron steamship to do so.[67] Fyfe continued as her engineer, as she traded between Sydney and the Hunter River.[68] In 1845 he was appointed superintendent engineer on shore, followed by other roles, and, on the formation of the Engineering Association of New South Wales in 1870, was its first President.[69] Practical engineering skills, like Fyfe's, were exported from Britain as an adjunct to engineering artefacts. In this the Scottish diaspora was prominent.

Some of Fairbairn's fitters and erectors were peripatetic. In 1848 the first limited company in Sweden commissioned Fairbairn in connection with a cotton mill and 40ft-diameter waterwheel at Strömsbo, Gefle.[70] He visited Gefle in mid 1850 and again in June 1851, and had erectors on site in 1851 and 1852.[71] At least twenty English fitters and instructors from various firms, some with their families, lived in the *Engelska Byggningen* at Strömsbro during the building and starting up of the mill, some from Fairbairns. These men are significant because there is a record of their wage rates, apparently including a premium for 'working away'. They included Joseph Claughton, there for forty-four weeks at 10*s*. per day,[72] Joseph Oddy, paid 10*s*.6*d*. per day, Charles Oddy paid 10*s*. per day; and Alfred Larkum, paid 8*s*. per day.[73] Joseph Oddy is of particular interest because he was active in at least three locations. From March to May 1850 he was at the

Brenneriveien Weavery at Oslo, setting up a steam engine.[74] From there he went to Germany, and then to Strömsbro.[75] Another peripatetic erector was Lambert. In July 1855 he was one of four European erectors, probably from Fairbairns, working at the Bombay Spinning and Weaving Company.[76] He appears to have been in India until 1859 and it is likely that he also worked on the Oriental Spinning and Weaving Mill. After that Fairbairns sent him to Ontario to erect swing-bridges over the Desjardines and Welland Canals.[77] Peripatetic erectors, such as Oddy and Lambert, were permanent Fairbairn employees, moving from job to job. Resourceful and adaptable, engineering firms in Britain relied on such men to overcome the problems of getting structures erected and machines to work, sometimes in primitive and undeveloped locations.

The erection of three Australian bridges illustrates the sort of co-ordination problems that could be faced by overseas erectors. In 1853 Fairbairns received orders for tubular-girder bridges over the Barwon, near Geelong, and over the Yarra, at Melbourne. John Croll was sent to manage their erection. He arrived in April 1855, followed by the ironwork on six ships. In October 1855 seven erectors arrived, but local contractors did not have the piers at Geelong ready for nearly two years. The bridge was erected 1858–9 – the two spans of 210ft were the longest bridge spans in Australia until 1881. The Yarra Bridge followed.[78] An order for another tubular-girder bridge, over the Mariburnong, near Melbourne, was placed with Fairbairns. For this bridge Douglas Galton was Inspecting Officer in Britain, a role transferred to I.K. Brunel during fabrication. Brunel sent drawings to Melbourne showing the tubular-girders divided into parts for shipping, with identification markings. Unfortunately Fairbairns used different markings, plans of which were sent with the ironwork, but the plans were left on board when the ship sailed back to England![79] Larger projects required larger teams. On the 17-span Warehouse Point Bridge in Connecticut in 1865 about a hundred erectors were involved, 'many of whom came specially from England'.[80]

Locomotive engineers faced different challenges. Another order received by Fairbairns through Brunel was for three 2-4-0WTs for the Adelaide & Port Railway. Broken down into components for shipping, they were accompanied by John Clarke (1825–79).[81] In January 1856 the locomotives were unloaded and assembled under Clarke's supervision. He then trained their drivers and stokers. Clarke did not return home but continued with the Railway Company, setting up a substantial workshop, and becoming locomotive superintendent of South Australian Railways.[82] In 1856 his brother-in-law, John Tipping, accompanied two Fairbairn locomotives for the Australian Agricultural Company at Newcastle, NSW, where he was engaged by the Company on the recommendation of Fairbairns.[83] The order for the Adelaide & Port locomotives included extensive spares and tools, an indication of the lack of facilities for maintenance and repair in Australia at that time.[84]

Fitters in far-away places needed a generous endowment of initiative as they put products together and made them work. These men were 'Fairbairns' to the local people in many countries who worked with them and learnt from them. At a time when travel was difficult and communications slow, fitters and erectors, independent and resourceful, were in the front line of the worldwide spread of British technology.

Ancoats Alumni

FAIRBAIRN'S Ancoats works was the successor to Maudslay's in London, in training and providing experience for young engineers. There were four groups of young men who trained at Fairbairn's; craft or trade apprentices, often the sons of millwrights; premium apprentices or pupils; journeymen seeking wider experience; and Fairbairn's assistants. There is no doubt that he attracted some remarkable men to train or gain experience with him.

For those who could not afford pupillage, a seven-year craft or trade apprenticeship in an engineering workshop was the normal way of entry into the industry, as Fairbairn himself had entered it.[85] Few records remain about Fairbairn's craft apprentices and only one is known to have founded a substantial business. This provides support for the view that, for craft apprentices, social mobility rarely extended beyond middle management or the proprietorship of a small firm. One reason for the limited social mobility of craft apprentices may have been that a material proportion of them comprised the sons of millwrights, which was 'considered a sufficient guarantee for skill and industry' but often was not.[86] The one Fairbairn craft apprentice who founded a significant company was Manchester-born George Saxon (1821–79), apprentice c.1836–44, outside foreman 1845–51, after which he moved to Benjamin Goodfellow at Hyde. He commenced business on his own account as a millwright in1854 and started building steam engines in the 1860s, patenting several improvements.[87] His work was well respected and his firm continued into the 20th century.

After a craft or trade apprenticeship many young men travelled as journeymen to gain experience, or because there was no employment available for them where they had trained.[88] The ablest sought positions with leading firms, often only staying a year or so, before moving on to another firm. These were not aimless restless wanderings but a broadening of experience by moving between workshops.[89] Fairbairns was high on the list of ambitious young journeymen. Scots were particularly drawn to their fellow-countryman. Peter Fairbairn (1799–1861), William's brother, was an early one.[90] Peter Carmichael (1809–75) from Dundee was another. He moved on, to work for Peter Fairbairn at Leeds, before returning to Dundee, and working his way up to senior partner of Baxters, by then the largest flax firm in the world.[91] John Anderson (later Sir John) (1814–86), a Scot from Aberdeen, was at the Ancoats works at the end of the 1830s. He then gained experience elsewhere, before joining the Department of the Inspector of Artillery.[92] Fairbairn worked with him on the Royal Small Arms Factory, Enfield,[93] the Balaclava flour-mill,[94] and the government powder mills at Waltham Abbey.[95] Archibald Sturrock (1816–1909) from Dundee, became Locomotive Superintendent of the Great Northern Railway.[96] Robert Morrison from Inverness, founded Robert Morrison & Co. at Ouseburn in 1853 and was soon employing over 500 men manufacturing engines and steam hammers.[97] Less typical were Greenwood and William Craven, whose family came to Manchester from Yorkshire. Greenwood worked for Sharps for twelve years, and William spent some time with Bodmer, before both brothers moved to Fairbairns in 1852. The following year they established Craven Brothers, described by S.B. Saul as 'probably the finest firm in the world supplying heavy machine tools'.[98] *Table 6.1*

Name	Apprenticeship	With WF	Apex of Career
Sir Peter Fairbairn	John Casson, Newcastle	1821, 1823–4	Principal, Wellington Foundry, Leeds
Peter Carmichael	William Low, Monifieth, Dundee.	1832–3	Senior Partner, Baxters of Dundee
John Haswell	Claud Girdwood & Co., Glasgow	1835–40	Engineer, Wien-Raaber-Eisenhahn Locomotive Works, Vienna
Archibald Sturrock	James & Robert Stirling, Dundee	1838–40	Locomotive Engineer, Great Northern Railway
William Hopper	Trotter, Broomhouse Paper Mills, Berwick	1838–9	Principal, Shipock Iron Works, Moscow
Sir John Anderson	Gordon, Barron & Co., Aberdeen	1839	Engineer, Woolwich Arsenal etc.
Edward Taylor Bellhouse	Wren & Bennett, Manchester	1840 (Millwall)	Principal, E T Bellhouse & Co., Engineers, Manchester
William Marshall	a millwright, possibly in Gainsborough	c.1840–2 (Agent)	Principal, William Marshall & Co., Engineers, Gainsborough
Robert Morrison	Reid, Inverness-shire, and then in Glasgow	1841	Principal, Robert Morrison & Co., Ouseburn, Newcastle upon Tyne
Greenwood Craven	Sharp, Roberts & Co., Manchester	1852–3	Principal, Craven Brothers, Engineers, Manchester
William Craven	Johann G Bodmer, Manchester	1852–3	Principal, Craven Brothers, Engineers, Manchester

Table 6.1: Some Journeymen who worked for Fairbairns and went on to successful careers

lists the places of their apprenticeships and apex of the careers of some of the successful journeymen who gained experience at Fairbairns.

One man who does not fit the description 'journeyman' but who gained his experience with Fairbairns and went on to found a major engineering consultancy was C.S. Allott (1842–1907). After a pupilage on the Ringwood and Christchurch Railway, Allott joined Fairbairns as a draughtsman in 1862, progressing to Chief Draughtsman and then to Assistant Manager. Major works of which he had charge included the roofs of the Royal Albert Hall and Liverpool Street Station and numerous bridges for the Intercolonial Railway of Canada. Following the Fairbairn liquidation, Allott started his own practice as a consulting civil engineer, initially supervising the completion of the Cornbrook Viaduct in Manchester for Fairbairns' liquidators.[99] He was engineer for many railway bridges, Stuart-Street generating station for Manchester Corporation, and cotton mills in Brazil and China.[100] In1897 the firm became C.S. Allott & Son and in the 20th century, as Allott & Lomax, it was the largest consulting civil engineering practice in Manchester with worldwide commissions, particularly in the power generation sector. In 2000 Allott & Lomax was taken over by Babtie, which, in 2004, was taken over by Jacobs, one of the largest consultancies in the world with over 70,000 employees by 2013.[101]

Even more remarkable than the journeymen were Fairbairn's premium apprentices or pupils – the terms are used interchangeably.[102] In the 19th century premium apprenticeships, pupilages or articles were the usual way of training

Name	Father's occupation	Prior education	Years at Fairbairns	Height of career
Albert Escher (1807–1845)	Cotton Spinner, Industrialist, Zurich		1824–5	Head of Engineering Department, Escher Wyss, Zurich
Andrew Murray (1813–1872)	Advocate, Sheriff of Aberdeenshire	Edinburgh Academy; Edinburgh University	1832–43	Chief Engineer, Portsmouth Dockyard
George Birkbeck	Professor of Natural Philosophy; Physician		1837–41	Tonge & Birkbeck, Patent Agents and Engineers, London
John Patchett			1838–9	
Frank Haydon (1822–1887)	Artist		1839–40	Public Record Office
Thomas Chappe de Leonval(1824-1895)		Private schools at Southport, Tutbury, Worksop and Paris	1840–6	Cotton mill proprietor, Waterloo Mills, Bolton
John England (c.1823–1870)	Solicitor		1840–7	Engineer, Tokyo to Yokohama Railway
— Rulph		Halifax	1841–6	
Hutton Vignoles (1824–1889)	Civil Engineer	Menars, France	1842–3	Consulting Engineer – railways and tramways
James Thomson (1822–1892)	Professor of Mathematics	Glasgow University, BA and MA in Mathematics	1843–5	Professor of Engineering, Glasgow
Edmund Edwards			1847–9	Consulting Engineer, London
Guilford Molesworth (1828–1925)	Anglican Vicar of Rochdale	King's College, Canterbury; Putney College (engineering); Articled to engineer L&BR	1848–52	Engineer to the Indian Railways
Suliman Effendi	Turkish Diplomat		1849	
Fleeming Jenkin (1833–1885)	Captain RN rtd; Coastguard	Edinburgh Academy; Genoa University, MA; One year in Genoa workshop	1851–4	Professor of Engineering, Edinburgh
William Anderson (1834–1898)	Banker, St. Petersburg	Commercial High School, St Petersburg; King's College London (App.Sci.)	1853–5	Director General of Ordnance Factories
James Cotterill (1836–1922)	Anglican Vicar	Brighton College	c.1854–7	Professor of Applied Mathematics, Royal Naval College, Greenwich
Laurence Grayson (c.1839–1916)		Manchester Academy	c.1854–8	Commissioner of Public Works, Australia
George Low 1833–1894	Banker, Stockbroker, Railway Chairman	Tutor at home	1854–9	Drawing office manager, F R & F Turner, Ipswich
H P Higginson		Collegiate School, Leicester	mid 1850s	Engineer and Manager, Wellington Gas Co., New Zealand
Robert Billington (1845–1904)			1859–63	Locomotive Engineer, London, Brighton & South Coast Railway
Harry Harman (1846–1895)	General Manager, Fairbairn Engineering Co.		1860s	Chief Engineer, Rownson, Drew & Co. (structural engineers)
Thomas F Pigot (1837–1910)	Chief Baron of the Exchequer in Ireland	Private tutor; St Geneviève College, Paris; University of France; University of Bonn; Trinity College, Dublin; Ecole des Ponts et Chaussées, Paris	1861	Professor of Descriptive Geometry, Mechanical Drawing, Surveying and Engineering, Royal College of Science for Ireland, Dublin

Table 6.2.
Known Fairbairn Pupils

Dates	Trading name	No. of known Premium Apprentices	No. of known journeymen who went on to very successful careers
1817–1832	Fairbairn & Lillie	1	1
1832–1842	William Fairbairn	9	8
1843–1854	William Fairbairn & Sons (from 1846)	8	2
1854–1864	Fairbairn & Company	4	0
1864–1875	The Fairbairn Engineering Co. Ltd	0	0

Table 6.3. Fairbairn's Premium Apprentices and successful Journeymen by Dates.

those who were to go on to manage engineering firms, become senior railway engineers, or partners in practices of consulting engineers. Competition for premium apprenticeships in the leading firms was stiff.[103] Ambitious fathers sought to place their sons with men of standing. Often there was a personal connection. Albert Escher, Hutton Vignoles, George Low, and probably Suliman Effendi were sons of Fairbairn's clients. Fairbairn worked with Molesworth's father in the formation of The Manchester Association for the Prevention of Smoke.[104] Frank Haydon's father was involved with Fairbairn in the founding of the School of Design. The calibre of many of the premium apprentices suggests Fairbairn took care in selecting them: 'it is only a few amongst the many that attain distinction ... unless the taste for knowledge and the power to apply it be there, it is vain to think of ascending the scale', he said.[105] Replying to his friend T. Romney Robinson who had referred a 'Dr Jones and his son' to him, Fairbairn wrote, 'Everything will however depend upon the young man himself, as provided he commences his career with the determination to *work hard* and *spend little*, I should have great hopes of his success'.[106] It is not known if Jones became a pupil, the probability is that he did not. There is no record of how many pupils Fairbairn had. Twenty-two have been identified, as listed in *Table 6.2*, inevitably including those who subsequently had distinguished careers. Young men had to be from a financially secure background to enter the profession in this way. Premiums were typically between £100 and £300 a year, on top of which were the costs of accommodation and living expenses.[107] Of those whose fathers' occupations are known, they are all from the middle-class, as *Table 6.2*.

Fairbairn had more pupils and successful journeymen in the years when he had become well-known beyond the textile industry, with reducing numbers in the years of the firm's decline, as shown in *Table 6.3*.

What did Fairbairn's premium apprentices do? It is clear that besides traditional millwright work in the factory, they assisted with experiments, were involved in site work and in at least some cases had a spell in the drawing office. Fleeming Jenkin told the Samuelson Committee that there was no difference from the artisan apprentice other than 'a certain amount of social consideration'.[108] This is perhaps unfair given that besides 'filing and chipping vigorously in a moleskin suit ... infernally dirty', and spending six weeks doing nothing but polishing brass valve boxes, Jenkin spent time in the drawing office, was in charge of the pattern shop for a time, and sometimes went out on site, for example to regulate a

Fig.6.1: James Thomson (1822–92)

governor at Saltaire. By the time he left Manchester he had become 'a fair average workman' and reckoned he was capable of building a locomotive.[109] Andrew Murray worked on the steamships *Railway* and *L'Hirondelle*, assisted with experiments and was involved in boilermaking.[110] John Patchett assisted Hodgkinson with his experiments.[111] Thomas Chappe assisted in experiments at both Ancoats and Millwall, and was involved in the erection of a corn mill in Cheshire.[112] William Anderson was engaged in the installation of machinery in Wales and Ireland. James Thomson worked on marine engines at Millwall and later on steam-engine building in the fitting shop at Ancoats. Guilford Molesworth was first engaged in millwright work in the factory and then sent to Portugal to help erect an engine for a cotton mill, followed by work on the pumping engine at Astley.[113]

In later years at least three Fairbairn's pupils looked back on their training. Jenkin wrote:

> The young professional engineer does not simply learn in the works how to file and chip. He learns the time required for all manner of jobs, the finish required in each class of work, the way the various parts are handled, the forms which are convenient, the routine of the shop, the character of the men – the system of storage, the materials and sizes to be bought in the market, and hundreds of other facts, which can only be made his own after contact with manufacture on a full scale. We cannot imitate this in college.[114]

He also advocated that premium apprentices should first study maths, physics and chemistry at university or technical college.[115] Some of Fairbairn's premium apprentices had done this but only a minority, as *Table 6.2* indicates. Anderson also emphasised the need for both the academic *and* the practical.[116] Molesworth recorded that

> the advantages gained by workshop experience are incalculable to the civil engineer ... The habits of precision, the thorough knowledge of the principles and practice of good work, the acquaintance with details and material ... as well as the knowledge of men, gained by mixing with them, all combine to make such training generally useful ... Practical experience ... must be gained in the school of life ... by keeping workmen's hours and by performing tasks that have to pass the criticism of the foreman, standing the test of actual work.[117]

Little is recorded about the life of premium apprentices outside work. Jenkin found time for social life, becoming friendly with the Gaskells – indeed he married a friend of the Gaskell girls, who it is claimed was his intellectual equal.[118] Unwin and Cotterill, having met at Ancoats, remained lifelong friends.[119] George Birkbeck was either resident with the Fairbairns at the Polygon or a regular visitor there, for his father wrote:

> We all rejoice in the effects of my son George's residence under your superintendence. His feelings are better regulated in consequence of the influence of occupation, under kind and friendly control, and he has acquired a taste for industrious pursuits ... He speaks in the highest terms of yourself and Mrs Fairbairn, and the rest of the family.[120]

Fig.6.2: Fleeming Jenkin (1833–85)

Fairbairn's pupils went on to work in varied fields of engineering. Two, Guilford Molesworth and William Anderson, were knighted for their work as engineers. Molesworth became Mechanical and Locomotive Superintendent to the Ceylon Railway.[121] Returning to England he wrote his *Pocket Book of Engineering Formulae* which ran to twenty-nine editions.[122] He then became Engineer to the Indian Railways, building many lines – one at the rate of a mile and a half per day – and undertaking reconnaissance for the line to Kandahar during the Afghan War, and from Rangoon to Mandalay in war-torn Burma.[123] Anderson joined Easton & Amos, which became Easton & Anderson, undertaking sugar mills, gun-mountings, waterworks and hydraulic lifts. He resigned in 1889 to become Director General of the Ordnance Factories. There he introduced an eight-hour day for 17,000 employees, apparently without loss of production. Fluent in Russian, he translated several engineering works into English. He was much involved in local education in Erith, where he lived, was a Fellow of the Royal Society and awarded an honorary doctorate by Durham University.[124]

Andrew Murray, after Millwall, moved to Woolwich Dockyard, and in 1846 was appointed Chief Engineer at Portsmouth Dockyard where he remained for nearly twenty-three years, being honoured with the Companionship of the Bath.[125] In 1844 Murray had won a Telford Medal from the ICE for a paper entitled 'On the Construction and Proper Proportions of Boilers for the Generation of Steam'.[126] At the Great Exhibition in 1851 he made notes of several exhibits among which was the Fairbairn crane.[127] Fairbairn cranes were subsequently installed at Portsmouth.[128] He wrote the entry on 'Shipbuilding' in the eighth edition of *Encyclopaedia Britannica,* and the entry was also published as a separate book.[129]

The abilities of at least some of the pupils whom Fairbairn took on are dramatically emphasised by the four who became Professors – James Thomson, Fleeming Jenkin, James Cotterill and Thomas Pigot – as did Fairbairn's assistant, William Unwin. Of these Thomson, Jenkin and Pigot had previously attended university, Cotterill went on to study mathematics at Cambridge, and Unwin obtained an external London degree whilst he was at Ancoats. The extent to which Fairbairn's constant involvement in engineering experimentation, his commit-

Fig.6.3: Guilford Molesworth (1828–1925)

ment to the education of engineers and the visits of leading engineers and scientists to the Ancoats factory, were factors that influenced his pupils is impossible to determine, but that his factory should produce five Professors of Engineering, in the early days of engineering as an academic discipline, must surely be beyond coincidence.

Specific influence can be identified in the cases of Cotterill and Pigot. In Cotterill's case there is a direct link from Fairbairns to an academic career. Whilst Cotterill was at Ancoats, William Hopkins, 'senior wrangler maker', made regular visits from Cambridge in connection with his experiments.[130] Hopkins' influence led to Cotterill reading mathematics at Cambridge.[131] His area of research was the use of energy functions in the theory of structures and in 1873 he was appointed Professor of Applied Mathematics at the Royal Naval College, Greenwich.[132] In Pigot's case he sought and took advice from Fairbairn which influenced his subsequent career. Educated to degree level in France, on Fairbairn's advice he spent three years at the Ecole des Ponts et Chaussèes in Paris, before coming to Ancoats for a year. After that he was involved in the construction of railways and in 1867 was appointed Professor at the Royal College for Science in Ireland, where he remained for twenty-three years.[133]

Thomson and Jenkin were outstanding academics. Thomson – the younger brother of William Thomson, later Lord Kelvin – was initially at the Millwall shipyard, directly involved in marine steam-engines, providing the setting for a significant correspondence with his brother. The letters show James's concern to improve marine engine efficiency, and demonstrate his grasp of the ideas of Clapeyron and Carnot on the motive powers of heat. He introduced the notion of the sea as a sink of lost work, both for water running into it in rivers and for heat diffusing itself to the temperature of the sea, a notion crucial to William's formulation of the second law of thermodynamics.[134] Intensely conscious of the need for accurate measurement of the mechanical effect of heat he prepared drawings at Fairbairn's in May 1844 for a modified Morin dynamometer, suggesting to Thomas Fairbairn that the instrument might be used for measuring the exact quantity of mechanical effort used by anyone who took power from the main engine at the shipyard. The dynamometer would thus serve not only for testing purposes, but to meter costs.[135] After leaving Fairbairns James developed the vortex turbine with Williamsons of Kendal and became Professor of Engineering at Belfast, with a particular interest in centrifugal pumps. He went on to succeed Rankine at Glasgow,[136] producing many academic papers.[137]

Jenkin, after leaving Fairbairns, worked on submarine cables, became Professor of Engineering at University College London in 1866 and moved to Edinburgh two years later. He filed thirty-five British patents and published some forty papers, many on telegraphic matters.[138] A polymath, he also produced papers on political economy such that Schumpeter described him as an economist of major importance;[139] and a critique of Darwin's *Origin of Species* using thermodynamics to counter Darwin's acceptance of 'an indefinite if not infinite' age of the earth.[140]

Fig.6.4: William Anderson (1834–98)

As Fairbairn sought high standards in his engineering, he seems to have had the insight to identify and appoint young men of high ability as his pupils, as journeymen seeking experience, and as assistants. The evidence suggests that the ethos of Fairbairn's works, with its emphasis on high quality engineering, experimentation and education, consistently rubbed off onto the lives of those who were pupils, or who sought experience there – at least five became Fellows of the Royal Society.[141]

It is much more difficult to trace any consistent influence in the approach to works organisation and industrial relations, because so little is actually known about the detailed organisation of the Fairbairn works, or about the subsequent organisations of those who trained or gained experience. The Committee for the Revision of Dockyards was full of praise for Andrew Murray in 1848 in a statement which on the face of it is unsupportive of the mismanagement case made above for the demise of the Millwall shipyard, but may indicate an emphasis on micromanagement at the expense of the overall financial control:

> Mr Murray ... having had the benefit of long experience in Private Trade, as the working Partner in the House of Messrs Fairbairn of Manchester, had assimilated all his own Regulations to those in use in that Establishment. In justice to this gentleman, we must express our satisfaction at the method, and ability, displayed in these arrangements. The labour account appeared to us to be perfect. There was not a man, out of 553 men employed, whose work upon any particular day, or half day, could not be at once ascertained and accounted for.[142]

Fairbairn's experiences of his pupils were good and he was keen to have them – differing from Nasmyth who had no enthusiasm for pupils. In this Nasmyth followed his mentor, Maudslay, who had ceased to take pupils because they caused 'much annoyance and irritation'.[143] Most of the famous graduates from the Maudslay 'nursery' went there as journeymen – Roberts, Napier, Clement, Whitworth – and Nasmyth was Maudslay's 'assistant'.[144] I.K. Brunel was also unenthusiastic about premium apprentices: 'I take pupils but rather as the exception than the rule: that is, I do not seek but rather raise difficulties than encouragement, and amongst other difficulties which I create is the premium I charge'.[145] Fairbairn was intensely proud of the successes of his pupils and ever ready to help them. Albert Escher, he wrote, 'having received mechanical instruction at the works in Canal Street, immediately on his return to Zurich commenced a large machine establishment, which ... stood high in the estimation of the Swiss, Italian and German public'.[146] He wrote to Robert Stephenson, 'Permit me to introduce to you ... Mr Rulph of Halifax, a highly respectable young man and a pupil of mine'. Rulph had been with Fairbairn for five years and the letter of introduction was in glowing terms.[147] Fairbairn put forward his pupil, Molesworth, for a job at Woolwich Arsenal.[148] When Jenkin moved to Penn's drawing office, it was with a letter of introduction from Fairbairn, who later proposed him for membership of the ICE and supported his election to the Royal Society.[149] In his *Treatise on Mills and Millwork* Fairbairn devotes nine pages to Thomson's vortex turbine, proudly referring to the inventor as 'one of my own pupils'.[150] A letter to Unwin sums it up. 'It is a high gratification to me to witness my young friends and pupils rising to distinction in their scientific and professional pursuits.'[151]

Conclusion

IN his commitment to education, investigation and the spread of useful knowledge Fairbairn personified Mokyr's 'Industrial Enlightenment'. That educational institutions, in whose foundations he played a significant role, continue today is testimony to his and his colleagues' foresight. Engineering machinery and structures could not have been exported without the accompanying knowledge of how to erect, operate and maintain them. This knowledge was provided by way of fitters and erectors through whom technology was diffused, often in remote places, by those who were the human face of Fairbairns. The achievements of Fairbairn's pupils were outstanding. These engineers have never previously been collectively identified and their identification provides evidence pointing to Fairbairn's Ancoats works as the successor to Maudslay's 'nursery' in providing training and experience for some of the ablest engineers and teachers of engineering of the next generation.

Notes for Chapter 6

1. Unpublished study by the author.
2. *Life*, p.157; Fairbairn, *UIfE3*, p.68.
3. R A Buchanan, *The Engineers: A History of the Engineering Profession in Britain 1750-1914*, (1989), p.163.
4. W Fairbairn, 'On Labour - Its Influence and Achievements', in Fairbairn, *UIfE3*, pp.51-2. He was speaking in 1865.
5. *The Practical Mechanic and Engineer's Magazine*, 1, 1842, 121.
6. Fairbairn, *UIfE*, pp.110-1.
7. Fairbairn, *UIfE*, pp.102-3.
8. W Fairbairn, 'Presidential Address', *Transactions of the Scientific and Mechanical Society, Manchester*, (Opening Meeting 1873); *Life*, p.420.
9. Fairbairn, *UIfE2*, p.160 (Fairbairn's italics).
10. A Prentice, *Historical Sketches and Personal Recollections of Manchester*, (1851), p.248. Roberts 'has been called the most versatile mechanic of the Industrial Revolution' (J Mokyr, *The Gifts of Athena*, (2002), p.65n55).
11. M Tylecote, 'The Manchester Mechanics' Institution' in D S L Cardwell (ed.), *Artisan to graduate. Essays to commemorate the foundation in 1824 of the Manchester Mechanics' Institution, now in 1974 the University of Manchester Institute of Science and Technology*, (1974), p.57.
12. *MG*, 10 April 1824; R H Kargon, *Science in Victorian Manchester: Enterprise and Expertise*, (1977), p.20; See Chapter 2.
13. M Tylecote, *The Mechanics' Institutes of Lancashire and Yorkshire before 1851* (1957), p.129.
14. Tylecote, 'Manchester Mechanics' Institution', p.58.
15. M Jacob, *The First Knowledge Economy: Human Capital and the European Economy, 1750-1850*, (2014), p.109 : 'the Manchester [Mechanics' Institute] proved to be a resounding success for several decades'. See also M A Walker, '"A Solid and Practical Education within reach of the Humblest Means": the Growth and Development of the Yorkshire Union of Mechanics' Institutes 1838-1891', (University of Huddersfield PhD Thesis, 2010) – 'questions the generally accepted view that mechanics' institutes made little contribution to adult working-class education'.
16. R G Kirby, 'An early experiment in workers' self-education: the Manchester New Mechanics' Institution, 1829-35' in Cardwell, *Artisan to Graduate*, pp.87-98.
17. Kargon, *Science in Victorian Manchester*, p.24.
18. Tylecote, 'Manchester Mechanics' Institution', p.62.
19. 'Report from the Select Committee on Education in England and Wales', (1835), (465), VII, appendix 2, quoted in Tylecote, 'Manchester Mechanics' Institution', p.62.
20. J Wheeler, *Manchester: its political, social and commercial history, ancient and modern*, (1836), p.414.
21. Tylecote, 'Manchester Mechanics' Institution', pp.75-6.
22. Fairbairn, *UIfE2*, pp.147-9.
23. *Manchester Examiner and Times*, 26 June 1852.
24. C Hartwell, *Manchester*, (2001), pp.196-7.
25. Manchester City Council, 'Whitworth Street Conservation Area', at http://www.manchester.gov.uk/info/511/conservation_ areas/970/whitworth_street_co (accessed 21 March 2016).
26. Fairbairn, *UIfE*, p.98; W Fairbairn, The Rise and Progress of Manufactures and Commerce and of Civil and Mechanical Engineering in Lancashire and Cheshire', in T Baines, *Lancashire and Cheshire, Past and Present*, (1869), p.xv.
27. G W Wood to H Brougham, 22 August 1826, quoted in Tylecote, *Mechanics' Institutes of Lancashire and Yorkshire*, p.139; W. Fairbairn, *Address of William Fairbairn, Esq. President, to the Members of the Lyceum, Great Ancoats Street; with the Report of the Provisional Directors, and Proceedings at the General Meeting, held on the 24th of January, 1839*, (1839), p.10, quoting Manchester Statistical Society, *On the State of Education in Manchester in 1834*, (1835); *MG*, 18 December 1839; S Nicholls, *In Memoriam. A Selection from the Letters of the Late John Ashton Nicholls, FRAS, &c.*, (1862), p.4; M Hewitt, *The Emergence of Stability in the Industrial City: Manchester, 1832-67*, (1996), p.135.
28. B Heywood and W Fairbairn, *Lyceum, Great Ancoats Street: Proposals for the Erection of a Building for the Purposes of the Institution.* (1838); A Kidd, *Manchester* (1993), p.58. Ancoats was soon followed by Chorlton-on-Medlock and Salford. (W E A Axon, *The Annals of Manchester: A Chronological Record from the earliest times to the end of 1885*, (1886), pp.204, 206).
29. [B Love], *Manchester As It Is*, (1839), p.108; Alan Godfrey Maps *Manchester (Piccadilly) 1849*, (2009 reprint - Manchester Sheet 29).
30. Fairbairn, *Address*, p.21;
31. J Morley, *The Life of Richard Cobden*, (One-volume Edition 1903), pp.175-6.
32. http://www.spartacus.schoolnet.co.uk/IRheywood.htm (accessed 2 June 2011).
33. http://webcache.googleusercontent.com/search?q=cache:RNegyIP2m3wJ:www.manst (accessed 02.06.11).
34. Fairbairn, *Address*, p.21; M Beetham, 'Heywood, Abel (1810-1893)', *Oxford DNB on-line*, (2004); J Johnson, *Clever Boys of our time who became Famous Men*, (nd, c1863), pp.180-97.
35. The occupations of the 715 initial members were: Principals, engaged as merchants, manufacturers and machinists 10; Professional men 4; Schoolmasters 6; Shopkeepers, Master Tradesmen, and their Assistants 87; Warehousemen and Book-keepers 132; Mechanics, Millwrights, Engineers, Moulders & Smiths 137; Engravers and Pattern Designers 7; Spinners, Weavers, and other Mill Hands 102; Other trades connected with the manufactures of the town, as Dyers, Calico Printers, Fustian Cutters, &c. 22; Building trades 37; Sundry Handicraft trades 85; No profession 7; School Boys 22; Females 57. (Fairbairn, *Address*, p.18).
36. A Gordon, rev. A C Howe, 'Nicholls, John Ashton (1823-1859), *Oxford DNB on-line*, (2004); *MG*, 20 January 1847; A B Reach (ed. by C. Aspin), *Manchester and the Textile Districts in 1849*, (1972), pp.35-6.
37. *MG*, 31 January 1849 and 6 February 1850; *Manchester Times*, 13 March 1852.
38. *Manchester Times*, 13 March 1852.
39. Nicholls, *In Memoriam*, p.15.
40. Kidd, *Manchester*, p.58.
41. D Cardwell, *James Joule: A Biography*, (1989), pp.27-8.
42. Kargon, *Science in Victorian Manchester*, pp.38-41.
43. S Macdonald, *The History and Philosophy of Art Education*, (2004), pp.84, 114.
44. W B Pope (ed.), *The Diary of Benjamin Robert Haydon*, (1963), p.454. Pope confuses James Heywood and Benjamin Heywood (p.454 n.1). Haydon says of Heywood that 'He was in the House at D'Israeli's maiden speech' (p.455), at which time James was a Member but Benjamin had ceased to be. Darcy follows Pope (C P Darcy, *The Encouragement of the Fine Arts in Lancashire 1760-1860*, (1976), p.113).

45. Pope, *Haydon*, p.457.
46. *MG*, 21 February 1838 and 24 February 1838.
47. *MG*, 3 October 1838; S Macdonald, 'The Royal Manchester Institution' in J H G Archer (ed.), *Art and Architecture in Victorian Manchester*, (1985), p.40.
48. J Thompson, *The Owens College: Its Foundation and Growth; and its connection with the Victoria University, Manchester*, (1886), pp.19-20; Kargon, *Science in Victorian Manchester*, p.154.
49. Thompson, *Owens College*, pp.25-7.
50. Kargon, *Science in Victorian Manchester*, pp.157-8.
51. *MG*, 26 October 1867.
52. Thompson, *Owens College*, pp.295-6; Kargon, *Science in Victorian Manchester*, p.184; H Lamb, rev. R H Kargon, 'Reynolds, Osborne (1842-1912), *Oxford DNB on-line*, (2004).
53. Walker, *Unwin*, pp.54-5.
54. Thompson, *Owens College*, pp.315-6
55. Thompson, *Owens College*, pp.391-2; *Life*, p.428.
56. [Love], *Manchester As It Is*, p.213 [Love's italics].
57. J Tann, 'Marketing Methods in the International Steam Engine Market: The Case of Boulton and Watt', *Journal of Economic History*, 38.2, 1978, 374-6.
58. [Love], *Manchester As It Is*, pp.212-3.
59. E J Molesworth (ed.), *Life of Sir Guilford L Molesworth, KCIE, 'The Nestor of the Engineering Profession'*, (1922), p.22; *MPICE*, 120, 1895, 352.
60. [D Scott], *The Engineer and Machinist's Assistant*, (1850), p.244.
61. www.biographiea.ac.at/oebl_2/206.pdf (accessed 6 October 2010). Haut says Haswell's position with Fairbairns was in 'the ship-building yard'. As the Millwall yard started up in 1835 it is possible he went there, but given his subsequent work this seems unlikely. *L'Hirondelle*, a packet boat for the Humber, was still being built at Ancoats in 1835, and also engines for some of the steamers for the European lakes, so this may have led to Haut's statement. (F J G Haut, 'The Centenary of the Semmering Railway and its Locomotives', *TNS*, 27, 1949-50, 19-29; *CE&AJ*, 1841, 147).
62. Austrian Imperial Railways – Exhibition Catalogue, (2009), pp.4-5.
63. Including the *Duplex* and the *Steierdorf* at London in 1862 (*ILN*, 1 Nov 1862, 479-80; *The Engineer*, 13, Jan.-June 1862, 265; *MPIME*, 1863, 97-9 and Plates 27-31).
64. *The Sydney Herald*, 14 August 1840.
65. *CE&AJ*, 1841, 148.
66. *Minutes of the Engineering Association of New South Wales*, 1889, 213-6. This refers to his having early married a Rutherglen girl, and they were together for nearly seventy years.
67. *MM*, 38, 1840, 400; D Gregory, *Australian Steamships Past and Present*, (1928), pp.35-7.
68. *The Australian*, 27 May 1841; *The Maitland & Hunter River General Advertiser*, 22 June 1844.
69. *Minutes of the Engineering Association of New South Wales*, 1889, 214-6.
70. *MG*, 31 July 1850; Carlberg, 'Personal Contacts', 57-8. The wheel was 18ft wide. Carlberg gives its cost as £3,132, whereas Söderberg gives it as £3,422 (T Nilson, *"The Invisible Hand of Knowledge". Market Forces and the Diffusion of Technology and Knowledge from Britain to Sweden during the Eighteen-fifties: The Case of the Swedish Textile Industry*, EBHA Conference Paper 2001: Business and Knowledge, A4: Industrial Diffusion, p.17, reprinted in *A Case of Identities. Festschrift to Martin Peterson*, (Göteborg University 2006)); Fairbairn, *UIfE2*, pp.229-31. Gefle is now named Gävle. See Chapter 4.
71. *Life*, p.365; *MG*, 31 July 1850; S Söderberg, 'Förstaden vid Testaboån', *Gefle Dagblads Hembygdsbibliotek*, XI, 1966, 73.
72. Nilson, *"Invisible Hand"*, p.19; Carlberg, 'Personal Contacts', pp.57-60. Claughton appears to have still been there in the late 1860s, with his family, possibly as works engineer. (J Haslingden, *The History of Hannah Haslingden and her family*, (nd but c.1900) p.3). It is unclear if he brought his family from England, or married a local girl.
73. Carlberg, 'Personal Contacts', p.58.
74. K Bruland, *British Technology & European Industrialization: The Norwegian textile industry in the mid nineteenth century*, (1989), pp.114, 156, and Appendix D. Bruland also refers to William Oddy (p.87) but this may be an error. Brenneriveien was a pilot plant for the much larger Hjula Weavery which followed and for which around thirty-six workers and managers were recruited from England - at least five of them by Fairbairn (Bruland, *British Technology*, pp.42, 87, 115).
75. Bruland, *British Technology*, p.114, Appendix D.
76. *Bombay Telegraph and Courier*, quoted in S D Mehta, *The Cotton Mills of India 1854-1954*, (1954), pp.13-17.
77. *Hamilton Spectator* 22 March 1860; *The Engineer*, 9, Jan-June 1860, 255.
78. P Alsop, 'Bridging the Rivers, Part 5', *The Investigator*, 4.2, 1969. 84-101; G Vines, *Metal Road Bridges in Victoria, Part 1 – History of Metal Road Bridges in Victoria*, (rev. ed. 2010), pp.19-20.
79. W Humber, *A Complete Treatise on Cast and Wrought Iron Bridge Construction including Iron Foundations*, (3rd ed. 1870), pp.197-9 and Plate 35; http://209.85.229.132/search?q=cache:L7izsE7qFJOJ:vhd.heritage.vic.gov.au/places/h (accessed 11 December 2009); B Harper, 'The true history of the design of the Melbourne, Mount Alexander and Murray River Railway', *Australian Journal of Multi-disciplinary Engineering*, 3.1, 2004, 88-9. The Mariburnong River was previously known as the Salt Water River. The bridge was not designed by Brunel as suggested by A Buchanan, *Brunel: The life and Times of Isambard Kingdom Brunel*, 2002, p.100.
80. H M Burt, *Burt's Illustrated Guide of the Connecticut Valley*, (1867), p.60.
81. *South Australian Register*, 18 January 1856. John Clarke was the younger brother of Richard Clarke, manager of Fairbairns' locomotive department.
82. R Stewein with M Thomson, *A History of the South Australian Railways. Volume 1: The Early Years*, (2007), pp.150, 220-1; *The South Australian Advertiser*, Adelaide 14 April 1860.
83. Australian Agricultural Company, 'Minutes of Court of Directors', June 1855; Australian Agricultural Company, 'Despatch to Directors', 11 May 1857. I am indebted to John Broadley for these references.
84. Strewein with Thomson, *South Australian Railways*, pp.211-2.
85. A Guagnini, 'Worlds apart: academic instruction and professional qualifications in the training of mechanical engineers in England, 1850-1914' in R Fox and A Guagnini (eds.), *Education, technology and industrial performance in Europe, 1850-1939*, (1993), p.25; Fairbairn, *M&MWI*, p.vi.
86. Fairbairn, *M&MWI*, p.vi.

87. D A Collier, *A Comparative History of the Development of the Leading Stationary Steam Engine Manufacturers of Lancashire c.1830-1939,* (PhD Thesis, Manchester University,1985),Vol.1,pp.221-4; D A Collier, 'The impact of other engine-makers on Oldham and district', in D Gurr and J Hunt, *The Cotton Mills of Oldham,* (3rd ed. 1998), pp.40-1; A S Roberts, 'Arthur Roberts' Engine List' (1921), at http://oneguyfrombarlick.co.uk/forum_topic.asp (accessed 2 October 2010) – gives details of several Saxon engines.
88. 'Select Committee on Scientific Instruction' [Samuelson Committee], *Parliamentary Papers,* 1867-8, [c.432] XV.I, p.123.
89. G Cookson, 'Early Textile Engineers in Leeds 1780-1850', *Thoresby Society,* 2nd Series 4, 1994, 50.
90. G Cookson, 'Fairbairn, Sir Peter (1799-1861), *Oxford DNB on-line,* (2004) (accessed 9 January 2014); *Fortunes Made in Business,* pp.252-6.
91. E Gauldie, (ed.), *The Dundee Textile Industry 1790-1885, from the papers of Peter Carmichael of Arthurstone,* (1969), pp.xiii, 48.
92. *MPICE,* 86, 1886, 346-53; G Roberts, 'Sir John Anderson, 1814-86: The Unknown Engineer who made the British Empire possible', *TNS,* 78, 2008, 261-91; O F G Hogg, *The Royal Arsenal: Its Background, Origin and Subsequent History,* (1993),Vol.1, pp.720-3..
93. *Life,* p.327; see Chapter 4.
94. Fairbairn, *M&MW2,* pp.255-9; see Chapter 4.
95. Fairbairn, *M&MW2,* pp.132-8; see Chapter 4.
96. T Vernon, *Archibald Sturrock: Pioneer Locomotive Engineer,* (2007).
97. *MPICE,* 31, 1871, 220-2.
98. R E Green, 'Craven Brothers (Manchester), Ltd. Vauxhall Works, Reddish, Stockport, *Machinery,* 3 February 1965, 237-48; S B Saul, 'The Market and the Development of the Mechanical Engineering Industries in Britain, 1860-1914', *Economic History Review,* 2nd Series 20, 1967.
99. W T Pike (ed.), *Contemporary Biographies,* (1899), p.187.
100. *MPICE,* 1907, 383-4; *MG,* 1 March 1907.
101. http://www.heraldscotland.com/sport/spl/aberdeen/jacobs-seals-babtie-deal-engineeri... (accessed 29 May 2012).
102. There is little secondary literature about premium apprentices despite their importance in management training. C More, *Skill and the English Working Class, 1870-1914,* (1980), pp.45, 104-6, and Guagnini, 'Worlds apart', both have useful, but short, descriptions; and there is a paper specific to the subject, but not related to engineering - Jonathan Reinarz, 'Fit for Management: Apprenticeship and the English Brewing Industry, 1870-1914', *Business History,* 43.3, 2001, 33-53.
103. Guagnini, 'Worlds apart', p.24.
104. S Mosley, *The Chimney of the World: A History of Smoke Pollution in Victorian and Edwardian Manchester,* (2001), pp.119-20; Kargon, *Science in Victorian Manchester,* pp.111-2.
105. Fairbairn, *UIfE2,* p.152.
106. Cambridge University Library, Stokes Papers, Add 7342, TR24 (Fairbairn's emphases).
107. Guagnini, 'Worlds apart', p.24. Sums paid by only two of Fairbairn's pupils have been found - £100 paid by Thomson's father (C Smith, 'Thomson, James (1822-1892)' *Oxford DNB on-line,* (2004), (accessed 9 April 2012)); and £300 paid by Jenkin's father ('Select Committee on Scientific Instruction', p.138). Reasons for the difference are not known, nor is it clear whether these were annual or one-off sums.
108. 'Select Committee on Scientific Instruction', p.123.
109. G Cookson and C A Hempstead, *A Victorian Scientist and Engineer: Fleeming Jenkin and the Birth of Electrical Engineering,* (2000), pp.21-2, 112.
110. *MPICE,* 36, 1872-3, 272; A Murray, *The Theory and Practice of Ship-building,* (1861), pp.v-vi; *MM,* 29, 1838, 348.
111. *Journal of the Franklin Institute,* 28, 1839, 239.
112. *MPICE,* 120, 1895, 352-3.
113. Bexley Council Local Studies Note No. 66, *Sir William Anderson 1834-1898,* (n.d. but c.2000); C Smith, 'Thomson, James (1822-1892)', *Oxford DNB on-line* (accessed 9 April 2012); J Larmor and J Thomson (eds), *Collected Papers in Physics and Engineering by James Thomson…,* (1912), pp.XXIV-XXV; Molesworth, *Molesworth,* pp.19-22.
114. S Colvin and J A Ewing, (eds.), *Papers Literary, Scientific, &c. by the Late Fleeming Jenkin, FRS, LLD,* (1887), Vol. 2, pp.187-8.
115. Colvin and Ewing, *Papers,* Vol.2, p.175.
116. W Anderson, 'The Interdependence of Abstract Science and Engineering', *MPICE,* 114, 1893, 256.
117. Molesworth, *Molesworth,* pp.159-60.
118. Cookson and Hempstead, *Fleeming Jenkin,* pp.21-9.
119. Walker, *Unwin,* pp.59, 162.
120. *Life,* p.191.
121. Molesworth, *Molesworth,* pp.39-51.
122. Molesworth, *Molesworth,* pp.30, 40-1.
123. Molesworth, *Molesworth,* pp, 51-3, 58-97, 99-100; J Rainbow, 'Guilford Molesworth in India', *The British Overseas Railways Journal,* 38, 2013/4, 25-8.
124. Bexley Council, *Sir William Anderson*; *MPICE,* 135, 1899, 321-3; *MPICE,* 72, 1883, 24.
125. *MPICE,* 36, 1872-3, 271-3.
126. A Murray, 'On the Construction and Proper Proportions of Boilers for the Generation of Steam', *MPICE,* 3, 1844, 331042; *MPICE,* 36, 1872-3, 272; Fairbairn, *UIfE,* p.90.
127. D Evans, *Building the Steam Navy: Dockyards, Technology and the Creation of the Victorian Battle Fleet 1830-1906,* (2004), p.89.
128. Fairbairn, *UIfE2,* p.296.
129. A Murray, *The Theory and Practice of Shipbuilding,* (1861).
130. *Life,* pp.287-308. See Chapter 5.
131. A D D Craik, *Mr Hopkins' Men: Cambridge Reform and British Mathematics in the 19th Century,* (2007), p.215.
132. T M Charlton, H*istory of theory of structures in the nineteenth century,* (1982), pp.118-22, 177; *The Times,* 18 January 1922.
133. *MPIME,* May 1910, 782-3; *Transactions of the Institution of Civil Engineers of Ireland,* 37, 1910-11, 191-3.
134. C Smith and M N Wise, *Energy and Empire: A biographical study of Lord Kelvin,* (1989), pp.283, 289-90.
135. Smith and Wise, *Energy and Empire,* p.292.
136. Smith, 'Thomson, James (1822-1892)'.
137. Larmor and Thomson, *Collected Papers by James Thomson.*
138. Cookson and Hempstead, *Fleeming Jenkin*; Colvin and Ewing, *Papers Literary, Scientific, &c.*
139. [H C] F Jenkin, *The Graphic Representation of the Laws of Supply and Demand, and other Essays on Political Economy,* (1931 ed.); J A Schumpeter, *History of Economic Analysis,* (1954), p.837.
140. S W Morris, 'Fleeming Jenkin and *The Origin of Species*: a reassessment', *The British Journal for the History of Science,* 27.3, 1994, 313-43; D L Hull, *Darwin and his*

Critics: the reception of Darwin's theory of evolution by the scientific community, (1973), p.302.
141. Anderson, Unwin. Thomson, Jenkin, Cotterill.
142. Evans, *Building the Steam Navy,* p.83.
143. S Smiles, (ed.), *James Nasmyth, Engineer, an Autobiography,* (1895 ed.), pp.218, 121.
144. J Cantrell, and G Cookson (eds.), *Henry Maudslay and the Pioneers of the Machine Age,* (2002).
145. R A Buchanan, *Brunel: The Life and Times of Isambard Kingdom Brunel,* (2002), pp.164-5. By 1846 Brunel was charging £1,000 for a four-year apprenticeship.
146. *Life,* p.129.
147. W Fairbairn to R Stephenson, 29 September 1846, No.110, Stephenson Papers, Institution of Civil Engineers.
148. Molesworth, *Molesworth,* pp.24-8.
149. Cookson and Hempstead, *Fleeming Jenkin,* pp.25-8, 99, 127.
150. Fairbairn, *M&MWI,* pp.166-174.
151. Walker, *Unwin,* pp.56-7.

Sources, Credits and Acknowledgements for Illustrations in Chapter 6.

Fig.6.1 J. Larmor and J. Thomson (eds), *Collected Papers in Physics and Engineering by James Thomson* (1912), Frontispiece.

Fig.6.2 S. Colvin and J.A. Ewing, (eds), *Papers Literary, Scientific, &c. by the Late Fleeming Jenkin, FRS, LLD* (1887), Vol.I, facing page lxviii.

Fig.6.3 E.J. Molesworth (ed.), *Life of Sir Guilford L Molesworth, KCIE, 'The Nestor of the Engineering Profession'* (1922), p.23.

Fig.6.4 http://www.gracesguide.co.uk/File:Im1898En V86-p587.jpg (accessed 1 June 2016).

Chapter 7

'Retirement'

1854–1874

Fig 7.1:
C.A. Du Val, *William Fairbairn* (mid 1860s – detail).

Introduction

WILLIAM Fairbairn never 'retired' in the conventional use of the term. Having ceased to be active in the day-to-day management of the company from 1854 he remained at the height of his profession for the next two decades, which he devoted to experimental work, engineering challenges, participation in learned societies, boiler safety, patent law, and writing books and papers. The breadth and disparate nature of these many activities, during the two decades in which Fairbairn was no longer running the firm he founded, raises difficulties in identifying common themes beyond incessant engineering-centred activity. His standing was such that he was called on from all sides for advice and reports. Some of his reports followed boiler explosions – experiences which led to his founding the Manchester Steam Users' Association for the Prevention of Steam Boiler Explosions, a model for similar associations. His stamp is to be found on boiler and patent legislation enacted some years after his death. Engineering experiments continued apace providing useful and reliable knowledge. In these and his books from 1855 to 1862 he was assisted by William Unwin. Fairbairn has been seen as a transitional figure from cast iron to wrought iron but this is a limited view, in that wrought iron was itself transitional, forming the bridge between cast iron and steel.

These were fulfilling years for Fairbairn. Mowery and Rosenberg maintain that 'if one had to choose any fifteen-year period on the basis of the density of scientific breakthroughs, it would be hard to beat the years 1859–1874'[1] – the last fifteen years of Fairbairn's retirement. But scientists and engineers are no different from others, whatever age they live in, and these years were also tinged with sadness for Fairbairn, with the deaths of daughter-in-law Marianne and son Peter in 1859, brother Sir Peter in 1861, and son John in 1867.[2]

Challenge and Innovation: Bridges and Buildings

THROUGHOUT his career Fairbairn had built mills and bridges, and he continued to do so in 'retirement'. He relished engineering challenges as illustrated by the Dinting and Mottram Viaducts, and by alterations to McConnel and Kennedy's Sedgwick Mill. They reveal a lot about his character and reputation. First they show the respect in which his engineering skills were held – it was to him that novel problems were referred, often by peer engineers. Secondly, they demonstrate his ingenuity and boldness, providing solutions that aroused the interest and admiration of other engineers. Thirdly, his sheer fascination with engineering is demonstrated by these challenges, undertaken well after normal retirement age when he already had ample money to live very comfortably.

The challenge accepted by Fairbairn at the Dinting and Mottram Viaducts was to replace unsafe timber arches, dating from the early 1840s, with wrought-iron beams. C.B. Vignoles and Joseph Locke were the original engineers for the timber arches,[3] which were repaired by A.S. Jee in 1855.[4] The spans were up to 150ft, with the railway up to 120ft above the ground. Fairbairn was given the stipulation that the seventy trains a day should not be interrupted. He achieved this with typical simplicity of method. Iron tubular girders were constructed on the existing bridges and lowered between the timber ribs. Fairbairn personally supervised the work, which was the subject of a paper to the Institution of Civil Engineers.[5]

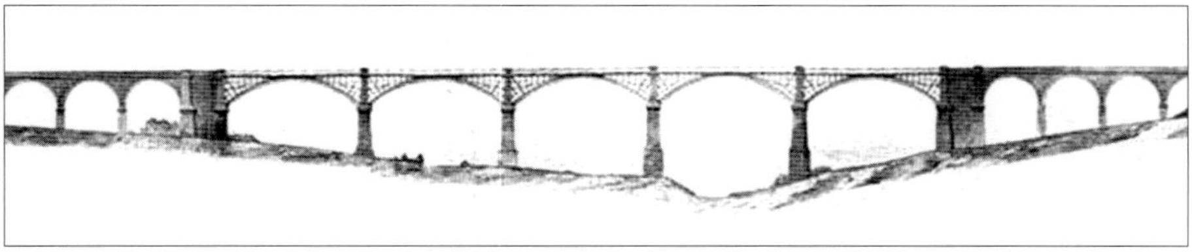

Fig.7.2: Dinting Viaduct: *(top)* as originally built 1843–4; *(above)* with the timber arches replaced with tubular wrought-iron girders in 1859 by Fairbairn (photographed in 1916).

Vignoles visiting the work with Fairbairn, recorded his interest 'in visiting scenes of former labours' and noted that 'a capital job' was being done.[6] Some appreciation of Fairbairn's achievement can be gained by a comparison with the reconstruction of his 1849 tubular-girder bridge at Gainsborough with steel plate girders in 1992, when the method adopted was similar to Fairbairn's at Dinting and Mottram, but caused track closure for three months.[7] At some point Fairbairn's wrought-iron girders have been replaced with riveted steel and additional brick supports introduced.[8]

Another challenge accepted by Fairbairn was to move columns, each supporting around 90 tons, in an eight-storey mill with brick-arch floors, whilst the mill remained in production with 300 people continuing to work on the upper floors. This was at McConnel & Kennedy's Sedgwick Mill which half-a-century earlier had been Fairbairn & Lillie's second major commission. McConnel & Kennedy were fearful of what is known today as progressive collapse, bringing the whole mill down. Fairbairn's method was ingenious. A new column with a bracket was inserted; the floor beneath the old column was propped; the old column was cut out; the bracket on the new column was extended to pick up the loads previously carried by the old column; and the props were removed. At second floor level the six stories of new columns were carried on new tubular girders. Fairbairn was around 75-years-old when he undertook this work, which was the subject of papers to the Institution of Mechanical Engineers and the British Association, indicative of the ingenuity of the novel solution, as of the confidence of this elderly engineer.[9]

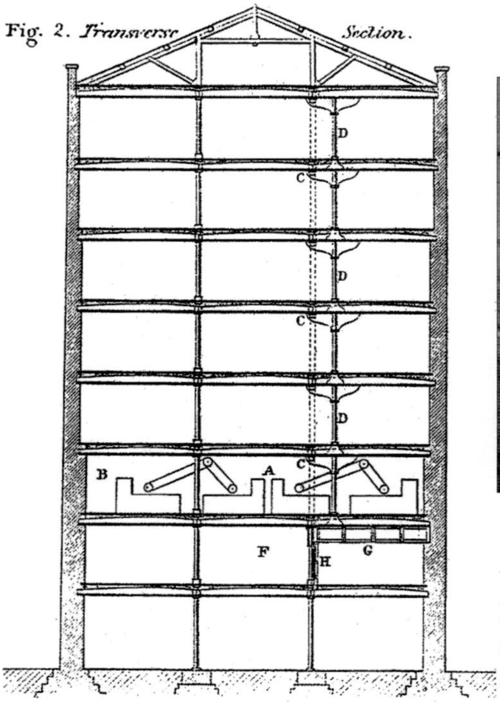

Fig.7.3: Sedgwick Mill – Moving of Columns.

(above right) Fig.7.4 : Sedgwick Mill – Altered Columns.

Others turned to Fairbairn with less hair-raising briefs, such as the young Manchester architect Alfred Darbyshire, commissioned to design an eight-storey sugar-refinery warehouse in Dublin. His clients required an experienced engineer to be involved, and in August 1862 Darbyshire 'entered the sanctum of the great man with diffidence and anxiety'. Examining the plans Fairbairn intimated that this was a building in which to introduce wrought-iron beams throughout, in lieu of the usual cast-iron.[10] They were built up from angles riveted to plates, as illustrated below. Innovation had not ceased with Fairbairn's 'retirement', and this was probably the first major building in Britain with wrought-iron beams throughout – one of the transitional steps between cast iron and the modern steel frame.[11]

Giedion drew attention to this building in the 1940s but wrongly dated it as 1845, incorrectly believing it to have concrete floors with permanent wrought-iron shuttering, of which he reproduced Fairbairn's illustration of 1857 – see *fig. 7.7.*[12] In fact the illustration related to 'a cotton or flax mill',[13] and was a sophisticated development of a similar floor using corrugated iron shuttering, patented by George Nasmyth in 1848.[14] Here was Fairbairn once again 'taking a hint', developing it and promoting it. Unlike his built-up beams at Dublin, Fairbairn appears to have envisaged rolled beams for his permanently-shuttered floor. At 16in. deep these joists would have been around twice the depth of any known rolled joist at that date. Following the collapse of Radcliffe's Mill in Oldham in 1844, on which Fairbairn and David Bellhouse reported, Thomas Cubitt called for the introduction of rolled-iron beams to replace cast-iron.[15] These were used in 'filler-joist' floors in Lancashire mills from the 1850s.[16] In America purpose-rolled wrought-iron floor joists were used in the mid 1850s.[17] There is no known example, surviving or demolished, of Fairbairn's permanently-shuttered floor with rolled-iron beams. If one was to be found, with 16in. rolled beams it would be the most advanced structural floor of its day.

Fig.7.5:
The Bewley Moss Sugar Refinery Warehouse, Dublin, in 1984.

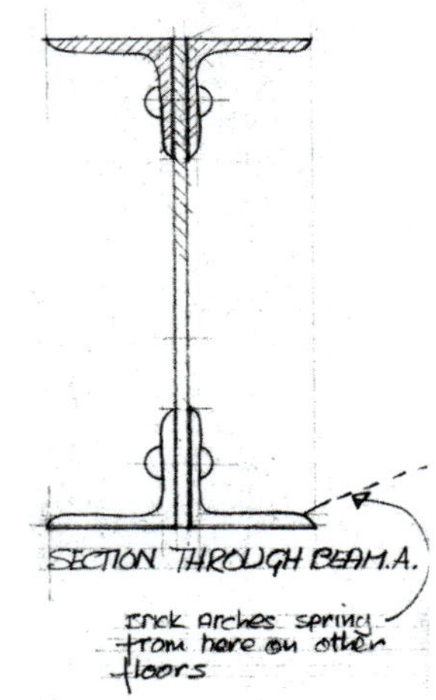

Fig. 7.6:
Wrought-iron beams, Dublin.

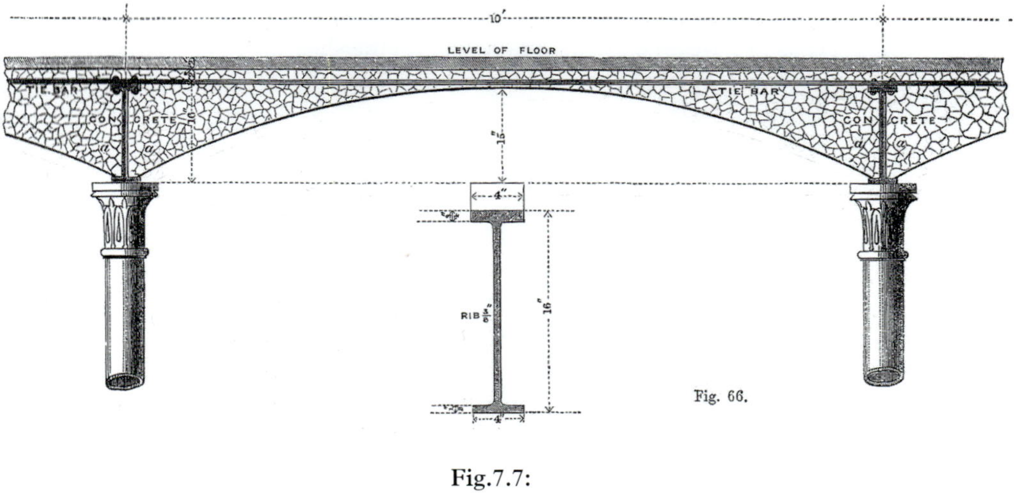

Fig.7.7:
Fairbairn floor comprising ¼"-thick wrought-iron permanent shuttering, or formwork, with concrete above, to receive floor finish.

William Unwin

WILLIAM Fairbairn was much assisted during seven years of his retirement, 1856–63, by William Cawthorne Unwin (1838–1933), on whom his mantle fell. After 'retirement' Fairbairn needed someone to provide assistance. On advice from his friend Thomas Tate, Fairbairn appointed the 18-year-old Unwin. Educated at City of London School, and in science at New College, St John's Wood, Unwin provided valuable assistance to Fairbairn with his research, including that on the strength of boiler flues, the properties of steam, the trials of continuous mechanical railway brakes, fatigue testing of wrought-iron girders, and armour-plating of warships. He assisted with Fairbairn's various books, at the same time achieving an external London University BSc.[18] Unwin left Ancoats to be works manager for Williamsons in Kendal, returning to Fairbairns in 1866 in a managerial role. It did not work out and he moved to London,[19] where Fairbairn was responsible for Unwin's first lecturing engagement, and wrote, 'do not hesitate to make use of my recommendation in any case where you think I can serve you'.[20] Unwin's inclination was towards teaching and he became continuously active in the advancement of engineering education and training. From 1872 Unwin was successively Professor at the Royal Indian Engineering College, the Central Institution of the City and Guilds, and London University. Like Fairbairn he became a Fellow of the Royal Society, President of the Institution of Mechanical Engineers and President of the Mechanical Science Section of the British Association. He was also President of the Institution of Civil Engineers. Accepting the Kelvin Medal of the British engineering institutions in 1921, he spoke of Fairbairn – 'his first chief'.[21] Fairbairn and Unwin had a profound respect for each other.[22] With nearly fifty years between their ages it resembled a father and son relationship, lasting eighteen years until Fairbairn's death. Fairbairn relied heavily on Unwin's assistance, writing of him in 1861, 'my assistant and secretary ... to whose assiduous attention and love of science I am greatly indebted'.[23] In 1872 he was writing to Unwin about experiments on fuel consumption of his five-tube boiler; and about rivets, 'I have an experimental paper on the shearing of rivets ... If I send it up to you could you find time to review and work out the law of shearing so as to render the experiments useful and satisfactory?'[24] One of his students recorded Unwin's lectures being enlivened by illustrations of his 'many experiences with Fairbairn'.[25]

Unwin's biographer refers to the breadth of Unwin's expertise:

> This is attributable directly to his early training and experience. Fairbairn's work ... covered the whole range of engineering ... bridges, mill framing, millwork, steam boilers and engines, water wheels and turbines, ships and masonry structures, all ... passed through his hands ... It was inevitable that the young man whose first job had been to make abstracts of scientific and technical data for Fairbairn should find his interests spread over an equally wide field.[26]

He might have specified just how extensive a field: machine design, hydraulics, bridges, structures, materials testing, hydro-electrics, engineering standards and the internal combustion engine. It was in Unwin's materials-testing that Fairbairn's influence was most evident. In 1876 he delivered a paper 'On the Resistance of

Boiler Flues to Collapse' in which he re-examined Fairbairn's results of 1857 – with which he had assisted – together with later work in Germany, which was based on Fairbairn's results.[27] Five years later he reported on riveted joints, referring back to Fairbairn's experiments on riveting as 'probably the first ever made'.[28] In 1887 he followed Fairbairn's papers on the strengths of cast iron and wrought iron at different temperatures with a paper on the 'Strength of Alloys at Different Temperatures'.[29] In 1903 he presented his paper, 'Tensile Tests of Mild Steel; and the Relation of Elongation to the Size of the Test-Bar'.[30] Unwin's standard work on *The Testing of Materials of Construction,* ran to three editions. In this work he was following his mentor as the country's leading authority on mechanical testing.[31] His *Elements of Machine Design,* running to many editions, and translated into French and German, placed on a scientific basis what was formerly derived from slowly accumulated experience embodied in empirical rules.[32] Writing only two decades after Fairbairn's *Mills and Millwork,* Unwin applied mathematics to the solution of engineering problems in ways that Fairbairn was not able to do. Unwin responded to a critic of his book, 'There may be practical engineers to whom the mathematical parts of the work will be difficult, but this is not a sufficient reason for omitting them'.[33] Theoretical engineering was advancing very fast.

Fig.7.8: Professor William Cawthorne Unwin, FRS, in 1911.

Research

FROM the early 1820s Fairbairn had been involved in testing, technical investigations and engineering experiments. Peer engineers also undertook experiments, but the extent and breadth of Fairbairn's work was exceptional. His work in this field may be differentiated into five groups, although inevitably there are overlaps and blurring of edges. First, throughout his career he undertook empirical testing of cast-iron beams for mills and bridges, to ensure their safety, providing an early example of systematic materials testing, now an important branch of engineering. Secondly, he experimented to solve specific problems where the requisite knowledge, theoretical or practical, did not exist, and it was a matter of proceeding by trial-and-error or parameter variation.[34] The experiments to increase the speed of canal boats were in this category,[35] as were those to meet the challenge of bridging the Menai Straits.[36] The former showed that fast canal travel was not feasible: the latter led to the Britannia Bridge. The investigations into smoke-prevention,[37] the practicality of working an incline by locomotive power,[38] and armour-plating experiments also fit here.[39] The third group of experiments are examples of early forensic engineering, comprising investigations and sometimes experiments to find the causes of collapses and explosions. For example, in connection with an explosion at LNWR's steam-shed at Longsight, Manchester, in 1853, in which five men died, Fairbairn and Richard Roberts gave evidence that the cause was the safety valve having been screwed down. The Government Inspector disagreed, blaming the state of the boiler stays. The jury accepted the inspector's evidence.[40] Fairbairn was not pleased and, typically, set about experiments on an engine of the same age and type as that which had exploded. His experiments vindicated his opinions and, again typically, he ensured that the results were widely known.[41] Forensic investigations led to experiments on boiler flues and the establishment of regular boiler inspections, reducing the number of explosions.[42] They provide a good illustration of the interaction of the sociological and the technological. Fourth, there were comparative product tests, notably tests on railway brakes,[43] and various submarine cable samples.[44] The fifth category of experiments, the largest, exemplified the British Association's understanding that science was based on hard-won experimental results, which provided the only secure basis for mathematical generalisations.[45] Fairbairn undertook many tests on the properties of various cast and wrought irons, alloys and steels. Insofar as these involved comparisons they overlap with the previous category. Systematic tests were also undertaken on the effects of prolonged loading, temperature, and re-meltings, on the strength of wrought-iron plates and riveted joints,[46] on wrought-iron beams and trellis girders, on metal fatigue, boiler construction and various properties of steam.[47] These provided reliable data for engineers. An unusual example from this group was Fairbairn's experiments on trussed cast-iron girders which ensured their swift demise.[48] Work in this fifth category exemplifies Mokyr's 'Industrial Enlightenment', providing useful and reliable knowledge by systematic measuring, recording and analysing of key data in the tradition of Newtonian science.[49] Much of it also falls within the description of 'Humboltian Science',[50] notably for example the measuring of temperatures at descending depths at Astley Pit.[51] Fairbairn's experiments show that research, experimentation and investigation do not follow any one pattern, but vary with the circumstances.

Many of his experiments used 'Fairbairn's Lever' – a mechanical device to apply loading to beams and to apply pressure to materials. It was used, sometimes with modifications, in his investigations of strength of cast-iron beams, in determining the crushing strengths of various materials, in investigations on the effects on strength of temperature and of repeated re-melting, and on the strengths of different mixtures of iron. It was used for tests on the strength of wrought-iron plates and riveted joints, for tests on the strength of wrought-iron tubes, for fatigue tests on wrought iron, for investigating the influence of pressure on solidification, for experiments on the insulation of submarine cables, and for the testing of various types of steel. Initially it was used by Hodgkinson.[52] Beyond Fairbairn's lifetime, it was used by Unwin.[53] A feature of Fairbairn's research from the mid 1820s to the mid 1860s, 'Fairbairn's Lever' was a piece of experimental apparatus to which he gave his name. However the idea was far from original and once again Fairbairn developed and publicised an existing concept, this time from Galileo in the 17th century, developed by the Dutch physicist, van Musschenbroek, (1692–1791) and, in 1817, by George Rennie (1791–1866).[54] Many, more sophisticated, lever testing machines were developed in the second half of the 19th century. They are described by Unwin.[55]

'I confess', wrote William Fairbairn, 'that nature had endowed me with a strong desire to distinguish myself as a man of science'.[56] But not all awarded him that accolade. Woodcroft observed, 'facility in dealing with algebraic processes is a far more searching and sure test of the highest order of scientific mind than is perhaps any other enquiry' and Fairbairn had admitted that he found algebra difficult and ultimately had to give it up.[57] R. Angus Smith was unstinting in some aspects of

Fig.7.9: **The Spey Bridge.**

his praise but was forced to conclude, 'we cannot say that he had scientific knowledge of an exact kind, indeed he was sadly deficient in it'.[58] Fairbairn was well aware that he was no mathematician. He knew when he needed to turn to others for help. In earlier years that was to Eaton Hodgkinson, in later years to Thomas Tate and to William Unwin.[59] Yet his peers recognised Fairbairn's achievements, giving him high acclaim including a Royal Society Gold Medal, membership of the French Académie des Sciences, appointment as a juror at the 1855 Paris Exhibition, and honorary doctorates from Cambridge and Edinburgh Universities.[60] Essentially Fairbairn was an experimental engineer, but in the 19th century Lord Wrottesley, and in the 20th century, A.I. Smith saw him as also a 'man of science'.[61] Perhaps he is best seen as a forerunner of 'modern technology' with its 'scientists who "do" technology and technologists who function as scientists'.[62] Indeed this is how Fairbairn viewed the Mechanical Sciences Section of the British Association:

> [Its] meetings have afforded incalculable benefit to the practical engineer and mechanic. They bring all his projects and contrivances under the strict rules of science, and on the other hand they bring the abstract theorist to the test of practical experience.[63]

Metal Fatigue

As an experimental engineer, Fairbairn was the author of a famous paper in the history of fatigue testing.[64] It arose from controversy which enveloped the 230ft-span tubular-girder Spey Bridge, built shortly after Fairbairn had 'retired', but with which he was deeply involved. The bridge was constructed on timber staging, but deflected by between seven and eight inches, when the blocking was removed.[65] The Board of Trade Inspector refused to sanction its use. Fairbairns, having contracted to meet the Board of Trade's requirements, were compelled to strengthen it at their own expense. Re-supporting the bridge and adding strengthening plates reduced the sag by half, after which it remained in service, for single rather than the intended double track, until 1906. The problem appears to have been that the empirical formula which was used failed to take account of the effects of increased stresses in longer spans.[66] The Treasury granted £150 for 'experiments to ascertain the durability, and the measure of strength to be allowed, of wrought-iron bridges,

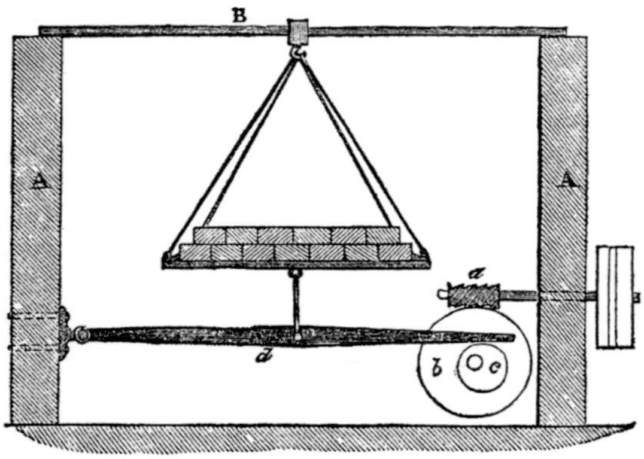

Fig.7.10: Repeated Loading Apparatus.

subjected to changes, shocks and vibrations of a continued and variable load', which Fairbairn undertook from March 1860 onwards. He set up apparatus in an Ancoats basement where, by means of Fairbairn's Lever and a waterwheel, a load was continuously imposed and removed from a built-up wrought-iron beam.[67]

The fatigue of materials – the drop in durability under repeated variable loads – is of major importance to today's industry, generating many research papers. The description '*fatigue de méteaux*' is attributed to Poncelet in the 1830s.[68] It first appeared in Britain in 1854 in a paper by F. Braithwaite, at which Fairbairn was present and contributed to the discussion.[69] Only a few mid-19th-century engineers – Rankine, Hodgkinson, James, Galton, Fairbairn – recognised that there is a difference between the bearing powers of a beam exposed to changes of pressure and one which has to sustain a static permanent load.[70]

The subject was addressed by the Railway Structures Commissioners whose Report included experiments on continued impact by Hodgkinson (a Commissioner) by repeatedly dropping an iron ball onto a cast-iron bar, and by James and Galton by passing weights over bars at different velocities.[71] Fairbairn gave evidence to the Commissioners, following which he wrote to them, referring to his impact testing of the first tubular-girder bridges at Blackburn, and suggesting three distinct types of experiment – to measure the effects of repeated stress, of vibration and of repeated blows. Details, with illustrations *(figs 7.10 & 7.11)* of the apparatus, were included in the Report.[72] These proposals were not acted upon.[73]

Over a decade after the Commissioners' Report the Spey Bridge experiments were put in hand. A built-up wrought-iron beam 22ft long by 16in. deep was subject to seven or eight loads, imposed and withdrawn, per minute. Deflection was regularly measured. The experiment continued for two years with three million cycles. Fairbairn found that girders were 'not safe when submitted to violent disturbances equivalent to one-third the weight that would break them' but exhibited 'wonderful tenacity when subjected to the same treatment with one-fourth the load'. The latter would accept twelve million changes without injury and thus with 100 trains a day would require 328 years before security was impaired.[74]

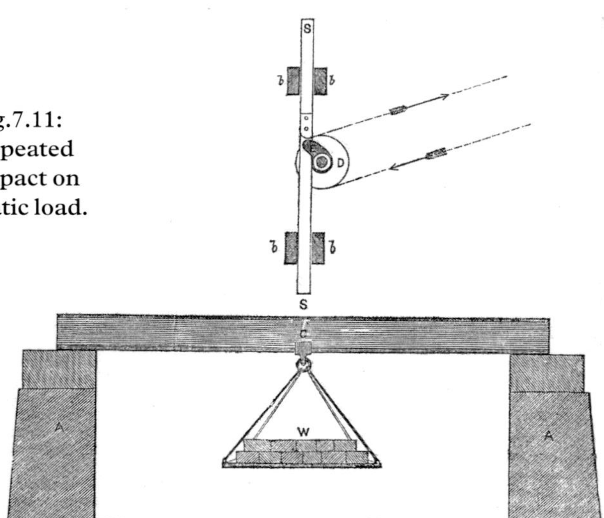

Fig.7.11: Repeated impact on static load.

This work was one of only a very

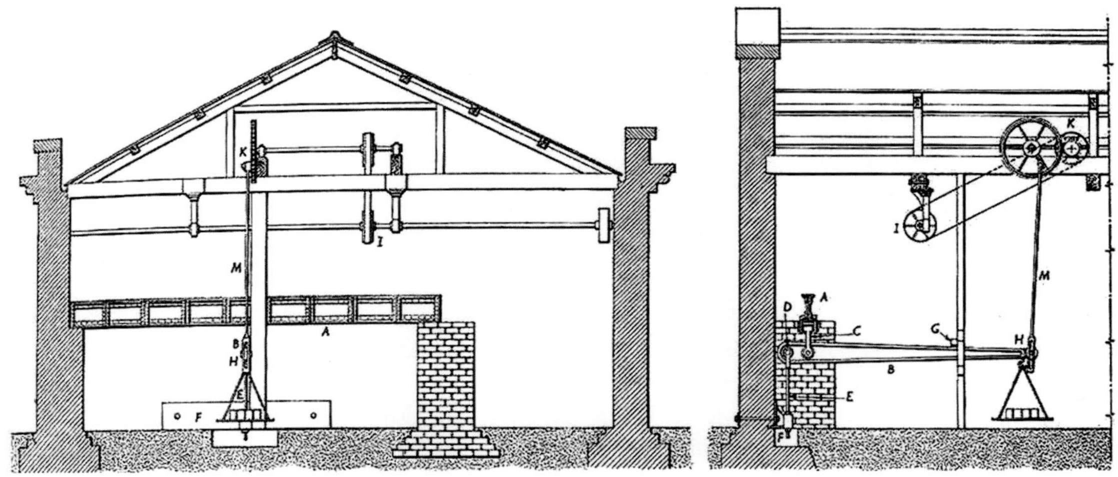

Fig.7.12: Fairbairn's Fatigue Testing Apparatus.

small number of tests on built-up structural components. The earlier experiments by James and Galton had a bearing on Fairbairn's work and he summarised their results in his paper;[75] but the vast majority of early work on fatigue, apart from Wilhelm Albert's 1837 paper on conveyor chains, related to railway axles, following the Versailles rail accident in 1842.[76] Fairbairn's findings were discussed by Benjamin Baker in 1873, but thereafter the British contribution to literature on fatigue was negligible until the start of the 20th century.[77] The first book on fatigue, written in 1924 by H.J. Gough, sixty years after Fairbairn's experiments, described them as 'deservedly famous' and his apparatus of 'great historic interest'.[78] The reason why full-scale fatigue tests on riveted bridge members are important is because they take into account the composite action of the different parts, showing the overall behaviour. Åkesson lists such tests chronologically, describes them, and compares their results with those from small specially fabricated specimens. He is aware of only nine full-scale tests.[79] The first is Fairbairn's in 1864 and the next Reemsnyder's in 1975.[80]

The immediate outcome of Fairbairn's fatigue experiments was to provide evidence that the Spey Bridge was satisfactory in respect of long-continued changes in load. The more important outcome was previously unavailable knowledge of metal fatigue, particularly important to bridge-builders.[81] Here was a case of failure, in this case the unexpected deflection, leading to new knowledge.

Fairbairn's writings were a combination of intuitive insight into the mechanism of fatigue and correct assessment of its practical significance. Six years before the commencement of his experiments he had written:

> [The] constant movement or sliding of the atoms or parts of crystalline, as well as fibrous bodies, is therefore the cause of breakage; and – assuming a change to take place in the molecules of the body – however slight the strain may be when first applied in one direction, and then changed to another direction, it only becomes a question of time how long the body will bear these continued repetitions before rupture takes place: sooner or later fracture must occur.[82]

This idea of fatigue failure being due to exhaustion of plastic ductility was later to be the basis of theories of fatigue in 1923 and 1939.[83]

As an interdisciplinary postscript on fatigue, in 1955 Louis Goldberg described Fairbairn's fatigue experiments which

> suggest that the notion that all long-term assets deteriorate physically in strict proportion to their physical use is somewhat naive and warrants careful reconsideration by accountants in conjunction with technical officers.[84]

Boilers and Boiler Safety

WILLIAM Fairbairn was, during the third quarter of the 19th century, the leading figure in the arena of boiler safety, with knowledge of boiler hazards said to be greater than any other living person.[85] That knowledge had a threefold base, his experience as a boilermaker over a quarter of a century, his research into the collapse of boiler flues, and his investigations of boiler explosions. His research is exemplified in four papers to the Royal Society. The first, 'Researches on the Resistance of Cylindrical Vessels of Wrought Iron to Collapse', relates to tubular flues passing through cylindrical boilers, containing the products of combustion to heat the water. As steam pressures continued to rise and boiler sizes increased, catastrophic explosions were frequent. By 1857 Fairbairn saw the necessity for testing cylindrical shells to obtain factual data on the strength of boilers for design purposes. The tests provided that data, but also demonstrated for the first time the collapse of cylindrical shapes due to *external* pressure. Before this work little or nothing was known as to the resistance of cylindrical vessels to an external collapsing pressure.[86] The axiom was that a cylindrical tube, such as a boiler flue, when subject to uniform external pressure, was equally strong whatever its length. Fairbairn had misgivings: hence the experiments. Wrought-iron tubes of different lengths, diameters and construction – forty in all – were inserted in turn into an 8ft-long cast-iron cylinder and surrounded with water. The hydraulic pressure was increased until collapse. The results showed that the resistance of cylindrical tubes to collapse from a uniform *external* pressure varies in the inverse ratio of their lengths. Further experiments showed that the length of wrought-iron tubes had only slight influence on their resistance to *internal* pressure. As well as an empirical formula based on his results, Fairbairn added a practical application, illustrating a 30ft boiler flue divided into three by simple curved T-iron ribs. The cost was minimal, but the strength of the tube was increased nearly threefold.[87] The results were so unexpected as to suggest similar experiments with other materials. Fairbairn, aware of the influence of directional properties and joints in plates, sought to confirm the results using a homogeneous material, glass. On this occasion he worked with mathematician Thomas Tate. They found that the resistance of glass cylinders to compression followed the same pattern as wrought-iron cylinders.[88]

These experiments were of major importance for boiler safety. They showed the cause of many unexplained explosions to be *external* pressure on the cylindrical fire-tubes, and provided a simple and inexpensive means of prevention, by way of the T-iron ribs which were rapidly adopted throughout the industry. Fairbairn's experiments and resulting formula were the basis of boiler design for nearly half a century, and the starting point for further debate in Europe and America, leading to improved empirical formulae. These included the formulae of Grashof, Love, Nystrom, Unwin, Belpaire and Wehage.[89] These developments were reviewed by C.R. Roelker in 1881,[90] some work was done in Danzig in 1887–92, and after that nothing further until Carman, and Stewart, both in America, in 1905–6.[91] Carman and Carr, writing in 1906, summarised the position:

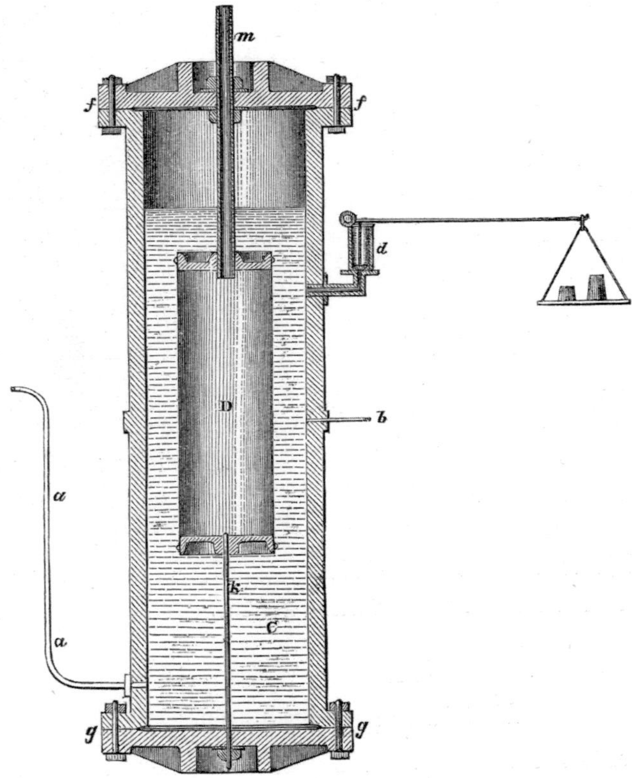

Fig.7.13: Fairbairn's Apparatus for Testing the Resistance of Tubes to Collapse.

> The first ... systematic experiments on the collapse of tubes were those of Sir William Fairbairn ... Fairbairn's formula has been the text for practically all the discussions on the collapse of tubes for the last fifty years. Aside from Fairbairn's experiments, there have been no systematic experiments described until within the last year. It is noteworthy that a problem so widely discussed, the solution of which has not only great scientific interest but also valuable technical applications, should remain for so many years with so little experimental work. The only reasons that can be assigned are the considerable expense involved in the experiments and the fact that the appliances needed are not commonly available in testing laboratories ... Immediately after the publication of Fairbairn's paper, one writer after another began to discuss the data for the purpose of deducing a more general and convenient empirical formula.[92]

In the 21st century Fairbairn's experiments on cylindrical vessels are still referred to in the historical introductions to major academic texts.[93]

Another area of Fairbairn's research exemplified in his papers to the Royal Society is the Bakerian Lecture of 1860, 'Experimental Researches to determine the Density of Steam at different Temperatures and to determine the Law of Expansion of Super-heated Steam'. This was again co-authored with Thomas Tate who provided the mathematics.[94] At the time there was a well-known formula that was the basis of the tables of the density of steam, on which calculations of the energy efficiency of steam-engines were founded. The formula was questioned by Joule, William Thomson and Rankine but there was no reliable experimental knowledge to verify the position. Fairbairn and Tate appear to have worked together on building the quite complicated apparatus, and in undertaking the experiments. Unwin was also involved with 'his untiring zeal and superior skill in manipulation'.[95]

The method involved a known weight of water in a glass globe of known capacity and devoid of air, uniformly heated in a steam bath. The temperature at which the whole of the water was just vapourised was recorded. Measurements of the volume of the steam were made by reference to a column of mercury in the stem of the globe. Then, knowing the weight, volume and temperature of the steam, its specific gravity could be calculated. The major result showed that the density of saturated steam was invariably greater than that which was devised from the previous formula. Tate provided a new formula.[96] No less an authority than Rankine wrote that Fairbairn and Tate had 'for the first time determined directly the density of steam through an extensive range of pressures and temperatures, thus making a contribution to physical knowledge of the highest order, both as to practical utility and scientific importance'.[97] The results enabled the designers of high-pressure steam-engines to base their calculations on more accurate and reliable data, enhancing both safety and economy. The part each man played in these experiments is unclear, except that Tate did the mathematics. The paper was translated into Italian.[98] Two years later there was a further paper providing data for steam at higher temperatures, the results being in line with the previous ones.[99]

Fairbairn's research had shown him that larger diameter boilers, for which there was a continual call, were more susceptible to explosion, particularly when used with high-pressure steam, of which he had long been an advocate. He gave practical effect to these issues by designing a new 'five-tube' high-pressure boiler, patented in 1870. It had two flues each closely surrounded by its own shell, with pipes connecting these shells to a third shell above. The three shells, of lesser diameters than a single shell, reduced the risk of explosion.[100] Its first use was for his friend Thomas Ainsworth at Cleator Moor, Cumbria, in 1872 when Fairbairn, aged eighty-three, was on site undertaking tests and experiments on the boiler.[101]

In 1874, with a view to settle the debated question of the relative merits of the inside-flue boiler and the French boiler with outside flues, a committee of nine members of the Société Industrielle de Mulhouse was appointed to test a double-flued Lancashire boiler against a French boiler. To these was added a Fairbairn five-tube boiler, which proved to be the most efficient, followed by the Lancashire boiler.[102] That of three leading boilers trialled in the heartland of the French textile industry, two were Fairbairn patents – and these two were the more efficient – highlights the importance of William Fairbairn in the field of steam boilers *(fig. 4.15)*.

With this success one might have expected the new boiler to have been universally adopted. It was not, apparently because it primed badly. Thomas Beeley, a boilermaker who had commenced business in Hyde in 1867, suggested fitting uptake pipes to aid circulation, and lengthening the lower shell, with smoke tubes connecting the rear of the fire-tube to the back-plate as in a multi-tube boiler.[103] How Beeley and Fairbairn came to be working together is unclear, but in 1873 they took out a joint patent for the changes – hence 'the Fairbairn-Beeley boiler'.[104] A prototype was built in 1874 for the Manchester Steam Users' Association to test to destruction by hydraulic pressure,[105] about which Fairbairn wrote his last letter on engineering matters on 8 June 1874, a few weeks before his death.[106] Beeley manufactured several hundred Fairbairn-Beeley boilers over the next twenty years, with exports to Russia, Finland, France and India,[107] but they never

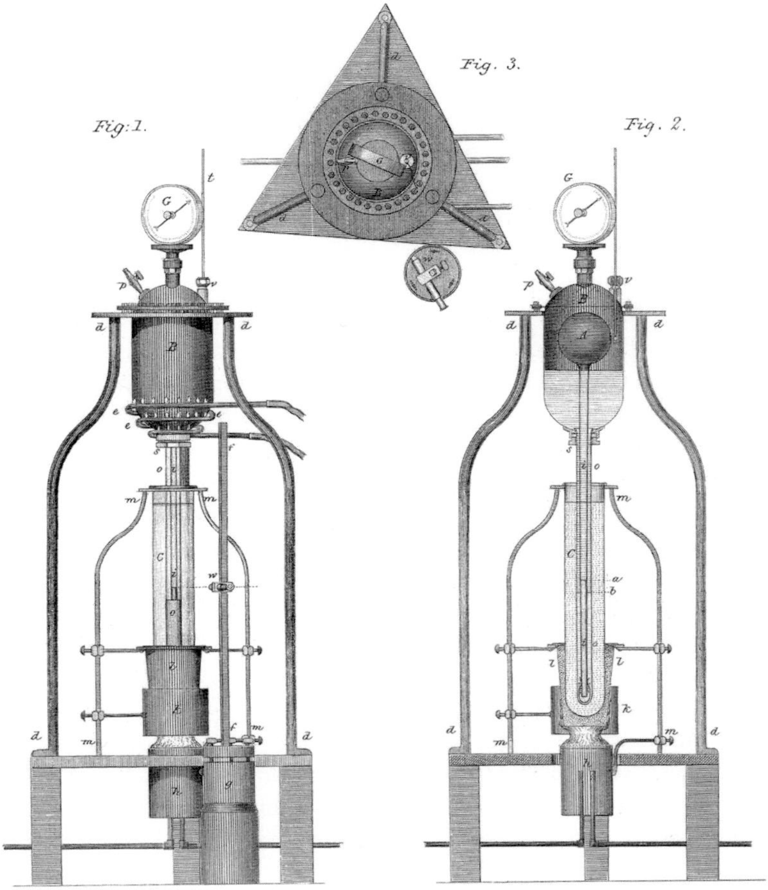

Fig.7.14: Apparatus for Experiments on the Density of Steam at Pressures from 15 to 70lb/sq.in.
– fig.1 Elevation
– fig.2 Section
– fig.3 Plan

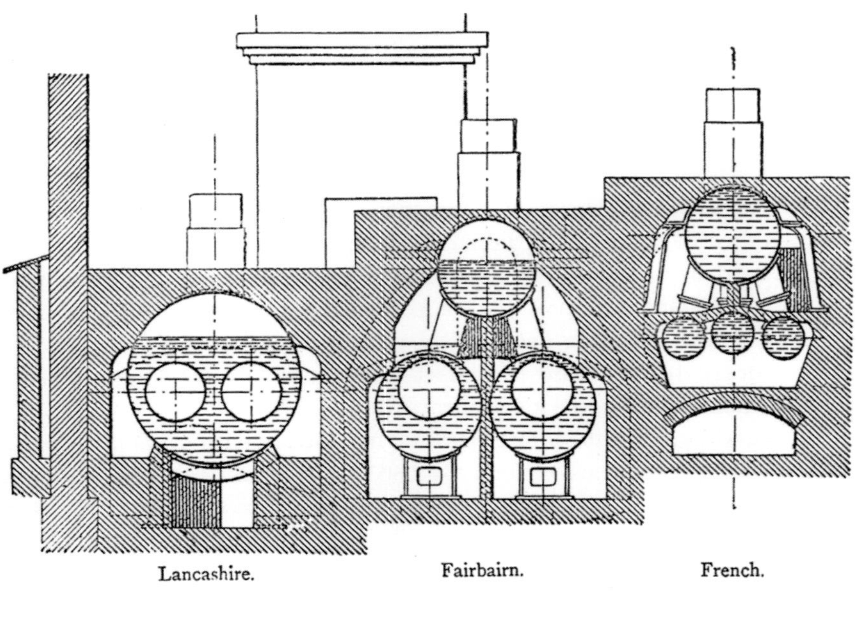

Fig.7.15: Experimental Boilers at Mulhouse 1874.

replaced Fairbairn and Hetherington's Lancashire boiler of three decades earlier, albeit with continuous incremental improvements, as the most popular boiler, particularly in the textile districts.

In the fifteen years 1856–70 there were 2,163 boiler explosions in Britain. Fairbairn told the Social Science Congress in Manchester in 1866, 'There is no mystery in steam boiler explosions; they are all traceable to causes', and elsewhere he expressed the view that such explosions were 'for the most part preventable'.[108] For many years he had been appalled at the fatalities from these explosions, a concern focussed by preparing reports for inquests.[109] One of these involved a locomotive built by Sharps which exploded whilst being tested, with nine fatalities. *The Engineer* published a wood-cut *(fig. 7.16)* taken from a photograph provided by Fairbairn.[110] It does not say who took the photograph or if was used at the inquest, nevertheless this was one of the earliest examples of the forensic use of photography, three years after the formation of the Manchester Photographic Society, of which Fairbairn was a Vice-President.[111] Soon after the Atlas Works explosion *The Engineer* carried an advert by a Committee headed by Fairbairn 'for raising a Fund to relieve the Widows and Orphans of those who were killed by the late melancholy accident at Sharp, Stewart, and Co.'s Works'. The subscription list included: 'Messrs William Fairbairn and Sons – £100.0.0', and 'William Fairbairn Esq – £10.10.0' – equivalent in 2016 to around £7,000 and £750 respectively.[112]

During inquests at Stockport and Blackburn in1853 Fairbairn recommended regular boiler inspections, but the catalyst for action was an explosion at Rochdale in July 1854 when ten people died.[113] He called for the establishment of associations whose members 'should appoint inspectors to take cognizance of the boilers within their respective precincts'.[114] With Henry Houldsworth and Joseph Whitworth, Fairbairn convened a meeting to establish a 'Steam Users' Association for the Prevention of Steam Boiler Explosions and the Economical Working of Boilers'. By April 1855 three inspectors had been appointed.[115] Houldsworth was President, followed in 1859 by Fairbairn who held the office and attended committee meetings assiduously until 1874. Apart from technical matters, two issues caused concern. The first was that the Association did not provide insurance, because it was thought insurance would induce carelessness. This caused loss of members to the Steam Boiler Assurance Company, a Manchester company founded 1858 and the first to insure boilers against explosion.[116] Fairbairn proposed converting the Association to a mutual insurance society; then he proposed insurance by an outside insurance company, based on the Association's inspections, but neither proposal was accepted. To counter the loss of business, Fairbairn suggested a 'guarantee system' for boilers under the Association's care. This found favour and thirteen committee members, led by their President, put up £1,000 each to underwrite a guarantee in respect of explosions arising following negligent inspections. Remarkably, given the very limited nature of this safeguard, 'the effect was electric'. New members joined the Association and an additional inspector was appointed.[117]

The second issue was that of legislation. The Association advocated prevention of explosions by regular inspections. By 1863, with an annual average of 1,600 boilers under inspection, only three accidents, involving two fatalities, had occurred during the eight years of the Association's existence. Fairbairn was against Government taking over the Association's role, believing this would inhibit progress.

Fig.7.16: Fatal Explosion at Sharp Stewart & Co., Atlas Works, 2 July 1858.

This view was not held by all.[118] A proposed Bill in 1869 would have transferred the Association's methods to the Board of Trade, but was unsuccessful.[119] The British Association set up a committee, which Fairbairn chaired, and of which several of his friends from Manchester were members.[120] It recommended that inquests should have reports from two competent engineers, and expressed a 'dread of any Government interference' believing it would be 'a barrier to progress'.[121] The Bill was re-introduced in 1870 and Fairbairn successfully lobbied for a Select Committee instead, but this did not take the matter much further.[122] The Manchester Association called for compulsory inspections controlled by boards elected by steam users. By 1879, Government having done nothing, Hugh Mason, who followed Fairbairn as President of the Manchester Association and had become an MP, successfully introduced the Boiler Explosions Act (1882). For the first time boiler explosions had to be reported, and Board of Trade surveyors had to investigate and report. The number of boilers continued to increase, but the number of deaths from explosions fell.[123] More than any other, Fairbairn had driven boiler safety in the third quarter of the 19th century and, eight years after his death, his approach was reflected in legislation which enhanced public safety with minimal government involvement.

Fairbairn's influence through the Manchester Steam Users' Association went beyond Britain, and not only in the field of boiler safety. In 1866, 'the most celebrated of [the Association's] offsprings came into existence' – the Alsace Association. By 1905 there were 57 similar associations in Continental Europe, and others as far afield as Calcutta.[124] In 1877 Fleeming Jenkin's campaign for healthier homes inspired the Edinburgh Sanitary Protection Association. It was set up to inspect domestic drains and modelled on Fairbairn's Steam Users' Association. Many other cities, including London and Glasgow, set up Sanitary Protection Associations.[125]

Lavington Fletcher, Chief Engineer of the Manchester Steam Users' Association from 1861, wrote about boiler legislation:

> it seems to me to open a very interesting and important question, and one that is being raised with reference to the conduct of other of the world's industries, such as mining, shipping, &c, the problem being how to control

> the careless without hampering the careful, and how to save human life and check recklessness without hampering progress. That questions of such interest should spring out of the simple subject of boiler inspection might not perhaps at first be anticipated.[126]

Railway Brakes

IN 1853 William Fairbairn had attended trials of Newall's 'break' [*sic*], the first effective continuous brake, applied by a continuous shaft over the roofs of the carriages.[127] He was enthusiastic and read a paper about it to the Institution of Mechanical Engineers.[128] A few years later Colonel Yolland of the Board of Trade reported on several braking systems, recommending Newall's. Fay of the Lancashire & Yorkshire Railway was aggrieved and it was decided that the two systems should be tested under Fairbairn's superintendence.[129] By that time both systems had brake-blocks on every wheel operated by continuous shafts beneath the framing. Initial tests took place on the 1 in 27 Oldham Incline. Similar trains descended by gravity and at a specific point detonated a fog-signal, at which point the speed was calculated, the brakes applied and the stopping distance measured.[130] Unwin travelled on the trains to ensure the brakes were applied at the right time and to register the speed and stopping distance for each experiment.[131] He also drafted the report, which Fairbairn signed.[132] They found little to choose between the two systems and were enthusiastic about both. *The Engineer,* summarising the findings, wrote, 'Railway travelling would be rendered incalculably safer by this summary control over the moving mass, a control which would have prevented ... nine-tenths of all the accidents which have ever happened on the railways of this country'.[133]

The Temperature in Italy – an Interlude

GEORGE Fairbairn suffered from ill-health.[134] Robust as his father's health was, William was suffering some rheumatism by his late sixties. Thus it was determined that father and son would spend two months in the warmth of Italy during the winter of 1857–8. They travelled via Paris, Marseilles and Nice, to Genoa, and visited most of Italy's major cities. They were surprised: expecting a warm climate, they found that in 'the whole of Lombardy and ... between the Alps and the Apennines, the cold in winter was generally intense, sometimes almost Russian in its character'.[135] Fairbairn set about recording temperatures and on his return read a paper to the Manchester Literary & Philosophical Society, 'On the Comparative Temperature of the Climate of England and some parts of Italy'.[136] This was hardly exact science but it was typical of the man and illustrates the influence of the British Association and Humboldt.[137]

Armour-Plating

SOMETHING of the background to iron warships was given in Chapter 3. When, as Pole puts it, having built the *Warrior,* Government 'began to enquire how they ought to have built her', the outcome was the appointment of a 'Special Committee on Iron'.[138] There were seven members – three military men, a metallurgist, Fairbairn and his subsequent biographer, Pole. The Committee was actively at work over four years, addressing the best material, the manufacture of

Fig.7.17: Inspection of the La Gloire Target after the firing, Shoeburyness, 1864.

plates, thickness, form, method of attachment and timber backing. Practical experiments were undertaken at Shoeburyness with targets of different construction, and with different guns and types of shot, as Armstrong and Whitworth vied with each other. The results pointed, contrary to the ideas generally prevalent, to plating of wrought-iron, as soft and tough as could be procured, to allow the energy of the shot to be expended in the indenting and battering of the plates without fracturing them, and with a timber backing.[139] There were two main aspects of Fairbairn's involvement, targets and technical research.

The *Warrior* had already been built, with 4½in. armour-plating on a backing of two 9in. layers of hardwood. A target using the same construction as the *Warrior* had been constructed and Fairbairn built an alternative target with elm in front of the plating. Trials took place in 1861 and the Fairbairn target failed, as did targets by Hawkshaw and Roberts.[140] By 1862 it seems that Fairbairn had become convinced – and the Committee agreed – that timber cladding or backing was not required, and an all-iron target was built at Ancoats and sent to Shoeburyness.[141] This 'Committee target' also failed, notably at the fixings of the plates. It is said that a friend of Fairbairn's 'proposed that the Southend hearse and mourning-coaches should be sent for to take the inventor and the committee to the railway'.[142] The target was repaired, and sheets of India-rubber or felt were introduced between the plates and iron skin, to reduce vibration.[143] Further trials of the *Warrior* and Committee targets took place in March. Both targets were disabled – the latter the more seriously.[144] Targets by Scott Russell and Samuda were no more successful.[145] Fairbairn played a leading role as *The Mechanics' Magazine* reported:

> In spite of wretched weather, and trampling in mud and slush, the members of the Special Committee were engaged in their useful labours from 10 to

> past 5 o'clock; Mr Fairbairn, who took the command, being most active and persevering up to the last moment, continually urging the artillery to new trials of strength in this remarkable combat.[146]

In September Fairbairn spoke about guns v. armour-plating at the opening of the Mechanical Sciences Section of the British Association – he was President of the Section. 'No vessel', he said, 'is able to carry armour-plates of sufficient thickness to resist such powerful ordinance as has been brought against them.'[147] This was endorsed by experiments in 1864 with Armstrong's 600-pounder gun firing steel shot.[148] In November of that year Fairbairn read a paper at the Society of Arts, 'Applications of Iron for the purpose of Naval Construction', with the Duke of Somerset, the First Lord of the Admiralty, presiding. In the ensuing discussion Somerset recalled that they

> were told that iron offered the best resistance, and that they should not mix iron and wood; but that was tried at Shoeburyness, and Mr Fairbairn gave in, and said they must put a certain quantity of wood behind the iron if they wanted resistance.[149]

Woodcroft sees the Committee target, 'constructed at the express desire of Fairbairn and in accordance with his designs ... the thick armour-plate being bolted directly upon the iron skin of the ship by bolts and nuts screwed up very tight, and without any elasticity in the construction of the washers' as 'a good instance of Fairbairn's defective conception of molecular forces'. When the first rounds broke half the bolts and many rivets, Fairbairn 'alleged in explanation that the bolts had been screwed up so hard as to produce a severe and unnecessary initial strain upon the iron, and partly by his own hands the nuts of such bolts as remained sound were slightly slackened; but the next round was attended with no better results'. Woodcroft is also highly critical of Pole for keeping these facts 'wholly out of view'.[150]

Clearly the hoped-for results were not achieved. But, as with the canal experiments over thirty years earlier, they provided a negative result which in itself was 'useful knowledge'. The journalist P. Barry wrote:

> The committee and Mr Fairbairn were, however, not to be caught napping, and, to the amusement of the Shoeburyness officers and others, coolly stated that the target had been invented for no other purpose than trying the value of the 17 inches of the Warrior teak backing.[151]

Indeed Fairbairn does say that the Committee's target was 'for the exclusive purpose of ascertaining whether or not the wood backing could be entirely dispensed with',[152] and Pole wrote that, 'it was said that ... the wood backing was not only unnecessary, but was absolutely prejudicial ... It was necessary to test this, and no less than three expensive targets were constructed for the purpose'.[153] But the ever-critical Barry is out of order in attributing pecuniary motives to Fairbairn, now well into his seventies and having retired from the management of the firm over seven years earlier.[154] In his *History and Progress of Iron Ship Building* (1865), Fairbairn commenced the chapter 'On the Properties of Iron Armour-Plates, and their Resistance to Shot at High Velocities' with the admission that 'The best mode

1861	On the Mechanical Properties of Specimens of the Iron and Steel plates subjected to Experiments with Ordnance, their Strength, Ductility, Resistance to Punching Force, &c.
1862	On the Resistance of different kinds of Shot to a Force tending to crush them, and the forms they assume under Pressure. Experiments on Punching.
1863	On the Mechanical Properties of Iron Plates. On the Manufacture of Armour Plates. Experiments to test the value of annealing Armour Plates. On the Tenacity of Cast-Iron Shot.
1864	On the Mechanical Properties of iron plates. On the Manufacture of Armour Plates. General Summary of Results.

Table 7.1: Fairbairn's Investigations for the 'Special Committee on Iron'.[159]

of employing armour-plates for the protection of the sides of ships is a problem yet to be solved'.[155]

The second main area of Fairbairn's involvement on the 'Special Committee on Iron' was in undertaking various investigations for the Committee. In this he was assisted by Unwin.[156] These were included in the Committee's Annual Reports for 1861–4, which were never published in full.[157] They are listed in *Table 7.1*. The only one that is readily available is the first, which is included in the 6th edition of Peter Barlow's *Treatise on the Strength of Materials* (1867).[158]

The Atlantic Cable

FAIRBAIRN brought his expertise to bear on two aspects of the great engineering achievement of the 1860s, the successful Atlantic cable. These were the choice of insulating material for the cable and the comparative strengths of various proposed cables. Following the failure of the 1858 attempt, Government set up a ten-strong Commission, chaired by Douglas Galton, which took evidence from forty people, and produced a comprehensive and optimistic Report.[160] One of its members was William Fairbairn.[161] The Report was later described as 'the most valuable collection of facts, warnings and evidence ever compiled concerning submarine cables',[162] which 'proved to be decisive in shaping the future progress of deep-sea cables'.[163] The Commission, needing to know the most effective insulating material for the cable, given the pressures involved, requested Fairbairn to investigate. Fairbairn's Lever was used with a modification similar to that employed by Hopkins.[164] Various insulators were tested. Dry samples of each

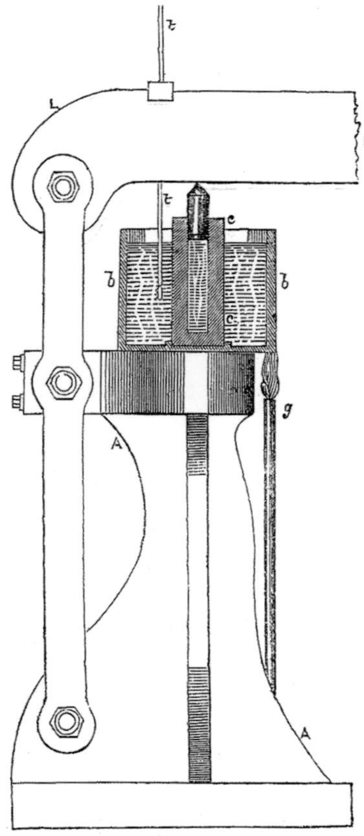

Fig.7.18: Apparatus for measuring absorption of water by insulators at higher temperatures.

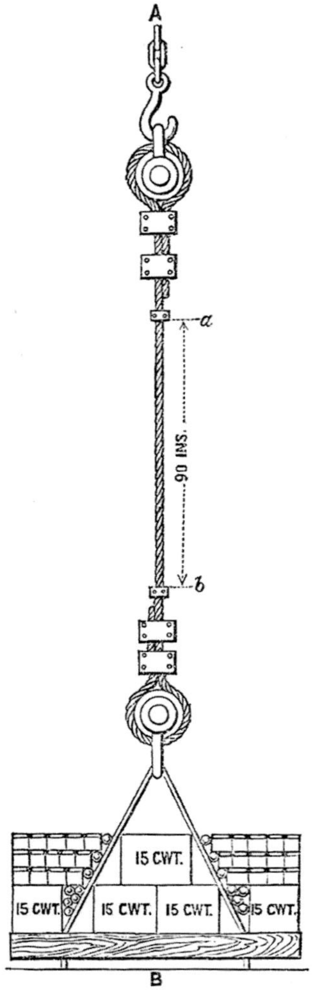

Fig.7.19: Apparatus to measure the breaking strain of cables.

were weighed, subjected to pressure in a cylinder of water, surface-dried, and re-weighed to determine the amount of water they had absorbed. Measurements were recorded at up to 20,000psi – equivalent to a depth of 8.72 miles – and for up to 450 hours. Raw India-rubber was found to absorb twenty-seven times as much water as Chatterton's compound. The experiments were repeated at higher temperatures, by means of the cylinder being immersed in a container of heated water. This showed that temperature had a marked effect on the amount of water absorbed. The best results were obtained from Chatterton's compound with gutta-percha.[165] Fairbairn then attempted a much more complex experiment, to compare the loss of electric charge through different insulators under pressure. This was fraught with difficulties but the results, whilst lacking conclusivity, were sufficient to endorse a system of alternate coats of Chatterton's compound and gutta-percha.[166] Four layers of each were used in both the 1865 and the successful 1866 cables.[167] The Commission's Report sets out details of Fairbairn's experiments.[168]

The Atlantic Telegraph Company set up its own Scientific Committee comprising Douglas Galton, William Fairbairn, Charles Wheatstone, Joseph Whitworth and William Thomson (later Lord Kelvin).[169] When tenders for a new cable were invited, about sixty different samples by Glass Elliot & Co and some by other firms, were delivered to Fairbairns' Ancoats works where Fairbairn and Whitworth instituted tests to determine the weight, breaking strain, and specific gravity of each cable, together with tests on each component of the cables. Samples of each cable, 7ft 6in. long, were subjected to loading with increments of one hundred-weight, with the elongation measured as each increment was added, until the sample broke.[170] Based on the results of these tests, the Committee recommended cable No.46 of Glass Elliot & Co.[171] However, Glass Elliot could not afford to finance the work including the gutta-percha insulation and, driven by John Pender, merged with the Gutta Percha Company to form the Telegraph Construction and Maintenance Co Ltd ('Telcon'), which entered a contract with Daniel Gooch's Great Eastern Steamship Co. to lay the 1865 cable.[172] The *Great Eastern* sailed from the Nore to Valencia, in south-west Ireland, where cable-laying was to commence. Fairbairn was on board as far as Valencia.[173] The attempt failed when the cable broke and could not be recovered.[174] The Atlantic Telegraph Company then found that its Act of Parliament would not allow it to raise money as it intended, by way of Preference Shares. Gooch and Pender therefore established a new company, the Anglo-American Telegraph Company, to take over the project. A new cable was obtained, and the *Great Eastern* left Greenwich on 30 June 1866 on the successful attempt – and the 1865 cable was recovered, a feat which Fairbairn considered 'one of the most successful triumphs of marine engineering'.[175] By this time William and Thomas Fairbairn and Pender were on the board of the newly-incorporated Fairbairn Engineering Co. Ltd, and in 1870 Thomas would join the Telcon board.[176]

Patents and Patent Law Reform

WILLIAM Fairbairn had no aspirations to become a Member of Parliament, or to become Mayor like his brother Peter who became Mayor of Leeds and Peter's son Andrew who became an MP. However where there was legislation afoot on engineering issues that affected him, he was a campaigner, as the boiler legislation issue has shown. The patent controversy had similarities. That an heroic ideology of invention continues to inform popular understanding partly results from the celebration of inventors stimulated by the outcome of the patent controversy during the third quarter of the 19th century. From 1837 to 1873 Fairbairn filed nine patents, as *Table 7.2.*

Historians have discussed the use of patent statistics as an indicator of invention, and as a starting point for tracking the transfer of innovations across sectors, but have faced the difficulty of not knowing how many ever got beyond the drawing-board, and of those that did, how many did not work.[178] Of Fairbairn's patents, the riveting machine, Lancashire boiler, tubular girder bridge and tubular crane were eminently successful. The five-tube boiler was only successful with Beeley's later improvements. The proposed mode of constructing metal ships failed, the mode of driving screw propellers appears to have been a non-starter and it is unclear what the improvements in marine engines were. This indicates a success rate of around 55 per cent.

Patents were expensive and procedures were archaic and cumbersome. The National Association for the Reform of Patent Laws was formed in 1847. Pressure mounted as exhibitors at the Great Exhibition feared unprotected exhibits would allow competitors free access to their designs.[179] In Manchester the Patent Reform Committee dating from 1828 was re-formed in December 1850 as the Manchester Patent Law Reform Association, with Fairbairn chairing the meeting.[180] Government introduced a Provisional Act and the matter was referred to a Select Committee which received evidence from thirty-three witnesses, including Fairbairn.[181] The outcome was indefinite as there was a strong lobby to

Table 7.2: **Fairbairn's Patents.**

Date	No.	Patentee(s)	Brief Description
1837	7,302	Robert Smith*	riveting machine
1841	9,072	William Fairbairn	improvements in marine steam engines – direct action
1842	9,409	William Fairbairn	construction of metal ships
1844	10,166	William Fairbairn John Hetherington	Lancashire boiler
1844	10,181	William Fairbairn	mode of driving screw propellers
1846	11,401	William Fairbairn Robert Stephenson	tubular girders
1850	13,317	William Fairbairn	tubular crane
1870	810	William Fairbairn	five-tube boiler high-pressure
1873	270	William Fairbairn Thomas Beeley	Improvements to high-pressure boiler

* paid for by Fairbairn.[177]

abolish the patent laws, including leading engineers I.K. Brunel, William Armstrong and William Cubitt.[182] The debate in some ways foreshadowed that at Harvard in the mid 20th century where Schumpeter distinguished between invention and the entrepreneurial function of 'innovation',[183] while others took the view that major innovations are built up of numerous incremental improvements, made over lengthy periods of time, the cumulative importance of which is of decisive economic significance.[184] There were also affinities to the debate in the 1990s between Christine Macleod and Gillian Cookson.[185] Brunel's belief was

> that the most useful and novel inventions and improvements of the present day are mere progressive steps in a highly wrought and highly advanced system, suggested by, and dependent on, other previous steps, their whole value and the means of their application probably dependent on the success of some or many other inventions, some old, some new.[186]

This was not the view of Fairbairn and the Manchester engineers who believed in a right to the ownership of intellectual property. Even in the citadel of free trade, they were not in favour of 'a fee trade in inventions'.[187] They pressed to retain legislation but with patents more easily obtainable, whilst appreciating that in return for a period of monopoly, a published patent allows others to build upon it. The reformers were largely successful and the Provisional Act was replaced by the Patent Law Amendment Act 1852.

Two years later the British Association set up a committee to review patent laws. Fairbairn presented its reports in 1858 and 1859.[188] There was a very public exchange in 1861 when the President of the British Association (Fairbairn) 'rebuked' the President of the Institution of Mechanical Engineers (Armstrong) 'for his unwarranted and groundless attack on the patent system'.[189] In 1864 Fairbairn was appointed to a Royal Commission on the Working of the Patent Law. It made relatively modest proposals, one of which Fairbairn dissented from – he believed that in some cases the time for which a patent was granted should extend beyond fourteen years.[190] In 1870 the Manchester Patent Law Reform Association, under Fairbairn's presidency, petitioned Parliament.[191] In 1871 there was another Select Committee.[192] This eventually led to the Patent Law Amendment Act 1883. The patent system had survived. The controversy had affirmed 'an individualistic understanding of invention'.[193] Fairbairn was one of the leaders of opinion who, along with Woodcroft and Brougham, had worked for this outcome. Nine years after his death, his views had prevailed by way of legislation.

The British Association for the Advancement of Science

IN both his boiler and patent campaigns, Fairbairn had utilised the British Association for the Advancement of Science. This was but one of the learned and professional societies in which he actively networked. In Manchester he was a member of the Literary and Philosophical Society and President 1855–60. Nationally he was a member of the engineering institutions and President of the Mechanicals 1854–5, and a Fellow of the Royal Society from 1850.[194] To all of them he read papers. Notable meetings of the Mechanical Engineers included that in Manchester in 1857 – to coincide with the Art Treasures Exhibition – when

Fairbairn unveiled the statue of his hero James Watt which he had first advocated in his proposals for Piccadilly in 1836.[195] Also in 1857 Longmans published the English translation of Arago's *Biographies of Distinguished Scientific Men* to which the publishers appended a 'Note by W Fairbairn'. This begins:

> In writing his historical eulogy of James Watt, the distinguished French philosopher has allowed his partiality for his own countrymen to overstep the boundaries within which an impartial writer should be restrained. To associate Dr Papin as a coadjutor of Watt in the discovery and invention of the steam-engine, is to give to the former a degree of prominence to which he is certainly not entitled.[196]

And at the Mechanical Engineers meeting at Newcastle in 1858 Fairbairn visited the scene of his apprenticeship at Percy Main.[197]

Fig.7.20: Watt, unveiled by Fairbairn, IMechE meeting, Manchester 1857.

The organisation to which Fairbairn made the greatest contribution was the British Association for the Advancement of Science, which met for an annual congress each summer in a different location. The Association influenced by the Whig spirit of reform, 'fulfilled Bacon's dream of applying knowledge to the benefit of mankind'.[198] The relationship between Fairbairn and the Association was symbiotic.[199] He gave much to it, reading papers, serving on committees and undertaking research projects. From the Association he received research grants, but by far the primary benefit was the opportunity to mix with men of scientific renown, absorbing their approach to science, and ultimately filling the Association's highest office, with all the accompanying prestige. During his 'retirement' he presented papers or reports in all but one year up to 1872, and from 1853 to 1870 he was President or a Vice President of the Mechanical Sciences Section in all but two years. Here he was amongst many of Britain's leading engineers.

Manchester had welcomed Fairbairn and he had become a through and through

Fig.7.21: J. Penn, R. Napier, J. Whitworth and W. Fairbairn, IMechE meeting, Manchester 1866.

(left)
Fig.7.22: Sir Peter Fairbairn (1799–1861).

(right)
Fig.7.23: John F. La Trobe Bateman (1810–89).

Mancunian. It was he who was the driving force in wooing the British Association to Manchester in 1861, only achieved after seven consecutive invitations.[200] For the first time, and only two years after Prince Albert had occupied the position, the President was an engineer without benefit of university education or aristocratic connections, or as *Chambers's Journal* put it:

> The selection of one of the *illustrissimi* of the town, Mr William Fairbairn, Mechanical Engineer, as President, was an interesting circumstance ... all felt how right and fitting it was that so admirable a specimen of the self-made men of industry should be put into the place of honour on such an occasion.[201]

His Presidential Address stressed the integration of science and technology – a theme on which he had spoken to the Manchester Mechanics' Institution a decade earlier: 'If the British Association had effected nothing more than the removal of the anomalous separation of theory and practice, it would have gained imperishable renown in the benefit thus conferred'. He referred to

> the present epoch as one of the most important in the history of the world. At no former period did science contribute so much to the uses of life and the wants of society. And in doing this it has only been fulfilling that mission which Bacon, the great father of modern science, appointed for it, when he wrote that 'the legitimate goal of the sciences is the endowment of human life with new inventions and riches.

Fairbairn may not have been a 'man of science', and may have expressed scepticism about some theoretical approaches, but he espoused and admired science and its application. It was 'engineering science' that had 'pre-eminently advanced the power, the wealth, and the comforts of mankind'. It was 'the application of science to the useful arts' which had resulted in, and would continue to result in 'the improvement of the condition of society'. He took his listeners on a tour of progress in science and engineering with accolades for James Watt, his friend George Stephenson and for Manchester engineers Richard Roberts and Joseph Whitworth. There was praise of his brother, Peter Fairbairn, with reference to flax machinery; and to his son-in-law J.F. Bateman with reference to Glasgow's water supply. He referred to the Britannia Bridge – but no mention of Robert Stephenson![202]

With over 3,000 attending, the Manchester meeting was a success.[203] Surprisingly one of the attendees was Karl Marx.[204] At the Polygon the Fairbairns entertained five former Presidents of the Association, Sir David Brewster, Revd Vernon Harcourt, Revd Dr T.R. Robinson, Major General Sabine and Lord Wrottesley.[205] His Vice Presidents included the cream of Manchester's science, engineering and commercial life – Joule, Whitworth, Sir Benjamin Heywood, Thomas Bazley MP, and would have included Hodgkinson had he not recently died.[206] In recognition Fairbairn was offered a knighthood, which, surprisingly for one with 'an appetite for distinction', was declined.[207] He did however accept a baronetcy eight years later. In the meantime he had unveiled city's grandest public monument, its memorial to Prince Albert – Matthew Noble's statue of the Prince beneath Thomas Worthington's magnificent Italian Gothic canopy or ciborium. Having not been successful in obtaining Victoria or a senior member of the royal family to unveil it, Thomas Goadsby, who had donated the statue, would have been the appropriate person to do so but he had died the previous year. Thus it fell to Fairbairn, 'one of the city's most senior public figures' to undertake the task.[208]

Fig.7.24: Prince Albert (1819–61) by Matthew Noble, beneath Thomas Worthington's magnificent ciborium, unveiled by William Fairbairn, Albert Square, Manchester, 23 January 1867.

Spreading Influence – the Written Word

FAIRBAIRN'S knowledge and experience was regularly diffused through papers to the British Association and the engineering institutions. His books and other publications were also a major channel for the diffusion of useful and reliable knowledge in the third quarter of the century, covering the fields of iron, the application of iron to construction, and the design of mills, millwork, prime movers and shipbuilding. He wrote ten books and was a significant contributor to at least five others, in addition to his many papers and reports.[209] To his writings he brought decades of practical engineering experience as he shared and showcased his knowledge and achievements. His writing was done late into the night in his library. He could not, he noted, 'suppress the desire I always had of giving to the world such information as I had collected in the varied forms and pursuits of my profession', and he was, he added, 'pleased to see myself in print'.[210] His books were generally well-received, widely read, and influential, at the same time raising the profile of their author.

Fairbairn's two most influential books were those related to construction. *On the Application of Cast and Wrought Iron to Building Purposes* (1854) was well-received and ran to four editions, becoming a standard text for the British construction industry.[211] It was translated into French and Spanish,[212] published in America,[213]

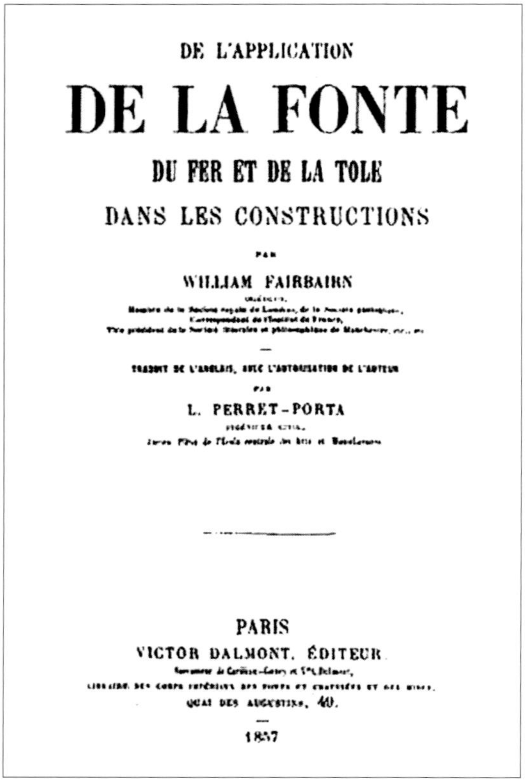
DE L'APPLICATION
DE LA FONTE
DU FER ET DE LA TOLE
DANS LES CONSTRUCTIONS
PAR
WILLIAM FAIRBAIRN
L. PERRET-PORTA
PARIS
VICTOR DALMONT, ÉDITEUR
QUAI DES AUGUSTINS, 49.
1857

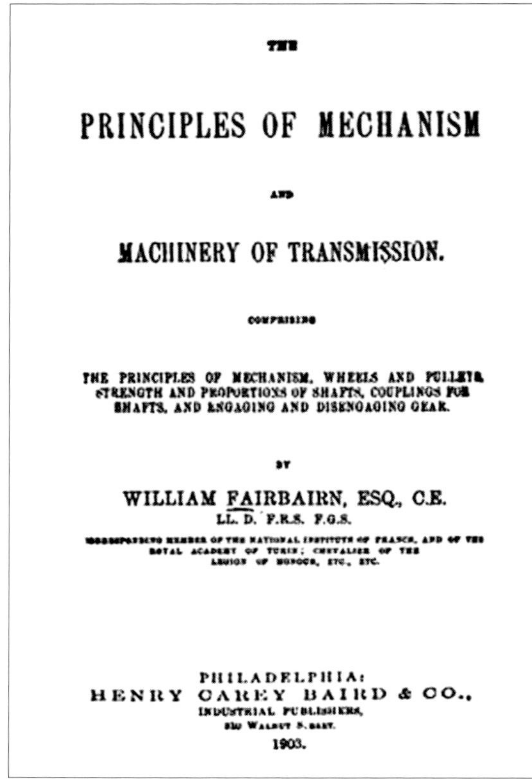
THE
PRINCIPLES OF MECHANISM
AND
MACHINERY OF TRANSMISSION.
COMPRISING
THE PRINCIPLES OF MECHANISM, WHEELS AND PULLEYS, STRENGTH AND PROPORTIONS OF SHAFTS, COUPLINGS FOR SHAFTS, AND ENGAGING AND DISENGAGING GEAR.
BY
WILLIAM FAIRBAIRN, ESQ., C.E.
LL. D. F.R.S. F.G.S.
PHILADELPHIA:
HENRY CAREY BAIRD & CO.,
INDUSTRIAL PUBLISHERS,
1903.

(above) **Fig.7.25: French Edition of *On the Application of Cast and Wrought Iron to Building Purposes*, Paris, 1857.**

(above right) **Fig.7.26: Final American printing of *Principles of Mechanism*, Philadelphia, 1903.**

and available in Australia.[214] His *Treatise on Mills and Millwork* was originally published in two Parts (1861 and 1863). Part I included water and steam power; Part II power transmission and descriptions of various types of mills. *The Mechanics' Magazine,* unenthusiastic about some sections of Part I – 'even a third- or fourth-rate man ... would have produced a much better history of mills' – nevertheless concluded that for engineers 'to neglect it is to fall behind the age'.[215] Part II fared better, being 'fully equal to anything Mr. Fairbairn has ever written'.[216] The books ran to four editions – the fourth, published in 1878, containing both Parts in one volume. There was a pirated American edition, with extracts from both Parts, published in 1867 in Philadelphia and reprinted at least five times, up to as late as 1903.[217] *Mills and Millwork* superseded Robertson Buchanan's earlier standard work on the subject and its continuing demand for over two decades was evidence of its wide recognition as a standard text.[218] Mokyr notes that 'Rankine did more than anyone in his time to bridge the gap between science and engineering by writing four textbooks that made the findings of the new science available to engineers'.[219] In the Preface to *A Manual of Machinery and Millwork,* (1869), which ran to six editions, Rankine included Fairbairn's *Mills and Millwork* in a list of four books that 'are so frequently referred to, that it may be desirable to mention them here'.[220] W.H.G. Armytage, choosing four books from around the mid 19th century 'which synthesized progress to date' in engineering included *Mills and Millwork* of which he wrote that it 'should be read by any student of the history of engineering'.[221]

Fairbairn's third most influential work was *Iron – Its History, Properties, and Processes of Manufacture,* an expansion of his article 'Iron' in the eighth edition of *Encyclopaedia Britannica.* It ran to three editions, each revised and enlarged, and was translated into French.[222]

Trouble-shooting

THE competition for the design of Bolton's new market hall was won in 1853 by George Robinson. In May 1854 the columns were in place and the roof in course of erection when the undercroft, described in Pevsner as 'highly impressive' started to move.[223] Fairbairn was called in. Robinson's undercroft had shallow-groined vaults carried on square stone piers. The movement was caused by unequal loading resulting from the wider spans over the roadways. Fairbairn inserted sixty brick arches springing from near the floor.[224] In places Robinson's square stone columns can still be seen, with Fairbairn's brick arches keyed into them.[225]

In 1858 the construction of the roof of Deane and Woodward's Oxford Natural History Museum was proceeding when a collapse occurred. Fairbairn was sent for. It turned out that the structure had been prepared for a lighter roof and that two years earlier Woodward had changed the specification of the glazing to a much heavier glass. Fairbairn put forward proposals, advising that if they were followed, 'they would ensure ample security and permanency to the structure'.[226]

Not all Fairbairn's interventions were successful. Called in a decade earlier to a recently completed 400ft chimney in Wigan, which was showing signs of movement, he recommended that it be lowered. Work on this was put in hand and about 80ft had been taken off the top, but the deviation from the perpendicular continued to increase and a series of stays were in course of construction to support the base. But all was without avail. The chimney fell.[227]

Public Health

In 1833 J.P. Kay delivered a lecture to the Manchester Statistical Society, 'Plans and Estimates for Public Swimming Baths for the use of the Operative Population'. Thereafter Manchester dragged its feet. The first establishments were a small one in Miller Street, Ancoats, which did not cater for swimmers, nor meet the demand for laundry facilities, and Miles Platting Baths, financed by Benjamin Heywood.[228] A public meeting was called in December 1854, following a paper by E.T. Bellhouse to the Statistical Society,[229] 'of all friendly to the formation of a joint-stock company for the construction of baths and laundries'. The concept of a philanthropic company dedicated to the advancement of social projects for a limited return on investment was not unusual at that time.[230] Fairbairn attended the meeting, was a member of the provisional committee (which also included his friend Joseph Whitworth) and became a major shareholder of the Manchester and Salford Baths and Laundries Company.[231] Thomas Worthington was appointed architect and the pioneer project was Greengate Baths, Salford, comprising two swimming pools, the largest then built, a laundry and public wash-house. Opened in August 1856, it was an instant success. Fairbairn told the first annual meeting that Greengate Baths were 'likely to be an example to other cities'. The London establishments he had visited were not to be compared with the Manchester Company's first venture.[232] Whilst, unlike boiler safety and patent reform,

Fairbairn was not in the driving seat for baths and wash-houses, his involvement provides further evidence of his social concern.

Evidence of Fairbairn's environmental concerns (in addition to smoke abatement already referred to)[233] was his call for the cleaning of polluted streams and rivers. This he believed could be achieved by the use of settling tanks. It would be a great relief to many towns, and the residents on the banks of many rivers, to have the waters pure and transparent instead of a large black ditch coloured with indigo, or coal-tar, and almost every other abomination floating on its surface.[234]

But his greatest contribution to public health and safety was unquestionably in the reduction of boiler explosions.[235]

Looking Ahead – Steel and Aeronautics

Fascinated as he was with engineering history, on which he wrote and lectured,[236] Fairbairn's experimental engineering in 'retirement' looked forward. Besides his experiments assisting the building of safer and more powerful boilers, on metal fatigue relevant to the safety of bridges, on submarine cables and armour-plating, there was work in two newer fields of engineering – steel and aeronautics. If the world was waiting for one key invention in the mid century, it was a way of mass manufacturing malleable steel, and famously Henry Bessemer found it. Fairbairn was at the British Association meeting in 1856, when Bessemer announced, to considerable scepticism, his new system for the manufacture of malleable steel.[237] Both Bessemer and Fairbairn had been searching for this – Fairbairn by the fusion of malleable scrap with the pig iron, in a cupola furnace, but this had failed. Bessemer was aware that this was so and wrote, 'In my experiments I avoided the difficulties inseparable from Fairbairn's method, by employing a reverberatory furnace'.[238] Bessemer was by no means the first to make steel, but prior to his process its hardness and brittleness rendered it 'unfit to take the place of [wrought] iron for general purposes'.[239] Fairbairn celebrated Bessemer's achievement. In September 1858 he was the recipient of 'a first sample order' of Bessemer steel boiler-plates, and in 1861 he devoted a whole chapter of the first edition of *Iron* to 'Bessemer's Process'.[240] In 1865 Bessemer estimated steel production would be 6,000 tons per week, up from 400 tons per week fifteen years earlier.[241] Steel was *the* new material, and engineers were crying out for reliable knowledge about it.[242]

The British Association's 1866 Committee appointed Fairbairn and Tate 'to test the improvements in the manufacture of iron and steel'.[243] The following year Fairbairn presented the results of 1,915 experiments on 52 steel samples, from nine Sheffield manufacturers, including Bessemer. These tabulated the transverse, tensile and compressive strains of each sample. Tate produced the formulae in the paper and the results were widely disseminated.[244] Fairbairn and Tate were reappointed and at the 1869 BAAS meeting Fairbairn reported on another 45 samples, plus tests on a steel girder.[245] There was little that was novel about these experiments, apart from the material. They followed a similar pattern to earlier experiments on cast iron and wrought iron. But, with the flow of steel increasing and replacing wrought iron, Fairbairn and Tate served the engineering industry by making useful and reliable knowledge of its properties available, thus further accelerating steel's widespread adoption. Their experiments with steel undermine the thesis that technology did not advance by the application of knowledge derived from science by way of systematic experimental investigation. They also show Fairbairn as a transitional figure, not just from cast iron to wrought iron, but from cast-iron to the advent of steel, via wrought-iron of which he was an acknowledged master.[246]

Fairbairn remained at the forefront of the transport revolution. In 1862 he was appointed to the British Association's Balloon Committee. Grants enabled a programme of scientific ascents on a scale never previously attempted.[247] But it is not for his part in this, that Fairbairn is remembered in the context of aeronautics. The Aeronautical Society of Great Britain was founded in 1866 and Fairbairn joined its Council later that year. The first Aeronautical Exhibition was held in 1868, when the Jurors' Report stated that 'with respect to the abstract question of mechanical flight … we are still ignorant of the rudimentary principles which should form the basis and rules of construction'.[248] The minutes of early meetings of the Society show that Fairbairn urged experiments to provide data of the reactions and forces on surfaces in atmospheric currents.[249] He did not think it would be difficult to obtain this data.[250] An Experimental Committee was formed and by the end of 1870 the first wind tunnel in the world had been built at Penn's works at Greenwich.[251] It was designed by another of its advocates, Francis Wenham.[252] The Society's 1870 Report referred to the inauguration of systematic experiments on the connection between the pressure and velocity of air, which, it was believed, was the only data 'on which a true science of aeronautics can be founded … and Sir William Fairbairn, FRS, will afford all the aid in his power' to these tests.[253] The following year Fairbairn's age and health forced his retirement from the project.[254]

Those Less Fortunate

IN 1856 *The Times* referred to Henry Cort (1741–1800) as 'the father of the iron trade'.[255] In the same year in *Encyclopædia Britannica,* Fairbairn had written:

> It would be difficult to enumerate all the services rendered by Mr Cort to the industry of this country, or sufficiently to express our sympathy with the descendants of a man to whose mechanical inventions we owe so much of our national greatness.

Cort had been ruined in the carrying out of his inventions, and died a ruined man. In 1859 Fairbairn wrote to *The Times* and got up a memorial to the Government signed by over 130 iron manufacturers. This resulted in pensions to Cort's surviving children. In addition he organised a not very successful private appeal which raised a little over £500, of which Fairbairn contributed £100.[256]

Richard Roberts (1789–1864) arrived in Manchester from Maudslays in 1816 and before long met Fairbairn with whom he was one of the founders of the Manchester Mechanics' Institution. Roberts has been described as 'a universal mechanical genius'.[257] From the mid 1820s until 1843 he was a partner in Sharp, Roberts & Co., and from 1843 in Roberts, Fothergill & Dobinson but this was not profitable and was wound up in 1852. Thereafter he practised as a consulting engineer, moving to London in 1860.[258] In March of that year Roberts wrote to *The Engineer* maintaining that it was he, and not Fairbairn, that was due the credit for recognising, as far back as 1848 and in his patent of May 1852, 'the necessity of assimilating the construction of a ship to that of a beam or girder of equal length' and recognising the advantages of 'adopting the cellular system (as applied in the construction of the Conway Tubular Bridge) to the strengthening of ships'.[259] Fairbairn did not respond – it seems both men were pursuing similar ideas. On the balance of probabilities Roberts' claims are unfounded taking into account the example of the grounding of the *Prince of Wales* prior to the 1845 tubular bridge experiments,[260] the references to cellular construction in Fairbairn's Royal Society paper in 1850,[261] and his references to both considering a vessel as 'a huge hollow beam or girder' and to cellular construction in his lecture in Manchester in April 1852.[262] C.F. Beyer visited Roberts in London in 1861 and found him financially distressed. Fairbairn in his capacity as President of the British Association wrote, unsuccessfully, to Palmerston seeking a civil list pension for him. He then 'agitated for a private subscription to be set up' and a Committee was formed: but it all took too long and Roberts died in poverty in March 1864. At a meeting at the Society of Arts the following month Fairbairn proposed a motion which was passed unanimously...

> This meeting is of the opinion that the eminent services which Mr Richard Roberts has rendered to the manufacturing industry of this Country, and to the World at large, by his useful inventions, demand some substantial and permanent record; and that the most appropriate tribute to his memory would be, in the first instance, to provide for the independence of his only daughter.[263]

He reported that the Manchester Committee had raised £1,500 with a further £1,000 promised. The total raised is not known. However, in 1866 Eliza Mary Roberts was granted a civil list pension.[264]

The End of a Full Life

In October 1873, Fairbairn, in his capacity as a Governor, played host to the Duke of Devonshire at the opening of Owens College – the future Manchester University. It was a cold day and Fairbairn caught bronchitis from which he never recovered. In the following June he went to Brighton 'to try to get relief by a change of air', but to no avail. In July he visited his son William Andrew in London, from where he wrote his last letter, to Major Maxwell, the fiancé of his granddaughter Maggie Bateman

> to express Lady Fairbairn's and my own deep regret that the extremely infirm state of my health prevents us being present [at the wedding] ...You will pardon me for adding more [sic], as I am scarcely able to read or write, and can only add our united blessing, and prayer for your future happiness.[265]

In July, soon after the marriage, he travelled to Moor Park, Farnham to be with his daughter, Anne, and son-in-law, J.F. Bateman, to whom he was closer than to any of his sons. There, on 18 August 1874, William Fairbairn died, aged 85.[266] The funeral was in Manchester. The cortege from the Polygon passed through Piccadilly and down Market Street to Prestwich Parish Church, where the tomb remains.[267] That tens of thousands of Mancunians turned out for the funeral is witness to his standing in the city he had made his home for sixty years.[268] His national and international repute and influence were attested to by the many obituaries, including portraits in *The Illustrated London News, Allgemeine Deutsche Polytechnische Zeitung,* and on the title page of *Scientific American,* which described him as 'the author of experiments and works which have changed the whole practice of iron construction: to few men do the engineering profession and the great metal industries of the world owe so large a debt'.[269]

Fig.7.27: Moor Park, where Fairbairn died.

William and Dorothy had been married for 58 years, and the Polygon had been their home for 35 of them. His personal estate was 'sworn under £120,000', equivalent to around £7.5M in 2015. This figure may be misleading as much of his wealth would previously have been in the firm, which he had passed to his sons. His will made provision for Dorothy: £500, use of the Polygon and the interest from £25,000 for life. He left £18,000 to the surviving sons, William Andrew and Adam, £16,000 to his daughter Anne, and £6,000 to his grandson Edward Cleather Fairbairn (whose father, Peter, had died whilst Edward was an infant) with the residue of the estate to his eldest son Thomas.[270] The family engaged William Pole to write a biography.[271] The Fairbairn Memorial Committee commissioned a statue by the relatively unknown E.E. Geflowski.[272]

Conclusion

ALL the indications are that Fairbairn thoroughly enjoyed his retirement, meeting engineering challenges, experimenting and writing, without the day-to-day pressures of contracting and management. The chronology of this phase of his career, and in particular that Fairbairn was active in retirement for twenty years, has not been adequately appreciated. Thus Musson implies that a factor in the demise of the firm was that 'in later life' Fairbairn 'was perhaps too much concerned with being a public figure and establishing his status as a scientist and writer', apparently not appreciating that in these later years he had no significant role in the management of the firm.[273]

Fig.7.28: The Fairbairn Grave, Prestwich.

Fairbairn's achievements in retirement were more than most people achieve in a lifetime and his influence was never greater than during these two decades.

His advice was sought, his publications were read, his talks were listened to, and his peers heaped honours upon him. The breadth of his interests was wide, encompassing buildings, bridges, boilers, cables, steam, ordnance, steel, patents, public health and education, yet all linked together by the common thread of engineering. His research provided useful and reliable knowledge for boiler-makers, steam-engine builders, and bridge builders; and there were important contributions to submarine cables and the properties of Bessemer steels. Fairbairn occupies 'a unique place' as a 19th-century experimental engineer.[274] His scheme for boiler inspections was widely adopted, and his approach was incorporated into legislation after his death. Similarly his approach to patent law influenced the Patent Law Amendment Act of 1883. Having ascended to the height of his profession by the time of his 'retirement', Fairbairn remained there for the rest of his life, as he remained within walking distance of his Ancoats works. Effusive worldwide obituaries and the crowds of Mancunians witnessing his funeral were evidence of the renown in which he and his diverse engineering achievements were held.

Notes for Chapter 7

1. D C Mowery and N Rosenberg, *Technology and the Pursuit of Economic Growth,* (1989), p.22.
2. See Chapter 1.
3. L G Booth, 'Laminated Timber Arch Railway Bridges in England and Scotland', *TNS,* 44, 1971, 11-2; *CE&AJ,* 11, 1848, 35-6 and Plate 4. For other illustrations of the original bridges see: Dinting: N W Webster, *Joseph Locke: Railway Revolutionary,* (1970), Plate 5; Mottram (or Etherow): *ILN,* 2, 1843, 52.
4. *CE&AJ,* 18, 1855, 112.
5. J Rapley, *The Britannia and other Tubular Bridges,* (2003), p.123; W Fairbairn, 'On the Reconstruction of the Dinting and Mottram Viaducts', *MPICE,* 22, 1862-3, 237-5.
6. C B Vignoles diary, 10 August 1860. I am grateful to John Vignoles for this reference (e-mail 22 May 2015). See also K H Vignoles, *Charles Blacker Vignoles: Romantic Engineer,* (1982), pp.90-1.
7. A C G Hayward, 'The Construction of Railway Bridges Then and Now', *The International Journal for the History of Engineering and Technology,* 84.1, 2014, 78-9.
8. O Ashmore, *The Industrial Archaeology of North-west England,* (1982), pp.85-6.
9. W Fairbairn, 'Description of the Removing and Replacing of the Iron Columns in a Cotton Mill', *MPIME,* 17, 1866, 181-5 + figs.; *CE&AJ,* 30, 1867, 127-8 and Plate 2; *BAAS1866,* pp.141-3. See also I Miller, C Wild and S Little, 'The Development of Ancoats: Archaeological Case Studies' in I Miller and C Wild, *A & G Murray and the Cotton Mills of Ancoats,* (2007), pp.53-4, where a less successful modification to the Sedgwick West Wing is also discussed. There was correspondence between McConnel and Fairbairn about this (McConnel & Kennedy Archive, John Rylands Library, MCK2/2/23 3 Sept. 1868, 17 Nov 1868, 30 Jan 1869, 22 Feb 1869, 29 Feb 1869). However it is unlikely that this was Fairbairn's work as his sketch of 3 Sept 1868 was not followed, the work was not put in hand until well into 1869, at the earliest, by which time Fairbairn would have been over eighty, and the detailing lacks his normal robustness. S Little, to whom I am grateful for source references, writes elsewhere that these alterations were 'probably by Fairbairn but there is no certainty on the issue' (S Little, 'The Mills of McConnel and Kennedy, Fine Cotton Spinners', *Transactions of the Lancashire and Cheshire Antiquarian Society,* 104, 2008, 50).
10. T A Lockett, *Three Lives,* (1968), pp.24-45; A Darbyshire, *An Architect's Experiences,* (1897), pp.65-8.
11. Nathaniel Beardmore's concrete floor of 1848 had used built-up wrought-iron joists (T Potter, *Concrete: Its Uses in Building from Foundations to Finish,* (3rd ed. 1908p.176). There had been buildings with wrought-iron beams, of wider spans, in America since the mid-1850s - beams of box section 'constructed on ... the principles developed in the thorough experiments of Messrs Stephenson, Hodgkinson and Fairbairn' (Cooper and Hewitt to unknown recipient, 1 September 1854, reproduced in C E Peterson, 'Inventing the I-beam: Richard Turner, Cooper & Hewitt and Others', *Bulletin of the Association of Preservation Technology,* 12.4, 1980, 20-1).
12. S Giedion, *Space, Time and Architecture: the growth of a new tradition,* (3rd ed. 1953), p.192. The incorrect date is also found in R Sheppard, *Cast Iron in Building,* (1945), pp.55-60, K Hudson, *Building Materials,* (1972), p.55, and P Guedes, *The Macmillan Encyclopaedia of Architecture and Technological Change,* (1979), p.102. There was a possible precursor in a system using cast iron plates, patented by Samuel Wyatt in 1800 and, as far as is known, used for the first and only time by his brother, James Wyatt, at George III's new palace at Kew in 1801. (B Cherry and N Pevsner, *Buildings of England. London: South,* 1983 ed.), p.507).
13. Fairbairn, *Application,* pp.153-8. Richard Sheppard incorrectly states that this was 'common building practice' (Sheppard, *Cast Iron in Building,* pp.57-61).
14. Potter, *Concrete,* p.177, Fig.40; Patent No. 12,260, 4 September 1848. George Nasmyth was the elder brother of James Nasmyth.
15. W Fairbairn and D Bellhouse, 'Report on the Causes of the Fall of the Cotton Mill at Oldham in October, 1844' in Fairbairn, *Application,* pp.274-9; *MM,* 43, 1845, 221. See Chapter 3.
16. R N Holden, 'Concrete filler-joist floors and the development of Lancashire cotton spinning mills', *IAR,* 34.2, 2012, 116; Fairbairn, *Application,* (3rd ed.), p.170.
17. Peterson, 'Inventing the I-beam' 15-24; *The Builder,* 12, 1855, 455.
18. Walker, *Unwin,* pp.30-51.
19. The indications are that it was William Fairbairn who encouraged Unwin to return to Ancoats, and it seems likely that he left because he was unable to work with Thomas Fairbairn who by this time was chairman of the limited company. However, in the absence of hard evidence, this can be no more than speculation.
20. Walker, *Unwin,* pp.52, 55.
21. Walker, *Unwin,* pp. 82, 104, 126, 141-2, 150-2, 159-62.
22. Walker, *Unwin,* p.65.
23. Fairbairn, *M&MWI,* p.xi.
24. Walker, *Unwin,* p.60; W Fairbairn to W C Unwin, 21 November 1872, Imperial College Archives, W C Unwin/107.
25. J S Wilson, 'Forward', *The Central,* January 1934, quoted in Walker, *Unwin,* p.210.
26. Walker, *Unwin,* p.178.
27. W C Unwin, 'On the Resistance of Boiler Flues to Collapse', *MPICE,* 46, 1876, 225-241; W Fairbairn, 'On the Resistance of Flues to Collapse', *Philosophical Transactions of the Royal Society,* 1858, 389-413, and 2 Plates.
28. W C Unwin (reporter), 'First Report to the Council of the Institution of Mechanical Engineers on the Form of Riveted Joints', *MPIME,* 1881, 301-368.
29. W C Unwin, 'Strength of Alloys at Different Temperatures', *BAAS1889,* 746-7. Cf. W Fairbairn, 'On the Strength and other Properties of Cast Iron obtained from the Hot and Cold Blast', *BAAS1837,* 396-406; W Fairbairn, 'Appendix to a report on the "Strength and other Properties of Cast Iron obtained from the Hot and Cold Blast"', *BAAS1842,* 88-92; W Fairbairn, 'On the Tensile Strength of Wrought Iron at Various Temperatures', *BAAS1856,* 405-22.
30. W C Unwin, 'Tensile Tests of Mild Steel; and the Relation of Elongation to the Size of the Test-Bar', *MPICE,* 155, 1904, 170-233.
31. W C Unwin, *The Testing of Materials of Construction,* (1888, 2nd ed.1899, 3rd ed. 1910); A I Smith, 'William Fairbairn and Mechanical Properties of Metals: Effect of Temperature on Strength', *The Engineer,* 216, 1963, 1017. Other publications by Unwin included *The Elements of Machine Design* (1877); 'Hydraulics' and 'Bridges' in *Encyclopaedia Britannica* (11th ed.); *On the Development and*

Transmission of Power from Central Stations, (1893 Howard lectures to the Society of Arts).
32. A Chatterton, in 'Introduction' to Walker, *Unwin,* p.xiii.
33. W Unwin, *The Elements of Machine Design, Part I, General Principles, Fastenings, and Transmissive Machinery,* (5th ed. 1890), p.vii.
34. Rosenberg and Vincenti, *Britannia,* pp.71-2.
35. W Fairbairn, *Remarks on Canal Navigation illustrative of the Advantages of the use of Steam, as a Moving Power on Canals,* (1831).
36. See Chapter 5.
37. W Fairbairn, 'On the Consumption of Fuel and the Prevention of Smoke', *BAAS1854,* p.s149; Fairbairn, *UIfE,* pp.48-92.
38. W Fairbairn, 'An Experimental enquiry into the Relative Powers of the Locomotive Engine, and the Resistance of Railway Gradients', *Memoirs of the Manchester Literary and Philosophical Society,* 14, 1850, 149-78; W Fairbairn, 'On Railway Gradients', *BAAS1845,* pp.93-4.
39. *Life,* pp.345-60, 488-9; W Fairbairn, 'On the Mechanical Properties of Specimens of the Iron and Steel Plates which had been subjected to experiment with Ordnance at Shoeburyness', *Transactions and Report of the Special Committee on Iron, between Jan. 21st 1861, and March 1862,* in P Barlow, *A Treatise on the Strength of Materials,* (6th ed. 1867), pp.215-32; W Fairbairn, 'On Guns and Iron-plated Ships', *BAAS1862,* pp.223-4; W Fairbairn, 'On the Properties of Iron and its resistance to Projectiles at High Velocities', *Proceedings of the Royal Institution,* 4, 1862, 491-502; *CE&AJ,* 25, 1862, 238-42; W Fairbairn, 'On the Law of Resistance of Armour Plates composed of one or more thicknesses', (in two parts) : Part 1, Reprint from *Transactions of the Institution of Naval Architects,* 1869; Part 2, *Transactions of the Institution of Naval Architects,* 11, 1870, 103ff; *Engineering,* 9, 1870, 285-6 and 307-8. For illustrations see *ILN,* 20 August 1864.
40. *MG,* 12 March, 19 March, 26 March 1853; *The Observer,* 14 March 1853; *MM,* 59, 1853, 284-6.
41. W Fairbairn, 'Experimental Researches to determine the Strength of Locomotive Boilers, and the Causes which lead to Explosion', *BAAS1853,* pp.52-7; Fairbairn, *UIfE,* pp.lix-lxxiv.
42. See below.
43. W Fairbairn, 'Experiments to determine the Efficiency of Continuous and Self-Acting Breaks (sic) for Railway Trains', *BAAS1859,* p.s76; *The Engineer,* 9, Jan-June.1860, 2-3; *Proceedings of the Manchester Literary and Philosophical Society,* 1859-60, 178-80; W Fairbairn, 'On the Efficiency of various kinds of Railway Breaks (sic), with Experimental Researches on their retarding Powers', *MPICE,* 19, 1860, 490-526, and 214, 1922, 235-57; *The Engineer,* 12, July-Dec.1861, 207; Walker, *Unwin,* pp.38-41.
44. See below.
45. J Morrell and A Thackray, *Gentlemen of Science: Early Years of the British Association for the Advancement of Science,* (1981), pp.271
46. See Chapter 4.
47. See below.
48. See Chapter 5.
49. J Mokyr, *The Enlightened Economy: Britain and the Industrial Revolution 1700-1850,* (2009), pp.1, 40, 42-3; R C Allen, *The British Industrial Revolution in Global Perspective,* (2009),p.241.
50. L Goldman, 'The Origins of British 'Social Science': Political Economy, Natural Science and Statistics, 1830-1835', *The Historical Journal,* 26.3, 1983, 602; K Olesko, 'Humboldt, Alexander von' and 'Humboldtian Science' in J L Heilbron (ed.), *The Oxford Companion to the History of Modern Science,* (2003), pp.383-7.
51. See Chapter 4.
52. S P Timoshenko, *History of Strength of Materials,* (1953), p.125.
53. A I Smith, 'William Fairbairn and Mechanical Properties of Metals: Effect of Temperature on Strength', *The Engineer,* 216, 1963, 1017.
54. M S Loveday, T Gray and J Aegerter, 'Tensile Testing of Metallic Materials: A Review', (TENSTAND –Work Package 1- Final Report, 2004), pp.55-6.
55. Unwin, *Testing of Materials,* pp.109-132.
56. *Life,* p.157.
57. [Woodcroft], X; *Life,* p.101.
58. R A Smith, *A Centenary of Science in Manchester,* (1883), pp.257-8.
59. *Life,* pp.211, 270,273-4, 360, 421; Walker, *Unwin,* pp.34-51; G Howson, 'Tate, Thomas (1807-1888)' *Oxford DNB on-line* (2004), (accessed 26 October 2013).
60. *Life,* pp.385, 371-2, 384, 241-51, 392-4. It is said that in the election to the Académie Fairbairn received 37 votes, Robert Stephenson three votes and Professor Willis one vote (W Fairbairn to T R Robinson, 17 May 1852, Cambridge University Library, Stokes Papers, TR16).
61. *Life,* p.385; Smith, *Contribution,* p.7.
62. E Layton, 'Conditions of technological development', in I Spiegei-Rösing and D de Solla Price (eds.), *Science, Technology and Society: A Cross-Disciplinary Perspective,* (1977), p.210, quoted in T J Pinch and W E Bijker, 'The Social Construction of Facts and Artifacts: Or How the Sociology of Science and the Sociology of Technology Might Benefit Each Other', in W E Bijker. T P Hughes and T Pinch (eds.), *The Social Construction of Technological Systems: New Directions in the Sociology and History of Technology,* (2nd ed. 2012), p.14.
63. Fairbairn, *UIfE,* p.114.
64. W Fairbairn, 'Experiments to determine the effect of Impact, Vibratory Action, and long-continued Changes of Load on Wrought-Iron Girders', *Philosophical Transactions of the Royal Society,* 1864, 311-25, and 2 Plates.
65. Fairbairn, *Application,* pp.246-55; *Engineering,* 4, July-Sept.1867, 321, 323; Walker, *Unwin,* p.43.
66. D Ross, *The Highland Railway,* (2005), p.31; J Mitchell, *Reminiscences of my Life in the Highlands,* (1884), p.186; A I Smith, 'William Fairbairn and Mechanical Properties of Materials: The Effect of Repeated Loading on Strength', *The Engineer,* 26 June 1964, 1133-6.
67. Walker, *Unwin,* p.211. This waterwheel is more likely to have been in a basement at the Ancoats works rather than at the *Polygon,* as suggested by Walker.
68. L Tóth and S Y Yarema, 'Formation of the Science of Fatigue of Metals, Part I. 1825-1870', *Materials Science,* 42.4, 2006, 673.
69. F Braithwaite, 'On the Fatigue and consequent Fracture of Metals', *MPICE,* 13, 1854, 463-74.
70. Smith, 'Effect of Repeated Loading', 1133.
71. *Report of the Commissioners appointed to inquire into the Application of Iron to Railway Structures,* (1849), pp.x-xi, 4, 16-19, 215-49.
72. *Report of the Commissioners,* pp.324, 403-7, 410-1. On the tests at Blackburn see Chapter 4.
73. Sendeckiyj appears to be unaware of Fairbairn's evidence and communications regarding fatigue in the

appendices to the *Report of the Commissioners* (G Sendeckiyj, 'Early Railroad Accidents and the Origins of research on fatigue of metals', Appendix A in T Nicholas, *High Cycle Fatigue: A Mechanics of Materials Perspective,* (2006), pp.485-8).
74. Fairbairn, 'Experiments to determine the effect of Impact', 322; *BAAS1860,* pp.45-8; *BAAS1861,* pp.285-9.
75. Fairbairn, 'Experiments to determine the effect of Impact', 313-5.
76. W Schütz, 'A History of Fatigue', *Engineering Fracture Mechanics,* 54.2, 1996, 263-300; R A Smith, 'The Versailles Railway Accident of 1842 and the Beginnings of The Metal Fatigue Problem', in H Kitagawa and T Tanaka (eds.), *Proceedings Fatigue 90: The Fourth International Conference on Fatigue and Fatigue Thresholds,* (1990), Vol.4, pp.2033-41.
77. B Baker, *Long-span Railway Bridges,* (Rev. Ed. 1873), pp.105-11; Smith, 'Effect of Repeated Loading', 1136.
78. H J Gough, *The Fatigue of Metals,* (1924), pp.3-4.
79. B Åkesson, *Fatigue Life of Riveted Steel Bridges,* (2010), pp.7-9.
80. H S Reemsnyder, 'Fatigue Life Extension of Riveted Connections', *Journal of the Structural Division, ASCE,* 101.12, 1975, 2591-2608.
81. *BAAS1860,* pp.45-8; *BAAS1861,* pp.286-9; Fairbairn, 'Experiments to determine the effect of Impact', 311-25, and 2 Plates; Smith, 'Effect of Repeated Loading', 1133.
82. Fairbairn, *Application,* p.55.
83. Smith, 'Effect of Repeated Loading', 1136; H J Gough and D Hanson, 'The Behaviour of Metals Subjected to Repeated Stresses', *Proceedings of the Royal Society A,* 104, 1923, 538-65; E Orowan, 'Theory of the Fatigue of Metals', *Proceedings of the Royal Society A,* 171, 1939, 79-106.
84. L Goldberg, 'Concepts of Depreciation', *The Accounting Review,* 30.3, 1955, 473.
85. W H Chaloner, *Vulcan: The History of One Hundred years of Engineering and Insurance 1859-1959,* (1959), p.8.
86. Unwin, 'Resistance of Boiler Flues', 225
87. W Fairbairn, 'On the Resistance of Tubes to Collapse', *BAAS1857,* pp.215-9; W Fairbairn, 'On the Resistance of Tubes to Collapse', *Philosophical Transactions of the Royal Society,* 1858, 389-413 and 2 Plates; Fairbairn, *UIfE2,* pp.1-45; *MM,* 68, Jan.-June1858, 5-7, and 70, Dec.1858-June 1859, 8-9; *The Engineer,* 7, Jan.-June 1859, 2-4.
88. W Fairbairn and T Tate, 'On the Resistance of Glass Globes to Collapse from external pressure; and on the Tensile and Compressive Strength of various kinds of Glass', *Philosophical Transactions of the Royal Society,* 1859, 213-247; Fairbairn, *UIfE2,* pp.46-95.
89. F W Grashof, 'Fairbairn's Versuche über den Widerstand von Röhren gegen Zusammendrückung', *Zeitschrift des Vereines deutscher Ingenieure,* 3, 1859, 234-43; G H Love, *Mémoire sur la Loi de Résistance des Conduits Intérieurs a Fumée dans les Chaudières à Vapeur Dédeite des Expériences de M. W. Fairbairn,* (Paris, 1859); G H Love, 'Sur la Résistance des Conduits Intérieurs à Fumée dans les Chaudiéres à Vapeur', *Memoirs Compte Rendu des Traveaux, Societe des Ingenieurs Civils de France,* 12, 1859, 471-500; J W Nystrom, *A New Treatise on Steam Engineering,* (1876), pp.106-7; Unwin, 'Resistance of Boiler Flues', 225-6; T Belpaire, 'Note sur la Résistance des Tubes Pressées de l'Extérieur', *Annales du Genie Civil,* 1879, 177-87; H Wehage, 'Zur Berechnung der Flammröhre von Dampfkessela', *Dingler's Polytechnisches Journal,* 242, 1881, 236-43.
90. C R Roelker, 'Experimental Investigation of the Resistance of Flues to Collapse', *Van Nostrand's Magazine,* 34, 1881, 208-220.
91. C Bach, 'Die auf der kaiserlichen Werft in Danzig von 1887-1892 ausgefürten Versuche Über die Widerstandsfähiceit von Flammrohen', *Zeitschrift des Vereines deutscher Ingenieure,* 38.23,1894, 689-96; A P Carman, 'Resistance of Tubes to Collapse', *Physical Review,* 21, 1905, 381-7; R T Stewart, 'Collapsing Pressure of Bessemer Steel Lap-Welded Tubes, Three to Ten Inches in Diameter', *Transactions of the American Society of Mechanical Engineers,* 27, 1906, 730-822.
92. A P Carman and M L Carr, 'Resistance of Tubes to Collapse', *University of Illinois Engineering Experiment Station Bulletin,* 5, June 1906, 1-5.
93. J Singer, 'Experimental Studies in Shell Buckling', in N F Knight, Jr and N P Nemeth (eds.), *Stability Analysis of Plates and Shells: A Collection of Papers in Honor of Dr Manuel Stein,* (1998), p.21; J Singer, 'Stiffened Cylindrical Shells' in J G Teng and J M Rotter, *Buckling of Thin Metal Shells,* (2004), p.287; J Singer, J Arcocz and T Weller, *Buckling Experiments: Experimental Methods in Buckling of Thin-Walled Structures: Volume 2, Shells, Built-up Structures, Composites and Additional Topics,* (2002), p.624.
94. W Fairbairn and T Tate, 'Experimental Researches to determine the Density of Steam at different Temperatures and to determine the Law of Expansion of Super-heated Steam', *Philosophical Transactions of the Royal Society,* 1860, 185-222 and 3 Plates. A similar paper was read to BAAS at Oxford (W Fairbairn, 'On the Density of Saturated Steam and on the Law of Expansion of Superheated Steam', *BAAS1860,* p.s210; *CE&AJ,* 23, 1860, 230).
95. Fairbairn and Tate, 'Experimental Researches', 192.
96. Fairbairn and Tate, 'Experimental Researches', 220.
97. W J M Rankine, 'Fairbairn, Sir William', in *The Imperial Dictionary of Universal Biography,* Vol.2, (1863).
98. W Fairbairn, 'Ricerche sperimentali per determinare la densita' del vapour a tutte le temperature e la legge di dilatazione del vapour soprariscaldato', *Il Nuovo Cimento,* 12.1, 1860.
99. W Fairbairn and T Tate, 'On the Law of Expansion of Superheated Steam', *Proceedings of the Royal Society of London,* 12, 1862, 53-7.
100. Patent No. 801, 18 March 1870; *The Engineer,* 33, Jan-June 1872, 273-4, 276; *Engineering,* 13, Jan-June 1872, 298 and Plate,
101. Walker, *Unwin,* p.60; Fairbairn Engineering Co Ltd, *The Fairbairn Five-Tube Boiler,* (nd but c1873), pp.5-6.
102. *Bulletin de la Société Industrielle de Mulhouse,* 1875, 241; D K Clark, *The Steam Engine: A Treatise on Steam Engines and Boilers,* (1891), Vol.1, pp.228-32; D K Clark, 'The Evaporative Performance of Steam Boilers', *MPICE,* 46, 1875-6, 261-2.
103. J P Glithero, 'The history of engineering in Hyde to 1914', (PhD thesis, UMIST, 2002), p.4.3.
104. Patent No. 270, 23 January 1873; Fairbairn Engineering Co Ltd, *Fairbairn Five-Tube Boiler.*
105. *The Engineer,* 41, 1876, 213-6.
106. *Life,* p.281. Fairbairn's last letter was to his granddaughter Maggie Bateman's fiancé – see below.
107. Glithero, 'Engineering in Hyde', p.4.10; *The Engineer,* 75, 1893, 43-6, 51; E B Ellington, 'Hydraulic Power Supply in Towns: Glasgow, Manchester, Buenos Aires, etc.', *MPIME,* 1895, 356 and Plate 89.

108. W Fairbairn, 'Casualties Arising from Some Boiler Explosions', *Transactions of the National Association for the Promotion of Social Science, Manchester Meeting, 1866,* (1867), pp.605-6; W Fairbairn, 'On Steam Boiler Legislation', *Quarterly Journal of Science,* 8 (old series), 1871, 214-27; *MG,* 7 October 1866.
109. Including at Manchester, Bolton, Halifax , Rochdale, Darwen, Huddersfield, Bacup and Blackburn.
Manchester: *MG,* 13 November 1841; *MG,* 29 March 1851; *Observer,* 21 April 1851.
Bolton: *Mining Journal,* 16, 1846, 10.
Halifax: *The Times,* 16 December 1850; *MG,* 18 December 1850; *The Engineer and Machinist,* 2, August 1850-March 1851, 343-4.
Rochdale: Fairbairn, *UIfE,* pp.lxxv-lxxxii; *MG,* 22, 26 and 29 July 1854; *Manchester Times,* 29 July 1854; *Mining Journal,* 9 September 1854; A McEwan, *Historic Steam Boiler Explosions,* (2009), pp.15-19.
Darwen: *MG,* 2 February 1857.
Huddersfield: *Observer,* 6 December 1857.
Bacup: McEwan, *Historic Explosions,* pp.36-9.
Blackburn: *MG,* 16 November 1853.
110. *The Engineer,* 6, July-Dec.1858, 37-8.
111. *MG,* 27 July 1855.
112. *The Engineer,* 6, July-Dec.1858.
113. Anon., *A Sketch of the Foundation and of the Past Fifty Years' Activity of the Manchester Steam Users' Association for the Prevention of Steam Boiler Explosions and for the Attainment of Economy in the Application of Steam, together with an account of the Jubilee Celebration on the 14th February, 1905,* (1905), p.5: *MG,* 16 November 1853; McEwan, *Steam Boiler Explosions,* pp.15-9. (The date was July not 9 September as stated in Chaloner, *Vulcan,* pp.5-6, following Anon., *Sketch of the Foundation,* p.5.) Fairbairn passed his fee to the Fund for the Relief of the Sufferers (*MG,* 29 July 1854), as he had done at Blackburn (*MG,* 16 November 1853).
114. *MG,* 29 July 1854; *Mining Journal,* 9 September 1854; Fairbairn, *UIfE,* p.lxxx. It is said in that the idea initially came from Richard Johnson in a letter to Fairbairn (Anon., *Sketch of the Foundation,* p.5).
115. *MG,*16 September 1854, 20 September 1854; Anon., *Sketch of the Foundation,* pp.22-3.
116. Anon., *Sketch of the Foundation,* pp.26-8; Chaloner, *Vulcan,* pp.7-9. It was followed by the National Boiler Insurance Company (1864) and British Engine Insurance Company (1878). See Anon., *100 Years of British Engine Insurance Ltd. Founded in Manchester 12 November 1878,* (1978).
117. Anon., *Sketch of the Foundation,* pp.34-9.
118. To a French engineer Fairbairn wrote,' I am quite aware of the regulations which exist in France as regards the construction of boilers, but in this country we prefer that the owners of boilers should be responsible in every instance where loss of life or injury to the person arises from any neglect on their part. This is the extent of the English law, and every man is at liberty to make any description of boiler he pleases, but he must be answerable for the results. It is true, and we admit it to be the duty of every Government to afford protection to life and property, but not to interfere so as to cramp the energies and enterprise of individuals in their pursuit of knowledge, and the advancement of industrial resources' (W Fairbairn to C Combes, 28 March 1863, quoted in *Life,* pp.275-6).
119. Anon., *Sketch of the Foundation,* p.52.
120. Joseph Whitworth, John Hick MP, Hugh Mason, Lavington Fletcher and C F Beyer (*BAAS1869,* p.47; *BAAS1870,* p.1).
121. *BAAS1869,* pp.53-4; *The Engineer,* 28, July-Dec.1869, 238.
122. *Life,* pp.27-8; *The Engineer,* 30, July-Dec.1870, 206.
123. Anon., *Sketch of the Foundation,* pp.53-5 and Fold-out chart 2.
124. Anon., *Sketch of the Foundation,* pp.67-9.
125. G Cookson and C A Hempstead, *A Victorian Scientist and Engineer: Fleeming Jenkin and the Birth of Electrical Engineering,* (2000), pp.165-6.
126. L E Fletcher to W Pole 5 July 1876, in *Life,* p.284.
127. *MG,* 12 November 1853.
128. W Fairbairn, 'On the Retardation and Stoppage of Railway Trains', *MPIME,* 1853.
129. W Fairbairn, 'Experiments to determine the Efficiency of Continuous and Self-acting Breaks for Railway Trains', *BAAS1859,* pp.76-80.
130. W Fairbairn, On the Efficiency of various kinds of Railway Breaks; with Experimental Researches on their Retarding Powers', *MPICE,* 19, 1860, 490-526.
131. Walker, *Unwin,* p.39.
132. 'Report on the Fay and Newall Brakes', dated 30 June 1859. Typescript Copy, Imperial College Archives, Ref. WCUnwin/1.
133. *The Engineer,* 12, July-Dec.1861, 207.
134. *MG,* 7 December 1868.
135. *Life,* pp.380-1.
136. W Fairbairn, 'On the Comparative Temperature of the Climates of England and some parts of Italy', *Memoirs of the Manchester Literary & Philosophical Society,* 2nd series 15, 1858, 45-8.
137. Morrell and Thackray, *Gentlemen of Science,* pp.512-3.
138. *Life,* p.350.
139. *Life,* pp.350-8.
140. P Barry, *Shoeburyness and the Guns: A Philosophical Discourse,* (1865), pp.180-2.
141. *MG,* 29 January 1862.
142. Barry, *Shoeburyness,* pp.186-7.
143. *MM,* 7, Jan.-June 1862, 244-5; *MG,* 11 April 1862.
144. *MM,* 7, Jan.-June 1862, 245.
145. Barry, *Shoeburyness,* pp.187-9.
146. *MM,* 7, Jan.-June 1862, 245.
147. *MG,* 4 October 1862; *BAAS1862,* p.181.
148. *The Observer,* 20 March 1864.
149. *MG,* 25 November 1864. This accords with Barry's view (Barry, *Shoeburyness,* pp.186-7).
150. [Woodcroft] IX.
151. Barry, *Shoeburyness,* p.187.
152. W Fairbairn, *History and Progress of Iron Ship Building,* (1865), p.182
153. *Life,* p.357.
154. Barry, *Shoeburyness,* p.43.
155. Fairbairn, *Iron Ship Building,* p.153.
156. Walker, *Unwin,* 48-9, 161; *Life,* p.360.
157. *Life,* pp.351, 359.
158. Fairbairn, 'On the Mechanical Properties of Specimens of the Iron and Steel Plates' in Barlow, *Treatise on the Strength of Materials,* pp.216-32.
159. *Life,* p.359.
160. 'Report of the Joint Committee appointed by the Lords of the Committee of the Privy Council for Trade and the Atlantic Telegraph Company to inquire into the Construction of Submarine Telegraph Cables; together with

the Minutes of Evidence and Appendix', *Parliamentary Papers,* 1860[2744], p.xxxvi.
161. G Saward, *The Trans-Atlantic Submarine Telegraph: A Brief Narrative of the Principal Incidents in the History of the Atlantic Telegraph Company,* (1878), pp.44-5. Robert Stephenson died and James Wortley became ill, leaving eight. In the same year, 1861, that the Committee presented its Report, the Saturday evening soirée at the British Association meeting in Manchester – Fairbairn's presidential year – was devoted to 'electric telegraphs' (*MM,* 6, 1861, 125).
162. 'Obituary of Sir Douglas Galton', *The Electrician,* 1900, p.80.
163. Cookson and Hempstead, *Fleeming Jenkin,* p.53.
164. See Chapter 4.
165. Fairbairn, *UIfE3,* pp.250-63. Chatterton's compound was constituted from three parts gutta-percha, one part rosin and one part Stockholm tar. Gutta-percha is natural latex produced from the sap of a tropical tree, *palaquium gutta,* native to southeast Asia and northern Australasia.
166. Fairbairn, *UIfE3,* pp.264-7.
167. Fairbairn, *UIfE3,* pp.273-4.
168. 'Report of the Joint Committee', pp.xx, 342-9.
169. Saward, *Trans-Atlantic Submarine Telegraph,* p.48; *UIfE3,* p.275.
170. W Fairbairn, 'Preliminary Investigation of the Mechanical Properties of the proposed Atlantic Cable', *BAAS1864,* pp.408-15; Fairbairn, *UIfE3,* pp.276-289.
171. 'Report of the Scientific Committee appointed to consider the best form of Cable for submersion between Europe and America', (21 October 1863), reproduced in Saward, *Trans-Atlantic Submarine Telegraph,* pp.50-1; Fairbairn, *UIfE3,* pp.276-89; *MM,* July-Dec.1864, 204.
172. G Cookson, *The Cable. The Wire That Changed the World,* (2003), pp.137-40. Cookson mentions neither Fairbairn's experiments on insulating materials nor his tests on the cable samples.
173. W H Russell, *The Atlantic Telegraph,* (1865), pp.42-3.
174. W Fairbairn, 'On Some of the Causes of Failure of Deep Sea Cables', *BAAS1865,* pp.178-84; Fairbairn, *UIfE3,* pp.317-25; *CE&AJ,* 28, 1865, 298.
175. *The Observer,* 1 July 1866; Fairbairn, *UIfE3,* pp.324-5; Cookson, *The Cable,* pp.145-9.
176. See Chapter 8.
177. *Life,* p.164.
178. J Tann, 'Borrowing Brilliance: Technology Transfer Across Sectors in the Early Industrial Revolution', *The International Journal for the History of Engineering & Technology,* 85.1, 2015, 96-8; C MacLeod, *Inventing the Industrial Revolution: The English Patent System, 1660-1800,* (1988), p.5.
179. H I Dutton, *The patent system and inventive activity during the industrial revolution 1750-1852,* (1984), pp.58, 66n10.
180. *MG,* 11 December 1850.
181. 'Report of Select Committee of House of Lords to consider Bills for Amendment of Law touching Letters Patent for Inventions', *Parliamentary Papers,* (1851), XVIII, pp.169-182.
182. Dutton, *Patent system,* pp.58-65.
183. V W Ruttan, 'Usher and Schumpeter on Invention, Innovation, and Technological Change', *The Quarterly Journal of Economics,* 73.4, 1959, 597; J A Schumpeter, *Business Cycles,* (1939), Vol.1, p.84; N Rosenberg, *Perspectives on Technology,* (1976), p.67. On the time-lag between invention and innovation see the table reproduced in Rosenberg, *Perspectives,* pp.69-70, from J Enos, 'Invention and Innovation in the Petroleum Refining Industry', in *The Rate and Direction of Inventive Activity: Economic and Social Factors,* (National Bureau of Economic Research Conference Papers, 1962), pp.307-8.
184. F Redlich, 'Innovation in Business: A Systematic Presentation, *American Journal of Economics and Sociology,* 10.3, 1951, 290; N Rosenberg, *Schumpeter and the Endogeneity of Technology: Some American Perspectives,* (2000), p.61.
185. C MacLeod, 'Strategies for Invention: the diffusion of new technology in nineteenth-century British industry', *Economic History Review,* 45.2, 1992, 290; G Cookson, 'Innovation, diffusion, and mechanical engineers in Britain, 1780-1850', *Economic History Review,* 47.4, 1994, 751-2.
186. I Brunel, *The Life of Isambard Kingdom Brunel, Civil Engineer,* (1870), p.492.
187. W Fairbairn, 'Presidential Address', *BAAS1861,* p.lxvi.
188. *BAAS1858,* pp.164-7; *BAAS1859,* pp.191-3.
189. *MM,* 6, 1861, 141.
190. 'Report of the Commissioners appointed to inquire into the Working of the Law relating to Letters Patent for Inventions', *Parliamentary Papers,* (1864), XXIX, p.17.
191. *MG,* 17 March 1866, 15 August 1868, 26 January 1870; *English Mechanic and Mirror of Science,* 20 May 1870, pp.202-3.
192. *Engineering,* 11, Jan-June 1871, 173,191-2.
193. C MacLeod, 'Concepts of Invention and the Patent Controversy in Victorian Britain', in R Fox (ed.), *Technological Change: Methods and themes in the History of Technology,* (1996), pp.141-53.
194. *Life,* pp.130, 241-2, 371, 383-4; R H Parsons, *A History of the Institution of Mechanical Engineers 1847-1947,* (1947), p.287.
195. *The Engineer,* 3, Jan.-June 1857, 511; *Life,* pp.377-8; T Wyke with H Cocks, *Public Sculpture of Greater Manchester,* (2004), pp.117-8; W Fairbairn, *Observations on Improvements of the Town of Manchester,* (1836), pp.16-7.
196. W Fairbairn, 'Note' in F Arago, *Biographies of Distinguished Scientific Men,* (English Translation by W H Smyth, B Powell and R Grant, 1857), p.604.
197. *The Engineer,* 6, July-Dec.1858, p.172.
198. Morrell and Thackray, *Gentlemen of Science,* p.266; J Mokyr, *The Enlightened Economy: Britain and the Industrial Revolution 1700-1850,* (2009), p.53.
199. T Tuovin, 'The Life and Work of Sir William Fairbairn; A Case study in the History of the Engineering Profession in 19th Century Britain', (Pro Gradu Thesis, University of Jyväskylä, 2000), p.61.
200. L Miskell, *Meeting Paces: Scientific Congresses and Urban Identity in Victorian Britain,* (2013), pp.63-4.
201. Morrell and Thackray, *Gentlemen of Science,* p.410; *Chambers's Journal of Popular Literature, Science and Art,* 28 September 1861, p.208.
202. Fairbairn, 'Presidential Address', pp.li-lxiv. Did Fairbairn attend Robert Stephenson's funeral at Westminster Abbey? Julia Elton suggests not (J Elton, 'Robert Stephenson in Society' in M R Bailey (ed.), *Robert Stephenson – The Eminent Engineer,* (2003), pp.236-7, probably following the list of attendees in *The Observer,* 24 October 1859 which does not include him. However he does appear in the list of attendees in *The Engineer,* 8, July-Dec.1859, 309).
203. *Chambers Journal of Popular Literature,* 28 September 1861, p.208.

204. R and E Frow, *Karl Marx in Manchester,* (1985), p.22.
205. *MG,* 3 September 1861.
206. *BAAS1861,* p.xxviii; Morrell and Thackray, *Gentlemen of Science,* p.410.
207. Fairbairn, *UIfE3,* p.68; *Life,* pp.390-1.
208. T Wyke, 'The Albert Memorial and the making of civic Manchester', in C Hartwell and T Wyke (eds.), *Making Manchester: Aspects of the history of architecture in the city and region since 1800. Essays in honour of John H G Archer,* (2007), p.106; *MG,* 14 January 1867; *Manchester Examiner and Times,* 24 January 1867; Wyke with Cocks, *Public Sculpture,* pp.11-6; A J Pass, *Thomas Worthington: Victorian Architecture and Social Purpose,* (1988), pp.40-52. The sub-committee which arranged the form of the memorial comprised Thomas Goadsby, James Prince Lee (Bishop of Manchester). William Fairbairn, Thomas Bazley, Algernon Egerton, Thomas Ashton, S J Stern, Ernest Reuss and Malcolm Ross. Worthington was himself a member of the Memorial Committee. (Wyke, 'The Albert Memorial', pp.95-6).
209. See Bibliography, 1.
210. *Life,* pp.468, 157.
211. C Condit, 'Buildings and Construction' in M Kranzberg, and C W Pursell Jr, (eds), *Techology in Western Civilisation,* (1967), Vol.1, p.376.
212. W Fairbairn, *De l'Application de la fonte, du fer et de la tôle dans les constructions,* (translated by L Perret-Porta, 1857); W Fairbairn, *Aplicacion del hierro a las construcciones,* (translated by D E Saavedra, 1857), with *Suplemento,* (1859).
213. The American edition was published in New York by J Wiley.
214. *Sydney Morning Herald,* 28 July 1858; *South Australian Register* (Adelaide), 6 April 1857.
215. *MM,* 19 April 1861, 273-4.
216. *MM,* 9 October 1863, 700.
217. W Fairbairn, *The Principles of Mechanism and Machinery of Transmission, comprising the principles of mechanism, wheels and pulleys, strength and proportions of shafts, couplings for shafts, and engaging and disengaging gear,* (1867) - comprising extracts from Part I and more extensively from Part II of Fairbairn, *M&MW.*
218. R Buchanan, *Practical Essays on Mill Work and other Machinery wth Notes and Additional Articles ... by Thomas Tredgold ... revised into a Third Edition by George Rennie,* (3rd ed. 1841).
219. J Mokyr, *The Gifts of Athena: Historical Origins of the Knowledge Economy* (2002), p.89n17.
220. W J M Rankine, *A Manual of Machinery and Millwork,* (6th ed. 1887), p.vi.
221. W H G Armytage, *A Social History of Engineering,* (1976), p.152.
222. W Fairbairn, *Iron - Its History, Properties, and Processes of Manufacture,* (1859); W Fairbairn, *Le Fer – Son Histoire, ses Propriétés at ses Différents Procédés de Fabrication,* (translation by G Maurice, 1864).
223. C Hartwell and M Hyde, *Pevsner Architectural Guide: Lancashire: Manchester and the South East,* (2004), p.141.
224. AHP, 'Architectural History Report, Bolton Market Hall, Appendix C', (2004), p.5.
225. *MG,* 14 April 1855.
226. F O'Dwyer, *The architecture of Deane and Woodward,* (1997), pp.263-5; W Fairbairn, 'Report on the Ironwork supporting the Roof of the New Museum at Oxford', (1858). This was the building in which, two years later at the British Association meeting, Samuel Wilberforce and Thomas Huxley debated about Darwin's *On the Origin of Species.* Fairbairn, as the next President of BAAS, must surely have been present.
227. *MM,* 46, Jan.-June 1847, p.117.
228. Pass, *Thomas Worthington,* pp.73-5.
229. E T Bellhouse, 'On Baths and Washhouses for the People', *Transactions of the Manchester Statistical Society,* 1854, pp.91-103.
230. A J Pass, 'Thomas Worthington' in J H G Archer (ed.), *Art and Architecture in Victorian Manchester,* (1985), pp.85-6.
231. *MG,* 13 January 1855; Pass, *Thomas Worthington,* p.79.
232. Pass, 'Thomas Worthington', p.87.
233. See Chapter 3.
234. W Fairbairn, 'On Public Health', in G W Hastings (ed.), *Transactions of the National Association for the Promotion of Social Science, London Meeting 1862,* (1863), p.63.
235. See above.
236. W Fairbairn, *Three Lectures on the Rise and Progress of Civil and Mechanical Engineering and on Popular Education,* (1859); Fairbairn, *UIfE2,* pp.187-243; Fairbairn, 'Presidential Address'; W Fairbairn, 'The Rise and Progress of Manufactures and Commerce and of Civil and Mechanical Engineering in Lancashire and Cheshire', forming Division IV of T Baines, *Lancashire and Cheshire, Past and Present,* (1869).
237. H Bessemer, 'On the Manufacture of Malleable Iron and Steel without Fuel', *The Times,* 14 August 1856; H Bessemer, *Sir Henry Bessemer, FRS. An Autobiography,* (1905), pp.154-61; Fairbairn, *Iron,* p.131; Smiles, *James Nasmyth,* pp.353-4.
238. Bessemer, *Bessemer,* pp.138-9. In a reverberatory furnace the metal is not in contact with the fuel, whereas in a cupola furnace it is.
239. Steel is an alloy of iron and a small amount of carbon, with sometimes other trace elements. Before Bessemer steel was relatively expensive and difficult to make. It was made in small quantities and used for tools, swords and cutlery. Krupp exhibited steel axles at the 1851 London Exhibition (E H Ahrons, *The British Steam Railway Locomotive 1825-1925,* (1927), p.165; Fairbairn, *Iron,* p.149).
240. Bessemer, *Bessemer,* pp.180, 220-1; Fairbairn, *Iron,* pp.131-68.
241. H Bessemer, 'On the Manufacture of Cast Steel, its Progress, and Employment as a Substitute for Wrought Iron', *BAAS1865,* p.172.
242. Available information included a few tests on Bessemer steel by Wilmot, which Fairbairn included in the first edition of *Iron* (Fairbairn, *Iron,* pp.213-4) and tests on axle steels by August Wöhler in Germany (Ahrons, *British Steam Railway Locomotive,* p.165).
243. *BAAS1866,* p.xliii.
244. W Fairbairn, 'Experimental Researches on the Mechanical Properties of Steel', *BAAS1867,* pp.161-274; *The Engineer,* 13, July-Dec.1867, 233-7; W Fairbairn, 'On the Mechanical Properties of Iron and Steel', *Quarterly Journal of Science,* 5, 1868, 10-14; Fairbairn, *Iron,* (3rd ed. 1869), pp.310-32 and Tables I - III.
245. *BAAS1867,* p.lxiv; W Fairbairn, 'Experimental Researches on the Mechanical Properties of Steel', *BAAS1869,* pp.96-150. The samples were from the Barrow Haematite Company, and the Heaton Steel and Iron Company. *The Engineer,* 28, July-Dec.1869, 130-6; *Engineering,* 8, July-Dec.1869, 152-3; W Fairbairn, *An Experimental Enquiry into the Strength, Elasticity, Ductility, and other properties of Steel manufactured by the Barrow*

Hæmatite Steel Company, (1869). Aged 80 one might have expected that this would be Fairbairn's last paper on steel. It was not. Following a fatal rail accident near Hatfield in 1870 caused by the breaking of a steel tire (sic), the inquest found that it was 'caused by the intensity of the frost'. It was not, and Fairbairn, in a paper read to the Manchester Literary and Philosophical Society in January 1871, set out the reasons why it was not and advised how such accidents could be avoided (W Fairbairn, 'On the Properties of Iron and Steel as applied to the Rolling Stock of Railways', *Proceedings of the Literary and Philosophical Society of Manchester,* 10, 1870-1, 86-91). This was endorsed by James Joule in a following paper (J P Joule, 'On the Alleged Action of Cold in rendering Iron and Steel Brittle', *Proceedings of the Literary and Philosophical Society of Manchester,* 10, 1870-1, 91ff).
246. H J Hopkins, *A Span of Bridges: An Illustrated History,* (1970), p.128, 'Fairbairn, the greatest of the Victorian Ironmasters'.
247. *BAAS1864,* pp.iv, 193-326; L T C Rolt, *The Aeronauts: A History of Ballooning 1783-1903,* (1966), p.192.
248. J L Pritchard, 'A century of British Aeronautics: the Royal Aeronautical Society, 1866-1966', *Journal of the Royal Aeronautical Society,* 70.661, 1966, 7-11.
249. Thirty years earlier he had constructed an apparatus fitted to the front of a train to measure air resistance (E Hodgkinson, 'Report on Experiments on the Resistance of Air', *BAAS1842,* p.211).
250. Pritchard, 'Century of British Aeronautics', 11.
251. Smith, *Contribution,* p.6; *Journal of the Royal Aeronautical Society,* 61, 1957, pp.160-1; R Hartree, *John Penn and Sons of Greenwich,* (2008), p.73. The Duke of Argyll was President of the Society from 1866-1895.
252. J D Anderson, *A History of Aerodynamics: And Its Impact on Flying Machines,* (1997), pp.120-6.
253. 'Annual Report of the Aeronautical Society of Great Britain, 1870', quoted in Smith, *Contribution,* p.6.
254. Smith, *Contribution,* p.6.
255. *The Times,* 29 July 1856.
256. *Life,* pp.31-2, 378-9.
257. Mokyr, *The Gifts of Athena,* p.65n55.
258. R L Hills, *Life and Inventions of Richard Roberts, 1789-1864,* (2002), pp.18, 36, 46.
259. *The Engineer,* 9, Jan-June 1860, 254.
260. E Clark, *The Britannia and Conway Tubular Bridges,* (1850), Vol.1, p.30.
261. W Fairbairn, 'An Experimental Inquiry into the Strength of Wrought Iron plates and their Riveted Joints as applied to Ship-building and vessels exposed to severe strains', *Philosophical Transactions of the Royal Society,* (1850), 705-6.
262. Fairbairn, *UIfE,* pp.130, 134-7.
263. Hills, *Roberts,* pp.50-1.
264. Hills, *Roberts,* p.51.
265. *Life,* pp.429-30. Sadly Maxwell, died within two years of the marriage (*Life,* p.429).
266. *Life,* pp.428-31. On Bateman, see P E Russell, 'John Frederick La Trobe-Bateman (1810-1899), Water Engineer', *TNS,* 52, 1980-1, pp.119-138.
267. Prestwich churchyard visited 2013. Why was Fairbairn buried in this graveyard? Firstly it was not unusual for those who lived in the city to be buried in suburban churchyards where there was available grave-space even if they were not members of the Church of England (the nearby St Paul's, Kersal, is a prime example). Secondly the first deceased commemorated on this tomb-chest was William's son Peter (d.1859), who was a member of the Church of England – he was a churchwarden at Manchester Cathedral in the mid-1850s (*MG,* 26 March 1856). The other sons who predeceased their father, John (d.1867) and George (d.1868) are also buried here.
268. *Life,* p.433.
269. *ILN,* 29 August 1874, p.205; *Allgemeine Deutsche Polytechnische Zeitung,* 12 September 1874; *Scientific American,* 5 September 1874, p.143.
270. *MG,* 2 October 1874. The evidence suggests that Edward Cleather in fact died a few months before his grandfather (http://fairbairn.lornahen.com/lineages/defairbairnjohnandersonhelen.htm (accessed 10 February 2014)).
271. *Life,* p.v.
272. *Life,* p.446. Pole incorrectly refers to Emanuel Edward Geflowski as 'G E Geflowski'. By commissioning Geflowski, after much acrimony, the Committee dealt a blow to Thomas Fairbairn, who wished his friend Thomas Woolner to undertake the commission. The Committee's decision appeared irregular and was surprising given that Hugh Mason, a Committee member, mobilised support amongst subscribers for Woolner. That Geflowski was to charge £840 compared with Woolner's £1,500 is unlikely to have been an issue as the Committee had raised £2,700 (Wyke, with Cocks, *Public Sculpture,* pp.41-2). Much more likely is that Thomas Fairbairn and his Pre-Raphaelite friends were still treated with suspicion by many in Manchester (see Chapter 8). Thomas had to make do with a plaque by Woolner in Cross Street Chapel (*Life,* pp.457-8). The plaque was destroyed in the blitz of WW2.
273. A E Musson, 'Introduction' to 1970 facsimile edition of *Life,* p.xx.
274. Smith, *Contribution,* pp.6-7.

Sources, Credits and Acknowledgements for Illustrations in Chapter 7

Fig.7.1 *Engineering,* 5, 1868, 59. The original is at the Institution of Mechanical Engineers to whom it was presented by Sir W.A. Fairbairn, Bt., in 1956, both parties believing it was by W. Holman Hunt (Institution of Mechanical Engineers, Annual Report 1956, p.303). Image by courtesy of Institution of Mechanical Engineers.

Fig.7.2 G. Dow, *The First Railway between Manchester and Sheffield* (1945); Postcard 1916: Dinting Arches, Glossop.jpg – Wikimedia Commons.

Fig.7.3 W. Fairbairn, 'Description of the Removing and Replacing of the Iron Columns in a Cotton Mill', *MPIME,* 17, 1866, fig.2.

Fig.7.4 Photograph, R.J. Byrom, 1961.

Fig.7.5 Industrial Development Authority Ireland, *Dublin inner city – renewal through enterprise,* (nd but c.1984), front cover. Image used by permission of IDA, Dublin.

Fig.7.6 Barry & Associates, Drawing 30 May 1983. I am indebted to Barry and Associates for copies of their drawings of the Dublin Warehouse.

Fig.7.7 Fairbairn, *Application,* p.156; reproduced in S. Giedion, *Space, Time and Architecture: the growth of a new tradition* (3rd ed. 1953), p.192.

Fig.7.8 Walker, *Unwin,* frontispiece.

Fig.7.9 K. Fenwick and N.T. Sinclair, *The Inverness and Aberdeen Junction Railway,* (2008), p.2. I am grateful to Messrs Fenwick and Sinclair, and to Anne-Mary Paterson for permission to reproduce this illustration.

Fig.7.10 *Report of the Commissioners appointed to inquire into the Application of Iron to Railway Structures* (1849), p.405.

Fig.7.11 *Report of the Commissioners,* p.406.

Fig.7.12 A.I. Smith, 'William Fairbairn and Mechanical Properties of Materials: The Effect of Repeated Loading on Strength', *The Engineer,* 26 June 1964, 1135.

Fig.7.13 W. Fairbairn, 'On the Resistance of Tubes to Collapse', *Philosophical Transactions of the Royal Society,* 1858, 391.

Fig.7.14 W. Fairbairn and T. Tate, 'Experimental Researches to determine the Density of Steam at different Temperatures and to determine the Law of Expansion of Super-heated Steam', *Philosophical Transactions of the Royal Society,* 1860, Plate VII. For a detailed explanation see pages 187–9 of 'Experimental Researches'.

Fig.7.15 D.K. Clark, *The Steam Engine: A Treatise on Steam Engines and Boilers,* (1891), Vol.1, p.229.

Fig.7.16 *The Engineer,* 6, July–Dec.1858, 37.

Fig.7.17 *ILN,* 20 August 1864.

Fig.7.18 Fairbairn, *UIfE3,* p.261.

Fig.7.19 Fairbairn, *UIfE3,* p.277.

Fig.7.20 Photograph 2016. Bronze copy by William Theed of Francis Legatt Chantrey's marble statue in Westminster Abbey.

Fig.7.21 C.F. Carter (ed.), *Manchester and Its Region,* (1962), Plate XIX (and reproduced elsewhere).

Fig.7.22 Engraving c.1859.

Fig.7.23 J.F. La T. Bateman, *History and Description of the Manchester Waterworks,* (1884), Frontispiece (also reproduced in P. Russell, 'John Frederick La Trobe-Bateman (1810–1889) Water Engineer', *TNS,* 52, 1980–1, 121).

Fig.7.24 Photograph 2016.

Fig.7.25 W. Fairbairn, *De L'Application de la Fonte du Fer et de la Tole dans les Constructions* (Paris 1857).

Fig.7.26 W. Fairbairn, *The Principles of Mechanism and Machinery of Transmission* (Philadelphia 1903 ed.).

Fig.7.27 P.E. Russell, 'John Frederick La Trobe Bateman FRS, Water Engineer 1810–1889', (MSc Thesis, UMIST, 1980), fig.57.

Fig.7.28 Photograph, R.J. Byrom.

CHAPTER 8

Thomas Fairbairn and the Demise of a Great Company

Fig.8.1:
William Holman Hunt, *Thomas Fairbairn* (1873).
(Fairbairn Family Collection)

Introduction

THIS book is about William Fairbairn and the company he founded. One of the questions addressed is why the company failed only twenty-one years after Fairbairn retired from its management, and within a year of his death. In order to answer this it is essential to understand his son Thomas Fairbairn. Secondary material shows no awareness that by 1859 Thomas was the sole owner of the Fairbairn company, nor after his triumph of the Art Treasures Exhibition of his rejection by Mancunians, nor of his early use of the 1862 Companies Act to realise capital from the company to become a country landowner. While a number of reasons have been put forward for the demise of the Fairbairn concern, none of them refer to the role played by Thomas. What little has been written about him is almost all confined to his important roles in the art world for which he was offered a knighthood at the age of 34.[1]

We have seen that Thomas was denied a university education by the firm's financial problems in the early 1840s. The claim has also been noted that Thomas proved himself by 'bringing the disastrous Millwall concern to a close'.[2] He spent eight years there, excepting ten months in 1841–2 which he spent in Italy. We know nothing about that Italian visit, of where he went or whom he met, but it appears to have been immensely formative.[3] His subsequent love of art and his organisational brilliance are both illustrated by the Manchester Art Treasures Exhibition.

This chapter describes the activities of the Fairbairn company from William's retirement at the beginning of 1854 to its voluntary liquidation in 1875 and final winding up in 1899. From his role as partner in the family business, Thomas successively occupied the roles of sole proprietor, chairman of the limited company and joint liquidator. The chapter highlights Thomas's roles in the art world and his 'gentlemanly capitalism', which is necessary to understand the decline and demise of the company. At first sight this was a classic case of second-generation failure, but the story is not so simple, as rather than just retire to the life of the landed gentry on the country estate, Thomas pursued a life of 'gentrified capitalism'. The tragic demise of a great firm was mirrored by his own tragic demise.

Exeat the Family

THOMAS returned to Manchester in 1848 with his new wife, Allison Callaway, daughter of a Chiselhurst surgeon.[4] They first lived at Irwell Bank, Broughton Park,[5] in fashionable North Manchester, not far from William Andrew at 6 Montague Terrace. Thomas was soon active in Manchester life, but that included neither Cross Street Chapel nor the Literary and Philosophical Society.[6] By 1857 the Thomas Fairbairns had moved to Northwood, Sedgley Park – a villa far more grandiose than his parents' home at the Polygon.[7] They had nine children. The first seven were born in Manchester, and the last two after their move south. Mary Eleanor, their first, born in 1850, was baptised at Manchester Cathedral – the Unitarian roots had been cut.[8] Two years later she died.

At the Ancoats works Thomas appears to have looked after the locomotive department. By the late 1840s the family firm was successful and prosperous, with steam engines and tubular-girder bridges to the fore, and the Millwall losses made good. When William Fairbairn stood down from the management of the firm in favour of his sons Thomas, William Andrew and George at the beginning of 1854, the firm was on the crest of a wave of success and renown.

For the following twenty years William retained his office at the works and for most of those years had a personal assistant who was based there. There is no evidence of his intervening in the running of the company during those years, although he clearly took an interest in the engineering work – for example in 1862 he tested a model of a bridge to cross the Tay at Dalguise,[9] and 1869, when the roof of the Albert Hall was pre-assembled in Fairbairns' yard, he is said to have 'modified some of the details'.[10] The research and consultancy work William undertook did generate some commissions for the company, but they were mainly discrete one-off commissions such as the Dinting and Mottram viaducts. More important to the company was William Fairbairn's name and high profile, particularly through his Presidency of the British Association, giving potential customers the feeling of stability and reliability. When the company was floated in 1864 his name was high on the list of directors (despite his reservations about limited liability)[11] but he played little part in its management. When, aged eighty-three, he retired from the board, the company stressed that he had not retired 'from his interest in the establishment'.[12]

All the indications are that the relationship between William and Thomas was cool. Father and son, both endowed with great gifts, were very different in everything else. William invested in the latest machine tools; Thomas, with acute perception, in avant-garde art. William attended Cross Street Chapel; Thomas did not. William lived all his life within walking-distance of his works; Thomas moved to a landed estate in Hampshire. William's friends were leading engineers and scientists; Thomas's friends included bohemian artists. William conducted experiments and wrote books; Thomas did neither. It was perhaps in their approach to business that father and son differed most significantly. Whereas William had developed his business through innovative engineering, which had made and lost fortunes, Thomas's approach was that of a trader who sought to ensure that money, once made, was not thrown away. But his was a short-term approach which left the company starved of innovation and investment, which boded badly for the future.

By the mid 1850s all was not well at Ancoats. In 1856 George resigned from the partnership. He was only twenty-six, but was beset by poor health.[13] In 1859 William Andrew retired to become a 'private gentleman' in London. He was aged only thirty-five.[14] It was also the year in which his wife Marianne died, leaving him with four children.[15] There is no hard evidence why these resignations occurred but Thomas's virtual absence from the business for the two years 1856–7 may have caused more than a little tension. The presumption is that Thomas bought out George and William Andrew. In any event, by 1860 the Fairbairn company had but one owner, Thomas Fairbairn.

Thomas chose not to engage in hands-on management at Ancoats. In 1860 he appointed Henry Harman (1815–75) as Manager, a post which Harman held until ill-health forced his retirement in 1873.[16] He had served an apprenticeship with Maudslay & Field and was a competent engineer, but no entrepreneur.[17] Prior to joining Fairbairns he had been Chief Engineer of the Manchester Steam Users' Association.[18]

Fig.8.2:
W. Holman Hunt,
The Awakening Conscience,
1853.

Art and Exhibitions

THOMAS'S activities during his two-year virtual absence from the firm in 1856–7 provide evidence of his outstanding organisational and entrepreneurial abilities.[19] They also show where his interests and priorities lay. Thomas had developed a thirst for art without which it is unlikely that either the Manchester Art Treasures Exhibition or his fruitful friendship with John Pender would have occurred. Visiting the Royal Academy in 1853 he encountered Holman Hunt, resulting in his commissioning *The Awakening Conscience* – which excited comment then, as now – and a lifelong friendship with Hunt.[20] Having lived with the picture, Thomas prevailed on Hunt to repaint the girl's face in time for him to loan it to the Art Treasures Exhibition. It was further modified in 1879 – perhaps Allison did not wish her home decorated with a picture of someone with Annie Miller's reputation.[21] The other Hunt which Fairbairn loaned to the Exhibition was *Two Gentlemen of Verona – Valentine rescuing Sylvia from Proteus* which he had bought in 1855.[22] These were a small part of Thomas's growing collection.[23]

While staying with the Fairbairns at Northwood, to visit the Art Treasures Exhibition, Hunt extolled the abilities of the Pre-Raphaelite sculptor, Thomas Woolner, with the result that the Fairbairns commissioned Woolner to sculpt a marble group of their two oldest surviving children, Constance and Arthur, both without hearing and speech.[24] It was exhibited at the 1862 London Exhibition and widely admired.[25] Also commissioned were a bust of William,[26] a medallion of Thomas,[27] and a bust of the swashbuckling adventurer, James Brooke.[28] Thomas retained links with Woolner. William Bell Scott the painter, dropping in at Woolner's in 1864, 'came upon a large party over their wine after dinner. Tennyson, Holman Hunt, the two Palgraves, Fairbairn, Spedding, and a lot more'.[29] It is easy to forget that the work of Hunt and his friends was considered avant garde and by no means universally accepted. Thus when Hunt referred to them as Pre-Raphaelites, Thomas pulled him up, 'Let me advise you, when talking to Manchester people about the works of your school, not to use that term; they are disposed to admire individual examples, but the *term* has to them become one of such confirmed ridicule that they cannot accept it calmly!'[30] That four paintings from Thomas's collection are in the Tate today, and others are displayed in major galleries, is a telling testimony, as Holman Hunt put it, to his 'immense instinct for what is good in Art'.[31]

At the Dublin Exhibition in 1853, financed (at personal loss) by William's friend William Dargan,[32] Fairbairns exhibited a high-pressure engine and a tubular crane,[33] and Thomas loaned J.B. Pyne's *View of Palanza on the Lake Maggiore*, which featured a Fairbairn iron steamboat.[34] At Dublin Thomas met J.C. Deane, an Assistant Secretary to the Exhibition.[35] It was Deane who had the vision of an exhibition of the Art Treasures of England, identified Manchester as the venue, and Thomas Fairbairn as the man to make it happen. On 10 February 1856 he wrote to Thomas urging an exhibition of 'the Art wealth of England' which was 'of greater value than any other country of the world'. Thomas, with Thomas Ashton Joseph Heron and others convened a meeting of 'influential gentlemen' in the Mayor's parlour.[36] Deane attended, as did Peter Cunningham, author of *The Lives of the Most Eminent British Painters, Sculptors and Architects,* and together they presented a paper. 'Manchester was the place to bring these art treasures together'.[37] Thomas took up the challenge. Accompanied by one or two friends he 'waited upon the leading citizens of Manchester' and in a few weeks £74,000 was guaranteed. Thomas, his father William, brother William Andrew and brother-in-law J.F. Bateman guaranteed £500 each.[38] He obtained estimates for the building. He was part of a deputation that sought and obtained Prince Albert's support.[39] On 20 May 1856 the guarantors gave the go-ahead and appointed an Executive Committee, which appointed Thomas Fairbairn as its Chairman and J.C. Deane as General Commissioner.[40]

They faced a daunting task – the Exhibition was to open in fifty weeks. A site was obtained in Old Trafford. By July building work had started, and requests for loans of works of art were in hand. There was another visit to the supportive Prince Albert who wrote advocating 'a chronological and systematic

Fig.8.3: T. Woolner, *Brother and Sister*, 1858–62.

Fig.8.4: Louis Haghe, *Thomas Fairbairn handing over the address to the Prince Consort* (detail), at the opening of the Manchester Art Treasures Exhibition, 1857.

arrangement',[41] as Deane and Cunningham had done. It emanated from Gustav Waagan (1794–1868), director of the Royal Museum in Berlin, author of *Treasures of Art in Great Britain,* and cataloguer of Albert's collection. Waagen, held in high esteem in the art world, was particularly helpful to the Committee. No other art historian of the day exerted greater influence on the organisation of the exhibition.[42] Thomas Fairbairn wrote to many owners of works of art, exploiting to the full the Queen and Prince Albert's example in loaning exhibits.[43] There was an overwhelming response. The building took shape.[44] The 16,000 exhibits arrived.[45] Thomas Fairbairn's particular interest was the Modern Masters and here he and the Executive Committee kept to themselves the choosing and obtaining of the works. Thomas's friend, the artist Augustus Egg, was engaged to oversee the hanging of the Modern Masters,[46] and Thomas loaned nine works by contemporary painters.[47] A spectacular acquisition for the exhibition, and for the nation, was the Soulages Collection, which the Committee bought at £13,500 'out of their own seven purses'.[48]

On 5 May 1857 Prince Albert opened the Exhibition with great ceremony. Thomas, in the uniform of deputy lieutenant read an address to the Prince.[49] At Northwood he and Allison hosted the Duke of Newcastle, the Earl and Countess of Yarborough and Lady Sophia Pelham, the Earl of Lincoln and Lord Ward,

Fig.8.5: A. Brothers, *William Fairbairn*, 1856.

whilst at the Polygon William and Dorothy entertained the Earl and Countess Granville, and Lord Wrottersley, the President of the Royal Society.[50] At the end of June the Queen, Albert and four of their children arrived in Manchester to visit the Exhibition, hosted by the Earl of Ellesmere at Worsley Hall. Lord Palmerston, the Prime Minister, and Lady Palmerston arrived the same evening and were the guests of Thomas and Allison at Northwood.[51] Thomas read an address to the Queen. William, who must have been very proud of his son, was presented to 'kiss her Majesty's hand'. The Queen knighted the Mayor, James Watts, but Thomas declined the offered knighthood.[52]

Many employers arranged for their work-forces to attend, including 2,500 from Saltaire, paid for by Titus Salt,[53] and 700 Fairbairn employees.[54] 400 Prestwich schoolchildren were guests of Thomas and his neighbour the Countess of Wilton, 100 of them travelling in Wilton carriages.[55] However the Exhibition made no concessions to the working man's ability to appreciate art. There were no descriptive notices adjacent to the exhibits because these might reduce the sales of catalogues. The *Art Journal* was rightly critical.[56] When the Exhibition closed on 17 October, it had been visited by 1,335,915 people,[57] and showed a small surplus on a turnover of £110,588.[58]

William Fairbairn's interest in art was meagre but he was supportive of his son and, as usual, brought his engineering skills to bear. At the Dublin Exhibition he had been called in to check the wrought-iron trellis girders supporting the galleries. He undertook tests, found them wanting and introduced modifications.[59] The same happened at the Manchester Exhibition where he found the girders 'weak and imperfect'.[60] Appropriately for 'the photographic capital of the world',[61] the Exhibition included a photographic section – one of the first photographic exhibitions. This included a photograph of William Fairbairn, a Vice-President of the Manchester Photographic Society, by one of the Society's leading photographers, Alfred Brothers.[62]

On the basis of the wealth of exhibits, attendance figures, and the reception it received, the Manchester Art Treasures Exhibition was an innovative and stunning success. To achieve, in such a short space of time, what F.M.L. Thompson has described as 'an exhibition of international importance, larger and of higher quality than anything previously staged in Britain', was an amazing feat.[63] The Exhibition also marked the start of the Hallé orchestra, which was formed to play there.[64] Thomas loved art and had a determination to introduce it to every section of society, with the naively utopian belief that art would change industrial communities.[65] Nevertheless he and his committee established the enduring principle of universally accessible art,[66] such that Elizabeth Pergam sees American art museums as 'the permanent heirs' to the Manchester Art Treasures Exhibition.[67]

Thomas Fairbairn was the entrepreneurial force behind that Exhibition, but there was a heavy cost.[68] On 30 December, at a déjeuner for the Executive Committee and others (including Fairbairn's art-collector friend John Pender) the new Mayor, Ivie Mackie, had some observations:

> The time that these gentlemen had devoted to the undertaking was not, perhaps, sufficiently known to many members of the community; but he knew for himself, that Mr. Fairbairn and others had devoted their whole time to the work throughout; and most of those present knew enough of the active management of a large business to be able to appreciate the sacrifice involved in being absent from it so long and continuously.

It is unlikely that the Mayor appreciated just how pertinent his words were. Apart from everything else, the Executive Committee had 327 formal meetings.[69]

Set-backs and Opportunities: Thomas Fairbairn leaves Manchester

IN the wake of the success of the Art Treasures Exhibition, Thomas experienced three setbacks in the following three years. First there was the controversial adventurer, James Brooke.[70] Born in Benares in 1803 and educated in England, he served in the Bengal Army from 1819 to1825 when he was wounded and returned to England. In 1835 he inherited from his father and sailed to Indonesia on a voyage of enquiry. In 1841, after military and political conflicts, he seized power in Sarawak obtaining appointment as Rajah from the Sultan of Brunei. Visiting England in 1847 he was knighted, and became an icon of early Victorian imperialism.[71] He returned as British Consul in Borneo but by 1850 Cobden,[72] Hume, Bright,[73] and others were expressing concern about the way the coastal waters were freed from pirates. Thus Cobden, 'We cannot go before the world with clean hands on any other question if we are silent spectators to such atrocities'.[74] The British Government distanced itself from Brooke, and he was forced to resign his British offices. Yet P.F. Tidman, who knew Brooke in Sarawak, believed he was 'one of the really great men of his time' who 'came to a disorganised crowd of savages, and left them a compact nation' with 'peace in their borders'.[75] Brooke was back in England again in 1858, bereft of money and campaigning for Government recognition.

How Thomas Fairbairn had met Brooke remains unclear. A possible link is the Earl of Ellesmere who chaired a meeting in London at which Brooke spoke.[76] Another link is Angela Burdett-Coutts, said to be the wealthiest woman in England, who was friendly with Brooke and also visited the Art Treasures Exhibition on several occasions. But there is another twist to what Barley calls 'an extraordinarily complex melodrama'. Brooke had an illegitimate son who, in the first half of the 1850s, worked in the stables of Lord Ward.[77] Was it mere coincidence that in 1857 Ward stayed with the Fairbairns at Northwood,[78] and in January 1858 the Fairbairns visited Ward,[79] or is there a hidden story here?

In April 1858 a public dinner was held in honour of Brooke. William and Thomas Fairbairn and Thomas's friend, John Pender, were on the organising Committee, chaired by the Mayor.[80] It is clear that Thomas was the instigator. At Northwood the Fairbairns hosted Brooke and Admiral Keppel,[81] who had assisted Brooke against the Borneo pirates,[82] and at the dinner Thomas proposed the toast to Brooke.[83]

It was the business opportunities which particularly interested Manchester merchants and manufacturers, and an outcome of the dinner was a petition urging that Sarawak be brought 'under the direct control of the Colonial Department ... for the benefit of British commerce'.[84] In October Brooke was back in Manchester to attend an Athenaeum Soirée at which Lord John Russell spoke. Again Brooke was a guest at Northwood. Towards the end of the Soirée he suffered a mild heart attack, and was at Northwood for some time convalescing.[85] Brooke's former Secretary wrote, 'Thomas Fairbairn, knowing the wretched state of the Rajah's finances, proposed to a few intimates that something should be done to relieve him'. A fund was set up but 'it did not succeed': it raised less than half what had been hoped for.[86]

In November William and Thomas Fairbairn, with John Pender, joined others from Liverpool and Glasgow at an interview with Lord Derby. Thomas spoke for Manchester and presented the petition. Lord Derby was unmoved.[87] Brooke, with help from Burdett-Coutts returned to Sarawak in 1859, coming back to retire to Devon in 1863,[88] in which year Thomas and Allison's fourth son was named James Brooke – the present sixth baronet (born 1930) is also Sir James Brooke Fairbairn. Brooke visited the Fairbairns at Brambridge in 1867,[89] and died in 1868.[90] Thomas was a Trustee under his will,[91] and involved with the Rajah Brooke Memorial Fund.[92] The dynasty Brooke founded ruled in Sarawak for over a century. But Manchester had not been sympathetic to Thomas Fairbairn's call for financial support for the Rajah. Perhaps that was inevitable in a liberal city.

Thomas's second set-back in Manchester was to his aspiration to become a Member of Parliament. Despite the reputation for radicalism acquired through the Anti-Corn-Law league, Manchester Liberal politics retained conservative undercurrents and a concern for economic self-interest. Thus in 1857 the town continued to be represented by Liberals – Sir John Potter topped the poll – but the Manchester electors dispensed with the services of its radical MPs, including John Bright who had opposed the Crimean War.

By mid-1858 Potter's health was causing concern. The London correspondent of the *Belfast News-Letter* was one of the first to have got wind of this: 'The honourable and pinguid member for Manchester is going to resign on the grounds of ill-health, and an impudent attempt is about to be made to get Mr Thomas Fairbairn, the boiler-maker, elected in his stead', said his despatch on 6 July.[93] Others followed, referring to Fairbairn and to Thomas Bazley, 'who would certainly be brought forward'.[94] Fairbairn announced he intended to contest the borough when a vacancy occurred. His opinions were 'truly liberal' but he was not 'a radical'. He was opposed to the ballot and did not desire universal or household suffrage, but would 'adopt machinery by which many professional and educated men, at present disqualified, should be entitled to vote'.[95] The reference to the ballot is to secret voting, called for to 'crush and extirpate electoral corruption'.

On 4 August, Liberal electors in Chorlton-on-Medlock unanimously passed the motion: 'That this meeting, having read the address of Mr. Thomas Fairbairn, deems him to be unfit to be representative of the City of Manchester in Parliament'. It then made a requisition to Bazley that he allow himself to be put in nomination as a candidate.[96] The Liberal electors of Collegiate Church Ward, chaired by Abel Heywood, passed a similar resolution.[97] On 25 October Sir John died, aged only 43.[98] Bazley's supporters put him forward as candidate.[99] He was a strong contender. He had established Deane Mills and community in Bolton, had been active in the Anti-Corn-Law League, was Chairman of the Manchester Chamber of Commerce, involved in education and temperance, a churchwarden of Stowell's evangelical Christ Church, Salford, and a Commissioner for the Great Exhibition.[100]

Sir John was buried with due pomp. Thomas Fairbairn, as an executor, was to the fore.[101] Three days later Thomas announced that he had 'determined not to contest the representation of Manchester'.[102] Bazley was elected unopposed.[103] Manchester may have moved from the radicalism of Cobden and Bright, but it had not moved very far. Thomas's political aspirations were dead: Manchester had delivered a second rejection.

At a meeting when testimonials were being given to members of the Art Treasures Executive, Thomas announced:

> I hope yet to see the inauguration of an institution to be dedicated to the Arts which shall be worthy of the wealth and importance and enlightenment of this great city ... let it be a fitting receptacle for the display of the most costly and the most meritorious works of Art ... let it be opened absolutely and entirely free to all.[104]

Receiving encouragement, in January 1860 he wrote to the Manchester press advocating a Free Art Gallery and Museum for Manchester, stressing:

> The influence of art ... upon character and public morality; its authority over our passions; and the mysterious connection which subsists between beauty and virtue.[105]

Thomas explained his proposals at a meeting in the Mayor's Parlour, at which those speaking in support included John Pender, Elkanah Armitage, William Gaskell, Thomas Wrigley and Edmund Potter.[106] At the Royal Academy banquet, with Palmerston, John Russell and most of the Government present, the President, Sir Charles Eastlake, proposed, 'Success to the Manchester Gallery and the health of Mr. Fairbairn'.[107] In Manchester Thomas published a pamphlet, *Free Art Gallery and Museum for Manchester,* with contributions from himself, Gaskell, Pender and others. The gallery was to attain 'a national importance'.

By September only half of the figure set to be reached before the scheme was taken forward, had been raised. Thomas wrote a exceptionally strongly-worded letter to the local press, with a morally imperialistic tone: 'These friends of the working man see in such an institution ... a temple of art [which] would be cherished and protected by the working classes as an institution peculiarly their own.' 'Why, then,' he asks, 'should the subscription list have hung fire?'[108] It was pointed out that 'many of our first-class men' were 'conspicuous by their absence' from the list.[109] They no doubt included those of more radical liberal views who were not too keen to be associated with a project driven by one with whose views on James Brooke, the ballot, and suffrage, they did not agree. Evangelical Christians, as distinct from people with latitudinarian or Unitarian views, were certainly lukewarm, convinced that for cultural change 'something more than mere amusement is required'.[110] And there was the wide awareness that neither the Mechanics' Institutes nor the Lyceums had been embraced by 'the working man'. The scheme was abandoned.[111] In spite of his triumph with the Art Treasures Exhibition, Thomas had lost his support in Manchester, as he met this third negative response in three years.

Hard on the heels of these rejections came three very attractive lures to London. One of the consequences of the Art Treasures Exhibition was the London Exhibition of 1862. When Wentworth Dilke, Chairman of the Society of Arts, visited the Art Treasures Exhibition Deane spoke with him, proposing that a Great International Exhibition of Arts should be held in London in 1861.[112] A memorandum followed and was submitted to the Society's Council. Encouraged by the forceful civil servant, Henry Cole (with whom Thomas had already had dealings over the Soulages Collection) the Society took the proposal on board, and appointed a committee, and then five Trustees, Earl Granville, The Marquis of Chandos (later Duke of Buckingham), Thomas Baring, C. Wentworth Dilke and Thomas Fairbairn. The Franco-Austrian war set things back a year. A site was agreed at South Kensington and Cole put plans in hand using Francis Fowke from his department to design the building. There was a skirmish at the Society of Arts with demands to know how Fowke had been appointed and why there had not been a competition. Cole repulsed the attack. William Fairbairn and William Baker (engineer L&NWR) were engaged to give advice on the proposals and reported on 22 December 1860 that 'by great exertions' the plans could be realised in the time available, but the costs would be far more than the budget. Tenders were vastly over budget and the building was hastily redesigned and a contract negotiated. The Queen appointed the five Society of Arts Trustees as Commissioners for the second London International Exhibition of Industry and Art. Work on site commenced on 9 March 1861. The building was dominated by two domes, each larger than any previous domes.[113] In April a five-man building committee was formed of Lord Chandos, Thomas Fairbairn, William Fairbairn, William Baker and Lord Shelburne (later Marquis of Lansdowne). Relationships between this committee and Cole, the contractors and Fowke became strained. Notwithstanding, the contractors' progress was truly amazing – they had eleven months in which to complete the building, and they did it. Finding Fowke in difficulties with the decor, the building committee, at the end of January 1862, turned to J.G. Crace who had decorated the interior of the Manchester Exhibition. Crace produced in what the Commissioners called 'an incredibly short time' relatively elaborate decorations which were generally well received.[114]

At the same time Cole's team, surpassing the 1851 Exhibition and the 1855 Paris Exhibition in scale, assembled 29,000 exhibits from 37 countries – it would have been higher if it had not been for the American Civil War. With so many exhibits it was impossible to take in anything like the full scope of the Exhibition within a reasonable time, and it became clear that the optimum size for an event of this type had been exceeded.[115] Reports in the Court Circular in *The Times*

indicated that the Commissioners met two or three times a week.[116] Prince Albert was to open the Exhibition on 1 May 1862, but on 14 December 1861 he died. Preparations had gone too far to postpone the event.

Thomas was responsible for the Fine Art Gallery but the selection of the artists was done by a committee of presidents of several British art societies. Nor was Thomas able to influence the placing of the works of his friends.[117] However Thomas did have influence, as Hunt indicates:

Fig.8.6: Burton Park, Petworth, West Sussex. Built c.1830.

> One night in his smoking room, when Woolner and I were his fellow guests, he spoke of the need for an authorised handbook on the works of art in the Exhibition, and asked whether we knew of any writer competent to undertake the guide. Woolner ardently assured our host that he knew the very man, Francis Turner Palgrave … [however] the author's prejudice against all other sculptors but Woolner was rampant, and his admiration of him riotous.[118]

Hunt rightly foresaw trouble. In *The Times* of 15 May there was a storming letter from the campaigning journalist, Matthew Higgins, under his *nom de plume,* Jacob Omnium. Not only had Woolner been unduly praised but his chief rival, Carlo Marochetti had been severely denigrated.[119] Worse followed: Palgrave lived at the same address as Woolner. The Commissioners set in motion the book's withdrawal.

Hunt wrote, 'Some of Millais' and my pictures, and, for the first time, several of Woolner's works in marble, were exhibited ... he had a dozen fine examples of his marble carving in the Exhibition ... Thomas Fairbairn was proud to have been one of his early patrons'.[120]

The 1862 Exhibition turned out something of a family affair for the Fairbairns. Thomas was a Commissioner and loaned five paintings, and three sculptures by Woolner, including the bust of William.[121] Fairbairn & Co. exhibited a six-wheel Goods Engine 'of excellent workmanship', designed for the Midland Railway by its Engineer, Matthew Kirtley.[122] George Fairbairn exhibited a working model of a tank locomotive.[123] William was on the Building Committee and had undertaken tests (with William Baker) by placing 400 men on the galleries and measuring deflections.[124] He was President of the Jury for machine tools,[125] was on the committee for sanitary appliances in Class X,[126] and wrote a report on the Department of Machinery.[127] Thomas's cousin, Andrew, who was now the sole principal of Sir Peter Fairbairn & Co. of Leeds following his father's death in 1861,[128] exhibited a large selection of machine tools.[129] The Exhibition was inaugurated by a grand banquet given by the Lord Mayor at the Mansion House. Thomas and Allison Fairbairn were there, as was William.[130] On 1 May the Exhibition was opened with much pomp and ceremony.[131] Whilst inevitably there were criticisms, the Exhibition

Fig.8.7: W. Holman Hunt, *The Children's Holiday*, 1864.

overall was a success, and, although it did not make a profit, it covered its costs.[132] There were 6.1 million visitors.

The 1851 Exhibition had left a significant financial surplus and the 1851 Commissioners continued in office to administer the funds, as they do to this day, with new members added as necessary. In 1861 Lord Chandos and Thomas Fairbairn joined, amongst others, Thomas Bazley.[133] Prince Albert had chaired the Commission until his death on 14 December 1861, but not until April 1864 was the new chairman, Lord Derby, appointed. During his time as Prime Minister, 1866–8, the Commissioners met at 10 Downing Street. One of the Commission's tasks following Albert's death was to provide an appropriate memorial. The result was the Royal Albert Hall, to which the Commission made a significant financial contribution. Thomas was not on the 'working committee' which was just as well given that the Fairbairn Engineering Company Ltd constructed the iron roof and columns. The roof, 'of gigantic nature and novelty', was designed by Grover & Ordish. It was first assembled in Fairbairns' yard at Ancoats, where the 80-year-old William showed interest in the details.[134] Thomas was still going to meetings of the Commissioners in 1881 – on 12 July of that year he was at a meeting of the Commissioners at Marlborough House, presided over by the Prince of Wales.[135]

The third lure to London was the presence there of Thomas's Pre-Raphaelite friends. Spurned by Mancunians,[136] in 1861, with these appointments and friends in London, Thomas turned his back on Manchester and moved. Northwood was sold and the Fairbairns became the first occupants of the town house, 23 Queen's Gate, built on land the Commission owned.[137] Thomas also purchased Burton Park, near Chichester, a classical building of c.1830 set in parkland.[138] James Brooke and Alan Barttelot were born at Burton Park, and it was here that Hunt painted the large family picture, *The Children's Holiday,* to hang at the top of Burton Park's imposing staircase.[139] During a stay Hunt records another visitor was Edward Trelawny (1792–1881) to whom he talked about Keats and Byron – Trelawney had been a friend of Shelley and Byron: Thomas was moving in very different circles from his father.[140] The residual Calvinist culture of Scotland and the residual Puritan culture of Cross Street had fully dissipated.

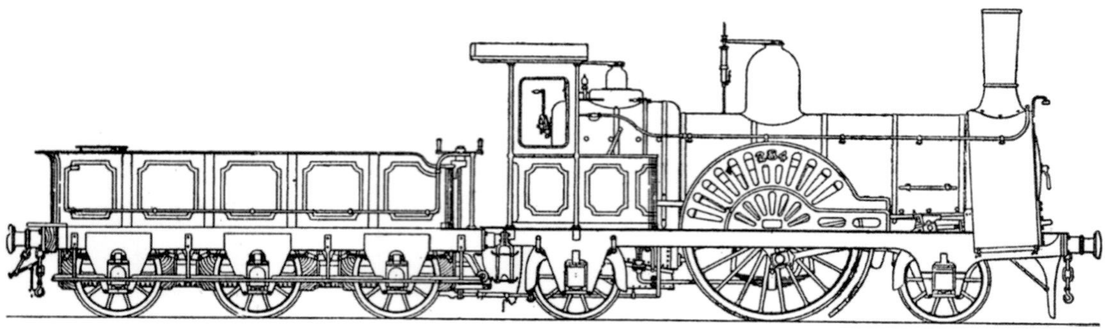

Fig.8.8: 2-2-2 express locomotive for the Great Eastern Railway. Designed by Robert Sinclair. Built by Fairbairns, 1862.

The Demise of the Locomotive Department

DURING the 1850s many fine locomotives were built, predominantly to the designs of railway company locomotive engineers, such as Clark of the Great North of Scotland Railway,[141] Kirtley of the Midland Railway,[142] and Sinclair of the Great Eastern Railway.[143] Engines were exported to Italy,[144] Australia,[145] Portugal,[146] India,[147] Canada,[148] Sweden,[149] Egypt,[150] and Spain.[151] There were also numerous engines for Ireland, mainly tanks.[152] But there was little in-house innovation, and investment in new machine tools was meagre. In 1856 an informed observer noted that the firm's machines were no longer up to date.[153] When Peter Carmichael had visited the north of England in 1837, Fairbairn's works had been on his itinerary: when he visited with his son and nephew in 1866, it was not, although Whitworth's, Platt's and Peter Fairbairn's at Leeds were.[154]

Thomas's approach, in contrast to his father's innovative engineering, was that of a trader. The L&NWR required twenty goods engines. Fourteen were to be built in-house and tenders were obtained for the other six. At £2,170 each Fairbairns' was the lowest tender and they were called to a meeting, at which point Thomas offered to reduce the price to £2,100 each for an order for twenty. His offer was accepted – all to be delivered that year.[155] The L&NWR also decided to invest in ten McConnell express passenger locomotives. Fairbairns did not submit the lowest price but won the contract by agreeing to take ten old engines at £450 each in part payment.[156] On another occasion Thomas wrote, albeit unsuccessfully, to the Lancashire & Yorkshire Railway seeking to persuade the company to increase its order for four 2-2-2 tanks to fifteen or twenty, as 'prices will rise in the next couple of years', and offering to accept part payment in L&Y's five-year bonds.[157] In 1860 when the Midland Railway accepted Fairbairn's tender for ten 0-6-0 goods engines, Thomas offered another twenty at £2,350 each, to be delivered within twelve months. Here he was successful, but there were delivery problems – thirteen of the thirty were up to 12 months late. The reasons for these major delays are not recorded. However, an unsuccessful request for payment on account may suggest there were cash-flow problems.[158] 1861 was a busy year, and the death of Richard Clarke at the start of 1860, after fourteen years managing the locomotive department, may have had a continuing adverse effect.[159] These first orders from the Midland Railway were also the last.

Company	Building locos in M/cr from:	Building locos in M/cr until:	No. of locos built in M/cr:
Sharp Roberts & Co. From 1847: Sharp Brothers From 1852: Sharp Stewart & Co.	1833	1888 [Moved toGlasgow]	3,442
William Fairbairn From 1846: William Fairbairn & Sons From 1859 Fairbairn & Company	1838	1863	c.500
Nasmyth Gaskell & Co. From 1850: James Nasmyth & Co. From 1860: Nasmyth Wilson & Co.	1838	1939	1,531
Beyer Peacock	1854	1966	c.8,000

Table 8.1. Major Locomotive Builders in Manchester (excluding railway companies)

On at least one occasion, it appears that Thomas may have resorted to a bribe. Sixteen locomotives had been built for the Great North of Scotland Railway. Although one had given trouble, John Ruthven, the locomotive superintendent, persuaded his directors to order three more. Fairbairns offered a fourth at a reduced price, and a contract was signed for four in August 1856. In December, Ruthven visited Manchester to inspect the work. As he was leaving, Thomas gave him a Christmas box of £50 – about £3,000 at 2015 values, which he accepted, but on returning to Scotland he handed it to his directors.[160] The GNSR ordered no further locomotives from Fairbairns.

In 1863 Thomas ceased locomotive building and disposed of this part of the Fairbairn business to Sharp Stewart & Co.[161] The Mill Street works was demolished and the materials sold.[162] The tools and machinery offered for sale included sixteen planing machines, thirty lathes, and twenty slotting and shaping machines.[163] So ended one of the 'best known locomotive builders' of the 1840s and 1850s.[164] The most likely reason for Thomas's closure of the locomotive department was an inadequate rate of return, caused by increasing competition, as more railway companies built their own locomotives and there was a growing number of Continental locomotive works.[165] The need for major investment in design engineers and the latest machine tools may have been an influence, as might having no railway access to the works, Clark's death, the somewhat erratic flow of orders,[166] and the loss of at least three important clients, the Great Northern Railway and the Midland Railway through late deliveries, and the Great North of Scotland Railway. That there were opportunities for locomotive builders is demonstrated by the continuation of the Manchester firms of Sharp Stewart, Nasmyth Wilson and Beyer Peacock as shown in *Table 8.1*.[167]

The Companies Acts of 1856 and 1862: The Fairbairn Engineering Co. Ltd

THE Joint Stock Companies Act of 1856, consolidated into the Companies Act of 1862, allowed the formation of companies on the principle that the liability of members was limited to the amount unpaid on their shares. Before these Acts business growth could be seriously restrained by discontinuity when a partner died, by difficulties in raising capital, and by each partner being liable for the debts of the partnership. This had tended to restrict businesses which involved more than minimal risk.[168] The Acts dramatically changed the scene. By floating a company, capital could not only be raised, it could also be realised. There were downsides – reckless risk-taking might be encouraged, the Act could be misused by the unscrupulous, owners were increasingly distanced from the operations of a business, and loyalty to long-serving employees might be reduced. William had no enthusiasm for limited liability companies.[169] Fairbairns was the only company of which he is known to have been a director, although he must have had many invitations to join company boards.

It has been claimed that the limited liability company is 'the greatest single discovery of modern times',[170] and it has been referred to as a 'Victorian invention that … changed the world'.[171] Nevertheless the move to limited liability was slow. After twenty years fewer than ten per cent of firms had become limited liability companies.[172] In January 1864, Thomas floated The Fairbairn Engineering Company Limited. The signatories of the new Fairbairn company included Thomas Fairbairn (7,000 shares), Augustus Novelli (1,000 shares),[173] and James McConnell (100 shares)[174] who were to be directors, Henry Harman (250 shares) who 'for some years has been the principal manager of the business' and whose services 'have been secured', and four others.[175] The other two directors were to be William Fairbairn and John Pender. Notably William was not a signatory.

There were 25,000 £10 shares – £1 payable on application and £2 on allotment. It was not intended to call up more than £5 per share. 23,220 shares were taken up.[176] The notice advised that the Directors had agreed to purchase the premises

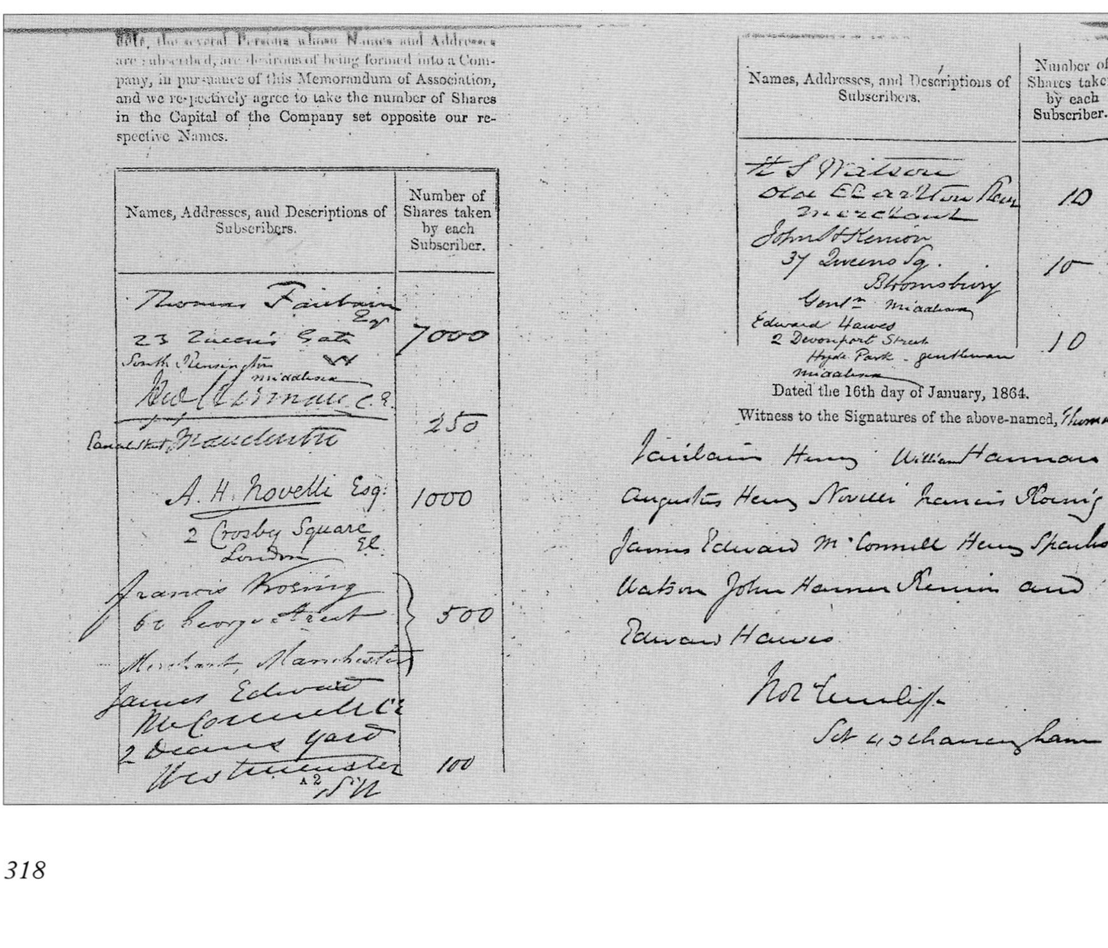
We, the several Persons whose Names and Addresses are subscribed, are desirous of being formed into a Company, in pursuance of this Memorandum of Association, and we respectively agree to take the number of Shares in the Capital of the Company set opposite our respective Names.

Names, Addresses, and Descriptions of Subscribers.	Number of Shares taken by each Subscriber.
Thomas Fairbairn Esq 23 Queen's Gate South Kensington W Middlesex	7000
[illegible] Manchester	250
A. H. Novelli Esq. 2 Crosby Square EC London	1000
Francis Koenig 62 George Street Merchant, Manchester	500
James Edward McConnell CE 2 Deans Yard Westminster SW	100

Names, Addresses, and Descriptions of Subscribers.	Number of Shares taken by each Subscriber.
H S Watson Old Elvaston Place Merchant	10
John H Kemion 37 Queens Sq. Bloomsbury Gent. Middlesex	10
Edward Hawes 2 Devonport Street Hyde Park - gentleman Middlesex	10

Dated the 16th day of January, 1864.

Witness to the Signatures of the above-named, Thomas Fairbairn Henry William Harman Augustus Henry Novelli Francis Koenig James Edward McConnell Henry Sparks Watson John Harman Kemion and Edward Hawes.

[illegible]

Sol 4 Salters [illegible]

Fig. 8.9: Signatures to The Fairbairn Engineering Company Limited Memorandum of Agreement.

and equipment of Fairbairn & Co 'on which ... large sums have been expended in the erection of new workshops and machinery'. In fact only two of the original four Fairbairn sites were transferred to the new company, the main Canal Street works and the Back Mather Street works.[177]

The flotation, together with the sale of the premises and equipment, enabled Thomas to realise capital which had been tied up in the company – even though he retained 28 per cent of the shares – and to invest the money where rates were higher, and in a landed estate. The total sum realised is not known but is likely to have been several million pounds in today's figures.

In the decade between William's 'retirement' and incorporation, the type of work undertaken at Ancoats was gradually moving from the firm's traditional base of millwork, steam engines and boilers, to concentrate on iron bridges and structural ironwork: there were some notable bridges including the River Barwon bridge in Australia,[178] the Dinting and Mottram viaducts,[179] and Keadby swing bridge over the Trent.[180] The list below gives a flavour of the type and scale of work undertaken by the company following incorporation and confirms its movement in the direction of structural engineering, notably lattice and plate-girder bridges, and roofs. These commissions represent only a small fraction of the work the firm must have undertaken during its eleven years as a public company but reports become less frequent as Fairbairns' work, with notable exceptions such as the roof of the Royal Albert Hall and the iron framework for the superstructure of the first two Spithead forts, was no longer at the leading-edge of engineering and thus less newsworthy. The roof of the Royal Albert Hall was an elliptical dome with diameters of 238ft and 272ft. It was formed of sixty arched iron girders.[181] Constructed 1870–72, the Spithead forts had an iron superstructure of 200ft diameter and 26ft high, designed to take 49 large guns in two tiers. The armour-plated cladding was by others. Fairbairns invested in new machine tools for the contract including several multiple drills capable of drilling thirty holes at once. The specification required all parts of the outer surface to be planed to a true circle. The manufacture, test erection at Ancoats and final erection at Spithead were all carried out under Harman and it was reported that the accuracy of construction was such that 'not a farthing of expense had to be incurred in corrections'.[182]

The following are some of the commissions undertaken by The Fairbairn Engineering Company Limited: dates are approximate – some contracts extended over several years.

1864	Plate girder bridge over Wyre Estuary: 16 spans.[183]
	Hauling engines for the São Paulo Railway.[184]
1865	Windsor Locks bridge, Hertford and Newhaven Railroad.[185]
	Logierat lattice bridge, Highland Railway.[186]
	Horbury Viaduct, L&Y Railway.[187]
	'Fairbairn' Crane, West India Dock.[188]
1866	Bridges for Indian railways.[189]
1867	Bridge over occupation road, Halifax.[191]
1868	Calder Bridge for the L&Y Railway.[192]
	'Fairbairn' Crane, Dover.[193]

	Ironwork for Manchester Police Courts.[194]
	Royal Albert Hall – roof and columns.[195]
	Crosland engine, Dewsbury. [196]
1869	Corn mill.[197]
	Steam engines for Jamaica.[197]
	Wrought-iron target.[197]
	Treadmill for Belle Vue Prison, Manchester.[197]
1870	Spithead forts and fortlets.[198]
1871	c.100 lattice and girder bridges for Intercolonial Railway of Canada.[199]
	Ironwork for basement of Manchester Town Hall.[200]
	Sugar Refinery, Egypt.[201]
1872	5-Tube boiler and 3-cylinder Crosland engine, Cleator.[202]
1873	Ironwork for incline at Oldham Road Goods Yard, Manchester, for L&Y Railway.[203]
1874	Ironwork for extension to Broad Street Station, London.[204]
	Ironwork for the roof of Liverpool Street Station, London.[205]
	Roof of the Library of Parliament , Ottawa.[206]
	Cornbrook Viaduct, Manchester.[207]

For the first two years the company paid a 10 per cent dividend. In 1867 the dividend dropped to 7.5 per cent and the orders on the books were 'much below the usual average.' A year later it was 6 per cent and in 1869 down to 5 per cent, with the Company facing a crisis.[208] Business 'scarcely exceeded one third' of its previous average and prices had been 'lower than in all previous experience'. Looking ahead, work in hand was meagre and remunerative rates 'almost impossible to secure'. The board forewent all remuneration and issued a profits warning: 'Shareholders must not be astonished if little or no profit be realised in the coming year'.[209] The Companies Act of 1867 permitted companies to reduce their capital and in 1869, as only half of the £10 shares had been called, the Company reduced the capital value of each share to £6.[210] By 1871 the dividend was back up to 5 per cent and the Company 'conceded to their hands the nine hours system'.[211] The 1872 dividend was increased to 10 per cent, with the firm 'fully and profitably employed, the works in progress being more extensive than at any time since 1866'. William, now 83, retired from the board, but not, it was stated, 'from his interest in the establishment'.[212] Clearly his name was still of marketing value to the company. The new profitability was short-lived. 1873 brought 'an enormous rise' in the costs of iron, coal, and all other materials which 'seriously diminished the profits of the year' and would probably 'be felt for some time to come'.[213] With Harman's failing health forcing him to retire, Charles Allott, chief draughtsman since 1862, was left to cope. The dividend slid back to 7.5 per cent in 1873. The next report referred to the unsettled state of the market during the year. The balance in hand was insufficient for a dividend to be paid, 'other than the 2.5 per cent distributed the previous October'.[214]

By 1874 Thomas had sold sixty per cent of his shares.[215] It appears he had lost confidence as well as interest in the firm. With no apparent attempt to replace Harman or promote Allott, and Thomas resident in Hampshire, the company was bereft of leadership. Thomas had shown by his leadership of the Art Treasures

Exhibition that he had the necessary administrative ability and drive, but he had chosen to distance himself from the smoke, the noise and the hassle of an Ancoats factory, and to pursue the more gentlemanly activity of non-executive directorships on London boards. The end was near.

Fig.8.10: Brambridge House, Hampshire. (Photograph by Peter Facey, 2005)

Gentrification in Hampshire

IN 1866 the Thomas Fairbairns, wealthy from the flotation, relinquished Burton Park and their London home at Queen's Gate. They moved to Brambridge House, Twyford, near Winchester, where they lived for twenty-five years until Thomas's death in 1891. Around the same time they also acquired a new London residence, 42 Wilton Crescent. An estate of around 500 acres has been identified as a prerequisite for entry into Burke's *Landed Gentry.*[216] With 600 acres, including four farms, Brambridge falls comfortably within this criterion.[217] Brambridge House was built in the 18th century. In 1872 it was damaged by fire, but the pictures were rescued. Thomas engaged the architect Matthew Digby-Wyatt who incorporated a large entrance-hall and staircase into the repairs.[218] The house then contained 'seven reception rooms and 23 bedrooms' with interior decorations 'of a refined character ... with no expense spared'.[219] The 1891 census showed eight resident staff.[220]

Thomas became High Sheriff for the County of Southampton in 1869.[221] The Assize sermons in Winchester Cathedral were preached by his brother, Adam.[222] On the Monday morning of the Assizes, the judges were conveyed to Court in Thomas's carriage, described in a local newspaper as:

> equipage ... of a very handsome kind; the coach has mounted panels, painted dark carmine, silver-beaded, and globe lamps; the wheels and underbed-work are in Chinese vermillion suitably picked out in blue and black; on the pannels [*sic*] are emblazoned the family arms ... The carriage was built by Peters and Sons, London. The servants' liveries, made by Poole, of London, were drab with scarlet facings, and the whole turn-out is very elegant.[223]

At the close of the Summer Assizes, there was a garden party at Brambridge for over a hundred local worthies to meet the Judge.[224] William, writing to his grandson, Arthur, was 'glad your papa is the High Sheriff but it costs a great deal of money'.[225] Children Arthur and Constance, without hearing or speech, were educated at Rugby College for the Deaf whereas Thomas Gordon, Reginald and James went to Eton and Cambridge.[226] When his father died in 1874, Thomas accepted the baronetcy.

The medieval church at Twyford required rebuilding. Thomas offered £1,000 towards the work on condition that either Waterhouse or Scott was appointed architect.[227] Waterhouse was chosen. Patronage of the established church marked another aspect of gentrification.[228] The new church building was Gothic Revival in style and the south transept was to form the Fairbairn chapel.[229] Culturally this was light-years distant from William and Cross Street Chapel.[230]

In 1885 *The Hampshire Advertiser* reported that Sir Thomas Fairbairn had seceded from the Liberal party, resigned from the Reform Club, and would be voting Conservative.[231] The cause was the matter of home rule for Ireland and on 22 May 1886 he attended the meeting in London from which sprang the Liberal Unionists,[232] and when the National Union Club was formed, Thomas was a vice-president.[233] Surely no example could better illustrate Martin Wiener's gentrification thesis for the decline of Britain than Thomas Fairbairn. Is he not the epitome of the businessman who 'increasingly shunned the role of industrial entrepreneur for the more socially rewarding role of gentleman'?[234] The answer is not so simple.

'Gentlemanly Capitalism' and John Pender's Companies

JOHN Pender, one of the foremost entrepreneurs of the second half of the 19th century, played a major role in Thomas Fairbairn's career.[235] Pender was a main driver of the great engineering achievement of the 1860s, the Atlantic Telegraph Cable. Moving from Glasgow, he became a textile merchant in Manchester, and friend of Thomas. He was one of the guarantors of the Art Treasures Exhibition, spoke in support of James Brooke, and supported Thomas's call for a Free Art Gallery. Pender founded numerous telegraph companies. By the time of his death in 1896, he controlled the largest and most successful international telegraph business, with 73,640 nautical miles of submarine cables – about one third of all the world's cables.[236] Like Thomas, Pender was an art collector; like Thomas he had a country estate and a London house;[237] but there the similarity stopped. Pender may fairly be described as a 'gentlemanly entrepreneur'.[238] To a meeting of Eastern Telegraph shareholders, he said,

> The greatest pleasure of my life is to attend to the interests and duties devolving on me as your chairman ... and I have made submarine telegraphy to a great extent my hobby.[239]

Some industrialists, like Pender and Thomas, did purchase country estates, but only a minority. Lancashire cotton masters preferred suburban mansions, within easy access of the mill.[240] F.M.L. Thompson identifies three possible relationships of the 'new men to the land'. First, the stereotypical successful businessman who, having purchased a country estate, ceased to be a businessman and became its antithesis, a landed gentleman of leisure. Second, the 'life-cycle' purchase – the rich man's version of acquiring a retirement home. The third case is 'a hybrid type of landed businessman', who sought to combine a landed position with continued activity in the business from which he had sprung.[241] This has some affinity with Thomas, but Thompson does not adequately allow for the profound effect of the introduction of limited companies. Not only did limited companies enable owners to realise tied-up assets and spread their risk more widely, but they facilitated investment from a distance without hands-on management. They allowed some to become company directors, as Thomas did, with the advent of railways making attendance at board meetings, usually in London, reasonably easy, and providing the opportunity to avoid the intense full-time involvement in industry which was incompatible with the gentlemanly ideal.[242] This does not resonate with Wiener's gentrification. It has some affinity with F.M.L. Thompson's 'pseudo-gentry' who made some show of having a gentleman's concern with farming and estate management but lived mainly on profits and dividends.[243] Thomas more accurately exemplifies the 'gentlemanly capitalism' described by Martin Daunton, which, with 'finance and commerce … removed from production'

> involved personal contact with members of one's own class rather than direct control of a large work-force; offered freedom from daily cares; were carried out in fashionable areas rather than in a grimy industrial environment; and were reconcilable with the gentry ideal.

A 'gentlemanly capitalist' did not despise the market economy, but he did hold production in low regard and avoided full-time work.[244]

Thomas joined the boards of several companies. He provides an early example of networking amongst directors of limited companies. In 1862 the board of the Bank of Manchester was augmented by Fairbairn, Pender and J. Aspinall Turner.[245] In 1863 it amalgamated with a London bank, forming the Consolidated Bank, with directors including Fairbairn, Pender, Turner, William Smith and Ivie Mackie.[246] Subsequently Thomas's friend, Augustus Novelli, was appointed. On 10 May 1866 panic followed the collapse of Overend Gurney, with the Bank of London sinking in the wake. The Consolidated Bank's directors announced that the Bank of London had transferred its customers' accounts to the Consolidated, with 'assets to cover the same'.[247] Four days later, a notice was published closing the Consolidated Bank following the discovery of 'serious errors' in the submitted list of assets.[248] However the problems were not fatal and the bank reopened. In the mid 1870s Thomas was still on the board of the Consolidated Bank,[249] where, unsurprisingly, the Fairbairn Engineering Company banked.

The directors of the Consolidated Bank were involved in the flotation of the wholesale drapery business of Morrison Dillon & Co – as the Fore-street Warehouse Company Limited – enabling the sons of one of the richest men in the country, James Morrison, to secure fortunes from their late father's drapery business, for which they had no enthusiasm.[250] In contrast to Fairbairns, two active partners from Morrison, Dillon & Co. joined the Fore-street board, with the other initial directors who were from the Consolidated Bank.[251] Later Thomas Fairbairn and Augustus Novelli were added.[252] The company traded until around 1940.[253]

In 1870 Thomas Fairbairn joined Pender's 'Telcon' Board.[254] Again unsurprisingly, the company banked with the Consolidated Bank.[255] Work flowed in, the majority from other Pender companies. There were many significant submarine cable contracts during the time Fairbairn was a director.[256] 'Telcon' was paying dividends of 25 per cent in the early 1870s, when railways were paying around 7 per cent.[257] In 1880 the Chairman wrote, 'our difficulty ... has been to keep the dividend down', and during the 1880s, it was a matter of 'our usual dividend of 20 per cent'.[258] For over two decades, 1870–1891, Thomas Fairbairn was a director of this hugely successful company which, until well into the 20th century, led the world in the manufacture and laying of submarine cables.[259] Other Pender companies of which Thomas was a director were The Eastern Extension, Australasia and China Telegraph Company Limited,[260] and The Brazilian Submarine Telegraph Company Limited.[261]

Thomas Fairbairn's second chairmanship was of Eley Brothers Ltd, a company founded in 1828 and famed for sporting ammunition.[262] In 1874 Eley Brothers incorporated to raise money to fund expansion.[263] In contrast to Fairbairns three of the former proprietors became executive directors.[264] The company's bankers were the Consolidated Bank.[265] The reports suggest Thomas was a prudent chairman of a successful company. At the 1879 AGM he announced 'the usual dividend' of 15 per cent for the year plus a bonus of the same amount. He referred to a reserve fund of £40,000 which he wanted to see invested in an immediately available security, to meet any contingency that might arise.[266] At the 1886 meeting, notwithstanding 'the utter stagnation of trade', the balance-sheet was as good as they had ever presented. He saw no reason why the company should not continue to be successful, as it still is.[267]

Table 8.2: Thomas Fairbairn and Directors' Linkages

	Thomas Fairbairn	John Pender	Augustus Novelli	J A Turner	William Smith	Ivie Mackie	Thomas Hankey	J C Pickersgill	William Fairbairn
Art Treasures Exhibition Guarantor	X	X	X	X		X			X
Bank of Manchester	X	X		X		X			
Consolidated Bank	X	X	X	X	X	X	X	X	
Fairbairn Engineering Co.	X	X	X						X
Fore-street Warehouse Co.	X	X	X	X	X		X	X	
Telegraph Const & Maintenance Co.	X	X							
Eastern Extension A&C Telegraph Co.	X	X							
Brazilian Submarine Telegraph Co.	X	X							
Eley Brothers	X				X				
Metropolitan Artizans' & Lab's Dwellings A.	X								

The striking linkages between the directors of these early public companies are shown in *Table 8.2*, providing an insight into the networking of the new genus of gentlemanly director. The roots of the network were amongst Thomas's friends who had supported the Art Treasures Exhibition, with the dual hubs of the Consolidated Bank, which held the accounts of most of the companies, and John Pender.

The Demise of the Fairbairn Engineering Co. Ltd

SIR William Fairbairn died in 1874. Nine months after his funeral, the Directors of The Fairbairn Engineering Company Ltd regretted that

> the years working had resulted in the loss of £9,874. This loss had been mainly incurred in connection with one large contract, which was now all but completed. Competition had been keener than ever ... The relations with the workmen, moreover, had in no wise improved ... Already foreigners were taking no inconsiderable portion of the work. Under these circumstances it would be for the consideration of the shareholders whether, before greater mischief be done, it would not be the wisest plan to abandon a business which the present Board do not see its way to carry on profitably. The Board recommended that this course should be adopted, and that the ... assets of the company should be realised as speedily as possible.[268]

On 25 May 1875 an Extraordinary General Meeting of the Company, held in London, resolved that the Company be wound up. Fairbairn and Novelli were appointed liquidators.[269] This was a terrible blow to long-term employees. The extent which realised assets were able to reimburse shareholders, if at all, is not known but, with prolonged difficulties in disposing of the Ancoats buildings and subsequent calls on shareholders, it is clear they lost money.[270]

Why did the company cease trading? What was the large loss-making contract to which the directors referred? Gordon Biddle identifies it as the roof of Liverpool Street Station in London.[271] This may have contributed but it was probably the Intercolonial Railway bridges contract in Canada. This contract was beset by delays, not of Fairbairns' making, and in 1878 the liquidators sent an agent to Ottawa to seek to obtain a settlement. After several months he was constrained to accept an offer which was very much less than the liquidators felt was due.[272] Whilst a loss-making contract may have caused financial difficulties, it did not bankrupt the company, and may have been a pretext. Three other reasons were given by the directors: keen market conditions, difficult labour relations and foreign competition. These too appear to be excuses. Of thirty-two leading North-west firms of mechanical engineers active during the years of the Fairbairn firm, 1817–75, only one, Fairbairns, ceased trading during the 1870s. Twenty-two (almost 70%) continued into the 20th century. These included older firms such as Whitworths, Nasmyths, Hicks, Galloways, Hetheringtons and Lairds, as well as firms founded in the 1850s such as Daniel Adamson, Beyer Peacock, Craven Brothers (two of whom had worked for Fairbairns)[273] and George Saxon (who had been an apprentice at Fairbairns).[274] These firms covered a wide range of work – boilers, steam engines, locomotives, machine tools, textile machinery, cranes and shipbuilding. A more poignant parallel is the Leeds engineering firm founded by William's brother Peter which, on Peter's death in 1861, was inherited by his son, Andrew, and continued well into the 20th century.[275]

The decision to major on structural ironwork did not help Fairbairns, but the causes of the demise were more fundamental and stretched back over two decades. They were fourfold and interwoven: succession, leadership, lack of innovation and lack of investment. Chapter 4 discussed William bringing in his sons as successors in what he must have hoped would be a dynastic company. John disappeared to India. George and William Andrew retired very early, but neither had the ability to lead a major company. Thomas certainly did have the ability, but he had no interest in engineering. There were men who could have taken Fairbairns on into the next century. John Haswell, who set up the Vienna locomotive works for Fairbairn and stayed on to become one of the best known locomotive builders in Europe, was one.[276] J.F. Bateman ,William's son-in-law was possibly another, and William Unwin might perhaps have been another. Three years after Unwin had moved to be works manager for Williamson Brothers, William Fairbairn initiated his return to Ancoats as Manager of the Engine Department. However Unwin stayed less than a year, leaving without a job to go to. The probability is that Unwin and Thomas were incompatible.[277] The desirability of bringing extra-familial talent to substitute for absent or withdrawing talent is illustrated by Thomas's cousin, Andrew Fairbairn, in Leeds. He formed Fairbairn, Naylor & Kennedy, majoring on flax machinery and machine tools with on-going success, allowing Andrew to pursue his interests as Mayor of Leeds and Member of Parliament.[278] In contrast to William Fairbairn, the two other members of the Manchester triumvirate, Whitworth and Nasmyth, had no sons to enter their firms. They brought in partners and their firms continued into the 20th century.

Succession and leadership go hand in hand. An absent partner, an absent proprietor, and then an absent chairman, with no executive manager on the board, are unlikely to nurture a successful company. Thomas was absent for the most part of two years for the Art Treasures Exhibition, absent in London from 1862, and absent in Hampshire from 1866. With a 600-acre estate and shrieval duties at the other end of the country, plus meetings of commissioners and boards in London, he had little time for Ancoats, and he failed to bring in executive directors to drive forward the day-to-day business of the company.

Coupled with the lack of leadership, a primary cause of the Fairbairn company's demise was the lack of innovation after William's retirement. Thomas never took out a patent, he undertook no research, and he read no papers to learned societies. Gone were the days when William had innovated in whatever field he was working. Now as windows of opportunity for a product closed, new windows were not being opened. Two pairs of examples illustrate William Fairbairn's achievements, their subsequent decline, and the absence of replacements.

William Fairbairn brought the waterwheel and flour milling with millstones to their zeniths – the culmination of two thousand years of development. But, following work by Poncelet and Fourneyron in France, and Francis in America, greater efficiencies were being achieved from turbines.[279] Unwin had left Fairbairns to build turbines designed by Fairbairn's former pupil, James Thomson.[280] The engineering division of Escher Wyss, developed in Zurich by Fairbairn's former pupil Albert Escher, manufactured turbines from 1839.[281] Yet Fairbairns continued to build only vertical waterwheels, for which demand decreased and gradually petered out. So too with millstones. Before the end of the 19th century millstones were finished. Roller milling was faster, used less power and provided a better product. There were experiments with roller mills in Switzerland as early as 1825 and roller milling was used in Budapest in 1839, with roller mills on a larger scale established there in 1874.[282] In Britain Fisons at Ipswich used roller milling for semolina in 1862, and several mills introduced the Hungarian system.[283] This was followed by a system developed by Daverio in Zurich and Henry Simon in Manchester.[284] The Daverio-Simon system was first used in England at McDougalls, a couple of miles from Ancoats.[285] Yet Fairbairns took no part in the rise of roller-milling.

The tubular-girder bridge and the tubular crane, in contrast to waterwheels and millstones, were innovations in the 1840s. Tubular-girders were used for some of the longest bridges but, by the mid 1860s, lattice and trussed girders had become the preferred options. Whilst Fairbairns built lattices they were no longer leading designers and the Intercolonial Railway's choice of pin-jointed trusses from a firm in Philadelphia for long spans was understandable.[286] Fairbairn cranes were manufactured over a slightly longer period of time. For heavy loads they were the preferred crane for docksides for twenty-five years from 1850. But by the mid 1860s lattice jibs were becoming the norm for all but the largest cranes, which had to wait until the mid 1870s for more radical solutions by such as Armstrong Mitchell and James Taylor.[287] It was not just the structure of the cranes that was changing. Their means of power were changing too. As early as 1846 William Armstrong had introduced hydraulic cranes and, in the last quarter of the century, manufacturers such as Stahl were building electrically operated cranes.[288] Fairbairns continued to build decreasing numbers of the same cranes they had built since 1850, with the same manual or steam powered operation. They were left behind.

It was not that there were no new opportunities. They were legion. A section of Chapter 7 focuses on William Fairbairn's involvement with steel. It was soon in use for railway lines, axles, boilers, beams, ships and cables, as it flowed from the new Bessemer furnaces, replacing wrought-iron. But the Fairbairn company was not involved in the manufacture or use of steel. Only two miles from Ancoats, at Knott Mill Ironworks, Fairbairns' boiler-making competitors, W. & J. Galloway & Sons, were securing orders for Bessemer steel-making plants for firms which were to operate them under licence from Bessemer. The Galloway sons were active partners in the business, which did not become a public company until 1899, but continued until 1932.[289] Musson identifies many new opportunities in this period – steelmaking, turbines, marine engines, hydraulic machinery, machine tools, rotary and composing machines in printing, the Siemens tank-furnace in glass-making, roller flour milling, the combine harvester, and the sewing machine.[290]

Innovation and the seizing of opportunities usually require considerable capital investment, in machines and in design engineers. Steel-making, turbines, roller-milling, trussed girder bridges, hydraulic machinery and a new generation of cranes could not be introduced without high calibre design engineers – and to secure their services would require either very high salaries or a share in the business. In the absence of company accounts there is little hard evidence, but all the indications are that once Thomas was in control there was meagre investment. This is consistent with Thomas investing his money elsewhere.

Musson suggests five reasons for Fairbairns' demise – loss of pioneering drive, 'too many irons in the fire' (following Ahrons), the onset of a serious trade depression, William Fairbairn being too much concerned about being a public figure coupled with establishing his status as a scientist and writer, and that the sons were less able men.[291] Hayward's reasons for the demise are the loss of Harman in 1873, the lack of proximity to a railway, unsuitable multi-storey buildings, high port-charges at Liverpool, and primarily the economic depression extending to 1896, visible in the run-down state of Ancoats.[292] Collier's reasons are lack of pioneering drive (following Musson),[293] too much diversification across general mechanical engineering (following Ahrons), increasing local competition, the general down-turn in trade, the retirement of Harman, and management problems.[294] Chaloner attributes the winding-up to 'a depression in trade'.[295] There are no references to the role of Thomas or the lack of investment.

The loss of pioneering drive is absolutely right, as is the reference to management problems, but what was the cause of that loss and those problems? The other reasons are questionable. By the time of the demise, Fairbairns had ceased shipbuilding and locomotive building and were concentrating on steam engines and, particularly, structural ironwork. By then there was only one son in the business and it had been a public company for over ten years. The 'serious trade depression' did not dislodge other leading North-west engineering firms in the 1870s. Saul believes the 'Great Depression' to be a myth,[296] and Emma Griffin has provided evidence that there was a two per cent annual growth rate in both manufacturing output and GDP between 1873 and 1913.[297] Crouzet simplifies the position to 'two long phases, on each side of a watershed situated around 1865–75 – one of "rapid growth" and one of "slower growth"'.[298] In an earlier paper Musson wrote that it was not expansion of production that had ceased, but that the annual rate of expansion of production had declined.[299] It is true that for William Fairbairn fame was a spur, but it spurred the innovations and projects that drove his ascendency. It does not appear to be have been appreciated that William was 65 when he withdrew from the day-to-day management of the firm, and that it was two decades later that the demise of the firm took place – the two decades of William's 'retirement' during which he wrote most of his books and took on his more important public roles. This illustrates a not uncommon failure to appreciate Fairbairn chronology. Staffing, location, buildings, transport charges, competition and the general economic climate were matters faced by almost every on-going company – and the North-west's other leading mechanical engineers survived.

Hayward considers the firm was right in the early 1860s to concentrate on structural iron fabrication and erection because it did not require a large design staff, even though 'it turned out to be an unfortunate decision'.[300] It was the wrong decision. Structural ironwork was an area where consulting engineers were increasingly involved in design. For most contracts the work was relatively simple to fabricate and erect. The result was that many fabricators were able and eager to price the work. This is illustrated by Fairbairns' tender of £11,500 in 1868 for ironwork for a bridge at Nottingham. Of twenty-four tenders, twenty were lower, the lowest being £7,072.[301] Fairbairns appear to have only come into their own on large, unusual and complex projects such as the Royal Albert Hall,[302] and the iron frameworks for two Spithead Forts.[303] In earlier days Fairbairn's ascendency was fuelled by innovative designs, rather than the execution of the designs of others.

The tools, plant, machinery and stores were auctioned. The company's creditors were all paid in full. In January 1876 the land and premises were advertised for sale, but could not be disposed of at an acceptable price, and were retained.[304] Some of the premises were let to tenants. Novelli resigned as liquidator in 1882, and was replaced. There was a problem: the rental income was less than the outgoing chief rents. In January 1889 Fairbairn resigned as liquidator and was replaced; and in February the Back Mather Street works was mortgaged to him for £6,000, probably explaining his resignation as a liquidator. In December he advanced a further £500. A year later the Company made a call on shareholders of 10*s*. per share, and in February 1891 Fairbairn's mortgage was repaid. Six months later he died. The liquidators made another attempt to sell in April 1892. In October 1892 a further call was made of 5*s*.6*d*. per share. Due to the unusually long time that had elapsed since the company went into liquidation, some of the shareholders were dead, with insolvent estates, some had disappeared, some were abroad, and some declined to pay. In August 1893 the liquidators went to Court and established that in the event of non-payment of a call they could sell a share or make it forfeit.[305] Finality came at a meeting on 26 April 1899. Resolutions were passed that the account of the winding up be received and adopted and that 'the books accounts and documents of the Company and of the Liquidators be destroyed by burning or otherwise'.[306]

Fairbairns was not unique – there were others with some similarities – Marshalls of Leeds is the classic example, of whom Rimmer writes:

> having made their fortunes, Marshall's sons thought in the 1840's that they could safely neglect the mills and, instead of searching methodically for profit and improvement as their father had always done, they collected whatever the mills had to offer, regarding profits with the fatalism of a peasant awaiting the harvest.[307]

Yet more often than not in family businesses, the second, and sometimes the third, generation rose to the challenge. Howe points out that by 1830 the nucleus of the cotton industry was 'relatively stable family firms'.[308] In the highest echelons of engineering, Robert Stephenson, I.K. Brunel, brothers George and John Rennie, John Penn II and the Lairds were all second generation engineers; and amongst the majority of Fairbairn's contemporary leading North-west engineering firms the immediate successors to the founders were hereditary.

An Unhappy Ending

IN June 1873 Thomas commissioned Holman Hunt to paint his portrait. The picture was exhibited at the summer Royal Academy exhibition. The background to the portrait of the overweight Thomas was perhaps intentionally suggestive, as it features the cabinet at the South Kensington Museum containing the Gubbio majolica drug-pot from the Soulages collection, purchased for the Art Treasures Exhibition in 1857.[309] Thomas's portrait was not well received. Why had Hunt 'selected a pose that is about the least graceful that he could have chosen'?[310] It was one of several pictures where Hunt was faced with expensive re-working due to adulterated orange vermillion pigment, bought on the advice of Thomas, and causing the flesh to turn black.[311]

The latter years of Thomas and Allison's life were marked with sadness. In August 1874 Thomas's father had died. There was the funeral to organise and letters of condolence to be replied to. Subsequently Thomas gave his father's library to the Manchester Mechanics' Institution. Then in November the Fairbairns' youngest child, Alan Barttelot died aged ten. Two months later their 17-year-old daughter, Florence, died. They were buried in the family vault at Twyford. The parents took a holiday in Biarritz to help them recover,[312] and the following year in the spa town of Carlsbad.[313]

Things appeared to improve. In 1877 there was a 'family' wedding when Thomas Gordon, the Fairbairns' second son, married his cousin Ada Maria, the youngest daughter of Thomas's younger brother, William Andrew, in a ceremony conducted by Adam Fairbairn, Vicar of Waltham St Lawrence, (the bride and bridegroom's uncle), assisted by the William Bateman, Vicar of Norwood, a cousin of the bride and bridegroom, and the eldest son of J.F. Bateman and Anne Fairbairn.[314]

In 1878 Allison and daughter Mary (May) attended the Queen's Drawing Room, and again in 1879.[315] At the Fairbairns' London house in Wilton Crescent, there were dinner parties and dances.[316] Marriage was in the air again in 1880 when third son, Reginald, married May Elizabeth Holt,[317] and his sister Mary married Archibald Lennox Napier, eldest son of Sir Robert J. Milliken Napier, Bart, in a ceremony conducted by her Uncle Adam, and attended by her 88-year-old grandmother, the Dowager Lady Fairbairn, and Mr and Mrs Holman Hunt, amongst others.[318] Five days later, Thomas and Allison left 'for Pau for the winter'.[319] Together with nearby Biarritz, Pau was a favoured destination of the English aristocracy in the latter part of the century, with packs of fox-hounds, tennis courts and golf courses.[320]

On 11 January 1882, Dorothy Fairbairn died aged 90, at Adam's vicarage.[321] She had quietly and loyally supported her husband over their long married life, entertaining his many guests. So there was another funeral at Prestwich. No longer were there any Fairbairns at the Polygon, or indeed, after seventy years, in Manchester. Thomas's last link with Manchester was cut.

The next family event was Arthur's marriage later in 1882 to Florence Long. Those who signed the register included Arthur's aunt, Anne Bateman, and among the guests, twelve years on from the death of James Brooke, were Spenser St John, his secretary, and Angela Burdett-Coutts, his friend and backer.[322] The marriage was not successful, and the couple soon separated.[323]

In 1887 Holman Hunt wrote to George Craik that Thomas was 'much dejected now',[324] and some of Fairbairn's 'Hunts' were sold at Christies.[325] Why was this? The answer lies with Thomas and Allison's second son, Thomas Gordon – he is on the right in *The Children's Holiday*. After Eton and Jesus, he became a partner in a London stockbroker. On 15 October 1887 the firm was 'hammered … Mr. Fairbairn's default being attended by painful circumstances'.[326] Those circumstances were unsuccessful speculation on the Stock Market using clients' money. The press reported:

> A private meeting of the creditors of the stockbroker Fairbairn who disappeared last week, was held to-day, and at it a worse state of affairs was disclosed than was generally imagined. It is now stated that £50,000 of the property of clients is missing, and that this money has not been refunded by the young man's family. He had had £20,000 from his father, but that appears to have gone long ago. In addition to these sums there are the 'differences' due in the Stock Exchange, which are put at £12,000. Altogether, therefore, the losses are upwards of £80,000.[327]

A warrant was issued for his arrest, but by then he had fled to France, and boarded a steamer to the United States.[328] *The Times* carried a notice, 'Twenty Pounds Reward – Wanted, Thomas Gordon Fairbairn ... Stockbroker, who absconded in the month of October last'.[329] He was declared bankrupt.[330] What happened to Ada Maria and their four children is unclear, nor is it clear what he did in America. Six years later Ada died, aged 37, and in 1899 Thomas Gordon married Jennie Davies. They lived in Boston, Massachusetts, and had four children.[331] Thomas Gordon and his father would never meet again. His parents must have been devastated, and Thomas's health cannot have been helped by this trauma. They fled the publicity, going to the spa at Pau, but in mid April, news was received that Thomas 'who has been in ailing health for some time past, was seized ... with a paralysis in the left side'.[332] He came home to be faced with the death of his youngest brother, Adam, in July 1888;[333] and the following year the death of his brother-in-law, J.F. Bateman.[334]

Thomas died on 12 August 1891 aged 68 and was buried at Twyford, in the family vault alongside Alan and Florence.[335] His funeral was modest and his decease was scarcely noticed in Manchester, where, at the time of the Art Treasures Exhibition, he had enjoyed fame and prestige. The *Manchester Guardian* devoted all of twenty lines of one column to his obituary,[336] and the *Manchester Times* a mere fourteen lines.[337] The contrast with his father's passing could hardly have been starker.

The Legacy of the Baronetcy

Hereditary honours are not without their critics, but they do draw attention to the fame of the original recipient. Thomas's estate was valued at £63,544 (c.£5m. in 2015). Family treasures were left to Arthur and provision made for Allison. There were bequests to the surviving children, except Thomas Gordon.[338] Arthur, without hearing or speech, inherited the baronetcy. Brambridge House and its estate were sold.[339] The family moved to 45 Brunswick Square, Hove.[340] In 1897 Arthur married again, to Gertrude Engledue, sixteen years his junior, at Westminster Registry Office.[341] What happened to Arthur's first wife, Florence Long, remains unclear – had they divorced, or was she deceased? Nor is it clear where Arthur and Gertrude lived during the first years of their marriage, but in 1909 they moved to Chichester.[342] Many of Thomas's pictures were sold in 1892 and 1895.[343] Others were retained by Arthur and passed to Gertrude, to be

Fig.8.11: Arthur Henderson Fairbairn (1852–1915), 3rd Baronet.

sold by her at Christie's in 1946.[344] *The Awakening Conscience* was sold anonymously at the same sale, presumably by her.

Of the thirty-five known grandchildren of William and Dorothy Fairbairn, the one who is best remembered is Thomas's eldest son, Arthur Henderson. This is on two counts. Firstly, amongst art historians as the subject, with Emma Constance, of Woolner's iconic sculpture of them as children; and secondly for his work on behalf of the deaf. He was a remarkable man who travelled extensively – Denmark, France, Germany, Italy, Norway, Russia and Sweden – and was an all-round sportsman, an accomplished equestrian, enthusiastic cyclist, and participant in field sports.[345]

Arthur and Constance were both deeply involved in many charities supporting those without speech or hearing, particularly those within the Church of England, where their Christian faith found active expression.[346] Arthur was Honorary Treasurer and a Trustee of the Royal Association in Aid of the Deaf and Dumb, with William Gladstone and the Duke of Westminster on the committee.[347] In 1912 he was chairman of a delegation to the Paris Congress of the Deaf.[348] Well respected, he was the subject of numerous articles.[349]

In 1904 Constance died aged 53. She and Arthur, brought together by their disabilities, had been close and lifelong friends. She was buried in Tunbridge Wells Cemetery. In 1907 Allison Fairbairn died, and she too was buried at Tunbridge Wells, as was James Brooke Fairbairn in 1947, and Arthur's wife, Gertrude, in 1952.[350] Herein lies a mystery – why was Allison not buried with her husband in the family vault at Twyford where there was ample space for her? Arthur died in 1915, aged 63. He was buried next to Constance, with the Woolner sculpture of them as children on a plinth above the graves, beneath a stone baldachin.[351] *The Times* reported Arthur's death on 4 June, referring to his marriage to Florence in 1882, but with no mention of Gertrude. It said that he 'leaves no issue, and is succeeded by his brother, Mr. James Brooke Fairbairn'.[352] With telegraphic communication now well established, the wires must have hummed as, five days later, *The Times* reported that 'Sir Arthur Henderson Fairbairn is succeeded in the title by his brother, Thomas Gordon Fairbairn, not James Brooke Fairbairn, as stated on Friday'.[353] In 1923 Thomas Gordon returned to England, to appear in the Bankruptcy Court in connection with the receiving order made against his estate thirty-five years earlier, but the criminal charges appear to have been dropped. The Official Receiver made him an allowance of £230 a year for his maintenance.[354] Thomas Gordon was now 68. Nine years later, in 1931, he died.[355] His only son with Ada Maria, Roderick Gordon, had died in 1909 and so his second son, William Albert, with an American mother and brought up in Boston, became the fifth baronet. He had come to England in the 1920s and married Christine Croft who achieved distinction in the gardening world.[356] In 1972 William Albert died and his son, the present baronet Sir James Brooke Fairbairn – William's great-great-grandson – succeeded to the title, the 6th Baronet Fairbairn of Ardwick.[357]

Conclusion

THE demise of the Fairbairn company was brought about by lack of leadership – initially an absentee proprietor, and then, after the flotation, by absentee directors. There was no executive on the board and the employed management, whilst sound engineers, were neither innovators nor entrepreneurs, and were almost certainly given no opportunity to be so. There was neither innovation nor sufficient investment for the future. The contrast with Pender's leadership of his companies is obvious and both Fore-street and Eley Brothers had previous partners as executive directors. The engineering industry was changing rapidly but Fairbairns was not moving with it. The many new opportunities, which would have needed investment in highly-paid design engineers and expensive machinery, were not taken. Whilst in previous decades William's emphasis was on engineering achievement, and profit often (but not always) followed, with Thomas's emphasis on profit, it tended to be more elusive. There were many engineering opportunities in the second half of the 19th century which other North-west engineering companies took, but which Fairbairns failed to take.[358] The most able premium apprentices and young journeymen no longer made their way to Ancoats. Burnham and Hoskins in their study of the iron and steel industry came to the view that 'if a business deteriorates it is of no use blaming anyone except those at the top'.[359]

Behind the immediate causes of demise there is the underlying question of why the succession failed, especially so as William, avoiding the common pitfalls, had retired and handed on the control of the company. The engineering industry was changing rapidly and rule-of-thumb had given way to a much more theoretical approach. Increasingly engineers were being trained at university level, sometimes in addition to a premium apprenticeship. The older Fairbairn boys experienced neither. They had some maths lessons from Hodgkinson and worked in the firm's office. All the indications are that Thomas was the only one of the older sons with the ability to undertake a university course, but financial circumstances at the relevant time apparently prevented this, and he had no interest in engineering beyond the money it brought him. William took his sons into partnership too soon – before they had had a thorough engineering education. His arrangements for succession included no recruitment of people of ability from outside the family, and the firm's subsequent demise endorses Rose's argument on the need for such leavening.[360] Thomas could have saved the firm by bringing in competent partners as his cousin Andrew did in Leeds, but he chose not to do so.

The effects of the 1862 Companies Act were far-reaching and possibly some were not anticipated when the Act was passed. 'The Act of Sixty-Two' may have changed the world, but not all benefited from it.[361] Thomas provides an early illustration of an owner realising capital through the use of the Act, a subject on which little has been written. Indeed Daunton does not mention this as a means of entry into the 'gentlemanly economy' when he describes the 'gentlemanly capitalist', removed from the grimy industrial environment to a landed estate, but living mainly on dividends, profits and fees arising from his shareholdings and directorships.[362] A better example than Thomas is unlikely to be found. Thomas also illustrates networking amongst directors of early limited companies – with in his case several links back to the guarantors of the Manchester Art Treasures Exhibition.

Gentrification *per se* was not necessarily the cause of company failures, as Pender showed by his gentrified entrepreneurship; nor was involvement in activities divorced from engineering, as was shown by Andrew Fairbairn. Similarly land ownership was not *per se* detrimental, as has been shown of some of the Greg family who combined successful business careers with the purchase of land, as a safe investment for surplus capital rather than for social prestige.[363] However Thomas Fairbairn is distinguishable from the Gregs in that one of his motives for the purchase of land *was* social prestige, and his purchase of a landed estate *was* made from realised capital, rather than from surplus profits.

Thomas may be viewed in at least two disparate and contrasting ways. First, that he was a shrewd and successful businessman. Finding his assets tied into a tired and declining business, paying a dividend of around seven per cent, and with the risk that vagaries of the market or a disastrous contract could precipitate bankruptcy, he took steps to use the new legislation to move his assets out of the company – an entrepreneurial exercise in itself. He then reinvested them, partly in relatively risk-free landed property and partly in companies paying dividends of 15–25 per cent. This enabled him to live in an attractive part of the country and receive dividends and director's fees for congenial work, which involved neither the grime of the foundry nor the inevitable hassles of an engineering business with a large labour-force. Here was the free market economy at work, with Thomas Fairbairn using it with considerable astuteness. It is an approach that William Fairbairn would have totally repudiated

The alternative view is that Thomas, with the privileges of wealth, had a duty towards those who had helped to generate that wealth, the skilled craftsmen and other employees, some of whom had worked for Fairbairns for decades; and a duty towards those who had placed their confidence in the firm's management by buying its shares. When William retired, the firm was far from being a tired and declining business, and opportunities in engineering were many. If by 1860 the business had started to decline it was because Thomas had failed to provide leadership and investment. For him to take out his money, rather than invest in management, machinery and innovation, was entirely selfish, showing a contemptuous disregard for employees and shareholders alike; and illustrates a wilful neglect of the adage that the first duty an employer of labour owes to those who work for him is to make his business succeed.

Technical innovation, economic circumstances and social change are factors which may force the closure of some businesses. Such closures may be inevitable adjuncts to a system that has brought, and continues to bring, such dramatically improved standards of living to so many. The primary cause of the closure of the Fairbairn works was not encompassed by these criteria. The gifted Thomas sought pleasure in art and the gentility of a stately home. He died sad and forgotten. His father, living for sixty years within walking distance of the noise and sweat of the foundries, forges and factories of Ancoats, achieved fulfilment and fame, as he single-mindedly followed his calling as a millwright and engineer. The Fairbairns do not mirror the Buddenbrooks but there are some similarities.[364]

Notes for Chapter 8

1. He is the subject of an entry in the *Oxford DNB* where the index describes him as 'art administrator and patron' (J Bronkhurst, 'Fairbairn, Sir Thomas, second baronet (1823-1891)', *Oxford DNB on-line*, (2004), (accessed 1 October 2015).
2. See Chapter 4.
3. *Life*, p.342.
4. J Bronkhurst, 'Fruits of a Connoisseur's Friendship: Sir Thomas Fairbairn and William Holman Hunt', *The Burlington Magazine*, 125, 1983, 587n9.
5. G J Eagling and A F Dimmock, *Sir Arthur Henderson Fairbairn 1852-1915*, (2006), pp.4-5.
6. He was on the Council of The Manchester School of Art (*Manchester Times*, 2 August 1864) and on the committee of The Patriotic Fund, to support dependants of servicemen killed in the Crimea (*MG*, 11 November 1854) and was a Magistrate (*MG*, 10 January 1855).
7. Almost opposite the main gate to Heaton Hall, it no longer exists but clearly was a splendid house, with lodge and drive, and extensive gardens. No photograph of it has been found.
8. Eagling and Dimmock, *Sir Arthur Henderson Fairbairn*, pp.4-5. Those born in Manchester were: Mary Eleanor (1850-1852); Emma Constance (1851-1904); Arthur Henderson(1852-1915); Thomas Gordon (1854-1931); Reginald (1856-1921); Florence (1858-1875) and Mary (1860-1936). The other children, James Brooke (1863-1947) and Alan Barttelot (1865-1874) were born after the move south.
9. Fairbairn, *Application*, (3rd ed. 1864), pp.123-36; Walker, *Unwin*, p.49.
10. *RIBA Sessional Papers, Session 1871-2*, 90.
11. *MG*, 19 January 1864; *MPICE*, 11, 1852, 475.
12. *The Times*, 17 April, 1872.
13. *London Gazette*, 15 August 1856; *Manchester Times*, 16 August 1856; *MG*, 7 December 1868. George had been married at the Cathedral the previous year by Canon Stowell, to Bessy Bagshawe, (*MG*, 13 July 1855) daughter of the Chaplain of the New Bailey Prison, yet George seems to have been the only one of the Fairbairn children to have retained links with Unitarianism (T Barker, *Memories of a Dissenting Chapel*, (1884), p.131) – in spite of having been married by the evangelical Canon. There were no children of the marriage. The following year William was suffering from rheumatism and George accompanied his father to Italy for two months (see Chapter 7). After this not much is known about George. In 1858 it appears that he was commissioned as Cornet (the fifth commissioned officer in a troop of cavalry, who carried the colours) in the Earl of Chester's Yeomanry Cavalry (*MG*, 7 August 1858). At the 1862 Exhibition he exhibited a working model of a tank engine (*1862 London Exhibition Catalogue: Class V*, No.1313). He died in 1868 at Offenbach in Germany aged only thirty-seven (*MG*, 7 December 1868).
14. *Life*, pp.330-1; G B Smith, *Leaders of Modern Industry, Biographical Sketches*, (1894), p.371.
15. www.thepeerage.com p.40247 (accessed 28 September 2013). William Andrew subsequently married Helen Birley, in 1865, and they had three children. From around 1870 they lived in Holland Park, London.
16. Hayward, 'Fairbairns', p.3.1; Anon., *A Sketch of the Foundation and of the Past Fifty Years' Activity of the Manchester Steam Users' Association for the Prevention of Steam Boiler Explosions and for the Attainment of Economy in the Application of Steam*, (1905), pp.32-3.
17. *MPIME*, January 1876, 20-1.
18. *MG*, 19 January, 1864.
19. B Bolton and J Thompson, *Entrepreneurs: Talent, Temperament, Technique*, (2nd ed. 2004), pp.16-7, where the definition of entrepreneur is broadened from the purely commercial.
20. Bronkhurst, 'Fruits', 588; W H Hunt, *Pre-Raphaelitism and the Pre-Raphaelite Brotherhood*, (1905), Vol.1, pp.347-8; C Arscott, 'Employer, Husband, Spectator: Thomas Fairbairn's Commission of the *Awakening Conscience*' in J Wolff and J Seed, (eds.), *The Culture of Capital: Art, Power, and the Nineteenth -century Middle Class*, (1988).
21. Bronkhurst, 'Fruits', 588, 594, 597. See also D Holman-Hunt, *My Grandfather, His Wives and Loves*, (1969).
22. Bronkhurst, 'Fruits', 588: *MG*, 12 September 1855.
23. This included E W Cooke's *View of Oneglia*, V Chavet's *Connoisseur*, W Wyld's *View of Nice*, a number of realistic landscapes such as John Brett's *Glacier of Rosenlaui*, and three by Edward Lear, *Comprehensive view of Petra by Sunrise*, *A View of Florence, taken from the Villa San Firenze near San Miniato* and *Ioánnina in Epirus and the Pindus Mountains*. There was Augustus Egg's *Beatrix Knighting Esmond*, R B Martineau's *The Knight's Guerdon*, and several depicting children including W F Witherington's *The Hop Garland*, Jean Baptiste Jules Trayer's *An Examination*, Désiré François Laugée's *Coming out of School* and James Archer's *Summer-time, Gloucestershire* – the last three being purchased at the 1862 International Exhibition.
24. Hunt, *Pre-Raphaelitism*, Vol.2 .p.160-2.
25. Known as *Brother and Sister.* Robert Browning wrote a poem about it and Woolner sent Browning's manuscript to Fairbairn. (A Woolner, *Thomas Woolner, R.A., Sculptor and Poet: His Life and Letters*, (1917), p.216). Tennyson wrote to the Duke of Argyll about it (C Y Lang and E F Shannon, Jnr. (eds), *The Letters of Alfred Lord Tennyson. Volume II, 1851-1870*, (1987), pp.304-5).
26. Pole, *Life*, pp.439, 441; Woolner, *Thomas Woolner*, p.195. Also exhibited at the 1862 London Exhibition. Original in private ownership in America.
27. B Read and J Barnes, (eds.) *Pre-Raphaelite Sculpture: Nature and Imagination in British Sculpture 1848-1914*, (1991), p.152.
28. Woolner, *Woolner*, pp.148, 158; Read and Barnes, *Pre-Raphaelite Sculpture*, p.151; Bronkhurst, 'Fruits', 591-2. On Brooke see below.
29. Lang and Shannon, *Letters of Alfred Lord Tennyson*, p.357. See also Hunt, *Pre-Raphaelitism*, p.225, 'One night in his smoking room, when Woolner and I were his fellow guests'.
30. Hunt, op.cit., Vol.II p.162.
31. The four in the Tate are: William Holman Hunt, *The Awakening Conscience;* Augustus Egg, *Beatrix Knighting Esmond;* W F Witherington, *The Hop Garland*; and John Brett, *Glacier of Rosenlaui.*
32. Fairbairn, *Application*, p.119n. On Dargan see F Mulligan, *William Dargan: An Honourable Life 1799-1867*, (2013).
33. *Official Catalogue of the Great Industrial Exhibition 1853*, (1853), p.48. The steam-engine was ordered by the committee - effectively Dargan - to drive exhibited machinery.

34. *MG*, 26.08.1848; http://www.gac.culture.gov.uk/work.aspx?obj=29747 (accessed 07.04.11).
35. Deane was the son of the Irish architect Sir Thomas Deane. He was a non-practicing barrister, who later became Secretary to the Anglo-American Telegraph Company and chronicler of the successful 1866 Atlantic Telegraph expedition. Described by W H Russell as 'a colourful character', he was a bankrupt for some years. He died in poverty in Italy in 1887 (S Roberts, 'John Connellan Deane', at http://atlantic-cable.com/CablePioneers/Deane/index.htm (accessed 11 April 2015)).
36. *Exhibition of Art Treasures of the United Kingdom held at Manchester in 1857. Report of the Executive Committee*, (1859), p.2. The Circular was signed by Thomas Bazley, J A Turner, William Entwistle, Oliver Heywood, Thomas Ashton, Joseph Heron (the Town Clerk), and Thomas Fairbairn.
37. *Report of the Executive Committee*, pp.2-5.
38. Thomas's art-collecting friends Augustus Novelli and John Pender guaranteed £1,000 and £500 respectively, Thomas Bazley £1,000. (*Report of the Executive Committee*, Appendix XX).
39. *Report of the Executive Committee*, p.6. The deputation to Prince Albert was James Watts (Mayor of Manchester), R N Philips (High Sheriff of Lancashire), Sir John Potter, Stephen Heelis (Mayor of Salford), Thomas Bazley, Joseph Heron (Town Clerk of Manchester) and Thomas Fairbairn. On Albert's involvement in the Exhibition see T Martin, *The life of His Royal Highness the Prince Consort*, (1875-80), Vol.4, pp.34-8.
40. *Report of the Executive Committee*, p.6.
41. *Report of the Executive Committee*, pp.16-7.
42. E A Pergam, *The Manchester Art Treasures Exhibition of 1857: Entrepreneurs, Connoisseurs and the Public* (2011), pp.33-5.
43. Pergam, *Art Treasures Exhibition*, pp.26-7.
44. *Report of the Executive Committee*, p.15, Appendix XXIII. The chosen design was by C D Young & Co of Edinburgh – iron framed with semi-circular roofs. The young local architect, Edward Salomons, was commissioned to design the facade and the organ case. Interior decor was by J C Crace. *The Builder* was uncomplimentary (9 May 1857, 264). *The Art-Journal* praised it (June 1857, 186).
45. *Report of the Executive Committee*, p.23.
46. Pergam, *Art Treasures Exhibition*, p.25. Pergam points out that Egg was in fact second choice, the artists Thomas Creswick and Richard Redgrave, who had overseen the hang at the 1855 Paris exhibition having been asked previously; also that, both Macleod and Cooper mistakenly give Egg credit for the choice of pictures (Pergam, *Art Treasures Exhibition*, p.44 n.80).
47. 323. Edward W Cooke, *View of Oneglia*. 381. William Gale, *An Angel*. 460. W Holman Hunt, *Two Gentlemen of Verona : Valentine rescuing Sylvia*. 496. F R Pickersgill, *Flight of the Pagan Deities on the Dawn of Christianity*. 550. W Holman Hunt, *The Awakened Conscience*. 551. Thomas Unwins, *The Saint Shop*. 563. Francis Goodall, *Episode in the Happier Days of Charles I*. 648. V Chavet, *The Connoisseur*. 669. William Wyld, *View of Nice*.
48. *The Times* 26 March 1857; U Finke, 'The Art-Treasures Exhibition', in J H G Archer (ed.), *Art and Architecture in Victorian Manchester*, (1985), p.116.
49. *The Observer* 10 May 1857.
50. *MG*, 5 May 1857.
51. *MG*, 30 June, 1857.
52. *MG*, 5 July 1857. One explanation was: 'It was not because he undervalued honours bestowed by the hand of royalty; but he held that such honours can answer their proper end only in being discriminative. At present knighthood is vulgarised in England. Gained by persons whose sole claim is money or mayoralty, and whose chief function in society is to vend this bane or the other, knighthood can be desired by none who are conscious of solid worth' (*Cassells Illustrated Family Paper*, New Series 2.31, 3 July 1858, 71-2). 125 years later Bronkhurst was tempted to suggest that his commissioning of Egg's *Beatrix Knighting Esmond* reflected 'Fairbairn's irreverence towards the honour he had declined' (Bronhurst, 'Fruits', 593).
53. *Manchester Times* 26 September 1857; *Art Treasures Examiner: a pictorial, critical, and historical record of the Art-Treasures Exhibition, at Manchester, in 1857; &c.* (1857), p.250.
54. *MG*, 24 August 1857.
55. *MG*, 24 August 1857, 22 September 1857; *Manchester Times*, 26 September 1857.
56. Finke, 'Art-Treasures', p.125; *Art Journal*, 1857, 362.
57. *The Times* 30 October 1857. The Tuesday of the last week saw the biggest attendance on a single day – over 27,000 (*Manchester Times* 17 October 1857).
58. *Report of the Executive Committee*, Appendix III.
59. Fairbairn, *Application*, pp.119-24.
60. Fairbairn, *Application*, pp.125-7.
61. D S L Cardwell, 'Two Centuries of the Manchester Lit.& Phil.', *Manchester Literary and Philosophical Society, Manchester Memoirs 1 (1980-81)*, 1982, 132; D S L Cardwell, *James Joule: A Biography*, (1989), p.239.
62. C Eshborn, *One Hundred Years of Photography – Manchester Photographic Society 1855-1955*, (1955), pp.12-3. Exhibit No.126 in the Photographic Section of the Manchester Art Treasures Exhibition, 1857.
63. F M L Thompson, *Gentrification and the Enterprise Culture: Britain 1780-1980*, (2001), p.116; Pergam, *Manchester Art Treasures Exhibition*, p.277; G Waterfield, 'A culture of exhibitions: the Manchester Art-Treasures Exhibition in context', *Bulletin of the John Rylands University Library of Manchester*, 87.2, 2005, 36. For a different and adverse view of the Exhibition see M T W Plampin, 'From Rio to Romola: Morality and Didacticism in the English Appreciation of Early Italian Art 1836-1863 (University of Plymouth PhD Thesis 2001), pp.172-295. Plampin's historical novel, *The Street Philosopher*, (2009) appears to confuse Thomas Fairbairn with his father. Thomas lived at Northwood, not The Polygon. Paintings were not loaned to the Exhibition from the Polygon, and William did not hold balls at The Polygon (p.140). Whilst this is only a novel, it is a travesty to associate The Polygon with 'immeasurable complacency' (p.258).
64. M Kennedy (ed.), *The Autobiography of Charles Hallé with correspondence and diaries*, (1972), p.136.
65. *The Times*, 30 October 1857.
66. K Layton-Jones, 'Re-presenting Manchester: printed and ephemeral images of the Art-Treasures Exhibition' in H R Leahy, 'Art, City, Spectacle: The 1857 Manchester Art-Treasures Exhibition Revisited', *Bulletin of the John Rylands University Library of Manchester*, 87.2 , 2005, 142.
67. Pergam, *Art Treasures Exhibition*, p.227.
68. Finke, 'Art-Treasures', p.124.
69. *MG*, 31 December 1858.

70. R H W Reece, 'Brooke, Sir James (1803-1868)', *Oxford DNB on-line* (2004), (accessed 12 December 2011); N Barley, *White Rajah: A Biography of Sir James Brooke,* (2002).
71. J H Walker, 'This Peculiar Acuteness of Feeling: James Brooke and the Enactment of Desire', *Borneo Research Bulletin,* 29, 1998, 169.
72. Cobden to Bright 6 December 1849, in J Morley, *The Life of Richard Cobden* (1903 ed.), pp.519-20; *Manchester Examiner & Times* 8 December 1849.
73. *Manchester Examiner & Times* 12 December 1849.
74. Morley, *Cobden,* p.520.
75. P F Tidman, 'Rajah Sir James Brooke', *The Monthly Packet,* 14 September 1874, quoted in Barley, *White Rajah,* pp.229, 249n19.
76. *Manchester Examiner & Times,* 26 July 1851.
77. Barley, *White Rajah,* pp.181-5.
78. *MG,* 5 May 1857.
79. Woolner, *Woolner,* p.140.
80. *MG,* 17 April 1858.
81. *MG,* 21 April 1858.
82. L G C Laughton, 'Keppel, Sir Henry (1909-1904)', *Oxford DNB on-line,* (2004), (accessed 22 December 2011).
83. *MG,* 22 April 1858.
84. *MG,* 22 April 1858.
85. *MG,* 23 October 1858.
86. S St John, *The life of Sir James Brooke, Rajah of Sarawak: from his personal papers and. Correspondence,* (1879), p.323.
87. *The Times,* 1 December 1858.
88. Barley, *White Rajah,* pp.202-3, 224.
89. *The Hampshire Advertiser* 20 July 1867.
90. Barley, *White Rajah,* p.228.
91. G Le G Jacob, *The Raja of Saráwak: An account of Sir James Brooke KCB LLD, given chiefly through letters and journals,* (1876), p.x; *The Times,* 25 July 1868.
92. *The Pall Mall Gazette,* 13 July 1868.
93. *Belfast News-Letter,* 6 July 1858.
94. *Birmingham Daily Post,* 23 July 1858; *Liverpool Mercury,* 30 July 1858.
95. *Daily News,* 2 August 1858.
96. *MG,* 5 August 1858.
97. *MG,* 11 August 1858.
98. *MG,* 26 October 1858.
99. *MG,* 26 October 1858, 27 October 1858.
100. A C Howe, 'Bazley, Sir Thomas, first baronet (1797-1885)', *Oxford DNB on-line* (accessed 6 December 2011).
101. *MG,* 1 November 1858.
102. *MG,* 3 November 1858.
103. *MG,* 16 November 1858.
104. *The Art-Journal,* February 1860, p.54.
105. *MG,* 1 February 1860.
106. *MG,* 6 March 1860.
107. *Manchester Examiner & Times,* 12 May 1860.
108. *MG,* 25 September 1860.
109. *Manchester Examiner & Times,* 20 September 1860.
110. *Manchester Examiner & Times,* 20 September 1860.
111. It was not until 1883 that Manchester had a Free Art Gallery, when the City Council took over the Royal Manchester Institution (*Manchester Examiner & Times* 29 August 1883).
112. H T Wood, *A History of the Royal Society of Arts,* (1913), p.416.
113. E Bonython, and A Burt, *The Great Exhibitor. The Life and Work of Henry Cole,* (2003), pp.202-4. On the building see also H Hobhouse, *The Crystal Palace and the Great Exhibition: Art, Science and Productive Industry. A History of the Royal Commission for the Exhibition of 1851,* (2002), pp.118-40; W Fairbairn, 'The Great Exhibition Buildings', *Popular Science Review,* 1863, 324-6.
114. F H W Sheppard, (ed.), *Survey of London: Volume 38: South Kensington Museums Area,* (1975), pp.137-147, at http://209.85.129.132/search?q=cache:HAdyTU3FgMAJ:www.british-history.ac.uk/r... (accessed 15 February 2009).
115. A Duncan, 'The 1862 London Exhibition' at http://www.mcgee-flutes.com/1862%20London%20Exhibition.htm (accessed 04 February 2012).
116. See, for example, *The Times,* 22 January 1861.
117. Bronkhurst, 'Fruits', p.592n62.
118. Hunt, *Pre-Raphaelitism,* p.225.
119. *The Times,* 15 May 1862.
120. Hunt, *Pre-Raphaelitism,* pp224-5.
121. *Catalogue of Fine Arts at the International Exhibition 1862. The Awakening Conscience* was not amongst them.
122. D K Clark, *The Exhibited Machinery of 1862: A Cyclopaedia of the Machinery represented at the International Exhibition,* (1662), pp.14-5; D K Clark, 'The Locomotive Engines in the International Exhibition of 1862', *CE&AJ,* 27, 1864, 18.
123. http://www.gracesguide.co.uk/wiki/1862_London_Exhibition:_Catalogue:_Class_V... (accessed 16 March 2010).
124. W Fairbairn and W Baker, 'Report to the Commissioners of the International Exhibition' reproduced in *Popular Science Review,* 1863, 324-5.
125. *Life,* p.394.
126. *MM,* 5 July 1861, 13.
127. Fairbairn, *UIfE3,* pp.180-203.
128. E Connell, 'Fairbairn, Sir Andrew (1828-1901) Textile Machinery Manufacturer' in D J Jeremy, (ed.) *Dictionary of Business Biography. A Biographical Dictionary of Business Leaders Active in Britain in the Period 1860-1980,* Vol.2, (1984), pp.311-3.
129. http://www.gracesguide.co.uk/wiki/1862_London_Exhibition:_Catalogue:_Class_7.:_ (accessed 16 March 2010).
130. *MG,* 28 April 1862.
131. *MG,* 2 May 1862.
132. Hobhouse, *Crystal Palace,* pp.118-140.
133. Hobhouse, *Crystal Palace,* p.403.
134. *RIBA. Sessional Papers,* 1871-2, 84-90.
135. *The Morning Post* 13 July 1881.
136. M Hewitt and R Poole, (eds.), *The Diaries of Samuel Bamford,* (2000), p.13n.9, where the influence of Thomas's disappointments was first suggested; Bronkhurst, 'Fruits', p.523.
137. Bronkhurst, 'Fruits', p.593.
138. Architect Henry Bassett. Now Listed Grade 1.
139. Bronkhurst, 'Fruits', p.593. It is believed to have been influenced by Van Dyck's *The five eldest children of Charles I,* which Hunt had seen at the Art Treasures Exhibition.
140. Hunt, *Pre-Raphaelitism,* pp.240-42. Trelawney had had three wives – one reputed to have been only thirteen, had been divorced twice, and lived with other young women (P Cochran, 'Trelawney, Edward John (1792-1881)', *Oxford DNB on-line,* (2004), (accessed 12 November 2011)).
141. M Barclay-Harvey, *A History of the Great North of Scotland Railway,* (2nd ed. 1949), p.180.
142. S Summerson, *Midland Railway Locomotives: Volume 2 : The Kirtley Classes,* (2007), p.146; Clark, *Exhibited Machinery,* pp.14-5.

143. E L Ahrons, *The British Steam Railway Locomotive from 1825 to 1925,* (1927), pp.144-5.
144. Ahrons, 'Famous Firms', p.185; J Davies, 'Fairbairn Works List' (Typescript).
145. L Oberg, *Locomotives of Australia 1854-2007,* (4th ed. 2007), p.18; J Broadley, *Making Go – Australia's Early Locomotive Engineers,* (2010); Ahrons, 'Famous Firms', p.185.
146. 'Liste der portugiesischen Lokomotiv – und Triebwagenbaureihen' at http://66.102.9.104/search?q=cache:SyFUWPVjoJ8J:de.wikipedia.org/wiki/Liste_der (accessed 02.06.08).
147. H Hughes, *Indian Locomotives: Part 1 – Broad Gauge, 1851-1940,* (1990), pp.39-40.
148. *The Engineer,* 1, Jan.-June 1856, 324; Davies, 'Fairbairn Works List'.
149. *The Engineer,* 1, Jan.-June 1856, 324; *The Engineer,* 7, Jan.-June 1859, 403; E Sundström and R Sten, 'Anglok i Sverige tillverkade i utlandet' at historyiskt.nu/normalsp/.../okj_rullande.html (accessed 20 December 2010).
150. Ahrons, 'Famous Firms', p.185.
151. F Wais, *Historia de los ferrocarriles Españoles,* Vol.1. (1974), pp.210-4.
152. Appendix 4.1
153. *The Engineer,* 1, 1856, 324.
154. E Gauldie, *The Dundee Textile Industry 1790-1885, from the papers of Peter Carmichael of Arthurstone,* (1969), pp.68-9, 184. Note that the quotation on p.69 is incorrect.
155. H Jack, *Locomotives of the LNWR Southern Division,* (2001), pp.49-70, 177.
156. Jack, *Locomotives of the LNWR,* pp.189-93.
157. T Fairbairn to G Wilson, 7 March 1854, M20/1, Wilson Papers, Manchester Central Reference Library.
158. Summerson, *Midland Railway Locomotives,* pp.146-7.
159. *The Engineer,* 9, Jan.-June 1860, 52.
160. *Great North Review,* 6.22, 1969, cited by Hayward, *Fairbairns,* p.2.58.
161. J W Lowe, *British Steam Locomotive Builders,* (1975), p.168. Two 0-4-2s which were under construction for the Newry & Armagh Railway were finished by Sharps.
162. *MG,* 9 April 1864.
163. *MG,* 14 May 1863.
164. Ahrons, 'Famous Firms', p.184.
165. *CE&AJ,* 30, 1867, 240.
166. See Appendix 4.1, which is built up from many sources.
167. But note that Nasmyths built very few locomotives in the late 1850s and early 1860s (J Cantrell, *Nasmyth, Wilson & Co. Patricroft Locomotive Builders,* (2005), p.138).
168. S G Checkland, *The Rise of Industrial Society in England 1815-1885,* (1964), p.107.
169. Discussion following B Poole, 'The Economy of Railways as a means of transit, comprising the Classification of the Traffic, in relation to the most appropriate Speeds for the conveyance of Passengers and Merchandise', *MPICE,* 1852, 475. Whilst many liberals were unhappy with it, Cobden argued that limited liability would help the poor to set up businesses (J Micklethwait and A Wooldridge, *The Company: A Short History of a Revolutionary Idea,* (2003), p.56).
170. N M Butler, quoted in J Dine and M Koutsias, *Company Law,* (7th ed. 2009), p.1.
171. Micklethwait and Wooldridge, *The Company,* pp.2-3.
172. F Crouzet, *The Victorian Economy,* (English Translation, A. Forster, 1982), p.339.
173. Novelli was from the Italian immigrant family of successful merchants in Manchester and Thomas's closest friend.
174. McConnell was a locomotive engineer who, until two years previously, had been Locomotive Superintendent of the LNWR, Southern Region (Jack, *Locomotives of the LNWR,* pp.49-70).
175. Hayward, 'Fairbairns', pp.3.8-3.10.
176. *MG,* 19 January 1864, 23 January 1864.
177. Hayward, 'Fairbairns', pp.3.11-3.13.
178. See Chapter 5.
179. See Chapter 7.
180. *MG,* 16 May 1864; *MPIME,* 1863, 24-6; *MPICE,* 203, 1916, 98-9; *MM,* 11NS, Jan-June 1864, p338. http://crowie.org/?p=4068 (accessed 18 May 2016). See Chapter 5.
181. *Engineering,* 5, 1868, 176, and 8, 1869, 117-24; *The Engineer,* 31, 1871, 10,221; R W Clark, *The Royal Albert Hall,* (1958), pp.46-7; F H W Shepherd (ed.), *Survey of London, Volume 38,* (1975), Chapter XI (on-line at http://209.85.129.132/search?q=cache:zj_9vvBqHNEJ:www.british-history.ac.uk/rep... (accessed 15 February 2009). For some excellent photographs see 'Royal Albert Hall: Roof | Flickr – Photo Sharing!'. See Fig.5.32.
182. Hayward, 'Fairbairns', p.3.20; *Engineering,* 9, 1870, 72; *The Times,* 11 January 1872; *Manchester City News,* 17 June 1871; G Mitchell with A Cantwell and P Sprack, *Spit Bank and the Spithead Forts,* (2003 ed.).
183. *MM,* New Series 12, July-Dec.1864, 173.
184. *Engineering,* 1, 2 February 1866; P C da S Telles, *A History of Brazilian Railways. Part 1 – The First Railways,* (ET by P E Waters, nd), pp.41-4 and Plate; M Cooper, 'Railway Museums in Brazil: State Politics and the Rise of the Volunteer Movement', *Journal of the International Association of Transport and Communications Museums,* 28, 2009, 52; P Catchpole, *A Very British Railway,* (2003), p.182; T T dos S Cruz, 'Paranapiacaba: A arquitetura e o urbanismo de uma Vila Ferroviária', (MA Thesis, University of São Paulo, 2007), p.59; Darwin Correspondence project, Letter 4129 – Aubertin to Darwin 27 April 1863 at http://www.darwinproject.ac.uk/darwinletters/calendar/entry-4129.html (accessed 1 November 2008). One of these engines remains – the only confirmed Fairbairn stationary steam engine to survive.
185. *Engineering,* 5, 1868, 279; H M Burt, *Burt's Illustrated Guide to the Connecticut Valley,* (1867), pp.60-2.
186. J Mitchell, 'On the Construction and Works of the Highland Railway', *BAAS1867,* pp.154-5; *ILN,* 5 August 1865, 104.
187. J Marshall, *The Lancashire and Yorkshire Railway, Volume 1,* (1969), p.217.
188. British History Online, 'The West India Docks: Power and transport', at http://.102.9.104/search?q=cache:Vs8yWKEA3IUJ:www.british-history.ac.uk/repo (accessed 15 March 2007).
189. *The Engineer,* 21, Jan-June 1866, 179.
190. Number not used.
191. National Archives, RAIL 343/897.
192. Marshall, *Lancashire and Yorkshire Railway, Volume 1,* p.260.
193. A J Hazelfoot, *The Batsford Guide to the Industrial Archaeology of South-East England,* (1978), pp.37-8.
194. Hayward, 'Fairbairns', p.3.19.
195. See above.

196. D A Collier, *A Comparative History of the Development of the Leading Stationary Steam Engine Manufacturers of Lancashire, c.1800-1939,* (PhD Thesis, Manchester University 1985), Vol.1, pp.42,61.
197. J A Bennison, 'Prize Essay' (1869, Lancashire Record Office, DDX/184).
198. See above.
199. S Fleming, *The Intercolonial. A Historical Sketch of the Inception, Location, Construction and Completion of the line of Railway uniting the Inland and Atlantic Provinces of the Dominion,...* (1876).
200. Hayward, 'Fairbairns', p.3.19.
201. *Manchester City News,* 17 June 1871.
202. Walker, *Unwin,* p.60.
203. Marshall, *Lancashire and Yorkshire Railway, Vol. 2,* p.46; National Archives, RAIL 795/543. Probably not constructed when liquidation took place (See T Wray, 'Oldham Road Goods station, Manchester', *LYR Focus,* 74, 2012, 4-14).
204. *The Maitland Mercury,* 23 April 1874.
205. S Bradley and N Pevsner, *Buildings of England: London I – The City of London,* (1997), p.311; *The Times,* 9 July 1875; A T Walmisley, *Iron Roofs. Examples of Design. Description, Illustrated with Working Drawings,* (2nd ed. 1888), Plates 61-3; G Biddle, *Great Railway Stations of Britain: Their Architecture, Growth and Development,* (1986), pp.124-6.
206. http://www.canadianarchitect.com/news/parliamentary-briefing/1000206736/?&er=NA (accessed 27 December 2013).
207. W T Pike, *Contemporary Biographies,* (1899), p.187. Completed by the liquidators.
208. Hayward, 'Fairbairns', pp.3.15-3.16.
209. *The Times,* 26 April 1869.
210. *MG,* 3 July 1868; *The Times,* 1 February 1869.
211. *MG,* 15 April 1871, 27 October 1871.
'The workpeople intend to show their appreciation of the concession by walking in procession on Saturday afternoon from the Company's works to the residences of Sir William Fairbairn and Mr. Harman, in Ardwick and Longsight respectively'. William was clearly still held in some esteem by the workforce.
212. *The Times,* 17 April 1872.
213. *The Times,* 18 April 1873.
214. *MG,* 24 April 1874.
215. Hayward, 'Fairbairns', p.3.26.
216. Thompson, *Gentrification,* pp.46-7.
217. *The Standard,* 15 July 1893, 13 June 1896; http://www.parksandgardens.org/places-and-people/site/4973?preview=1 (accessed 20 October 2013).
218. *Hampshire Advertiser,* 17 February 1872; *MG,* 17 February 1872; http://www.britishlistedbuildings.co.uk/en-145762-brambridge-house-colden-common (accessed 24 Jan. 2012).
219 . *The Standard,* 15 July 1893; 13 June 1896; http://www.parksandgardens.org/places-and-people/site/4973?preview=1 (accessed 20 October 2013). The house is now listed Grade II*.
220. Eagling and Dimmock, *Arthur Henderson Fairbairn,* p.16.
221. *Morning Post,* 13 November 1868; *The Times,* 13 November, 1869.
222. *Hampshire Advertiser,* 26 February, 1870; 2 March 1870; A H Fairbairn, *An Assize Sermon preached in Winchester Cathedral, on the Second Sunday in Advent, 1870, before the Hon. Sir Robert Lush, Knight, one of the Justices of the Court of Queen's Bench,* (1870).
223. *Isle of Wight Observer,* 5 March, 1870.
224. *Hampshire Advertiser,* 23 July 1870.
225. William Fairbairn to Arthur Henderson Fairbairn, 24 February 1870. Photocopy in the author's possession, courtesy of Julia Elton. The 1870 Lent Assizes is the only occasion of which there appears to be any record of William visiting Brambridge. In contrast he regularly visited his daughter, Anne, and her husband, J F Bateman (*Life,* pp.430-1).
226. Eagling and Dimmock, *Arthur Henderson Fairbairn,* p.8.
227. S Crooks, *Alfred Waterhouse in Twyford,* (2003), p.11. This was towards an estimate of £6,000. The final cost was £9,000.
228. There is no evidence, and it is improbable, that Thomas ever became a communicant member of Twyford church, which would have involved Confirmation, and probably, given his Unitarian background, Trinitarian Baptism.
229. Crooks, *Alfred Waterhouse in Twyford,* pp.57, 72. The original plan showed what appear to have been additional choir pews in the south transept.
230. It provides an illustration of the wide influence of the Tractarians, whose 'ideas were taken over and translated into bricks and mortar by people who had no use for the theology that animated it' (H Schlossberg, *The Silent Revolution and the Making of Victorian England,* (2000), p.283).
231. *Hampshire Advertiser,* 2 December 1885.
232. *Birmingham Daily Post,* 24 May 1886.
233. *Morning Post,* 28 December 1886.
234. M J Wiener, *English Culture and the Decline of the Industrial Spirit, 1850-1980,* (2nd ed. 2004), p.97.
235. A McConnell, 'Pender, Sir John (1816-1896)', *Oxford DNB on-line,* (2004), (accessed 02 December 2011); Macleod, *Art and the Victorian Middle Class,* pp.215-21, 459-60.
236. McConnell, 'Pender'.
237. Footscray Place, Kent, and 18 Arlington Street, London.
238. Macleod, *Art and the Victorian Middle Class,* pp.216, 220-1.
239. H Barty-King, 'Pender, Sir John (1816-1896)', in Jeremy, *Dictionary of Business Biography* (referring to Cable & Wireless Archives, Report of AGM, 25 January 1894).
240. A Howe, *The Cotton Masters,* (1984), pp.252-3. Around 500 large country houses were built or substantially remodelled in Britain in the fifty-five years between 1835 and 1889, about half by men of new wealth (F M L Thompson, *The Rise of Respectable Society: A Social History of Victorian Britain 1830-1900,* (1988), p.155).
241. Thompson, *Gentrification* pp.84-5.
242. S Nenadic, 'Businessmen, the urban middle classes, and the "dominance" of manufacturers in nineteenth-century Britain', *Economic History Review,* 44.1, 1991, 79; P J Cain and A G Hopkins, 'Gentlemanly Capitalism and British Expansion Overseas: I. The Old Colonial System, 1688-1850', *Economic History Review,* 39.4, 1986, 505.
243. Thompson, *Respectable Society,* p.160.
244. M Daunton, '"Gentlemanly Capitalism" and British Industry 1820-1914', *Past and Present,* 122, February 1989, 125-6.
245. *The Times* 30 May 1862. J Aspinall Turner was a Manchester cotton manufacturer and merchant. He and John Potter were elected in 1857 as 'Palmerstonian Whigs', when tactical voting unseated Bright and Gibson. He was also a renowned etymologist (Obituary, *The Gentleman's*

Magazine, 28 September 1867). In 1829 the Bank of Manchester became the first joint stock bank in Manchester. It failed in 1842, and was re-formed in 1852.
246. *The Times,* 14 August 1863.
247. *The Times,* 24 May 1866.
248. *The Times,* 28 May 1866. Of the board *The Spectator* wrote: 'It would scarcely be possible ... to frame an imaginary board better or more trustworthy than that which governed this ill-fated concern.... All were solid, responsible men, rich and up to their eyes in large affairs, men, some of them who, like Mr. Smith, had managed millions, or like Mr. Pender, had built up first class fortunes by their own skill and daring.' (*The Spectator,* 39, 1866, 599-600).
249. *The Morning Post,* 19 January 1877; *The Pall Mall Gazette,* 16 January 1879.
250. C M Leaf, *Walter Leaf 1852-1927,* (1932), pp.112-3, quoted in S Chapman, *Merchant Enterprise in Britain From the Industrial Revolution to World War I,* (1992), p.218. One of the sons, Charles, left an estate larger than any ever previously left by a British businessman (W D Rubinstein, 'Morrison, Charles (1817-1909)', *Oxford DNB on-line,* (2006), (accessed 17 January, 2012)).
251. *The Times,* 16 March 1864. It is unclear who was chairman.
252. *The Times,* 16 January 1875, 16 July 1877, 20 January 1883.
253. http://cosgb.blogspot.com/2011/10/fore-street-warehouse-company-ltd.html (accessed 14 January 2012).
254. *The Times,* 2 March 1892.
255. *The Times,* 9 April 1864.
256. G L Lawfoot and L R Nicholson, *The Telcon Story 1850-1950,* (1950), pp.174-5. These included the following cables: Suez-Aden-Bombay; Madras-Penang-Singapore; Singapore-Hong Kong; Brazil (Pernambuco) via Madeira and St Vincent; another Atlantic cable: Australia-New Zealand; Rangoon-Penang; Aden-Bombay, East coast of Africa (Aden to Cape Town); renewal of the1866 Atlantic cable; and to the West coast of Africa.
257. *The Times,* 27 February 1872; R B Wilson (ed.), *Sir Daniel Gooch Memoirs & Diary transcribed from the original manuscript and edited with an introduction and notes,* (1972), p.206.
258. Wilson, *Gooch,* pp.270, 280, 296, 308, 319, 341.
259. K R Haigh, *Cableships and Submarine Cables,* (1968), p.96; K G Beauchamp, *History of Telegraphy,* (2001), pp.167-171.
260. *The Times,* 6 May 1873; Haigh, *Cableships,* pp.116-8. Cables laid whilst Fairbairn was a director included: duplicate cable Penang-Malacca-Singapore; duplicate cable Singapore-Darwin; Philippines - Hong Kong; duplicate cable Singapore-Batavia; Hong Kong -Shanghai; and a new cable Hong Kong-Singapore.
261. Haigh, *Cableships,* p.23.
262. http://www.eley.co.uk/en/history-of-eley/ (accessed 16 January 2012).
263. D Thomas, *The Eley Story 1828-1978,* (1978), p.10.
264. http://www.eley.couk/en/history-of-eley/ (accessed 16 January 2012); *The Morning Post,* 28 February 1874.
265. *The Morning Post,* 28 February 1874. In the press notice, Fairbairn's address was given as 1A, Crosby-square, London, the address of the Fairbairn Engineering Co's London office. Amongst the other directors were the three Eley Brothers and William Smith.
266. *The Morning Post,* 30 January 1879.
267. *The Morning Post,* 28 January 1886. A hundred years later Eley won a Queen's Award for technological achievement, and in 2011, 705 World Cup and European shooting medals were won with Eley ammunition (http://www.eley.co.uk/en/history-of-eley/ (accessed 16 January 2012)).
268. *The Times,* 4 May 1875; *MG,* 4 May 1875.
269. *London Gazette,* 18 June 1875.
270. Shareholders included Holman Hunt who had £750 (c.£45,000 in 2014) invested in the company (J Bronkhurst, *William Holman Hunt: A Catalogue Raisonné* (2006), Vol.1, pp.19, 38n148).
271. Biddle, *Great Railway Stations,* pp.124-5.
272. Chrysler & Bethune, Solicitors, 'Memorial in respect to the Unpaid Claim of Mr H Bingham Higginson, in connection with Iron Bridges on the Intercontinental Railway', (1897), p.8. Higginson was Fairbairns' subcontractor, who was left to pursue a claim in equity against the Canadian Government, still ongoing in 1903.
273. R E Green, 'Craven Brothers (Manchester) Ltd., Vauxhall Works, Reddish, Stockport', *Machinery,* 3 February 1965, pp.237-48.
274. D A Collier, *A Comparative History of the Development of the Leading Stationary Steam Engine Manufacturers of Lancashire, c.1800-1939,* (PhD Thesis, Manchester University 1985), Vol.1, pp.221-6.
275. *Fortunes Made in Business,* pp.284-8; Connell, 'Fairbairn, Sir Andrew', in Jeremy (ed.) *Dictionary of Business Biography, 1860-1980.* The firm majored on flax machinery and machine tools.
276. *Austrian Imperial Railways – Exhibition Catalogue,* (2009), pp.4-5; www.biographiea.ac/oebi_2/206.pdf (accessed 6 October 2010).
277. Walker, *Unwin,* pp.51-2.
278. Connell, 'Fairbairn, Sir Andrew', p.311; *Fortunes Made in Business,* pp.288-92. Andrew did, however, make sure he understood flax machinery, visiting the United States, and Hanover where he studied German and familiarised himself with flax mills. Later he investigated the industry across Europe.
279. T S Reynolds, *Stronger Than a Hundred Men: A History of the Vertical Water Wheel,* (1983), pp.338-47.
280. Fairbairn, *M&MWI,* pp.166-75; C Smith, 'Thomson, James (1822-1892)', *Oxford DNB on-line,* (2004), (accessed 9 April 2012). James Thomson was the brother of William Thomson, Lord Kelvin. Crosbie Smith summarises the differences between Thomson's turbine design and the French designs.
281. J Moser, *Escher Wyss, a Century of Water Turbines,* (1940); http://www.andritz.com/hy-history-andritz-hydro (accessed 27 February 2012).
282. H Simon, 'Modern Flour-Milling in England', *MPICE,* 70, 1882, 199-203.
283. Simon, 'Modern Flour-Milling', 215.
284. B Simon, *In Search of a Grandfather: Henry Simon of Manchester 1835-1899* (1997), pp.47-8.
285. Simon, 'Modern Flour-Milling', 216.
286. See Chapter 5.
287. See Chapter 5.
288. O Bachmann, H-H Cohrs, T Whiteman and A Wislicki. *The History of Cranes,* (1997), p.53. An existing Stahl electrically-powered 'Fairbairn' crane at Rorschach on the Bodensee was visited in 2012.
289. W H Chaloner, 'John Galloway, 1804-1894, engineer, of Manchester and his reminiscences', in D A Farnie and

W O Henderson (eds.), *Industry and Innovation. Selected Essays:W H Chaloner,* (1990), pp.116-7; R M Birse, 'Galloway, John (1804-1894)', *Oxford DNB on-line* (2004), (accessed 18 February 2013).

290. A E Musson, 'The Great Depression in Britain, 1873-1876: A Reappraisal', *The Journal of Economic History,* 19.2, 1959, 206-8.

291. A E Musson, 'Introduction to the 1970 facsimile Edition' of *Life,* p.xx; Ahrons, Famous Firms', 185.

292. Hayward, 'Fairbairns', pp.3.30-1, 3,36-7.

293. A E Musson and E Robinson, *Science and Technology in the Industrial Revolution,* (1969), p.489n1; Musson, 'Introduction' to *Life,* p.xx.

294. Collier, *Stationary Steam Engine Manufacturers,* pp.42-4.

295. W H Chaloner, 'The birth of modern Manchester', in C F Carter (ed.), *Manchester and its region: a survey prepared for the meeting held in Manchester August 29 to September 5, 1962,* (1962), p.142.

296. S B Saul, *The Myth of the Great Depression, 1873-96* (2nd ed.1985). See also W H Chaloner, 'Was there a Decline in the Industrial Spirit in Britain 1850-1939?', *TNS,* 55, 1983-4, pp.211-18.

297. E Griffin, 'Patterns of Industrialisation' in M Hewitt (ed.) *The Victorian World,* (2012)' pp.90-1.

298. F Crouzet, *The Victorian Economy,* (English Translation by A. Forster, 1982), p.63.

299. Musson, 'The Great Depression', 206-8.

300. Hayward, 'Fairbairns', pp.3,36-7.

301. *The Engineer,* 25, 1868, 162.

302. *RIBA Sessional Papers,* 1871-2, 84-90; F H W Sheppard (ed.), *Survey of London, Volume 38: South Kensington Museums Area,* (1975), pp.177-95; Hayward, 'Fairbairns', pp.3.19-3.20. At the time was believed to be the largest roof in the world without internal support.

303. *Engineering,* 9, 1870, 72; *The Times,* 11 January 1872.

304. *MG,* 18 September 1875;15 January 1876.

305. Hayward, 'Fairbairns', pp.3.32-3.

306. *London Gazette,* 24 March 1899.

307. W G Rimmer, *Marshall's of Leeds, Flax-spinners 1788-1886,* (1960), p.297. The Gregs and the Rathbones are also relevant (M B Rose, *The Gregs of Quarry Bank Mill: The rise and decline of a family firm 1750-1914,* (1986), pp.92-3; S Marriner, *Rathbones of Liverpool, 1845-73,* (1961), p.131). These were all originally Unitarian families.

308. Howe, *Cotton Masters,* p.311.

309. Bronkhurst, 'Fruits', p.594.

310. *ILN,* 13 June 1874, 567. See also *The Times,* 1 July 1874.

311. W H Hunt to J L Tupper, 9 March 1876, in J H Coombs et al (eds.), *A Pre-Raphaelite friendship: the correspondence of William Holman Hunt and John Lucas Tupper,* (1986), p.205; W H Hunt to F G Stephens, 11 December 1876, in Bronkhurst, *Catalogue Raisonné,* Vol.1, p.235.

312. *Morning Post,* 30 April 1875.

313. *The Observer,* 23 July 1876.

314. *The Times,* 20 July 1877. On W F Bateman, see M La Trobe-Bateman, (ed.), *Memories Grave and Gay of William Fairbairn La Trobe-Bateman,* (1927). His Grand Tour of Europe as a young man was financed by his grandfather William.

315. *Hampshire Advertiser,* 11 May 1878; *The Times,* 9 May 1879.

316. *The Observer,* 22 June 1879, 6 July 1879.

317. http://thePeerage.com/ 'Reginald Fairbairn' (accessed 24 January 2012).

318. *Morning Post,* 17 December 1880. The marriage was not successful: in 1901 Mary, now Lady Napier, appears to have been living with her widowed mother in Hove (Eagling and Dimmock, *Arthur Henderson Fairbairn,* p.19). There was a divorce in 1903 and in 1904 Mary married G Graham Kirklington. (http://thePeerage.com/ 'Mary Allison Fairbairn' (accessed 24 January 2012)). There were no children from either marriage.

319. *Morning Post,* 21 December 1880.

320. *The Times,* 22 February 1888. Thomas's cousin, Sir Andrew Fairbairn and his wife initially rented, and in 1890 bought, Villa Trios Fontaines at Biarritz, where they entertained lavishly (*The Observer,* 2 March 1890). The architects for Villa Trios Fontaines, originally for Lord Ponsonby, were Salomons & Wornum (*The Builder,* 19 November 1910, 632.). This was Edward Salomons who had been the architect for the front of the Manchester Art Treasures Exhibition building, and for its organ case.

321. *The Times,* 14 January 1882.

322. *Morning Post,* 7 July 1882.

323. Eagling and Dimmock, *Arthur Henderson Fairbairn,* p.12.

324. Bronkhurst, 'Fruits', 597.

325. *Morning Post,* 28 December 1886. *The Scapegoat* sold well, whereas *The Awakening Conscience* was bought in at £105. *Valentine rescuing Sylvia from Proteus* sold for less than Fairbairn had hoped.

326. *Daily News,* 15 October 1887.

327. *The Standard,* 19 October 1887.

328. *The Times,* 26 January 1923.

329. *The Times,* 8 February 1888.

330. *The Times,* 14 February 1888, 23 February 1888; 'No.21 of 1888'.

331. http://thePeerage.com/ 'Sir Thomas Gordon Fairbairn, 4th Bt.', (accessed 24 January 2012).

332. *Ipswich Journal,* 17 April 1888.

333. *The Times,* 26 July 1888.

334. P Russell, 'Bateman, John Frederic La Trobe (1810-1889)', *Oxford DNB on-line,* (2004), (accessed 25 January 2011).

335. *Hampshire Advertiser,* 26 August 1891.

336. *MG,* 15 August 1991.

337. *Manchester Times,* 21 August 1891.

338. *Manchester Guardian,* 23 October 1891. Note that probate valuations were of personalty only and excluded all real property (Thompson, *Gentrification,* p.59).

339. *The Standard,* 15 July 1893, 13 June 1896, 20 July 1896.

340. Eagling and Dimmock, *Arthur Henderson Fairbairn,* p.19.

341. Eagling and Dimmock, *Arthur Henderson Fairbairn,* p.17. She retained her own name and was never known as Lady Fairbairn. There were no children from either marriage.

342. Eagling and Dimmock, *Arthur Henderson Fairbairn,* pp.21-2.

343. Bronkhurst, 'Fruits', pp.593, 597n89; *Manchester Guardian,* 23 October 1891.

344. Bronkhurst, 'Fruits', p.593n64.

345. Eagling and Dimmock, *Arthur Henderson Fairbairn,* pp.16-7.

346. *Hampshire Advertiser,* 26 November 1887; Eagling and Dimmock, *Arthur Henderson Fairbairn,* pp.13-4, 25-6.

347. http://www.aim25.ac.uk/cgi-bin/vcdf/detail?coll_id=15701&inst_id=118&nv1=browse.. (accessed 24 January

2012); Eagling and Dimmock, *Arthur Henderson Fairbairn,* p.15.
348. Eagling and Dimmock, *Arthur Henderson Fairbairn,* p.20.
349. 'Sir Arthur Henderson Fairbairn, a Deaf-Mute Baronet', *The Silent Worker,* 11.1, 1898, 3; J W Gilbart-Smith, 'Sir Arthur Fairbairn and his surroundings: the story of a deaf and dumb baronet', *Cassell's Magazine,* 1904, 606-12; 'Well-known deaf persons: Sir Arthur Henderson Fairbairn, Bart.', *British Deaf Times,* 1, 1904, 17-8; *Mayfair,* 25 July 1912, p.910; W R Roe, *Peeps into the Deaf World,* (1917), pp.280-2; G C Braddock, 'Notable Deaf Persons: Sir Arthur H Fairbairn, Bart.', *British Deaf Times,* 38, 1941, 81-2; A F Dimmock, 'Sir Arthur Henderson Fairbairn (1852-1915): Britain's Deaf and Dumb Baronet', *Deaf History Journal,* 6.1, 2002, 32-4.
350. Eagling and Dimmock, *Arthur Henderson Fairbairn,* p.30.
351. The sculpture was removed in 1989, 'to a place of safety'.
352. *The Times,* 4 June 1915.
353. *The Times,* 9 June 1915.
354. *The Times,* 26 January 1923.
355. http://thePeerage.com/ ' Sir Thomas Gordon Fairbairn, 4th Bt.' (accessed 24 January 2012).
356. *Telegraph,* 1 March 2001.
357. I am very grateful to Sir James Brooke Fairbairn, Bt., for the help and encouragement he has given me in my research.
358. Such companies included: Galloways, Whitworth, Nasmyth Wilson, Sharp Stewart, Lairds, Beyer Peacock, Hicks, Daniel Adamson, Craven Brothers.
359. T H Burnham, and G O Hoskins, *Iron and Steel in Britain, 1870-1930: A comparative study of the causes which limited the economic development of the British iron and steel industry between the years 1870 and 1930,* (1943), p.271.
360. M B Rose, 'Beyond Buddenbrooks: the family firm and the management of succession in nineteenth-century Britain' in J Brown and M B Rose, *Entrepreneurship, networks and modern business,* (1993), pp.133-5.
361. 'All hail, astonishing Fact!
All hail, Invention New
The Joint Stock Company's Act
The Act of Sixty-Two.'
(W S Gilbert and A Sullivan, *Utopia Limited, or The Flowers of Progress,* (1893), Chorus – end of Act 1).
362. Daunton, '"Gentlemanly Capitalism"', pp.125-6. He does however note that prior to 1907 there was reluctance by the merchant sector to adopt limited liability (pp.143-4).
363. M B Rose, 'Diversification of Investment by the Greg Family, 1800-1914', *Business History,* 30, 1988, pp.79-81, 85-8; Daunton, '"Gentlemanly Capitalism"', p.131.
364. T Mann, *Buddenbrooks: The Decline of a Family,* (1901. English Translations by H T Lowe- Porter, 1924, and J E Woods, 1993). This was the principal novel for which Mann received the Nobel Prize for Literature in 1929. It was banned by Adolf Hitler.

Sources, Credits and Acknowledgements for Illustrations in Chapter 8.

Fig.8.1 W. Holman Hunt, *Thomas Fairbairn* (1873). I am grateful for permission from the Fairbairn Family Collection to include this portrait.

Fig.8.2 W. Holman Hunt, *The Awakening Conscience* (1853), Tate Gallery, London.

Fig.8.3 T. Woolner, *Brother and Sister* (1858–62), Photograph by W. Jeffrey, c.1860, Photographs and Papers of Thomas Woolner: Image courtesy of Leeds Museums & Galleries (Henry Moore Institute Archive). See also J. Lukitsh, *Thomas Woolner: Seeing Sculpture through Photography* (2005), p.13.

Fig.8.4 Louis Haghe, *Thomas Fairbairn handing over the address to the Prince Consort* (detail), (Manchester City Art Gallery but currently [2016] not on view).

Fig.8.5 C. Eshborn, *One Hundred Years of Photography – Manchester Photographic Society 1855–1955* (1955), p.13.

Fig.8.6 http://www.barbsweb.co.uk/history/burton.htm (accessed 22 October 2016).

Fig.8.7 W. Holman Hunt, *The Children's Holiday* (1864), Torre Abbey Gallery, Torquay. Depicting Allison and the five younger children.

Fig.8.8 E.L. Ahrons, *The British Steam Railway Locomotive from 1825 to 1925* (1927), p.144.

Fig.8.9 National Archives: BT 31/879/909c, reproduced in Hayward, 'Fairbairns', p.3.10.

Fig.8.10 P. Facey, 'Brambridge House, looking W', at http://www.geograph.org.uk/reuse.php?id=63994 (accessed 28 December 2013). Image Copyright Peter Facey. This work is licensed under the Creative Commons Attribution-Share Alike 2.0 Generic Licence.

Fig.8.11 *British Deaf Times,* 1, 1904, 17.

Conclusion

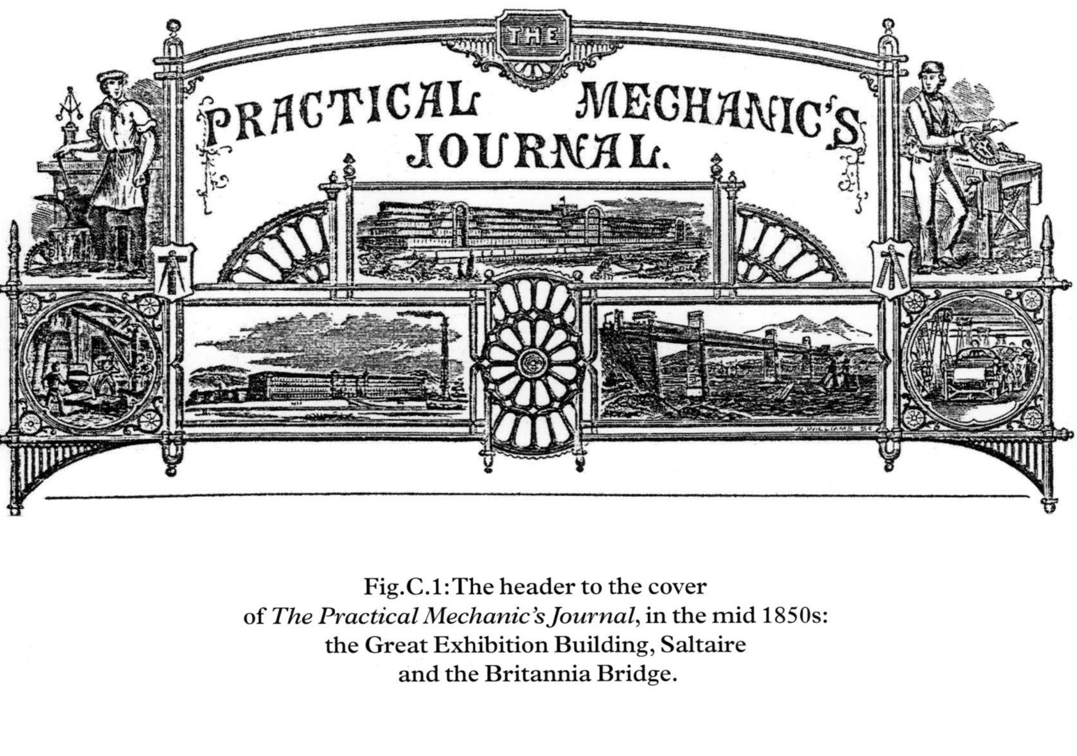

Fig.C.1: The header to the cover of *The Practical Mechanic's Journal*, in the mid 1850s: the Great Exhibition Building, Saltaire and the Britannia Bridge.

WILLIAM Fairbairn, an unknown apprentice millwright at a Newcastle colliery, became the best known living engineer of the 1860s. His importance is recognised today by leading historians of industrial development – thus Mokyr's description of him as 'a worthy heir to John Smeaton'[1] who was the first person 'to achieve distinction as an engineering scientist';[2] and Rosenberg's statement that the thesis could 'readily be defended' that 'Fairbairn contributed more than any other single individual to the emergence of mechanical engineering'.[3] It is therefore startling that so few today are aware of Fairbairn's work beyond his involvement in the Britannia Bridge, and that Fairbairn studies are so few.

This book has brought together Fairbairn's wide range of work. It has provided an overview of the engineer and his achievements – and failures – and in doing so has sought to answer the three questions posed at the outset: how did the unknown apprentice achieve this ascendency; in what areas was his work influential; and what was the cause of the demise, within a year of his death, of the major engineering company he founded?

Ascendency

OF the many factors that contributed to Fairbairn's ascendency the evidence indicates that there are nine that predominate – time and place, innovation, experimentation, optimisation, determination, hard work, networking, the British Association, and his innate abilities.

Settling in Manchester as a journeyman millwright, he was in the right place at the right time, with the right skills. During the 1820s, fuelled by the burgeoning cotton industry, the centre of mechanical engineering moved from London to Manchester. During Fairbairn's six decades in Manchester it became the centre of the industrial world at a time of which Checkland has written:

> It is probably not far from the truth to say that the period from 1815 to 1885 in Britain represents that range of human experience in which individual economic initiative had its greatest opportunity to operate upon men and things, and in so doing to re-make an ancient society.[4]

Important also was that 1817, when Fairbairn & Lillie commenced in business, was within a limited time-span when there was a call for millwrighting skills, but before the need for major capital outlay on machine tools, which would have been required only a decade later.

A factor that distinguished Fairbairn from most of his contemporary millwrights and engineers was his ability to see possibilities which others either did not see, or, if they did see, did not pursue. His frequent approach was that of Schumpeter's entrepreneurial 'innovator' who takes a 'hint' from something that is out there, but which no-one else has developed:

> I must candidly admit that whatever improvements I have effected in practical science have originated in some useful hint which I have applied, when ruminating on the subject, for the purpose I wished to attain.[5]

This was so in power transmission, ventilated waterwheel buckets, fan-driven ventilation, the riveting machine, the Lancashire boiler, drop valves and the tank engine, all of which brought him commercial benefits. The evidence shows that the radical improvements first used in the Murray and Sedgwick Mills – wrought-iron line-shafting of reduced diameters turning at much greater speeds – were attributable to Fairbairn, and were the foundation of Fairbairn & Lillie's success. He also had the ability to see and grasp the potential opportunities provided by the transfer of technology from one branch of engineering to another, as the use of marine engines in textile mills and coal mines; and by derivation from one engineering product to others, such as the tubular crane from the tubular-girder bridge.

Fairbairn was the great experimental engineer of his age. His involvement in experiments on the Forth & Clyde Canal led him into shipbuilding. His best-known experiments, those in connection with the Britannia Bridge, besides leading to the longest railway bridge of its day, were the earliest buckling tests on thin-walled tube structures. They have been recognised by Alec Skempton as 'perhaps the most significant single piece of research in structures in the nineteenth century',[6] and by Donald Cardwell as 'the most revolutionary advance in structural engineering' in that century.[7] These experiments in the late 1840s led to fame and, through their derivative tubular structures, to fortune.

The approach to much of Fairbairn's work was that of optimisation – bringing together the best from every source resulting in some of the most advanced machines and structures of his day, for example his waterwheels, the passenger lift and single-storey integrated textile mills. Optimisation is a term rooted in mathematics, but is increasingly found in economics and information technology. It is used in current engineering but is not generally found in the history of engineering where, in the sense of 'to make as perfect and efficient as possible' often bringing together elements from different sources, it provides a useful description of Fairbairn's work.[8] With optimisation, advanced products of high quality were produced, enhancing Fairbairn's reputation and attracting clients who had the vision and the money to procure the best.

Fairbairn was indefatigable. Through his Scottish background with its ethos of the Reformation and the Enlightenment, the residual Puritanism of Cross Street Chapel, and the Enlightenment *Sermons* of Hugh Blair, he was imbued with the work ethic. Time was never wasted: neither frivolity nor luxury attracted him. Motivated primarily by a striving for excellence and recognition as an engineer, pecuniary incentives were never to the fore: 'Fame with him was ever before money'.[9] He had many interests but, apart from reading some fiction, all were linked to the common cable of engineering.

Fairbairn's ascendency was also fuelled by his wide networking. He was drawn towards outstanding engineers – by 1824 he knew three of R.C. Allen's list of seventy-nine 'Great Inventors',[10] and from the 1830s, through the Institution of Civil Engineers and the British Association, he established a national profile, coming to know almost all the leading engineers in Britain. No less than six of his clients were sometime Presidents of the Institution of Civil Engineers.[11] Fairbairn & Lillie's most spectacular commissions at the Murray and Sedgwick Mills came from fellow-Scots with engineering backgrounds. The link with John Kennedy demonstrates the effect of the network of a satisfied patron. George Stephenson, whom Fairbairn had known from apprenticeship days a quarter-of-a-century earlier, was the catalyst for Fairbairn's cast-iron girder bridge at Manchester and later for the great wrought-iron Britannia and Conway bridges. Important in Fairbairn's ascendency was his involvement in the British Association from the mid 1830s. Here he met leading scientists and engineers. Here was a stimulus for his research and a platform for his papers – expressions of his Baconian understanding of progress for the benefit of mankind. In 1861 he received the Association's highest honour, its Presidency, two years after Prince Albert, and the first engineer to hold this position. Giedion writes of 'the instinctive prescience of genius' being shown by 'turning up wherever new things were being done', and Fairbairn always seemed to appear at important landmarks in engineering – the Liverpool and Manchester Railway, the first iron steamships to reach Australia, the Britannia Bridge, the most advanced factory of the time at Enfield, the Atlantic Telegraph, the first wind-tunnel.[12]

When all that has been said, the fact remains that Fairbairn, whilst not without weaknesses, had innate abilities beyond many, indeed beyond most, of his contemporary millwrights and engineers. With belief in a Creator, he saw the opportunities which he had 'turned to moderately good account' as 'under the blessing of Divine Providence'.[13]

Fairbairn lived through a transitional age and his career illustrates the technical transformation of the first seventy years of the 19th century. He was both an eye-witness of, and a participant in, that transition – in transport, from wooden sailing ships to iron ocean steamers, from horses and canals to a national rail network, from king's messenger to international telegraphy, from staircases to elevators, from balloons to the first wind tunnel; and in engineering, from water, wind and horse to steam power, from hand tools to machine tools, from rule-of-thumb to theoretical calculations, and from cast iron, via wrought iron, to steel.

Influence

FAIRBAIRN'S influence was extensive. In this section his six major areas of influence are summarised – power transmission, mill-building, boilers, tubular structures, shipbuilding, and experimentation – together with four means by which his influence was diffused – through structures and artefacts, through personal connections, through papers and books, and through employees and pupils.

The power transmission at the Murray and Sedgwick Mills provided the launch of the Fairbairn & Lillie partnership. It also established the pattern of power transmission within factories for over a century, until steam engines were replaced by electric motors.

That work at the Murray and Sedgwick Mills began half-a-century of Fairbairn's industrial buildings which incorporated cumulative improvements in layout, structure, appearance, prime-movers and building services, to provide examples at the cutting edge of innovation across five decades. No other engineer influenced the development of industrial buildings from 1820 to 1870 to the extent that Fairbairn did, from simple buildings in the functional tradition to the architecture of Saltaire and the automated single-storey factory, with an increasing emphasis on building services – heating, ventilation, lifts – and safety issues including fire protection, dust control, smoke prevention and boiler safety. The list of *existing* world *textile* sites prepared by The International Committee for the Conservation of the Industrial Heritage contains more mills on which Fairbairn worked than any other engineer, architect, contractor or entrepreneur.[14] His influence as a mill-builder was unmatched, and world-wide.

Fairbairn's introduction of the riveting machine in 1837 dramatically changed boiler construction, producing better boilers, more quickly, at a fraction of the cost. The Lancashire boiler, introduced by Fairbairn and Hetherington in 1844, became the most widely used boiler for the rest of the steam age. His later experiments on the 'Resistance of Cylindrical Vessels to Collapse', the first investigations of the conditions of rupture in vessels exposed to uniform external pressure, immediately led to the widespread introduction of strengthening rings to boiler flue tubes, reducing the risks of explosions. From these experiments elastic stability has developed into a major area of engineering science. Thus in 1998 Josef Singer, widely recognised as the leading authority on 'buckling', wrote of Fairbairn's work on cylindrical vessels:

> Fairbairn's reports were not only the first accounts of a comprehensive program [*sic*] in shell stability ... but they ... present a well documented discussion of a systematic and carefully planned study, that provides enlightening reading even today.[15]

Fairbairn was the driving force in boiler safety in the third quarter of the 19th century, instituting boiler inspections through the Manchester Steam Users' Association, which spawned similar associations in many cities.

The evidence introduced in Chapter 4 indicates that Fairbairn's shipyard was 'the first great iron ship-building yard in Britain'[16] – one of the yards which laid the keel of the British shipbuilding industry which by the end of the century had become the greatest in the world.[17] Fairbairn's major contributions to shipbuilding were his experiments on the strength of wrought-iron plates and riveted joints; and, following from the Britannia experiments, the cellular structure of ships' hulls, first used by Brunel and Scott Russell in the building of the *Great Eastern*.[18] Before that 'most of the ideas of the iron shipbuilder were obtained from the construction of wooden ships'.[19]

From the Britannia Bridge experiments engineers learned the importance of previously unknown buckling phenomena, how to use cells to avoid buckling in the compression flange, how to use stiffeners to avoid web buckling, and the behaviour of plates in compression and bending.[20] In Europe the Britannia Bridge influenced the shift from suspension-bridges to girder-bridges in the 1850s. It also influenced shipbuilding, and its derivatives included tubular-girder bridges and 'Fairbairn' cranes. Far from being a cul-de-sac in bridge design, Britannia was influential in its day and, as Åkesson has convincingly shown, a harbinger of the tubular bridges of the second half of the 20th century.[21]

Boilers and bridges were by no means the only fields of Fairbairn's experiments. His early work with Hodgkinson on cast iron provided knowledge widely relied on for the remainder of the 19th century. His sustained-load experiments, over an impressive seven years, were the first 'creep' tests undertaken in this country and no equivalent work was done for another century.[22] His investigation of the strength of wrought-iron plates and riveted joints disproved a widely held view that joints with a single row of rivets were stronger than the plates. The results increased marine safety and were relied on for thirty years. He undertook fatigue tests on wrought-iron beams, conceiving fatigue failure to be caused by exhaustion of plastic ductility, which would become the basis of subsequent 20th-century theories of fatigue.[23] By the mid 1860s, his experimental work had moved to steel. Once again he was to the fore in providing useful information as steel became more readily available.

The great bridge experiments and those on canal boats used trial-and-error or parameter variation in search of solutions, supporting the theory that technology was not the application of knowledge derived from science. But many of Fairbairn's experiments were in the Newtonian tradition where experimental observations provided the basis for mathematical generalisations – 'useful and reliable knowledge' on which engineers relied and which affected engineering practice – thus supporting the view that new knowledge *also* came from science and 'from the study of technology through scientific methods'. Here was 'the enlightened economy' of Mokyr and of Allen.[24] Fairbairn's experimental work was more extensive than that of any of his contemporary engineers, giving good grounds to endorse Macquorn Rankine's assessment, written during Fairbairn's lifetime, that he was 'one of the most eminent engineers and cultivators of mechanical science',[25] an assessment endorsed by the research engineer and scholar Alastair Smith in the 1960s.[26]

Fairbairn's influence was diffused in many ways, of which four stand out. First, the buildings and machinery he constructed were visited, and often copied – Schumpeter's 'imitation'. Examples include the waterwheels at Deanston and Wesserling, transmission shafting, drop valves, and tubular cranes. Mills, including the Murray and Sedgwick Mills, Orrell's Mill, Saltaire and the Royal Small Arms Factory at Enfield, attracted a steady stream of visitors from Britain and the Continent.

Personal connections were important in the diffusion of Fairbairn's influence. Clients from Switzerland, Turkey, Germany, Sweden and elsewhere visited the Ancoats works. Fairbairn made many overseas visits, including to all these countries, as well as to France, Russia and, on many occasions, to Ireland.

A third major route by which Fairbairn's influence was diffused was by his many papers and books. He delivered more papers than any of his contemporary peer engineers, and wrote more books. His papers were largely about his experiments and his engineering projects. They were generally widely reported. His major books became standard works, running to several editions with several translated into European languages. Engineers of the calibre of Rankine frequently referred to them.[27] Fascinated by engineering history, Fairbairn's approach was Baconian, emphasising the benefits brought by science and technology.

Fairbairn was an active advocate of education. He attracted many very able young engineers to Ancoats. Those who trained or obtained experience with him have never previously been collectively identified. With five who became professors of engineering and four engineers who were knighted, Fairbairn's Ancoats works was the successor to the Maudslay 'nursery'.[28] When these men moved on, something of Fairbairn's commitment to optimisation, quality, investigation and education moved with them.

An aspect of influence which is often overlooked is longevity. There are many who might be remembered today if their lives had not been cut short at a relatively early age. Some who achieved much, might have achieved more – Robert

Stephenson and I.K. Brunel both died in their fifties. Fairbairn lived until he was eighty-five, with many achievements in his years of 'retirement' – years of influence and widespread acclaim.

Demise

IN any comprehensive study of Fairbairn's life it is inevitable that much of the emphasis will be on his successes. Yet historians increasingly appreciate the importance of the study of failure, as well as of success.[29] Fairbairn encountered difficult times in childhood, as an apprentice and as a journeyman, which influenced his later life. The Fairbairn & Lillie partnership started with set-backs, and the dissolution of that partnership was stressful. Severe financial pressures at the end of the 1830s were followed by the demise of the Millwall shipyard. There were the disagreements arising from the Britannia and Conway Bridges, and failures to secure much-wanted commissions, such as the Cologne Bridge. Some of these difficulties and failures had positive aspects. If Fairbairn had continued in partnership with Lillie, it is likely he would have made a comfortable living, undertaking millwrighting work and, like many of his contemporaries, would have been scarcely remembered. Had he not resigned from the Britannia Bridge it is unlikely that his role in it would have been appreciated to anything like the extent that it is today. Fairbairn's response to set-backs and failures was 'an indomitable spirit and unflinching industry'.[30] This resonates with Bolton and Thompson's assessment that overcoming adversities is an entrepreneurial trait.[31] His deepest disappointment, although he never mentioned it, must surely have been the failure of his sons to take the company forward after he had passed its management to them – the shattering of the dynastic dream.

The decline and ultimate fate of the Fairbairn Engineering Company reinforces Rose's argument that succession 'forms the critical foundation upon which the future prosperity of a family firm rests'.[32] The latter years of the family business prior to Fairbairn's retirement were successful and profitable, with the company held in high renown. Thereafter things went awry. It was probably the entrepreneurial dynastic imperative, identified by Schumpeter and Perkin, which limited Fairbairn's later partners to his sons,[33] none of whom had studied engineering at university or served a premium apprenticeship. The demise of the company had its roots there, three decades before it actually ceased trading. Thomas's success in organising the Manchester Art Treasures Exhibition highlighted his abilities, while at the same time revealing interests, lifestyle and networks very different from those of his father. By 1859, six years after William's retirement, the firm was in the sole hands of Thomas who was quick to see and seize the opportunities provided by the 1862 Companies Act. The flotation of his company in 1864 provides an early example of the application of the Act by a proprietor eager to realise capital from his business, in this case to purchase a landed estate. This suggests perhaps a more immediate impact of this facet of the Act than has previously been appreciated. Thomas's action was not simple gentrification as envisaged by Wiener,[34] but rather Daunton's 'gentlemanly capitalism', where commerce was removed from controlling a large work-force in a grimy industrial environment to panelled boardrooms in the City.[35]

No other leading North-west engineering company ceased to trade during the

Fig.C.2:
E.E. Geflowski,
Sir William Fairbairn, Bart.
(1878),
Manchester
Town Hall.

1870s, and many continued into the 20th century. Contrary to the theories put forward by several scholars,[36] primary research into the career of Thomas Fairbairn points to the fact that it was largely his failure to provide effective management and leadership, coupled with his lack of both investment and innovation that destroyed the company.

Summation

THIS book has shown William Fairbairn to have been one of the most versatile engineers of his time, the great mill-builder and experimental engineer of the middle-quarters of the 19th century, a major contributor to iron shipbuilding during the critical decade 1835–44, and to wrought-iron bridge-building during two of the most formative decades in the history of bridge-building. His achievements read like entries in a book of records – the largest and most advanced factories of their day, the most powerful waterwheels, the tallest chimney, the first three-storey building without load-bearing masonry external walls, the first major experiments on creep, buckling, fatigue and the strength of cylindrical vessels, the first iron bridge on a main-line railway, the longest span railway bridge of its day, the longest-span bridge in Australia, the first iron steamship on the Central European Lakes, the first to cross the Atlantic and the first to reach Australia, the world's most powerful crane, the first railway locomotives in Brazil, the winding and pumping engines for the deepest coal-mine of its day, patents for the riveting machine and the Lancashire boiler – some achievement for a former apprentice colliery-millwright. *The Engineer*'s obituary included the apposite words: 'it is difficult to discover a branch of the art of mechanical engineering to which Fairbairn has not contributed something. His footprints may be found on every path which the engineer can tread'.[37]

Notes for Conclusion

1. J Mokyr, *The Lever of Riches: Technological Creativity and Economic Progress,* (1990), p.105; J Mokyr, *Twenty-five Centuries of Technological Change: A Historical Survey*, (1990), p.69.
2. AW Skempton (ed.), *John Smeaton, FRS,* (1981), p.1.
3. N Rosenberg,'Review of *The Life of Sir William Fairbairn, Bart.* Reprint with Introduction by A E Musson', *Economic History Review,* 25.4, 1972, p.712.
4. S G Checkland, *The Rise of Industrial Society in England 1815-1885,* (1964), p.103.
5. *Life,* p.106.
6. AW Skempton, *Proceedings of the Institution of Civil Engineers,* Vol.1.3, 1952, 405, cited in Smith, *Contribution,* p.4; AW Skempton, 'Discussion', following R J M Sutherland, 'The introduction of structural wrought iron', *TNS,* 36, 1963-4, p.83.
7. D Cardwell, *James Joule, A Biography,* (1989), pp.34-5. See also S P Timoshenko and J M Gere, *Theory of Elastic Stability, Second Edition,* (1961), p.432.
8. S S Rao, *Engineering Optimization: Theory and Practice,* (1996).
9. *Life,* p.466, quoting 'a friend who knew him in business'.
10. R C Allen, *The British Industrial Revolution in Global Perspective,* (2009), Appendix: A list of the great inventors [of the seventeenth and eighteenth centuries], pp.269-70.
11. Cubitt, Bidder, Vignoles, Hawkshaw, Fowler and R Stephenson.
12. S Giedion, *Space Time and Architecture,* (3rd ed. 1954), p.599. He was writing about the French Architect, Le Corbusier.
13. *Life,* p.72.
14. TICCIH, *The International Context for Textile Sites,* (2003).
15. J Singer, 'Experimental Studies in shell buckling', in N F Knight and M P Nemeth (eds.), *Stability Analysis of Plates and shells: A Collection of Papers in Honor of Dr Manuel Stein,* (1998), p.21.
16. S Smiles, *Industrial Biography,* (1863), pp.327-8.
17. LT C Rolt, *Victorian Engineering,* (1970), p.78.
18. *The Artizan,* July 1856, p.145.
19. E C Smith, *A Short History of Marine Engineering,* (1938), p.109.
20. J Singer, J Arbocz and T Weller, *Buckling Experiments: Shells, Volume 2, Shells, Built-up Structures, Composites and Additional Topics,* (2002), p.623.
21. B Åkesson, *Understanding Bridge Collapses,* (2008), pp.179-92.
22. A I Smith, 'William Fairbairn and Mechanical Properties of Materials – 50,000 hours sustained-load tests on iron bars', *The Engineer,* 216, July-Dec.1963, pp.543-5; L Haworth, 'The Importance of Properties of Materials to the designer', in T F Roylance (ed.), *Engineering Design: Papers given at the University of Nottingham Summer School of Engineering Design, September 1964,* (1966), p.114.
23. B Åkesson, *Fatigue Life of Riveted Steel Bridges,* (2010), pp.7-9; A I Smith, 'William Fairbairn and Mechanical Properties of Materials: The Effect of Repeated Loading on Strength', *The Engineer,* 26 June 1964, pp.1133-6.
24. J Mokyr, *The Enlightened Economy: Britain and the Industrial revolution 1700-1850,* ([2009] 2011 ed.), p.40; Allen, *British Industrial Revolution,* p.239.
25. W J M Rankine, 'William Fairbairn', in *The Imperial Dictionary of Universal Biography,* Vol.2. (1863).
26. A I Smith, 'William Fairbairn – Experimental Engineer', in E G Semler (ed.), *Engineering Heritage: Highlights from the History of Mechanical Engineering, Volume 2,* (1966), p.20, where Smith also wrote that 'It is to … Fairbairn and Rankine … that we owe, perhaps more than to any others, the establishment of modern engineering science. Rankine made unique contributions on the theoretical side, and Fairbairn on the practical and experimental'.
27. W J M Rankine, *A Manual of Machinery and Millwork,* (1869), p.vi.
28. J A Cantrell and G Cookson (eds.), *Henry Maudslay and the Pioneers of the Machine Age,* (2002).
29. P. Fridenson, 'Business Failure and the Agenda of Business History', *Enterprise and Society,* 5.4, 2004, 'There are … only a few significant cases in which historians have attempted to deal with both the birth and death of firms.' (p.565). T J Pinch and W E Bijker, 'The Social Construction of Facts and Artifacts: Or How the Sociology of Science and the Sociology of Technology Might Benefit Each Other', in W E Bijker, T P Hughes and T Pinch (eds), *The Social Construction of Technological Systems: New Directions in the Sociology and History of Technology,* (2nd ed.2012), p.16; R Morris, *Time's Anvil: England, Archaeology and the Imagination,* (2012), pp.19, 319.
30. *Life,* pp.149, 340-1.
31. W K Bolton and J L Thompson, *Entrepreneurs: Talent, Temperament, Technique,* (2000). p.20.
32. M B Rose, 'Beyond Buddenbrooks: the family firm and the management of succession in nineteenth-century Britain' in J Brown and M B Rose (eds), *Entrepreneurship, networks and modern business,* (1993), pp.133-5.
33. J A Schumpeter, *The Theory of Economic Development,* (1934), pp.93-4; H Perkin, *The Origins of Modern English Society 1780-1880,* (1969), pp.83-5.
34. M J Wiener, *English Culture and the Decline of the Industrial Spirit, 1850-1980,* (2nd ed. 2004).
35. M J Daunton, '"Gentlemanly Capitalism" and British Industry 1820-1914', *Past and Present,* 122, 1989, 125.
36. See Chapter 8.
37. *The Engineer,* 38, 1874, p.154.

Sources of Illustrations in Conclusion

Fig.C.1: *Practical Mechanic's Journal,* Front cover mid 1850s.

Fig.C.2: E.E. Geflowski, *Sir William Fairbairn, Bart.* (1878), Manchester Town Hall. Photograph 2016.

Appendices

Appendix 1.1
William Fairbairn – The Journeyman Years

Dates	Duration	Travel	Place	Employer	Work	Notes
1811 26 March			Newcastle	Percy Main Colliery		Seven year apprenticeship completed.
1811 April	Few weeks		Newcastle		Erection of sawmill. 20s. per week	Met David Hogg (from Coldstream).
1811 May – October	Six months		Bedlington	Mr Norval	24s. per week	With Hogg. Met Dorothy Mar
1811 November	Few weeks			unemployed		
1811 December	2 weeks +	On board a collier, to London.				With Hogg. 'North-east gale for nearly a week'
1811 December – January 1812	1 month		London	unemployed		With Hogg. Cold and hungry.
1812 January		Walked to Cheshunt via Hertford				Rain and sleet; melted snow underfoot.
1812 January	2 weeks		Cheshunt		Erection of windmill	With Hogg
1812 January		Walk to London				
1812 January – October (?)	12(?) months		London	Mr Grundy, Patent Ropery, Shadwell	Construction of rope works	With Hogg
1812 October – December	Few weeks		London	John Penn, Greenwich	Millwright work at Stratford Mill	
1813			London	Jointly with Revd Hall	Model of a digging machine for Society of Arts competition	£20 lost on this venture
1813			London	Pork butcher, Tottenham Court Road	Sausage-making machine	A successful venture
1813 April		'outside place' on coach to Bath				Overnight, snow, 'suffered dreadful cold'
1813 April – May	Six weeks		Bathgate		Millwright work	
1813		Walked to Cardiff,				Via Bradford-on-Avon, Trowbridge, Bristol, Newport.
1813	4 days	Sloop to Dublin				Arrived penniless and hungry.
1813	c.4 months		Dublin	Mr Robinson, Phoenix Foundry	Nail-making machinery	
1813 October	2 days	Packet to Liverpool				
1813 October		Coach to Manchester				
1813 October			Manchester	Adam Parkinson	Millwright, calico printing machinery	

1813 November – 1814 February	c.3 months		Manchester	Laid up with scarlet fever.		
1814 March – 1816 (?)	c.2 years		Manchester	Adam Parkinson	Millwright, calico printing machinery	
1816 June		Coach to and from Bedlington				Married Dorothy Mar on 16 June at Bedlington
1816 July	c.8 months		Macclesfield	Thomas Hewes (of Manchester)	Erecting waterwheels and millwork	First married home – lodgings in Macclesfield.
1817 March			Manchester			First child born 30 March. Had to borrow to pay medical bills.
1817 March – November	c.8 months		Manchester	Thomas Hewes	Draughtsman	
1817 November			Manchester		Millwright	Commenced own business. Joined by James Lillie

Appendix 1.2
The Known Children and Grandchildren of William and Dorothy Fairbairn

Name	Dates	Marriage	Children	
Anne	1817-1894	1841 – J F [La Trobe] Bateman	Mary 1842-1899 Margaret 1843-1923? William 1845-1926 Anne 1847-1910? Agnes 1849-1910?	Frederick 1853-1889 Lee 1859-1938
John	1821-1867	?	?	
Thomas	1823-1891	1848 – Allison Callaway	Mary 1850-1852 Emma Constance 1851-1904 Arthur Henderson 1852-1915 Thomas 1854-1931 Reginald 1856-1921	Florence 1859?-1875 Mary (May) 1860-1936 James Brooke 1863-1947 Alan 1865-c.1875
William Andrew	1824-1910	(1)1849 Marianne Murray d.1859 (2)1865 Helen Birley	William Murray 1850-1929 Emily 1853-1935 Edith 1854-1928 Ada 1856-1893 Francis 1865-1945 Edward 1866-1931 George 1869-1901	
Margaret	1826	—	-	
Peter	1828-1859	1854 Maria Cleather	Edward Cleather 1855-1873	
George	1830-1868	1855 Bessie Bagshawe		
Adam Henderson	1836-1888	1864 Anna Fuller-Maitland	Frances 1865-1951 Adam Maitland 1866-1921 Hilda 1868-1917 William 1869-1930 Thomas 1870-1944 Rosamund (Rosie)1872-1948	Helen (Dolly)1873-1914 Hugh 1874-1944 Ida 1876-1940 Alan 1878-1960 David 1882-1950

Appendix 2.1
The main Mill Buildings with which William Fairbairn was involved

Notes

1. The following list is largely restricted to occasions on which mills were built, or the power transmission installed. Occasions where work was limited to the prime mover and/or boilers are omitted – see further appendices below.

2. Mills where Hodgkinson beams are known to have been used, but where there is no other evidence of Fairbairn involvement, are omitted.

3. There is little doubt that there were many other mills by Fairbairn, particularly in industrial Lancashire.

Date	Location	Owner	Purpose	Type	Fire-proof	Prime mover	Archi-tect	Exit-ing	Notes	Main references
1818	Murray Mill, Ancoats, Manchester	A & G Murray	Cotton Spinning	Multi-storey	Yes	Steam	N/A	Yes	Fairbairn & Lillie renewed the power transmission	I Miller and C Wild (eds), *A & G Murray and the Cotton Mills of Ancoats*, (2007), pp.77-8.
1818–20	Sedgwick Mill, Ancoats, Manchester	McConnel & Kennedy	Cotton Spinning	Multi-storey	Yes	Steam (Boulton & Watt)	No	Yes	Fairbairn prepared drawings. Fairbairn & Lillie undertook the millwork. Altered by WF, mid 1860s.	Miller and Wild, *A & G Murray*, pp.51-54; S Little, *Report on the McConnel & Co. Mills. An Archaeological Survey on the Royal Mill Complex, Ancoats* (2007); S Little, 'The Mills of McConnel & Kennedy, Fine Cotton Spinners', *Transactions of the Lancashire & Cheshire Antiquarian Society*, 104, 2008, 35-59; C H Lee, *A Cotton Enterprise 1795-1840: a history of M'Connel & Kennedy fine cotton spinners* (1972).
[1823]	Stalybridge, Cheshire	Thomas Harrison	Cotton spinning and weaving	Multi-storey	Yes	Steam (Boulton & Watt)	No	No	Although there is no record of Fairbairn's involvement, Tann points out that it resembles Wood's factory in Bradford	J Tann, *The Development of the Factory* (1970), p.24.
1824 and later	Ancoats, Manchester	Own works	Engineering, incl. foundry	Various	In part	Manual then Steam	No	No		*Life*, pp.112-17; Hayward, 'Fairbairns', pp.1.12–1.15, 2.2–2.6.
1824–25 & 1828	Bean Ing Mills, Leeds	Benjamin Gott	Wool Spinning and weaving	Multi-storey extension	Yes	Steam (Boulton & Watt)	No	No		Letters, Fairbairn to Gott, Gott Papers, Brotherton, Library, Leeds; Fairbairn, *Application* , p.5; R Fitzgerald, 'The Development of the Cast Iron frame in Textile Mills to 1850', *IAR*, 10,2, 1988, 138; D T Jenkins, *The West Riding Wool Textile Industry 1770-1835. A Study of Fixed capital Formation* (1975), p.60.
1824	Caledon, Ireland	Earl of Caledon	Corn Mill	Multi-storey		Water. Steam later	No	No	Extent of Fairbairn & Lillie's involvement unclear	M Lake,*Caledon Conservation Area* (2003), p.18; W A McCutcheon, 'Water Power in the North of Ireland', *TNS*, 39,1966-7, 93; W A McCutcheon, *The Stationary Steam Engine in Ulster* , (reprinted from *Folk & Farm.* No date), pp.8-9, 23 and Plate 5.

1824–5	[Gorton Mills, Manchester]	A Lees	Cotton Spinning and weaving	Multi-storey with later single-storey additions		Steam	No	No	Circumstantial evidence only. Farnie dates it 1825. Similarities with Orrell's Mill	D A Farnie, *John Rylands of Manchester* (1993), p.12, Fig.3; F Wightman, 'Gorton Mills', (1980). Copy held at G M Record Office).
1825	Bradford	John Wood	Worsted spinning	Multi-storey	Yes	Steam (Boulton & Watt)	No	No		Fitzgerald, 'Development of the Cast Iron Frame', 139; Tann, *Development of the Factory*, p.40; Jenkins, *West Riding Wool Textile Industry*, pp.106-8; J T Ward, *The Factory System: Vol. 1: Birth and Growth*, (1970), pp.131–8
1825-28	Havelock Mill Manchester	Vernon Royle	Silk spinning	Multi-storey	No	Steam	No	No	Machinery by Fairbairn & Lillie	Ure, pp.249, 266, 272-5; *M&MWII*, pp.213-6; M Williams, 'Havelock Mill, Manchester: A case-study in the emergency recording of a large urban mill complex', *IAR*, 16.1, 1993, 100-10.
1826	Portlaw, Ireland	Malcolmson Brothers	Cotton spinning	Multi-storey	Yes	Water	No	Yes, in part		T Hunt, *Portlaw, County Waterford 1825-1876. Portrait of an Industrial Village and its Cotton Industry*, (2000), pp.46-8; A Bielenberg and J M Hearne, 'Malcolmsons of Portlaw and Clonmel: some new evidence on the Irish cotton industry 1825-50', *Proceedings of the Royal Irish Academy*, 106C, 2006, 342, 346-7.
Pre-1827	Pollard St, Ancoats, Manchester	J Pollard	Cotton spinning			Steam	N/A	No	Fairbairn & Lillie were responsible for the shafting	*MG*, 17 November 1827.
1829	Albion Corn Mill, Tib St, Manchester	J Fletcher	Corn milling			Steam	N/A	No	Fairbairn & Lillie were responsible for the machinery and millwork	Auction poster 1833, in the possession of Capes Dunn, Auctioneers, Manchester. I am indebted to Terry Wyke for this information.
1830	Egerton, Bolton	Ashworths	Cotton spinning, weaving and dyeing	Multi- and single-storey		Water and steam	No	No		R Boyson, *The Ashworth Cotton Enterprise. The Rise and Fall of a Family Firm 1818-1880*, (1970), pp.19-23, Plate IV; *Life*, pp.146-8; J H Longworth, *The Cotton Mills of Bolton: A Historical Directory*, (1987), pp.20-1.
1831	Macclesfield, Cheshire	J P Swanick (lessee).	Corn Mill	Multi-storey	It is believ-ed so.	Steam	No	No	This was almost certainly the first mill with Hodgkinson beams	T Bonson, 'Macclesfield Mill', *Wind and Water Mills*, 23, 2004, 30-6; N Brindley, 'Francis Brindley and the Marple Brindleys' at http://www.marple-uk.com/marple_brindleys.htm (accessed 26.03.13), p.3 of 12; *MG*,22 May 1830.
Pre-1832	Knott Mill, Manchester		Cotton	Multi-storey		Steam	N/A	No	Fairbairn & Lillie's work was probably only the shafting.	*MG*, 9 August 1834.
1833	Park Mill, Brereton, Cheshire.	John Howerd	Corn Mill	Multi-storey		Water and Steam	No	Yes		T Bonson, *Driven by the Dane : Nine Centuries of Waterpower in South Cheshire and North Staffordshire*, (2003), pp.175-83.

1834-37	Stockport	Ralph Orrell	Cottton spinning and weaving	Multi-storey + weaving shed	Yes	Steam	No	No		Ure, *Philosophy*, pp.33-4 and frontispiece: A Ure, *The Cotton Manufacture of Great Britain*, (1836), Vol.1, pp.296-304, 311-2 + Plates 1 and 2; Fitzgerald, 'Development of the Cast Iron Frame', 139-41.
1835	Hammer-stein, Germany	F A & W Jung	Cotton	Multi-storey		Water and steam				A Oehike, 'Das englische Vorbild: Die Einführung moderner Spinnereibauten und Textiltechnik aus Lancashire', in H Bönnighausen et al, *Cotton mills for the continent: Sidney Stott und der englische Spinnereibau in Münsterland und Twente* (2005), p.34n5; T C Banfield, *Industry of the Rhine embracing a view of the social condition of the rural and manufacturing population of that district, 2, Manufactures*, (1848), pp.145-6; 'Villa Hammerstein um 1850.jpg, Wikimedia Commons (accessed 26 September 2016
1835	Millwall, London	Own shipyard	Ship-building	Single-storey		Steam	No	No		*Life*, pp.335-42; R A Hayward, *The Story and Scandal of HMS Megaera*, (1978), pp.18-21; *CE&AJ*, 3, 1840, 397; *ILN*, June1848, 309; London Metropolitan Archives, Records of Sun Fire Office, MS 11936/555/1252067-8 and /562/1273798; H Hobhouse (ed.) Survey of London: Vols 34 and 44: Poplar Blackwall and the Isle of Dogs, (1994) pp.466-80, available at 'British History Online'.
1836	Stalybridge, Cheshire	Bailey Brothers	Cotton spinning and weaving	Two-storey		Steam	No	No	Side-lever engine	Ure, *Cotton Manufacture*, Vol.2, p.313; I Haynes, *Stalybridge Cotton Mills*, (1990), p.33; *M&MWI*, pp.246-8, Plate 8.
1836	Carlisle, Cumbria	Peter Dixon	Cotton spinning	Multi-storey	Yes	Steam	R'd Tatter-sall	Yes	Possibly first textile mill involving an architect	E Jones, *Industrial Architecture in Britain 1750-1939*, (1985), pp.58-9.
Pre-1836 and c.1860	Milford Mills, Carlow, Ireland		Corn Mills	Multi-storey		Water	No	Part	Reconstructed by Fairbairns c.1860.	*The Carlow Sentinel*, 8 October 1836; S C Hall and A M Hall, *Ireland: Its Scenery, Character, &c.* (1841), p.405; *Carlow Post*, 11 November 1862; *Carloviana: Journal of the Old Carlow Society*, 1993-4, p.12.
Pre-1837	Glossop		Cotton spinning and weaving	4 or 5-storey		Water and steam	No		Extent of Fairbairn's work unclear. Possibly gearing and shafting.	*MG*, 28 January 1837; *Manchester Times and Gazette*, 4 May 1844; *MG*, 9 November 1844.
1837	Scharfen-stein Germany	Fiedler & Lechla	cotton	Multi-storey		Steam	C F Uhlig			A Oehlke, 'Spinnmühlen in Sachsen – Technologietransfer und Architektinische Umsetzung einer neuen Bauaufgbe', *Mitteilungen des Chemnitzer Geschtsvereins*, Neue Folge VIII.69, 1999, 130-1.
1837	Northern Ireland		Flax and tow spinning			Water and steam				*MG*, 7 March 1846.

1837-38	St Petersburg Russia	Russian Government (per Alex. Wilson)	cotton							*MPICE*, 29, 1875, 259; M Chrimes, 'Wilson, Alexander' in A W Skempton (ed.), *A Biographical Dictionary of Civil Engineers in Great Britain and Ireland, Volume 1 1500-1830*, (2002), pp.785-6; *MPICE*, 30, 1870, 461-5
1838	Vienna		Locomotive Works			Steam				Austrian Imperial Railways Exhibition Catalogue (Vienna 2008).
1838	[Hunslet Mill, Leeds]	John Wilkinson	Flax	Multi-storey	Yes	Steam	No	Yes	Circumstantial evidence only for Fairbairn's involvement	M Binney, F Machin and K Powell, *Bright Future, The Re-use of Industrial Buildings* (1990), pp.42-50; C Giles and I H Goodall, *Yorkshire Textile Mills. The Buildings of the Yorkshire Textile Industry 1770-1930* (1992), various refs.
c.1840	Flax Street Salford		Flax	Multi-storey	Yes	Steam	No	No		F Wightman, 'The Flax Mill, Flax Street, off Silk Street, Salford' (1982), at https://www.flickr.com/photos/manchesterarchiveplus/9656443962/in/photolist-fHiN... and https://www.flickr.com/photos/manchesterarchiveplus/9662173775/in/album-721576... (accessed 1 April 2016).
	Leeds	Marshalls	Flax						Unclear what Fairbairn was involved with	Fairbairn, *M&MW*, p.198.
1840	Victoria Mill, Wigan.		Cotton	Two-storey	Yes	Steam	No	Yes		R Pollard and N Pevsner, *Buildings of England. Lancashire: Liverpool and the South-West*, (2006), p.675; E Crowe, *The Dinner Hour, Wigan* (1874, Manchester City Art Gallery).
1840	Brunswick Mill Manchester		Cotton	Multi-storey	Yes	Steam	No	Yes	Fairbairn's role is unclear	M Williams with D A Farnie, *Cotton Mills of Greater Manchester* (1992), pp.154-6.
1840	Constantinople		Corn Mill	Pre-fabricated Iron. 3-storey		Steam	No	No	An important building in the history of architecture	*M&MWII*, p.116; [D Scott], *Engineer and Machinist's Assistant*, (1850), Description of Plates, pp.91-8 and Plates 98-101; T C Bannister, 'Bogardus Re-visited, Part 1: The Iron Fronts', *Journal of the Society of Architectural Historians* (American), 15.4, 1956, 15; G Herbert, *Pioneers of Prefabrication: The British Contribution in the Nineteenth Century*, (1978), pp.41-2
Pre-1841	York Street, Chorlton-on-Medlock Manchester		Cotton (?)	Multi-storey		Steam	N/A	No	Fairbairn's work possibly limited to gearing and shafting	*MG*, 29 May 1841; *MG*, 18 March 1843; *Manchester Times and Gazette*, 25 November 1843.
1842-43	Robinwood Mill, Todmorden		Cotton	Multi-storey	No	Steam and Water	No	Yes	Acquired by Fieldens	*MG*, 19 October 1844; *Halifax Guardian*, 2 November 1844; Giles and Goodall, *Yorkshire Textile Mills*, various refs.

1842	Cassano d'Adda, Lombardy, Italy	Battaglia & Co.	Flax	Multi-storey		Water		Yes, in part	Factory designed by Benedict Albano, Civil Engineer, London	*CE&AJ*, 5, 1842, 138-9; 6, 1843, 143, 181, 191-2, Plate VII; S B Moody, 'Description of a Water-wheel constructed by Mr. W. Fairbairn, MInstCE, and erected in Lombardy', *MPICE*, 3, 1844, 66-7; *Archeologia Industriale. Notiziario della Società Italiana per l'Archeologia Industriale, Sezione Lombadia*, No.3, March 1979, 5; and No.7, March 1980, 6.
1843	Izmet Turkey	Turkish Govern-ment	Wool spinning and weaving	Single-storey with basement	Yes	Water	None			Fairbairn, *M&MWII*, pp.188-92 + Plates 17 and 18; W Fairbairn, 'Description of a Woollen Mill erected in Turkey', *MPICE*, 3, 1843, 125-6; *MG*, 22 March 1843, 20 December 1843.
1843	Turkey	Turkish Govern-ment	Furnaces, forges, rolling mills							*Life*, p.172.
1843	Turkey	Turkish Govern-ment	Silk Mill							*Life*, p.172.
1843	Turkey	Turkish Govern-ment	Cotton Mill							*Life*, p.172.
1843	Alexandria, Egypt	Turkish Govern-ment	Arsenal							British Library, Add. MSS, 37461, ff.378-87, 37462, f.61.
1846	La Foudre, Rouen, France	Lebaudy Peter et Cie	Flax	Multi-storey	Yes	Steam	Yes	Yes	First 'fireproof' mill in France	L'Association Mémoire et Patrimoine de Petit-Quevilly, 'Caserne Tallandier', (5-page leaflet, nd); http://thiriez.org/filaturesJN/LaFoudre.htm (accessed 16 September 2016) http://www.petit-quevilly.fr/histoire/la-foudre/ (accessed 16 September 2016)
1847-9	Whittaker's Mill, Aston-under-Lyne	John Whittaker & Sons	Cotton spinning and weaving	Multi-storey + weaving shed	Yes	Steam	No	No		Fairbairn, *Application* , pp.58-60; I Haynes, *Cotton in Ashton,* (1987), p.39.
1849	Tampere, Finland	Finlayson & Co.	Cotton			Water		Part	Extent of Fairbairn's involvement unclear	TICCIH, 'The International Context for Textile Sites', (2003), p.11; http://www.flickr.com/photos/36713050@N03/5248739189/in/set-72157625388834955 (accessed 19.03.13).
c.1850	Old Union Corn Mill, Birmingham		Corn Mill			Steam	None	No	Reconstruction and supply of machinery	Blackie & Son, *Engineer and Machinist's Assistant* (1850), Description of Plates, p.98 and Plate 106; *UIfE2*, pp.336-8.
1850	Invangorod, Russia	Alexander von Stieglitz	Flax Mill	Multi-storey	Yes	Water		Yes		Fairbairn, *M&MWII*, pp.198-211.

1850	Gefle (Gavle), Sweden		Cotton	Multi-storey	Yes	Water				Fairbairn, *UIfE2,* pp.229-32; P Carlberg, 'Personal Contacts between the Manchester Area and Gefle in Sweden a Hundred Years Ago: A Communication', *Transactions of the Lancashire and Cheshire Antiquarian Society,* 70, 1960, 58; S Söderberg, *Förstaden vid Testeboån: arkivanteckningar och traditioner om Strömsbros uppkomst och utveckling,* (1966), pp.59-63; *MG,* 31 July 1850; *Life,* p.365.
1851	Wolver-hampton	J & J Norton	Corn Mill	Multi-storey	Yes	Steam	No	No	Demolished 2015	R K Morriss, *The Old Steam Mill, Corn Hill, Wolverhampton. An Archaeological and Architectural Analysis,* (Report commissioned by Wolverhampton MBC, 1999).
1851-53	Saltaire, Yorkshire	Titus Salt	Spinning and weaving Alpaca	Multi-storey and single-storey	Yes	Steam	Lock-wood & Maw-son	Yes	This was the largest factory at the time it was built	*Life,* p.328; R Balgarnie, *Sir Titus Salt, Baronet: His Life and its Lessons,* (1877), pp.116-22; Fairbairn, *Application,* pp.165-76, Plates I-II; Fairbairn, *M&MWII,* pp.114-5; *ILN,* 1 October 1853, pp.287-9; *The Builder,* 12, 602, 1854, 437-9; Jones, *Industrial Architecture,* pp.95-7; J Reynolds, *The Great Paternalist: Titus Salt & the Growth of Nineteenth Century Bradford,* (1983).
1853-57	Royal Small Arms Factory, Enfield	Govern-ment	Rifles etc.	Single-storey		Steam	Lock-wood & Maw-son	Yes	This was the most advanced factory in Europe when it was built	*Life,* p.327; *MG,* 6 January 1855; *The Engineer,* 7, Jan.-June 1859, 204-5, 258, 294-5, 348-9, 422-3; *ILN,* 21 September 1861, p.298; *MM,* 6, 1861,110-1,127-8, 144-5; M Bowbelski, *The Royal Small Arms Factory,* (1977); E Ames and N Rosenberg, 'The Enfield Arsenal in Theory and History', *The Economic Journal,* 78.312, 1968, 827-42; N Rosenberg, *The American System of Manufactures,* (1969); D Pam, *The Royal Small Arms Factory, Enfield & Its Workers,* (1998); B Cherry and N Pevsner, *Buildings of England: London 4: North,* (1998), pp.45, 452-3; Jones, *Industrial Architecture,* p.140.
c.1854	Powder Mills, Waltham	Govern-ment	Gun-powder	Single-storey			No			Fairbairn, *M&MWII,* pp.255-61.
1854	Tardeo, Bombay.	Bombay Spinning and Weaving Co.	Cotton spinning and weaving	Probably single-storey		Steam	No		This was the second cotton mill in India. Machinery was by Hetherington	S D Mehta, *The Cotton Mills of India 1854-1954,* (Bombay 1954), pp.13-7.
1855	Bombay	Oriental Spinning & Weaving Co.	Cotton spinning and weaving	Single-storey		Steam	No		The third cotton mill in India.	Fairbairn, *M&MWII,* pp.178-87, Plates 15-16; Mehta, *Cotton Mills of India,* p.22; E Leigh, *The Science of Modern Cotton Spinning,* (3rd ed. 1875), Vol.1, pp.49-51, Plate 14 (where a similar mill is described as Bombay Spinning and Weaving Company, and the design attributed to Hetherington)

	Bombay		Cotton spinning and weaving	Single-storey		Steam	No		Fairbairn refers to having built 'several of these mills for the Bombay Presidency'	Fairbairn, *M&MWII*, pp.179.
1856	Hjula Weavery, Oslo	Halvor Schou	Cotton spinning and weaving	Multi-storey		Water	No		There was an earlier pilot mill, the Brenneriveien Weavery (1849)	K Bruland, *British Technology & European Industrialization. The Norwegian textile industry in the mid nineteenth century*, (1989), pp.42, 76, 87, 114-5, 170.
c.1860	Taganrog, Russia.	Russian Govern-ment	Corn Mill	Multi-storey	No	Steam		No	Possibly designed before the Crimean War	Fairbairn, *M&MWII*, 127-32, Plates 11-13.
1864	Dublin	Bewley Moss & Co,	Sugar Refinery Warehouse	Multi-storey.	Yes		Alfred Darby-shire.	Yes	An early, possibly the first, use of a wrought iron structural frame	Fairbairn, *Application*, 3rd ed. 1864), pp.167-8; S Giedion, *Space, Time and Architecture*, (3rd ed. 1953), p.192; A Darbyshire, *An Architect's Experiences*, (1897); T A Lockett, *Three Lives*, (1968), pp.24-45.
1871	Egypt		Sugar Refinery							*Manchester City News*, 17 June 1871.

Appendix 2.2
Known Fairbairn Waterwheels

Date	Location Client Purpose	No. of Wheels	River	Diam.	Width	HP	Existing	Notes	Main references
1822-5	Turton, Bolton H & E Ashworth New Eagley Mill (cotton spinning)	1	Eagley			45	No	Cost £880	R Boyson, *The Ashworth Cotton Enterprise 1818-80,* (1970), p.15.
1824-6	Zurich Escher Wyss & Co. (cotton spinning)	2	Limmat				No	Designed to be raised and lowered.	*Life*, pp.124-8; W O Henderson, *Britain in Industrial Europe 1750-1870*, (1965), p.205; W O Henderson, *Industrial Britain under the Regency,* (1068), p.5.
1824	Schinznach Baths, Switzerland	1					Leat and tailrace exist	Per J G Bodmer	*MPICE*,.28, 1865, 580; H Balmer, *Zur Geschichte von Bad Schinznach*, (1991) p.6.
1823	Caledon, Ireland Earl of Caledon (corn mill)	2	Blackwater	30ft 20ft			Part of tailrace exists	Circumstantial evidence only	W A McGutcheon, 'Water Power in the North of Ireland', *TNS*, 39, 1966-7, 93; http://www.caledon.org.uk/mill.php
c.1824	Portlaw, Co. Waterford. Malcolmsons (cotton spinning)	2	Clodiah	34ft 26ft	16ft 6in. 15ft			Cast in Cork. Fairbairn's role is unclear	T Hunt, *Portlaw, County Waterford 1825-1876. Portrait of an Industrial Village and its Cotton Industry*, (2000), pp.46-8; A Bielenberg and J M Hearne, 'Malcolmsons of Portlaw and Clonmel: some new evidence on the Irish cotton industry 1825-50', *Proceedings of the Royal Irish Academy,* 106C, 2006, 342, 346-7.
1825	Wesserling, Alsace. Gros, Davillier, Roman & Co. (cotton spinning)	1	Thur	7.45m	6.08m	60	No	Ure describes it as 'on the ventilating principle'	A Ure, *The Cotton Manufacture of Great Britain*, (1836), Vol.I, pp.lxvii-lxxii; Pole, *Life*, pp.77, 129; Drawings, 3 signed by Fairbairn, at 'Portefeuille Industriel', CNAM, Paris.
1825-8	The Vosges, Alsace, & other parts of France'	several							*Life*, p.129.
1825-7	Catrine, Ayrshire. James Finlay & Co. (cotton spinning)	2	Ayr	50ft	10ft 6in.	120 each	Weir, sluice gate, tunnel and reservoir remain	Demolished 1945	Fairbairn, *M&MWI,* (1861), pp126-30 + Plates 1-2; [C Brogan], *James Finlay and Company Limited 1750-1950,* (1951), pp.61-3; W Fairbairn, 'On Waterwheels with Ventilated Buckets', *MPICE,* 1849, pp. 57-8; G D Hay and G P Stell, *Monuments of Industry: an illustrated historical record,* (1986), pp.63, 65-6; R Stenlake, *A Lot o Genuine Folks and A Wheen o Rogues,* (2011), pp.42-6; http://yesterdayrules.wordpress.com/2010/09/23/catrine-cotton-mills/
1827-8	Mr. Cooke. Could be 'Cook', in which case see below	[1]							*Life*, p.452.

1827-8	Hagues, Cook & Wormald, Dewsbury, (blanket makers)	1					No	In use until well into the 20th century	F J Glover, *Dewsbury Mills, A History of Messrs, Wormalds and Walker Limited, Blanket Manufacturers, Dewsbury,* (PhD thesis, Leeds University 1959), pp.400-55; F J Glover, 'A Yorkshire blanketmaker's diary', *Journal of the Bradford Textile Society,* 1962-3, p.89 (The illustration on p.85 is NOT a Fairbairn & Lillie waterwheel).
1828	Handforth , Cheshire. Duckworth & Co. (print works)	1		15ft 6in.	18ft		No	Low-breast. Probably first ventilated wheel (without sole-plate)	Fairbairn, *M&MWI,* 133-4 + Plate I; Fairbairn, 'On Waterwheels', 51-4.; *Life,* pp.188-90.
1828	Linwood, Paisley, Scotland. Andrew Brown.	1					No	Openings in the sole plate	Fairbairn, *M&MWI,* 133-4 + Plate 3; Fairbairn, 'On Waterwheels', 51-4.; *Life,* pp.188-90; [D Scott], *The Engineer and Machinist's Assistant,* (1850), p.217.
1830	Deanston, Perthshire. James Finlay & Co. (cotton spinning & weaving)	2 (+2 by J Smith)	Teith	36ft		100 each	Lade exists.	Replaced by turbine for electricity for the Grid	Fairbairn, *M&MWI,* pp.130-4; [Brogan], *James Finlay,* pp.68-72; *Life,* p.314: B Byrom, *Old Doune and Deanston,* (2009), pp.30-9.
c1830	Egerton, Bolton. Ashworths et al. (cotton spinning and bleaching)	1		62ft		110–140	No	Designed by Bodmer. Controversy re ventilated buckets	Boyson, *Ashworth,* pp.19-22 + Plate I; B Jones, *Bolton's Industrial Heritage,* (2006), pp.47-9; *MPICE,* 28, 1868-9, 581; *The Engineer,* 5 October 1877, p.240.
	South of England. Silk Mill.	(1)		22ft	10ft			Circumstantial evidence only.	Fairbairn, *M&MWII,* pp.216-8.
	Park Mill, Stockport.. Peter Marsland. (cotton)	1				78	No	Also pumped water for Stockport Water Works	*MG,* 27 June 1847; P Arrowsmith, *Stockport: A History,* (1997), p.195.
1832	Probably Ashworths.	1	Eagley	32ft			No	Tailrace an inverted siphon in a tunnel	Fairbairn, *M&MWI,* pp.125-6.
1833	Hazelbank, Banbridge, N Ireland, Samuel Law. (flax spinning and bleaching	1	Bann	14ft	12ft		Leat remains	Replaced by turbine for electricity for the National Grid	E R R Green, *Industrial Archaeology of County Down,* (1963), pp.7,19; H D Gribbon, *The History of Water Power in Ulster,* (1969), p.21; J Smyth, 'On the Industrial Uses of the Upper Bann River', *BAAS1874,* pp.131.
1834	Seapatrick, Banbridge, N Ireland. F W Hayes. Flax spinning and weaving.	1	Bann	16ft	22ft	50	No		Green, *Industrial Archaeology of County Down,* pp.7, 18.

1834-8	[Lumbutts Mill, Calderdale, Yorkshire. Fieldens.] (cotton spinning).	(2+1)	Calder	31ft	6ft	54?	3 wheels, one above the other in a tower. The tower remains.	Attributed to WF on at least 3 websites. Law gives no attribution. May not be by Fairbairn	http://freepages.genealogy.rootsweb.ancestry.com/~todmordenandwalsden/lumbuttsm... (accessed 26 March 2013); http://www.geograph.org.uk/photo/361136 (accessed 26 March 2013); B R Law, *Fieldens of Todmorden: A Nineteenth Century Business Dynasty,* (1995), pp.60-1,94; J Glynn, *Rudimentary Treatise on the Power of Water,* (1853), p.82.
1835	Moylinney, Co. Antrim, N Ireland. William Chaine. (flax beetling)	1	Six Mile Water	17ft	12ft		No	Circumstantial evidence only	Gribbon, *Water Power in Ulster,* p.22.
Pre 1836	Milford Mills, Carlow, Ireland. John Alexander. (flour milling)	1 2 (?)	Barrow		22ft	120	Weir remains.	Replaced by turbine for the National Grid. Mill by WF	S C Hall and A M Hall, *Ireland: Its Scenery, Character, & Co.,* (1841), p.405; W Ellis, 'Milford Mills', *Carloviana: Journal of the Old Carlow Society,* 1993-4, 12-3.
c.1836	Park Mills, Brereton, Cheshire. John Howard. (flour milling).	1	Croco	18ft	6ft		Existing, in poor condition. Grade II*.	Mill turned into apartments.	T Bonson, *Driven by the Dane. Nine Centuries of Waterpower in South Cheshire and North Staffordshire,* (2003), pp.175-83; Original drawings held at Brereton Hall.
1835-9	Herdmans Mill, Sion Mills, N Ireland. (flax spinning)	2	Mourne				Weir and leat remain.	Replaced by turbine for electricity for the Grid.	Letter: Herdmans to Major Humphries 31 June 1835, held by Mrs C Ferguson, Sion Mills, to whom I am grateful for a view of this and other documents; Sion Mills Buildings Preservation trust at: http://www.sionmills.org/Default.aspx?tabid=126
1838-9	[Compstall, Marple, Cheshire. Andrew, Bruckshaw & Bradley.] (cotton spinning, weaving, & bleaching)	(1)	Etherow	50ft	17ft (?)		No	Almost certainly by James Lillie & Co.	R E Thelwall, *The Andrews and Compstall their Village,* (1972), p.14. J K Major believes that both Fairbairn and Lillie had access to the earlier Fairbairn & Lillie patterns (Letter, J K Major to R J Byrom, 20 February 1983).
Pre 1840	Gisors, France.	1						Used in trials with a turbine	*CE&AJ,* 3, 1840, 420-1.
Pre 1841	Buckley Brothers Todmorden, Yorks. (cotton and worsted)	1	Calder			24	No	Buckley Bros. were in bankruptcy in 1841	*Manchester Times and Gazette,* 22 October 1841.
c.1842	Cleator, Whitehaven, Cumbria. Thomas Ainsworth. (flax spinning)	1	Ehen	20ft or 21ft 6in	22ft or 24ft	100 or 130	No	Ventilated. Dimensions from Davies-Shiel differ	Fairbairn, *M&MWI,* p.137 + Plate 4; Fairbairn, 'On Waterwheels', 54-7+ Plate 2; [Scott], *Engineer and Machinist's Assistant,* pp.85-6 + plates 92-3; M Davies-Shiel, *Watermills of Cumbria,* (1979).
1842	Cassano, Lombardy, Italy. Battaglia & Co. (flax and hemp, spinning and weaving)	1	Adda	16ft	21ft or 24ft		No		*CE&AJ,* 5, 1842, 138; 6, 1843, 143,181, 191-2 + Pl.8; 7, 1844, 69-70; *MM,* 40, Jan.-June 1844, 356-8; S B Moody, 'Description of a Water-wheel constructed by Mr. W. Fairbairn, MInstCE, and erected in Lombardy', *MPICE,* 3, 1844, 66-8.

1843	Izmet, Turkey. Sultan Abdul Mecit I (woollen mill)	1		30ft	13ft			Mill by Fairbairn	Fairbairn, *M&MWII*, pp.190-2 + Plates 17, 18; *The Times*, 16 December 1843.
1843	Dublin J & R Mallet (lead rolling)	1		20ft	12ft		No	Never actually used for lead rolling	*Manchester Times and Gazette*, 6 July 1844.
1845	[Bergen, Norway. Rosendahl & Fane] (cotton spinning)	Not built					-	Unsuccessful tender	K Bruland, *British Technology and European Industrialisation: The Norwegian textile industry in the mid nineteenth century*, (1989), pp.61-2.
1845-7	Robinwood Mill, Todmorden. Fielden Brothers. (cotton spinning)	1	Calder				No	Fairbairn mill exists. Main power was steam	Law, *Fieldens of Todmorden*, p.285.
1846	Brough, Castleton, Derbyshire. Benjamin Pearson & Co. (cotton doubling)	1		16ft	6ft		No		*MG*, 16 June 1855; http://www.derbyshireheritage.co.uk/Menu/Archaeology/Mills.php.
1849	Tampere, Finland. Finlayson & Compagnie (cotton)	1	Tammer-koski Rapids						http://www.flickr.com/photos/36713050@No3/5248739189/in/set-72157625388834955
1850	Gefle (now Gävle), Sweden. Gefle Manufacturing Co. (cotton spinning and weaving)	1	Testeboån	40ft	20ft	150		Mill by Fairbairn	Fairbairn, *UIfE2*, pp.229-31; P Carlberg, 'Personal Contacts between the Manchester area and Gefle in Sweden 100 years ago: a Communication', *Transactions of the Lancashire and Cheshire Antiquarian Society*, 70, 1960, 56-9; S Söderberg, 'Förstaden vid Testeboån', *Gefle Dagblads Hembygdsbibliotek*, 11, 1966, 59-81; *Manchester Guardian*, 31 July 1850.
1851	Ivangorod, Russia. Alexander von Stieglitz. (flax mill)	1	Narva	24ft	20ft			Mill by Fairbairn	Fairbairn, *M&MW II*, pp.192-6, 204-5.
1851	Glasshouses Mill, Pateley Bridge, Yorks. Yorkshire. John & Geo. Metcalfe. (hemp spinning)	1	Nidd	24ft	21ft		Moved to and re-built at Styal Mill, Cheshire		C Giles and I H Goodall, *Yorkshire Textile Mills. The Buildings of the Yorkshire Textile Industry 1770-1930*, (1992), pp.126-7. Original documents and drawing, plus photographic record, held at the National Trust's Styal Mill.
1852	Midleton Distillery, Cork, Ireland	(1)		22ft	16ft		Yes	Detailling not typical WF. Without doc. evidence, provenance is doubtful	*Industrial Heritage Association of Ireland Newsletter*, 31 May 2008, 3, (where the wheel is said to have been 'installed by William Fairbairn').

1853	Stoke Mills, Sharnbrook, Bedford. Hipwell & Sons. (corn mill)	(1)	Great Ouse	14ft	10ft		No	Not entirely certain that it was built by Fairbairn	Original letter re Fairbairn's visit and subsequent estimate held at Bedfordshire and Luton Archives & Records Service, Bedford.
1853-6	Montreal Waterworks Thomas Keefer (engineer)	2		20-22ft	c.22ft	110 each	No		S M Ross, 'Steam or Water Power? Thomas C Keefer and the Engineers Discuss the Montreal Waterworks in 1852', *The Journal of the Society for Industrial Archaeology*, 29.1, 2003, 49-64.
1854	Hjula Weavery, Oslo. Halvor Schou	(1)				60		Estimate. Unclear if it was built by Fairbairn	Bruland, *British Technology*, p.76.
c.1856	Royal Gunpowder Mills, Waltham Abbey, Essex	1	Lea	14ft	12ft		Leat remains	To operate powder-mills	Fairbairn, *M&MWII*, pp.246-9.
c.1856	Royal Gunpowder Mills, Waltham Abbey, Essex	1	Lea				Existing wheel does not appear to be by Fairbairn	To operate hydraulic press etc.	Fairbairn, *M&MWII*, pp.246-9; J K Major, 'A Waterwheel by William Fairbairn at the Royal Gunpowder Mills Waltham Abbey', unpublished paper to Gunpowder and Explosives History Group, 7 May 2004.
1856-7	Belleek Pottery, Fermanagh, Northern Ireland	1	Erne			100	Axle remains on site	Installed by Mortimer Bros. of Derrylin	S Stajda, 'Belleek' at http://www.oldandsold.com/articles/article449.shtml .
1857-8	Melbourne, Australia	1		18ft	8ft 8in.			For sale, 'not yet installed'	*The Argus* (Melbourne), 31 July 1858.
1860	Ancoats, Manchester, Wm. Fairbairn. (fatigue testing)	1	Shooter's Brook (?)	small			No	Unlikely in cellar of Fairbairn's home – probably in a cellar at his works*	W Fairbairn, 'Experiments to determine the effect of Impact, Vibratory Action, and long-continued Changes of Load on Wrought-Iron Girders', *Philosophical Transactions of the Royal Society*, 1864, 318; Walker, *Unwin*, pp.43-5, 211*.
1863	Ancoats, Manchester. Wm Fairbairn. For sale	1		14ft	10ft				*MG*, 26 September 1863.
1864	Liverpool. For sale with flour mill machinery	1							*MG*, 9 July 1864.
1865	Liverpool. For sale. 'Never set up'	1		30ft	12ft			May be the same wheel as last above	*MG*, 29 June 1865.

Appendix 3.1
Some of the Stationary Steam Engines built by Fairbairns

Year	Client	Location	Purpose	Type	No.	nhp *	Notes	Main references
1828	Birley's	Chorlton Mills, Manchester	Cotton mill		1	6	£175. In 1830 Birleys were the 2nd largest employers in M/cr. Other work done for them	S Clark, 'Chorlton Mills and their Neighboura', *IAR*, 2.3, 1978, 220-1.
1831-2 (?)	Earl of Caledon	[Caledon Mills, N Ireland]	Corn mill	Single cylinder, condensing beam engine	1		May not be by Fairbairn – drive is off main shaft	W A McCutcheon, 'The Stationary Steam Engine in Ulster' in C Ó Danachair (ed.), *Folk &Farm: essays in honour of A T Lucas*, (Dublin, Royal Society of Antiquaries of Ireland, 1976), offprint of pp.8-12.
1833	William Haworth	Rochdale			2	60 each		Boulton & Watt MS 3147/3/448 H 80.
c.1834	John Howard	Park Mill, Brereton, Cheshire	Corn mill		1			A Bonson, *Driven by the Dane. Nine Centuries of Waterpower in South Cheshire and North Staffordshire*, (2003), pp.176-7.
1834		Kingston Mill, Stockport	Cotton mill	Side-lever engine.	2		First of this type. Based on marine engines	D A Collier, *A Comparative History of the Development of the Leading Stationary Steam Engine Manufacturers of Lancashire, c.1800-1939*, (Manchester University PhD, 1985), Vol.1, p.40
1834				In the course of the year, 'steam engines to 400 horses' power'	several			Ure, *Philosophy*, p.39.
c.1835	W Bailey & Brothers	Bailey St. Mill, Stalybridge, Gt. Manchester	Cotton mill	Side-lever engine	2	110 each	'At an early period of my own practice I introduced it on an extensive scale.' Fairbairn, *M&MWI*,p.247) £8,800 inc. mill gearing	Fairbairn, *M&MWI*, pp.246-8 + Plate 8; A Ure, *The Cotton Manufacture of Great Britain* … ,(1836), Vol.I, pp.313-4; J S Russell, *A Treatise on the Steam Engine from the Seventh Edition of the Encyclopaedia Britannica*, (1841), pp.vii-ix (Description of the Plates) + Plates 9-11; G Watkins, *The Textile Mill Engine*, (1970), Vol.I, p.42.
post-1835				Side-lever engines.	Several		' there are numbers now at work … giving entire satisfaction'.	Fairbairn, *M&MWI*, p.247.
1836	Ralph Orrell	Travis Brook Mill, Stockport	Cotton mill	Beam engines	2	80 each		Ure, *Philosophy*, pp.34, 109-12; Ure, *Cotton Manufacture*, Plates I and II.
1836-8	Watergrove Mine	near Foolow, Derbyshire	Lead mine pumping engine	High pressure, side-lever pumping engine	1		There were teething troubles. £2,900 inc. boilers.	N Kirkham, 'Steam Engines in Derbyshire Lead Mines', *TNS*, 38, 1965-6, 71-3; R A Hayward, 'The Watergrove Pumping Engine of 1838', *Bulletin of the Peak District Mines Historical Society*, 5.4, 1973, 200-14.

c.1839		Amiens, France	Flax spinning			80 each		*MM*, 32, Oct.1839-May1840, 255.
1839				'All sizes ... are frequently under hand, from ... 8hp to the enormous magnitude of 400hp ... this latter ... worth ... from £5,000 to £6,000.'				Love & Barton (publishers), *Manchester As It Is*, (1839), p.211.
1839-40	Biolly et Fils	Verviers, Belgium	Coal mine pumping engine	'on the Cornish principle'	1		Pit 720ft deep.	*Transactions of the Manchester Geological Society*, 1, 1841, 236 + Plate 8; *Mining Journal*, 9, 1839, 90; *Railway Times*, 2.90, 1839, 755.
c.1840		Norwich	locomotive shop'	Pair of columnar engines	2	10 each	In use until 1922	Science Museum, ref.1922-293 and neg.5882; Hayward, p.2.31; L T C Rolt, *Victorian Engineering*, (1970), Plate 21.
c.1840		Salford	Flax mill	Pair of beam engines	2		In use until c.1912	F Wightman, 'Flax Mill' at https://www.flickr.com/photos/manchesterarchiveplus/9662173775/in/album-721572... (accessed 1 April 2016).
c.1840				' the writer found ... upwards of sixty steam engines, many of them of great size, ... in various stages towards completion'.	60			Woodcroft], VIII; *The Engineer*, 12 October 1877, .253.
1840	Turkish Government	Constantinople	Corn mill	High-pressure columnar	1	12		Fairbairn, *M&MWII*, pp.118-26; [D Scott], *The Engineer & Machinist's Assistant*, (1850), Vol.I, pp.91-3, Vol.II, Plates 98-101.
1840s (?)	Samuel Kershaw	Turner Lee Mills, Glossop	Paper mill	condensing	1	20		*Manchester Courier*, 8 April 1848.
c.1843	7-storey mill	York Street, Manchester			1	20		*MG*, 18 May 1843.
1844	The Kingston and Dalkey Atmospheric Railway	Dalkey, Dublin	Pumping engine for atmospheric railway	40psi	1	110	36ft diam. flywheel.	*The Practical Mechanic*, 3, 1844, 281-2; C Hadfield, *Atmospheric Railways. A Victorian Adventure in Silent Speed*, (1967), pp.107-8 and Plate 4.
1845-6		Northern Ireland	Flax and tow spinning.		1	35		*MG*, 7 March 1846.
1845-7	Fieldens Robinwood Mill	Todmorden, Yorkshire	Cotton Mill					B R Law, *Fieldens of Todmorden, A Nineteenth Century Business Dynasty*, (1995), p.285.

1846	Duke Pit	Oak Colliery, Oldham	Coal mine pumping engine	Beam engine. $80^{1}/_{2}$ in. cylinder 10ft stroke	1		4 boilers. Raised 90gals per stroke from 780ft. Abandoned 1926.	G Fanning, *Oldham Coal*, (Northern Mine Research Society, 2001), pp.130-8.
1846	J & J Norton	Old Steam Mill, Corn Hill, Wolverhampton	Corn Mill				Mill destroyed by fire in 1851. Rebuilt by Fairbairn. Engine probably survived fire	*Wolverhampton Chronicle*, 14 January, 1846.
1847 (?)	John Whittaker and Brothers	Whittaker's Mill, Ashton-under-Lyne	Cotton spinning and weaving		2	150 each	Haynes says the engines were designed by Oldham Whittaker, but no source reference	Fairbairn, *Application*, (3rd ed. 1864), pp.170-1; I Haynes, *Cotton in Ashton*, (1987), p.39.
1848-9		Portugal	Cotton mill					E J Molesworth (ed.), *Life of Sir Gulford L Molesworth*, (1922), p.22.
1849 (?)	Royal Arsenal	Woolwich.	Foundry		1	30	with Fairbairn's double-beat equilibrium valves'	*The Times*, 14 April 1849.
Pre-1850	James Allen & Sons	Salford	Cotton spinning	'works expansively'	1	30		*MG*, 18 May 1850.
1850	Halvor Schou	Brenneriveien, Oslo, Norway	Weavery		1			K Bruland, *British Technology & European Industrialization. The Norwegian textile industry in the mid nineteenth century*, (1989), pp.114, 156, 179.
1851	F D P Astley	Astley Deep Pit, Dukinfield, Cheshire	Coal mine pumping engine	Cornish type side-lever single-acting, high-pressure expansive and condensing engine	1	160	At 2,059ft, believed to be the deepest pit in the world at the time	W Fairbairn, 'Description of a New Construction of Pumping Engine', *MPIME*, 6, 1855, 177-182 and Plates 33-6; *The Engineer*, 1, Jan.-June 1856, 183-4; 8, July-Dec.1859, 130; MG, 28 December 1855, 16 July 1858.
1851	Great Exhibition	London		Columnar engine	1	6		*Official Catalogue of the Great Exhibition of the Works of all Nations*, (1851), Class V, No.26.
1851		Chepstow	Corn mill	'vertical direct action'	2	30		*Gloucester Journal*, 28 August 1858.
c.1852	F D P Astley	Astley Deep Pit, Dukinfield, Cheshire	Coal mine winding engine	Direct-action with 'grass-hopper' arrangement. 60in. diameter cylinder; 8ft stroke. 30psi+. Two flywheels on which the wire ropes wind	1		Probably the most powerful winding engine of its day. Raised 32cwt of coal at a time, at 20+mph. On similar principle to marine engines for *Vulture*, *Odin* and *Dragon*	W Fairbairn, 'On a New Description of Winding Engine', *MPIME*, 4, 1853, 137-42 + Plates32-3; R H Parsons, *A History of the Institution of Mechanical Engineers 1847-1947*, (1947), pp.106-7.

1853	Titus Salt	Saltaire, Yorkshire	Alpaca spinning and weaving factory	Two pairs of condensing beam engines	4	100 each	Corliss valve gear fitted 1867 probably by others	Fairbairn, *M&MWI*, pp.235-42 and Plates 5-6; C Giles and I H Goodall, *Yorkshire Textile Mills: The Buildings of the Yorkshire Textile Industry 1770-1930*, (1992), p.154; R L Hills, *Power from Steam. A history of the stationary steam engine*, (1989), pp.177-8; Hayward, 'Fairbairns', p.2.27.
1853	Dublin Exhibition (Wm Dargan)	Dublin	To drive exhibited machinery	Direct-acting, single flywheel.	2	20 each	Fairbairn was a good friend of Dargan who financed the exhibition	*CE&AJ*, 16, 1853, 222-3.
c.1854	Oriental Spinning and Weaving Co.	Bombay	Cotton mill		2	80 each		Fairbairn, *M&MWII*, p.179 and Plates 15-16.
c.1855	Bombay Spinning and Weaving Co.	Tardeo, Bombay	Cotton mill	Pair of horizontal condensing steam engines	2	80 each		E Leigh, *The Science of Modern Cotton Spinning*, (3rd ed. 1875), Vol.1, pp.49-52+Plate 14.
1855	Paris Universal Exhibition	Paris	To drive exhibited machinery	Columnar engine	1			http://collections.vam.ac.uk/iten/O145806/annexe-west-south-side-steam-photograph... (accessed 26 March 2013).
1855	Board of Ordnance	Royal Small Arms Factory, Enfield	To drive rifle-making machinery	'supplied with Fairbairn's expansion gear'	2	40 (?) each	Fairbairn's contract for the structural ironwork, engines and shafting was reported to be about £60,000	*MG*, 6 January 1855; MM, 6, 1861, 110.
1856		Moscow			2	40 each		*The Engineer*, 1, Jan.-June 1856, 324.
1856		Spain			2	40 each		*The Engineer*, 1, Jan.-June 1856, 324.
1857	Thomas Ainsworth	Cleator, Cumbria	Flax mill	Beam engine				C Caine, *Cleator and Cleator Moor: Past and Present*, (1916), p.378.
1857	Nevsky Cotton Spinning Co. (Alexander von Steiglitz)	St Petersburg	Cotton mill	Pair. 60in. diameter cylinders; 8ft stroke.	2	150 each	Became the largest cotton-spinning factory in Russia	*MG*, 22 August 1857; *The Engineer*, 9, 1860, 286.
1857		Russia			several		'several other large engines for Russia [are] in hand'	*MG*, 22 August 1857.
1859-60	Fitzroy Iron Works	Mittagong, New South Wales, Australia	Blowing engine for iron works		1	100	The order included smelting furnace and refinery	*The Maitland Mercury & Hunter River General Advertiser*, 24 December 1859.
Not known	Wm. Fairbairn	Ancoats Works, Manchester	Engineering Works	'Two of Fairbairn's pillar high-pressure steam engines'	2		Sold when the works closed	*MG*, 25 September 1875.

Not known	Wm. Fairbairn	Ancoats Works, Manchester	Engineering Works	Two of Fairbairn's horizontal high-pressure steam engines	2		Sold when the works closed	*MG*, 25 September 1875.
Not known	North Seaton Colliery	North Seaton, Newcastle	Coal mine pumping engine	Condensing cylinder 78in. diameter. 7ft stroke			Malleable iron beam	*Transactions of the North of England Institute of Mining and Mechanical Engineers*, 34, 1886-7, 201-17.
Not known		Wallshaw Mills, Oldham		Pair of beam engines, with high-pressure cylinder 36in. and low pressure 40in. 6ft stroke	2	60		*Manchester Courier*, 7 December 1883.
c.1860	Corn Mill	Taganrog, Russia	Corn mill.	Pair beam engines with single flywheel	2	100 each	Probably designed before the Crimean War and built after it	Fairbairn, *M&MWII*, pp.127-8.
1860s (?)	Wm. Dargan	Chapelizod, Dublin	Flax and thread mills		2			F Mulligan, *William Dargan, An Honourable Life 1799-1867*, (2013). p.185.
1861	Wm. Degraves & Co.	Flinders Lane, Melbourne, Australia	Corn mill	Low-pressure beam engine	1	50		*The Argus*, (Melbourne), 17 April 1861; *South Australian Register*, 1 August 1873.
1862				Large horizontal engine (possibly next entry below)	2			Walker, *Unwin*, Plate opp. p.48
1863-5	São Paulo Railway	Santos to Jundiaí, Brazil	Hauling engines on four inclines	Pairs of horizontal engines with single flywheels.	2x4	150 each	One pair preserved at Paranappiacaba – the only known surviving stationary steam engine by Fairbairns	P C da S Telles, *A History of Brazilian Railways. Part 1, The First Railways*, (ET by P E Waters, 1987), pp.43-5; *Engineering*, 2 February 1866; P Catchpole, *A Very British Railway*, (2003), p.182; T T dos S Cruz, *Paranapiacaba : A arquitetura e o urbanismo de uma Vila Ferroviária*, (MA Thesis, Univ. of Sao Paulo, 2007), p.59.
c.1864 (?)	Joseph Westley	Nunn Mills, Northampton	Corn Mill		1	10	Vertical	*Northampton Mercury*, 7 May 1870.
1868	Calder Wharf Mill	Dewsbury, Yorkshire		Pair of three cylinder triple expansion engines, at 135psi. Designed by J Crosland	2		'the firm was one of the first to build a triple expansion engine for use on land'	Collier, *Stationary Steam Engine Manufacturers*, Vol.1, p.43; *The Practical Engineer*, 1890, 677.
1869		Jamaica			2	12		J A Bennison, *Essay*, 12 February 1869. – Lancashire CC Record Office DDX/184' p.15.
1869		Jamaica		'small steam engine'	1			Bennison, *Essay*, p.15.
1869				Horizontal engine	1		'in course of erection'	Bennison, *Essay*, p.16.

1871	Thomas Ainsworth	Cleator, Cumbria	Flax mill	Pair of Crosland engines with three cylinders – triple expansion at 150psi.	2		To supplement the beam engine of 1857	Caine, *Cleator and Cleator Moor*, p.378; Walker, *Unwin*, p.60.
1871-2				'other [Crosland] engines to similar design' (as last above)	Several			*The Practical Engineer*, 1890, 677.

Watt originally adopted the 'horse power' unit to measure the power of his engines. One unit is the power required to lift 33,000lb through 1ft in one minute. This was usually expressed as 'nominal horse power' (nhp) and was related to the size of the engine's cylinder. This often underestimated the actual or 'indicated horse power' (ihp) produced by the engine in practice. As steam pressure increased, the difference between nhp and ihp increased. Watt's nhp was based on a steam pressure of about 5–6psi whilst by 1850 most new engines had steam pressures of about 40psi or higher. It is difficult to ascertain the ihp of an engine whose capacity is given as an nhp figure, as there is no reliable method of conversion. However the use of nhp figures, as above, has some value for purposes of comparison.

For further discussion see
— Fairbairn, *M&MW I*, pp.239–41;
— R.L. Hills and A.J. Pacey, 'The Measurement of Power in Early Steam-Driven Textile Mille', *Technology and Culture*, 13, 1972;
— G.N. von Tunzelman, *Steam Power and British Industrialization to 1860* (1978), pp.25–7.

Appendix 3.2
Identified Ships and Marine Engines built by William Fairbairn

Fairbairn built circa 120 iron ships, plus engines for at least 10 ships built by others. So far information is only to hand as follows.

	Name of ship	B'lder	Built at	Built for	Mat	Plied	Type	Tons	Length /deck	Beam	Eng. B'ldr	Type of engine(s)	total hp	Cylin-ders	Drive	Diam P/w / screw	End and date
1830	*Lord Dundas [I]*	F&L	M/cr	Forth & Clyde Canal	iron	Forth & Clyde C	passenger	41	68'0"	11'6"	Steph'	locomotive	18		1 p/w		
1831	*Lord Dundas [II]*	F&L	M/cr	Forth & Clyde Canal	iron	Forth & Clyde C	passenger	44?	68'0"	15'0"	F&L	locomotive	20		p/w	11'0"	
1831/2	*Manchester*	F&L	M/cr	Forth & Clyde Canal	iron	Forth & Clyde C	cargo	70?	70'0"?	15'0"?	F&L		30		p/w		
1832	*Lancashire Witch*	F&L	M/cr	M/cr,Bolton & BCC	iron	Mcr, B&BC	passenger	10	60'0"	6'0"	-	none	-	-	horse	-	
1833	*La Reinede Belge*	WF	M/cr	Company in Bruges	iron	Ostend Bruges		64	73'0"	14'0"	WF		24		p/w		
1833	*(Vulcan) Minerva*	WF	M/cr	Caspar & Lämmlin	iron	Zurichsee	passenger/ cargo	108	98'0"	15'6"	WF	2 high-pressure	40		p/w		
1834	*Railway*	WF	M/cr	James Audus	iron	Selby-Hull	passenger/ cargo	164	110'0"	18'0"	WF		50		p/w		
1835	*L'Hirondelle*	WF	M/cr	James Audus	iron	Selby-Hull	passenger/ cargo	171	115'0"	18'0"	WF		60		p/w		
		E/W		Escher Wyss	iron	Zurichsee					WF				p/w		
1836	*Ludwig*	WF	M'w	Dampfboot AG	iron	L.Constance	passenger/ cargo	177	120'0"	17'0"	WF		40		p/w		sank 1870
					iron	L.Constance	passenger/ cargo								p/w		
					iron	L.Constance	passenger/ cargo								p/w		
					iron	L.Constance	passenger/ cargo								p/w		
1836	*Little Dreadnought*	WF	M/w		iron	R.Thames	cargo	14			-				p/w	-	
1837	*Sirius*	WF	M/w		iron	R.Rhone	passenger/ cargo	250	175'0"	17'1"	WF		70	24"	p/w		
1837	*Le Castor*	No'd			iron	Rouen-Havre					WF boiler	high-pressure			p/w		
1837	*Le Pollux*	No'd			iron	Rouen-Havre					WF boiler	high-pressure			p/w		
1838	*Inkerman*	WF	M/w	Russian Government	iron	Black Sea	packet	216	175'0"	18'0"	WF		80		p/w		Year
1838	*Nevka*	WF	M/w	Nicholas I of Russia	iron	S.Petersburg	Royal Yacht	231	150'3"	1838	WF	side-lever	70		p/w	15'6"	
1838	*Prussian Eagle*	WF	M/w	R. Marit. Soc.Berlin	iron	(R.Elbe)	passenger/ cargo	100	118'6"	14'0"	WF		50		p/w		sank 1838

1839	*Enterprise*	WF	M/w	Russian Government	iron	Black Sea	tug	92	76'6"*	16'0"	WF	side-lever	42	27 1/4" x 2'6"	p/w		
1839	*Concordia*	WF	M/w		iron	Upper Rhine	ferry	118	112'3"	15'0"	WF		36		p/w		
1839	*Shell*	WF	M/w		iron	R.Thames	barge	112	102'6"	15'3"	WF		30		p/w		
1839	*Woronzow*	WF	M/w	Russian Government	iron	Black Sea	tug	91	81'0"	16'0"	WF		40		p/w		
1839	*Pradpriatie*	WF	M/w	Russian Government	iron	Black Sea	tug	91	81'0"	16'0"	WF		40		p/w		
1839	*Dolphin*	WF	M/w	R'l Marit. Soc.Berlin	iron	R.Elbe	passage/ tug	107	114'6"	14'0"	WF	side-lever	42	28" x 2'0"	p/w	10'4"	
1839	*HMS Cyclops*	Ad'y		Admiralty	wood		6-gun frigate				WF				p/w		
1839 -40		WF	M/w	East India Company	iron	Bombay	passenger/ cargo	334	125'0"	24'0"	WF		80		p/w		
1839 -40		WF	M/w	East India Company	iron	Bombay	passenger/ cargo	334	125'0"	24'0"	WF		80		p/w		
1839 -40		WF	M/w	East India Company	iron	Calcutta	passenger/ cargo	334	125'0"	24'0"	WF		80		p/w		
1839 -40		WF	M/w	East India Company	iron	Calcutta	passenger/ cargo	334	125'0"	24'0"	WF		80		p/w		
1839 -40		WF	M/w	East India Company	iron	Calcutta	passenger/ cargo	334	125'0"	24'0"	WF		80		p/w		
1839 -40		WF	M/w	East India Company	iron	Calcutta	passenger/ cargo	334	125'0"	24'0"	WF		80		p/w		
1839 -40		WF	M/w	East India Company	iron	Calcutta	accom'ion	334	125'0"	24'0"	-	none	-	-	by tug	-	
1839 -40		WF	M/w	East India Company	iron	Calcutta	accom'ion	334	125'0"	24'0"	-	none	-	-	by tug	-	
1839 -40		WF	M/w	East India Company	iron	Calcutta	accom'ion	334	125'0"	24'0"	-	none	-	-	by tug	-	
1839 -40		WF	M/w	East India Company	iron	Calcutta	accom'ion	334	125'0"	24'0"	-	none	-	-	by tug	-	
1839 -40		WF	M/w	East India Company	iron	R.Hoogley	ferry	26	66'6"	9'0"	WF	oscillating	80		p/w		
1840	*Coquette*	WF	M/w		iron	R.Neva	ferry	163	150'0"	15'0"	WF		50		p/w	14'6"	
1840	*Iron Duke*	WF	M/w		iron	Demerara	barge	110	103'6"	15'0"	WF		24		p/w	10'0"	
1840	*Telegraph*	WF	M/w		iron	R.Weser	ferry	206	136'0"	18'0"	WF		45		p/w	12'3"	
1840	*(San Carlo?)*	WF	M/w		iron	L.Maggiore	ferry	38.5	81'0"	10'0"	WF		14		p/w		
1840	*Rose*	WF	M/w	Hunter R.St.Nav.Co	iron	E. Australia	passenger/ cargo	305	153'6"	20'6"	WF	side-lever	106	40 3/4" x 3'6"	p/w	17'0"	1872?
1840	*Thistle*	WF	M/w	Hunter R.St.Nav.Co	iron	E. Australia	passenger/ cargo	305	153'6"	20'6"	WF	side-lever	106	40 3/4" x 3'6"	p/w	17'0"	wrecked 1859
1840		WF	M/w	Fossdyke Navig'tion	iron	Fossdyke	dredger	54	65'0"	14'0"	WF		6				

1840		WF	M/w		iron		Canal boat	12	65'0"	6'0"	-	none	-	-	horse	-	
1840		WF	M/w		iron		Barge	32	78'9"	9'0"	WF				p/w		
1841	*Yarra Yarra*	WF	M/w		iron	E.Australia		93	96'0"	14'6"	WF		30		p/w		
1841	*Juno*	WF	M/w		iron	London-Hull	coaster	135	82'3"	19'6"							
1841		WF	M/w		iron	R.Thames	fire engine	68	60'0"	16'0"		none			48 men		
1841	*Aegir*	WF	M/w	Christ.VIII, Denmark	iron	Copenhagen	Royal Yacht	254	150'0"	19'0"	WF	side-lever	82	36 3/4 x 3'0"	p/w	16'0"	
1842	*Hamlet*	WF	M/w		iron	Denmark	tug	130	100'0"	17'0"	WF	side-lever	42	27 1/2" x 2'3"	p/w	11'0"	
1842	*Eagle*	WF	M/w		iron	L.Geneva	ferry	303	186'6"	18'6"	WF		100		p/w		
					iron	L.Lucerne	ferry										
1842	*Rhenus*	WF	M/w		iron	Upper Rhine	ferry		145'0*	18'0"	Fren'h		150		p/w	17'9"	
		WF	M/w		iron	Baltic											
		WF	M/w		iron	Baltic											
		WF	M/w		iron	Baltic											
		WF	M/w		iron	Baltic											
		WF	M/w		iron	Baltic											
		WF	M/w		iron	Baltic											
		WF	M/w		iron	Baltic											
		WF	M/w		iron	Baltic											
1842	*HMS Rocket*	WF	M/w	Admiralty	iron	Woolwich dockyard	tender	70	90'0"	12'8"	WF	2 oscillating	20	19" x 2'0"	p/w	9'8"	
1842	*HMS Firebrand*	Ad'y		Admiralty	wood						WF	2 direct ?			p/w		
1843	*Lady Burgoyne*	WF	M/w		iron	Loch Derg	ferry		130'0"	17'6"	WF	oscillating	92	39" x 3'0"	p/w	16'0"	
1843	*Der Pfalz Graf*	WF	M/w		iron	R.Rhine	tug	350	180'0"	20'3"	WF		160		p/w		
1843		WF	M/w	Russian Government	iron		dredger				WF				p/w		
1843	*HMS Mohawk*	WF	M/w	Admiralty	iron	Lake Ontario											
1843	*HMS Cormorant*	Ad'y		Admiralty	wood		1st class. sloop	1100	210'0"	36'0"	WF	2 direct	316	65 1/2" x 5'3"	p/w	26'8"	broken up 1853
1843	*HMS Vulture*	Ad'y		Admiralty	wood		2nd class. frigate	1190			WF	2 direct	476	80 5/8" x 5'9"	p/w	26'6"	broken up 1866
1844	*HMS Gladiator*	Ad'y		Admiralty	wood		6 gun frigate				WF	2 direct ?			p/w		

1845	*HMS Dragon*	Ad'y		Admiralty	wood		frigate				WF	2 direct	580	87½" x 5'9"	p/w	27'0"	
1845	*Pharos IV*	WF	M/w	Northern Lighthouse Bd	iron	E.Scotland		303	146'3"	21'0"	Penn	oscillating	152	47" x 4'3"	p/w	19'0"	broken up 1877
1845	*Sir H Pottinger*	WF	M/w	P&O	iron	India/China	passenger/ cargo	1225	218'0"	34'0"	M&R	oscillating	450	76" x 7'0"	p/w	30'0"	broken up c1867
1845	*HMS Grappler*	WF	M/w	Admiralty	iron		sloop	557	165'0"	26'6"	M'ley		440	40" x 4'6"	p/w	21'0"	1850
1846	*HMS Odin*	Ad'y		Admiralty	wood		1st class frigate	1310	208'0"	37'0"	WF&S	2 direct	560	87½" x 5'9"	p/w	26'6"	broken up 1865
1846-9	*HMS Magaera*	WF	M/w	Admiralty	iron		2nd class frigate	1298	196'0"	37'6"	Ren'ie	direct		49½" x 2'0"	screw	13'0"	wrecked 1871
1854	*HMS Algiers*	Ad'y		Admiralty	wood				218'7"	25'7"	WF&S	horizontal	430	61" x 3'3"	screw		broken up 1870

Appendix 4.1
Summary of Fairbairn Locomotives

Company	Type	Year																										Overseas	Ire	Sc	Eng.	Total
		38	39	40	41	42	43	44	45	46	47	48	49	50	51	52	53	54	55	56	57	58	59	60	61	62	63					
M/cr & Bolton Rly	0-4-0	1	3																												4	4
Gt N of E Rly	0-4-2		3																												3	3
	0-6-0		1																												1	1
Midland Counties Rly	0-4-0			2																											2	2
M/cr & B'ham Rly	2-2-2			2																											2	2
	0-4-2																6														6	6
M/cr & Leeds Rly	2-2-2			1	1																										2	2
	0-4-2					3	3																								6	6
	0-4-0									4	7	15																			26	26
Slamannan Rly	2-2-2				2																									2		2
London & S W Rly	rebuild				6r	5r																									[11r]	[11r]
	2-2-2						1			3	7																				11	11
	parts						4																									-
London & Brighton Rly	2-2-2				1	3																									4	4
Caledonian Railway	0-4-0											2																		2		2
Preston & Wyre Rly	0-4-0									1																					1	1
Lancs & Yorks Rly	2-2-2											5	16																		21	21
	0-4-2												3	7																	10	10
	2-4-0T																		4												4	4
Midland Gt W Rly Ire'd	2-2-2										6	6																	12			12
	2-2-2T														3														3			3
	0-4-2																							5	1				6			6
	2-4-0															1	1	2											4			4
Gt Northern Rly	0-4-0											3	3																		6	6
	0-6-0															4	5	1													10	10

East Lancs Rly	0-4-0											1	4																		5	5
South Staffs Rly	2-2-2												4																		4	4
	0-6-0												2									6									8	8
Shrewsb'y & B'ham Rly	2-2-2												1																		1	1
M/cr Sheffield & Lincs	2-2-2												1	4									4								9	9
	2-4-0																						3								3	3
North Western Rly	0-4-0													3	1																4	4
	2-2-2T													3			1r														4(1r)	4(1r)
Belfast & Cty Down Rly	2-2-2T													1	1														2			2
	0-4-2																						2						2			2
NewryWarren'pt & Ro	2-2-2T													1															1			1
Limerick & Ennis Rly	0-4-2																			1			1						2			2
Waterford & Kilkenny	2-4-0																		1										1			1
St Helens Rly	2-2-2T														1																1	1
London & N W Rly	0-6-0														2	8	12	4	4							5					35	35
	2-2-2															1	4	5													10	10
	0-4-2																	6													6	6
E de F Maua, Brazil	2-2-2T																3											3				3
S Devon Rly	4-4-0T															1															1	1
Waterford Limer & W	2-2-2T																		7										7			7
Waterford & T Dargan	2-2-2T																[3]												[3]			[3]
Hunter River Rly	0-4-2																		1	1	3							5				5
	2-2-2																				2							2				2
Waterford & Tramore	0-4-2																	1											1			1
	2-2-2T																		2										2			2
Dublin & Wicklow Rly	2-2-2T																2												2			2
	2-4-0																1							3					4			4
Gt N of Scotland Rly	2-4-0																	7	5		4									16		16
Furness Railway	0-4-0																	2	2		2				2						8	8
S FV Emanuele, Italy	2-4-0																	5										5				5

	0-4-0T																		6								6				6
Piedmont Railway	2-4-0																6										6				6
	2-2-2																				1						1				1
Birkenhead L & CJR	2-4-0																	2												2	2
	2-4-0T																	1												1	1
L'pool Crosby & S'port	2-2-2T																	4												4	4
G Thomson	2-4-0T																	1												1	1
BBC & PJ Rly	2-2-2T																	3										3			3
BBC & PJ R(Dargan)	2-2-2																	1										1			1
Belfast & Ballmena Rly	2-2-2T																							1				1			1
L'derry & Coleraine Rly	0-6-0																	1										1			1
	2-4-0																								1			1			1
Waterford & Kilkenny	2-4-0																	1										1			1
Dublin & Drogheda Rly	2-4-0																	1										1			1
Gt Southern & W Rly	2-4-0																	2										2			2
	0-4-2																		1									1			1
North British Rly	0-4-2																	4											4		4
Stamford & Essendine	0-4-2T																	1												1	1
C de F du Nord	2-2-2																	1									1				1
Adelaide & Port Rly	2-4-0T																	3									3				3
Australian Ag Co.	0-4-0T																	2									2				2
CCEP Portugal	2-2-2																	1									1				1
	2-2-2T																		2	2							4				4
East Indian Rly	2-2-2																	1									1				1
	0-4-2																	1									1				1
Köping-Hult ,Sweden	2-2-2T																	1									1				1
	2-2-2																			2	3						5				5
	2-4-0																							1			1				1
Gt W Rly of Canada	2-4-0																		6	4	2						12				12
France	0-4-0T																		6								6				6

Lancaster & Carlisle Rly	2-2-2																						5								5	5
	2-4-0																						7								7	7
Scottish Central Rly	2-4-0																				6									6		6
Ulster Railway	2-2-2																						1			2			3			3
Geelong-Ballarat																								1				1				1
Midland Railway	0-6-0																							10	14	6					30	30
West Midland Rly	0-6-0																								8	4					12	12
Egyptian Govt Rly	0-4-2																				2							2				2
	0-6-0																									3		3				3
GT Rly of Canada	4-4-0																								10			10				10
Tudela & Bilbao Rly	4-4-0T																								3	3		6				6
Asuncion & VR Rly	?-2-2																						1					1				1
Cork Brandon & S C	0-4-0T																									1			1			1
Gt Eastern Railway	2-2-2																									5					5	5
Newry & Armagh Rly	2-2-2T																						1						1			1
	2-4-0T																										1		1			1
	0-4-2																										[2]		[2]			[2]
Yearly totals		1	7	5	10	11	4	0	0	8	20	32	34	19	8	15	39	39	52	23	27	12	25	19	40	30	3	89	72	30	297	488
Year		38	39	40	41	42	43	44	45	46	47	48	49	50	51	52	53	54	55	56	57	58	59	60	61	62	63					

Appendix 5.1
Bridges with which Fairbairn is known to have been involved

An asterisk in the References column indicates inclusion of illustration(s) in the reference.

Date	Location	Type	Spans	Notes	Present State	Main references
1803	Kelso	Stone		Designed by John Rennie. Fairbairn's first job, labouring, ending with an accident after a few days	Existing	*Life*, p.65. Widely illustrated*
1817	Manchester: unsuccessful competition entry for bridge over the river Irwell	Iron		A stone bridge was built	Not built	*Life*, pp.105-6
	River Tame, Cheshire	Iron Bowstring			Demolished	*Report of the Commissioners appointed to inquire into the Application of iron to Railway Structures*, (1849), p.311
1830	Water Street, M/cr. Liverpool & Manchester Railway (G Stephenson, Engineer)	Cast-iron girders of Hodgkinson type	24ft 6in.	First use of Hodgkinson beams	Demolished	R S Fitzgerald, *Liverpool Road Station, Manchester: An historical and architectural survey*, (1980), pp.22-28*
1830	Kenyon & Leigh Junction Railway (?)	Cast-iron girders of Hodgkinson type		Circumstantial evidence only	Demolished	M R Bailey, 'The early Years' in M R Bailey (ed.), *Robert Stephenson – The Eminent Engineer*, (2003), p.26
1831	Forth & Clyde Canal	Design for iron swing bridge			Not known if built	Hayward, 'Fairbairns of Manchester', p.4.33
mid 1830s	21 bridges, London & B'ham Railway (R Stephenson, Engineer) + 'many more elsewhere'	Cast iron girders of Hodgkinson type		Extent of Fairbairn's involvement uncertain	Demolished	J Sutherland, 'Iron Railway Bridges' in Bailey (ed.), *Robert Stephenson*, p.304
1846-8	Railway Bridge, Conway. Chester & Holyhead Railway	Tubular	400ft	Two parallel tubes. The first Tubular Bridge	Existing. Strengthened	Fairbairn, B&CTB*; E Clark, *Britannia and Conway Tubular Bridges*, (1850)*; Rosenberg and Vincenti, *Britannia**; Sutherland, 'Iron Railway Bridges', pp.318-35*; J Rapley, *The Britannia and other Tubular Bridges and the men who built them*, (2003)*
1846-52	Britannia Bridge, Menai Straits, Chester & Holyhead Railway	Tubular	2 x 460ft and 2 x 230ft	Two parallel tubes. Longest span railway bridge when built	Destroyed by fire in1970. Towers exist, New steel arches	As Conway, immediately above.
1846	First proposal for bridge over the river Dee at Chester	Tubular-girder		Proposal before collapse of Stephenson's Dee Bridge	Not built	ICE Archive: Stephenson papers No.44*. Sutherland, 'Iron Railway Bridges', pp.324-5*

1846	Proposal for an aqueduct over the river Weaver	Tubular			Not known if built	R Rawlinson, '*Report on Liverpool's Water Supply*, (1846), p.25
1846-7	Patent for tubular-girder bridge			Granted 8 October 1846. Enrolled 8 April 1847		*CE&AJ*, 10, 1847, 143*
1846-7	Bridge over Leeds & Liverpool Canal, Blackburn, for Blackburn, Bolton & Darwen Rly	Tubular-girder	60ft	Three tubular-girders. The first significant tubular-girder bridge. C Vignoles, Engineer	Demolished	*CE&AJ*, 10, 1847, 376-7*; G D Dempsey, *Tubular and other Iron Girder Bridges*, (1850), pp.24-6
1846-7	Bridge over turnpike road, Blackburn, for Blackburn, Bolton & Darwen Railway.	Tubular-girder	66ft 6in.	Fairbairn undertook impact tests.	Demolished	*MM*, No.1392, 13 April 1850, 281-5*; Fairbairn, *Application*, pp.228-9; R Hunt (ed.), *Ure's Dictionary of Arts, Manufactures, and Mines*, (1867 ed.), Vol.3, p.957*.
1846-7	Two access bridges to Liverpool Landing Stage	Tubular-girder	154ft	Hinged to rise and fall with the tide	Replaced	Dempsey, *Bridges*, pp.27-30;* Life, pp.227, 318; *CE&AJ*, 18, 1855, Plate 17;* W Fairbairn, 'On the Rise and Progress of Manufactures and Commerce, and of Civil and Mechanical Engineering in Lancashire and Cheshire', in T Baines, *Lancashire and Cheshire, Past and Present*, (1869), pp.xcix-ci and Plate II.*
1846-7	Various railway bridges	Tubular-girder		Of the 200+ iron railway bridges with spans over 40ft reported in 1847 around 24 can be identified as wrought-iron tubular-girder bridges		National Archives, Kew, 'Returns and Plans of Iron Bridges, 1847', MT8/1; P Sibly, 'The Prediction of Structural Failures' (UCL, DPhil Thesis, 1977), Table 2.1; S Smith, 'The Development and Use of the Tubular Beam, 1830-1860', *History of Technology*, 14, 1992, p.115 and 132n42.
1847	Second proposal for bridge over the river Dee at Chester	Tubular-girder		Proposal after collapse of Stephenson's Dee Bridge	Not built	ICE Archive: Stephenson papers No.265*
1847	Two railway bridges, Bolton	Tubular-girder		Probably for James Thomson, Engineer of the Liverpool & Bury Railway	Demolished	Rapley, *Britannia*, p.120; W Fairbairn to R Stephenson, 21 July 1847, Nos 283-4 Stephenson Papers, ICE Archive
1847	Proposal for bridge over the Rhine at Basel	Tubular			Not built	*Life*, p.137
1847-8	Two railway bridges over Leeds & Liverpool Canal, Burnley	Tubular-girder			Demolished	J Marshall, *The Lancashire & Yorkshire Railway*, (1969), Vol.1, p.114
1847-8	Railway bridge over Haslingden Grane Road, Helmshore	Tubular-girder		On the East Lancashire Railway	Demolished	ILN, 30 September 1848, 203; Marshall, *Lancashire & Yorkshire Railway*, Vol.1, p.111
1847-8	Railway bridge over road, Wigan	Tubular-girder	65ft		Demolished	Marshall, *Lancashire & Yorkshire Railway*, Vol.2, p.131
1847-8	Railway bridge over Leeds & Liverpool Canal, Wigan	Tubular-girder	65ft		Demolished	Marshall, *Lancashire & Yorkshire Railway*, Vol.2, p.131
c.1848	Railway bridge, Ardwick, Manchester	Tubular-girder		Ardwick Branch Railway	Demolished	*Report of Commissioners*, Appendix 7;* P S A Berridge, *The Girder Bridge after Brunel and others*, (1969), p.42*

1848	Cowbridge Drain Bridge on the Firsby to Boston line of the East Lincolnshire Railway	Tubular-girder	42ft	Believed to be the oldest surviving tubular-girder bridge	Existing. Strengthened 1888 and 1905	I am grateful to Robert Thorne for information on this bridge (e-mails of 24 May and 10 June 2016). 'Report by Capt. R M Laffan' in 'Reports of the Inspecting Officers of Railways for 1848', National Archives MT29/7
1848	Two other bridges on the Firsby to Boston line	Tubular girder			Demolished	Ditto
1848	Stamp End Railway Bridge, Lincoln, over roads and river Witham	Tubular-girder	66ft 6 (main span) + 3 other spans		Main span in use (strengthened 1903); others replaced. Listed Grade II	Listing Notice, Lincoln, 1941-1/13/465; B M J Barton, 'Stamp End Bridge', *Lincolnshire History and Archaeology*, 33, 1998, 55-6.*
c.1848	Railway bridge for Messrs Peto (location unknown)	Tubular-girder	99ft 9in.			Original drawing held by E P Cannon, Galway*
1848-9	Railway bridge Gainsborough, to carry the MS&L Railway over the river Trent	Tubular-girder	2x154ft	John Fowler was the Company Engineer. When built this was the longest tubular-girder bridge.	Demolished 1992	*MG*, 18 July 1849; *Manchester Times*, 18 July 1849; Dempsey, *Bridges*, pp.30-2*; *CE&AJ*, 12, 1849, 255; Hunt, *Ure's Dictionary*, Vol.3, pp.949-54*; Rapley, *Britannia*, pp.120-4*.
1848-9	Railway Bridge, Worksop, to carry the MS&L Railway over a road	Tubular-girder	65ft 9in.		Demolished	*The Sheffield & Rotherham Independent*, 14 July 1849
1848-9	Railway Bridge to carry the Manchester and Altrincham Railway over the Mersey, near Stretford	Tubular-girder	c.70ft		Demolished	*MG*, 21 July 1849
1849	Rasbottom Street, Stalybridge, for Huddersfield & Manchester Railway	Tubular-girder			Demolished	*Manchester Times*, 18 July 1849
1849	Railway swing bridge, being part of the bridge over the river Tay at Perth (Dundee & Perth Railway)	Tubular-girder swing bridge (wrought iron)		One part of a multi-span timber viaduct on stone piers. Timber replaced by plate girders in 1864, but swing bridge remained	Replaced by fixed span	Rapley, *Britannia*, p.120; 'Perth, West Railway Bridge' at https://canmore.org.uk/site/79826/perth-west-railway-bridge [This website makes no reference to Fairbairn and Rapley gives no source]
1849-50	Proposal for a tubular bridge over the Rhine at Cologne	Tubular	2x570ft and 2 swing sections		Not built – lattice preferred	*Life*, pp.217-37; Fairbairn, *Application*, pp.261-72, 286-9 and Plate V.*
1849-50	Alternative proposal, for a tubular-girder bridge over the Rhine at Cologne	Tubular-girder	2x326ft and 2x244ft + 2 swing sections		Ditto	*Life*, pp.217-37; Fairbairn, *Application*, pp.261-72, 286-9 and Plate VI.*

1850	Torksey Railway Bridge , to carry the MS&L Railway over the river Trent	Tubular-girder	2x130ft	This bridge drew attention to the effects of continuity of beams over supports	Existing, in very poor condition. Listed Grade II*	W Fairbairn, 'On Tubular Girder Bridges' and Discussion following, *MPICE*, 9, 1850, 233-87 and Plate II;* Fairbairn, *Application*, pp.243-54; *CE&AJ*, 13, 1850, 142; *CE&AJ*, 15, 1852, 8-14 and Plate III; Rapley, *Britannia*, pp.81-7, 121-2;* K Keen, 'Torksey Railway Bridge', *Journal of the Railway & Canal Historical Society*, 197, 2007, pp.518-25.
1850	Railway bridge to carry the Dublin-Galway line over the river Suck at Ballinasloe	Tubular-girder			Still in use – strengthened	N V Torpey, 'Irish Railway Bridges', *Journal of the Irish Railway Record Society*, 22, 2004, 128.*
1850	Railway bridge to carry the Midland Great Western Railway at Lough Attalia, Galway	Tubular-girder	2x60ft and 2x40ft + swing section	After testing the swing bridge was never opened!	Existing. Swing section replaced by fixed span	W Humber, *A Practical Treatise on Cast and Wrought Iron Bridges and Girders …*, (1857), pp.43-9 and Plates 27, 28 and 29;* E Shepherd, *The Midland Great Western Railway of Ireland: An Illustrated History*, (1994), p.105
c.1850	Bridges in Norfolk, possibly including Vauxhall Bridge, Great Yarmouth	Tubular-girder		Vauxhall strengthened c.1890 with steel lattice arches above the three tubular-girders.	Vauxhall existing as strengthened. Listed Grade II	*The Builder*, 1850, 178; 'History of Vauxhall Bridge' at http://www.greatyarmouthpreservationtrust.org/media/uploads/History_of_Vauxhall_Bridge.pdf
1850	Proposal for the Liteiny Bridge over the river Neva at St Petersburg	Tubular		Proposal prepared after interviews with the Emperor	Not built	MPICE, 39, 1875, 259; O J Vignoles, *Life of Charles Blacker Vignoles*, (1889), p.352
1850s	Railway Bridge over canal and tramway on Sheffield goods Branch of the MS&LR	Plate girder	60ft and 96ft clear spans	Girders 168ft long		Humber, *Practical Treatise*, pp.76-8 and Plates 39 and 40.*
1851	Railway bridge over river Dodder, Dublin.	Tubular-girder (?)				Vignoles, *Life of Vignoles*, pp.186-7
1851	Proposal for the Boyne Viaduct, Droheda	Tubular-girder			Not built – lattice preferred	J Millar, *William Heap and his Company,1866*, (1976), pp.38-9; C O'Mahoney, 'Iron Rails and Harbour Walls: James Barton of Farndreg', *Journal of the County Louth Archaeological and Historical Society*, 22.2, 1990, 134-40; Smith, 'Development and Use of the Tubular Beam', 120-2.
1851-3	Cahir Viaduct to carry the Waterford and Limerick Railway over the river Suir	Tubular-girder	1x150ft and 2x50ft		Existing - repaired and strengthened	*The Engineer*, 1, 1856, 336;* W Fairbairn in J Scoffern et al, *The Useful Metals and their Alloys*, (1869), pp.456-64;* P McCarron and A McAdam, 'Preserving Ireland's iconic railway viaducts', *PICE*, Paper 1300020, 2013, pp.1-6.
1851-4	Most of the iron bridges on the Great Western Railway of Canada	Probably tubular-girders and plate-girders				Z Colburn, 'American Iron Bridges', *MPICE*, 22, 1863, p.556

c.1852	Road bridge over the river Aire, Saltaire	Tubular-girder	450ft in 4 spans		Demolished	Fairbairn, *Application*, pp.168, 273
	Two railway bridges, Gelderd Road, Leeds (L&NWR)	Tubular-girder		Girders under the track	Demolished	Information from R Fitzgerald
1854-5	Vertical lift bridge on the North Docks Branch of the L&Y Railway, Liverpool	Tubular-girder		Attributed to Fairbairn by Rapley, but no source given	Demolished	Rapley, *Britannia*, p.13;* Marshall, *Lancashire & Yorkshire Railway*, Vol.1, pp.143, 146.*
c.1855	Railway Bridges in Nova Scotia	Tubular-girders				J G James, 'Overseas Railways and the Spread of Iron Bridges, c.1850-70', (typescript, ICE, nd but c.1980), pp.26-7.
1856	Proposals for bridges in the Bois de Boulogne, Brest and elsewhere in France	Tubular-girders?		The Bois de Boulogne bridge, described as 'ornamental', may not have been a tubular-girder	Not built	*Life*, pp.376-7
1856	Railway Bridge at Valencia, Spain	Tubular-girder	c.100ft			*The Engineer*, 1, Jan-June 1856, p.324
c.1856	Railway Bridge over the river Sacavem on the Lisbon to Santarem line, Portugal	Tubular-girder	108ft			J S Valentine, 'Description of the Line and Works of the railway from Lisbon to Santarem', MPICE, 18, 1859, pp.14-5.
1856-8	Railway Bridge over the river Findhorn at Forres on the Inverness and Aberdeen Junction Railway	Tubular-girder	3x150ft		Existing. Still in use	R Paxton and J Shipway, *Civil Engineering Heritage: Scotland – Highlands and Islands*, (2007), p.139; K Fenwick and N T Sinclair, *The Inverness and Aberdeen Junction Railway*, (2008), pp.12-3*
1857	Four access bridges to Liverpool Landing Stage	Tubular-girders	113ft	Fairbairn was also involved in the construction of the landing stage.	Demolished	*MG*, 20 May, 23 May, 2 September 1857; *The Engineer*, 1, Jan-June 1856, p.324; 27, Jan-June 1874, pp.111-2,115,127,132.*
1858	Railway Bridge over the river Spey on the Inverness and Aberdeen Junction Railway	Tubular-girder	230ft	Important because of the fatigue testing to which it led	Replaced 1906	*Engineering*, 4, July-Dec.1867, 321,323;* Fairbairn, *Application*, pp.246-55;* A I Smith, 'William Fairbairn and the Mechanical Properties of Materials: The Effect of Repeated Loading on Strength', *The Engineer*, 217, 1964, pp.1133-6;* Fenwick and Sinclair, *Inverness and Aberdeen Junction Railway*, pp.2, 14-6, 61.* D Ross, *The Highland Railway*, (2005), pp.30-2, 133.*
1858-9	Road Bridge over the Barwon River, Geelong, Victoria, Australia	Tubular-girder	2x210ft	Erection team from England. Longest spans in Australia until Fitzroy River suspension bridge of 1881	Demolished 1924.	P Alsop, 'Bridging the Rivers: Part 5', *The Investigator*, 4.2, 1969, pp.84-101; G Vines, *Metal Road Bridges in Victoria, Part 1 – History of Metal Road Bridges in Victoria*, (rev.ed.2010), pp.19-20.*
1858-9	Railway Bridge over the Saltwater River on the Melbourne and Williamstown Railway	Tubular-girder	200ft clear	The river is now known as the Maribyrnong	Replaced 1911	W Humber, *A Complete Treatise on Cast and Wrought Iron Bridge Construction including iron Foundations*, (3rd ed. 1870), pp.197-9 and Plate 35;* B Harper, 'The true history of the design of the Melbourne, Mount Alexander and Murray River Railway', *Australian Journal of Multi-disciplinary Engineering*, 3.1, 2004, pp.83-90.

1858-60	Bridge to cross the river Yarra-Yarra near Melbourne	Probably tubular-girder				James, 'Overseas Railways', pp.20-1; J G James, Card Index , 'Victoria: Richmond', held in ICE Archive.
1859-60	Dinting Viaduct to carry the Manchester, Sheffield & Lincolnshire Railway	Tubular-girder	2x125ft	Replacement of timber arches, c.120ft above ground, without stopping the trains! For details of original arches see *CE&AJ*, 11, 1848, Pl.iv.	Still in use. 7 piers added in 1919. Girders replaced (Ashmore says with riveted steel)	W Fairbairn, 'On the Reconstruction of the Dinting and Mottram Viaducts', *MPICE*, 22, 1862-3, pp.227-33;* *Life*, p.331; G Dow, *The First Railway between Manchester and Sheffield*, (1945);* O Ashmore, *The Industrial Archaeology of North-west England*, (1982), pp.85-6.
1859-60	Mottram (or Etherow) Viaduct to carry the Manchester, Sheffield & Lincolnshire Railway	Tubular-girder	1x150ft and 2x125ft	Replacement of timber arches, c.120ft above ground, without stopping the trains!	Still in use. 3 piers added in 1919. Girders replaced	Fairbairn, 'On the Reconstruction of the Dinting and Mottram Viaducts';* *Life*, p.331; Dow, *The First Railway between Manchester and Sheffield*.
1859-60	'Most of the bridges for the Great Western Railway of Canada'			'but they are now making their own at Hamilton'		*MPICE*, 22, 1863, pp.540-56
1859-60	Railway Swing Bridge to carry the Great Western Railway of Canada over the Desjardines Canal, Hamilton, Ontario	Tubular-girder		Replaced a timber swing bridge which collapsed with c.70 fatalities	Demolished	*The Engineer*, 9, Jan-June 1860, p.225; Illustration at http://www.hamiltonparanormal.com/desjardines1.html (accessed 18 May 2016).*
1859-60	Swing bridge over the Welland Canal at Merritton, Ontario	Tubular-girder				*Hamilton Spectator*, 22 March 1860
1861	L&Y Railway Bridge over Rochdale Canal near Middleton Junction	Probably plate girder				Marshall, *Lancashire & Yorkshire Railway*, (1970), Vol.2, pp.14-5
1862	Railway swing bridge over the Caledonian Canal at Clachnaharry, Inverness	Tubular-girder	Girders 126ft, on turntable	The 1909 replacement is almost identical to the original but in steel. It is still operational. Listed Grade 'B'	Replaced 1909	*MM*, 21(NS), 1869, 84-5;* *The Engineer*, 24, July-Dec.1867, 223-4;* R and S Byrom, 'The Railway Swing Bridge at Clachnaharry' *Towpath Action Group Newsletter*, 83, 2012, p.15-7.*
1863-4	Bridges over the river Irwell and Gt Ducie St, Victoria Station, Manchester	Probably plate girders				*The Engineer*, 17, 1868, p.348; *MG*, 8 October 1864; Marshall, *Lancashire and Yorkshire Railway*, Vol.1, p. 187.
1863	Tay Viaduct at Dalguise, Perthshire, for the Inverness and Perth Junction Railway	Lattice girder	2 spans, 210ft and 141ft		Existing	Fairbairn, *Application*, (3rd ed. 1864), pp.123-32;* Paxton and Shipway, *Civil Engineering Heritage*, pp.300-1.* On the Scottish bridges generally see J Mitchell, 'On the Construction and Works of the Highland Railway', *BAAS1867*, pp.s151-61.
1863	Railway bridge over the river Dulnain on the Inverness and Perth Junction Railway.	Lattice girder	80ft		Demolished (abutments remain)	*MM*, 21(NS), 1869, 172;* *The Engineer*, 24, July-Dec.1867, pp.223-4;* *Engineering*, 4, July-Dec.1867, p.264.
1863	Railway Viaduct over the river Tilt at Blair Athol.	Lattice girder	150ft		Existing. Strengthened	*Engineering*, 4, July-Dec.1867, p.264; *ILN*, 19 September 1863, pp.287-9.*
1863	Railway Viaduct over the river Tummel at Ballinluig	Lattice girders	2x122ft and 2x35ft			Mitchell, 'Highland Railway', p.s155; *ILN*, 5 August 1865, 103-4.*

1863-4	Keadby swing bridge over the river Trent on the Trent, Ancholme and Grimsby Railway	Probably tubular-girder swing section and plate girders for the other spans	5 spans	Swing section gave 2 channels of 60ft each	Replaced with a rolling lift bascule bridge in 1916	*MG*, 16 May 1864; *MPIME*, 1863, 24-6; *MPICE*, 203, 1916, 98-9. For illustration see http://crowle.org/?p=4068 .*
1864	Road bridge over the river Wyre at Shard, on the Fylde	Plate girders	16 spans of 60ft			*MM*, 12(NS), July-Dec.1864, p.173
1864-5	Horbury Viaduct over the river Calder on the Wakefield to Goole line of the L&Y Railway					Marshall, *Lancashire & Yorkshire Railway*, Vol.1, p.216
1864-6	Railway Bridge over the Connecticut River for the New Haven, Hartford & Springfield Railroad	Plate girders for short spans; Whipple trusses for longer spans	17spans. Longest 177ft 4in. Total 1,524ft 6in.	Designed by J Laurie. Fabricated and erected by Fairbairns (2 spans sub-let). 'Nearly ¼ mile overall.'	Rebuilt 1903-4	*Engineering*, 2, July-Dec.1866, pp.83-5;* H M Burt, *Burt's Illustrated Guide of the Connecticut Valley*, (1868), pp.59-63;* G L Vose, *Manual for Railroad Engineers and Engineering Students*, (1873), pp.262-6.
1865	Railway viaduct over the river Tay at Logierait, Perthshire	Lattice girders	2x137ft and 2x41ft 6in.		Existing. Now a community-owned road bridge	Mitchell, ' Highland Railway', p.s155; 'The Logierait Bridge Company' at http://www.logieraitbridge.com/ (accessed 19 May 2016).*
1866	Bridges for Indian railways			'Very considerable contracts for bridge work, principally for Indian lines, are now being executed by the company.' *(Engineer)*.		*The Engineer*, 21, Jan-June 1866, 179. See also *The Engineer*, 9, Jan-June 1860, p.225
1867	Railway bridge over occupation road on the Halifax branch of the Lancashire & Yorkshire Railway	Plate girder				National Archives, Kew, Ref: RAIL 343/897
1867-9	Calder Bridge on the Halifax Branch of the Lancashire & Yorkshire Railway.					Marshall, *Lancashire & Yorkshire Railway*, Vol.1, p.260.
1871-5	c.100 Railway Bridges for the Intercolonial Railway of Canada	Lattice girders and plate girders	spans of 100ft and less.	61 spans of 100ft (some bridges had several spans), 10 spans of 80ft and 69 lesser spans		Anon., *The Intercolonial Railway: The Genesis of its Bridges*, (1875); S Fleming, *The Intercolonial: A Historical Sketch of the Inception, Location, Construction and Completion of the Lines of Railway Uniting the Inland and Atlantic Provinces of the Dominion*, (1876).*
1873-77	Cornbrook Viaduct, Manchester, for the Cheshire Lines Committee	Lattice girders		Designed by C S Allott. Contract completed by Fairbairns' liquidators	Existing. Used by Metrolink trams	'Allott' in W T Pike (ed.), *Contemporary Biographies*, (1899), p.187; For illustration see https://www.railscot.co.uk/locationnew.php?loc=Cornbrook%20Viaduct (accessed 19 May 2016)

Appendix 5.2
Known 'Fairbairn' Quayside, Shipyard and Railway Company Cranes, as at 6 January 2016

* Illustrated material. ** Jib curved but not tubular

Date	Location	Maker	Cap tons	Ht. (ft)	Diam. (ft)	Power	Present state	References
1850	Patent/ prototype	Fairbairn				manual	-	Patent No, 13317; *MM*, 1851, 381-3*.
1850	Keyham 1 (south basin)	Fairbairn	12	30	65	manual / steam later?	Dem.	D Evans, *Building the Steam Navy – Dockyards, Technology and the Creation of the Victorian Battle Fleet 1830-1906*, (2004), p.74; Fairbairn, *UIfE2*,pp.282-9; *Newton's London Journal of Arts and Sciences*, March 1858, 171-176, *ILN*, November 1853 .421*; R A Hayward,*The Story and Scandal of HMS Megaera*, (1978), p.57*; D Brewster, 'On a Wrought Iron Tubular Crane, designed by William Fairbairn, CE, FRS', *BAAS1850*, pp.s177-9; *The Engineer and Machinist* , 3, March-Dec. 1851, 174-6; K V Burns, *The Devonport Dockyard Story*, (1984), p.46.
1851	Great Exhibition	Fairbairn	5	15		manual	Dem.	L Haghe, *The Great Exhibition: Moving Machinery* (1851) (painting)*; *Official Catalogue of the Great Exhibition of the Works of all Nations* [1851] Class 5, pp.230-1*; *CE&AJ*, 14, 1851, p.294.
1851	St Petersburg	Fairbairn	12	30	65	manual		Fairbairn, *UIfE2*, p.290; *Life*, pp.366-7.
1851	St Petersburg	Fairbairn	10			manual		*Engineer and Machinist*, 3, 1851, p.179 and Plate XV*.
c.1851	experimental crane	Fairbairn	10			manual	Dem.	Fairbairn, *UIfE2*, p.290.
c.1851	experimental crane	Fairbairn	3			manual	Dem.	Fairbairn, *UIfE2*, p.290.
1852	Keyham 2	Fairbairn	12	30	65	manual / steam later ?	Dem.	As for Keyham 1, 1850, above.
1852	Keyham 3	Fairbairn	12	30	65	manual / steam later?	Dem.	As for Keyham 1, 1850, above.
1852	Keyham 4	Fairbairn	12	30	65	manual / steam later ?	Dem.	As for Keyham 1, 1850 above.
1852	Keyham 5	Fairbairn	12	30	65	manual / steam later ?	Dem.	As for Keyham 1, 1850 above.
1852	Keyham 6	Fairbairn	12	30	65	manual / steam later ?	Dem.	As for Keyham 1, 1850 above.
1852	Southampton	Fairbairn	12	30	65	manual	Dem.	Fairbairn, *UIfE2*, p.290; *ILN*, 21, 1852, p.165.
	Devonport	Fairbairn	12	30	65	manual	Dem.	Fairbairn, *UIfE2*, p.290; Burns, *Devonport Dockyard*, p.57.*
1853	Dublin Exhibition	Fairbairn				manual	Dem.	J Sproule (ed.), *The Irish Industrial Exhibition of 1853: A Detailed Catalogue of its Contents* (1853), p.178.

1853	Keyham (north basin)	Fairbairn	40			manual / steam later ?	Dem.	Evans, *Steam Navy*, p.93, 193; Burns, *Devonport Dockyard*, p.46.
c.1855	Birkenhead	Fairbairn	12	30	65	manual	Dem.	Fairbairn, *UIfE2*, p.290.
c.1855	Folkstone	Fairbairn					Dem.	*ILN*, 34, 1859, p.81.*
1857	Keyham	Fairbairn	60	60	106	manual 4-man, steam later	Dem.	*UIfE2*, p.290-4; W Fairbairn. 'Description of Large Tubular Wrought Iron Crane recently erected at Keyham Dockyard, Devonport' in *MPIME*, 1857, 87-96 + Plates*; *Newton's London Journal of Arts and Sciences*, 1858, pp.171-6; *ILN*, 23, 1853, 421-2; Evans, *Steam Navy*, p.98.
1857-8	West India Dock, London	Fairbairn	40	36		manual	Dem.	*British History on Line* http://66.102.9.104/search?q=cache:Vs8yWKEA3IUJ:www.british-history.ac.uk/repo... (accessed 15.03.07); *British History on Line* at http://www.britihistory.ac.uk/image.aspx?compid=46606&filename=figure0369-0 (accessed 09.10.07)*.
c.1857	Devonport	Fairbairn	12	30	65	manual	Dem.	Fairbairn, *UIfE2*, p.290.
c.1857	Devonport	Fairbairn	12	30	65	manual	Dem.	Fairbairn, *UIfE2*, p.290.
c.1858	'a Royal Dockyard'	Fairbairn	10	26	50	steam or manual	Dem.	Fairbairn, *UIfE2*, p.295*; *Life*, p.322*; Rosenberg and Vicenti, *Britannia Bridge*, p.50.
1859	Portsmouth (east wall of steam basin)	Fairbairn	60	60	106	steam	Dem.	Fairbairn, *UIfE2*, p.296; *The Engineer*, 7, Jan-June 1859, p.272.
1859*	Dublin**	Courtney, Stephens (W Anderson) (lattice jib)	6	18	36	manual	Dem.	W Anderson, 'Description of a Six-ton Crane, with Curved, Diagonally Braced Jib, at the Pigeon-house Fort, Dublin' in *Transactions of the Institution of Civil Engineers of Ireland*, 6,1863, 188-92 and Plate I*. See also B B Stoney, *The Theory of Strains in Girders and Similar Structures, with Observations on the Application of Theory to Practice and Tables of the Strength and other Properties of Materials*, (1869), pp.129-36.
	Russia	Courtney, Stephens (W Anderson)	20	69	63			Stoney, *Strains*, p.133.
	Sheerness						Dem.	
1859	Woolwich (proposal)	not built		120			-	http://www.photoship.co.uk/JAlbum%20Ships%20Misc/Ports%20Harbours%20Piers/... (accessed 4 December 2012).*
1861	Chatham gun wharf	Fairbairn	10 or 15		50	manual or steam	Dem.	*The Practical Mechanics Journal*, March 1861, p.329.
1861	Chatham gun wharf	Fairbairn	>10				Dem.	*The Practical Mechanics Journal*, March 1861, p.329.

1861	Middelburg Netherlands	De Prins van Oranje	10			manual	Exists. Conserved	Fairbairn Crane Documentation Group (fcdg), 'Middelburg, Rotterdamsekaai: Fairbairn crane' at http://www.flickr.com/photos/fairbairncrane/4735381348/in/pool-fairbairncranes* (accessed 7 August 2012); Industrieel Erfgoed Zeeland, 'Stadskraan Middelburg teruggeplaatst' at http://www.industrieelerfgoedzeeland.nl/stadskraan-middelburg-teruggeplaatst/* (accessed 03 August 2012); http://www.flickr.com/photos/fairbairncrane/4738856782/in/pool-fairbairncranes*(accessed 17 July 2012); 'wiez' Photograph November 2009 at http://www.flickr.com/photos/wiez/4738069101/in/pool-fairbairncranes*(accessed 07.08.10);* 'fcdg' photograph April 2010, at http://www.flickr.com/photos/fairbairncrane/4735381348/in/pool-fairbairncranes * {accessed 07 August 2012).
1861	Rotterdam	De Prins van Oranje	25			manual 1878 steam	Dem.	http://www.flickr.com/groups/fairbairncranes/ *(accessed 03 August 2012); Information from Prof. van Hooff (e-mail 19 August 2013).
1862*	Helsingør, Denmark	Fairbairn	10			manual	Exist.	'Cranes Kraner Krane', at www.my1287.dk/index.php?option=co... *(accessed 28.01/11); Ole Ryolf, photograph 2006, at http://www.flickr.com/photos/oleryolf/262244298/ *(accessed 07 August 2012); G Baviere, photograph 2012, at http://www.flickr.com/photos/84554176@N00/6941935342/in/pool-fairbairncranes/ *(accessed 17 July 2012); J Hallbæck-Andersen, photograph 2014, at https://www.flickr.com/photos/janhallback/14537775620/in/photolist-wWkXN1-uJg... (accessed 26 March 2016).
1863	Newcastle NSW	Fairbairn	15			steam	Dem.	*The Engineer*, 15, Jan-June 1863, 26; *Sydney Morning Herald* 21 November1862.
1863	Zulphen, Netherlands						Turntable exists	http://www.flickr.com/groups/fairbairncranes/ * (accessed 3 August 2012).
1865-6	West India Dock, London	Fairbairn	10	38			Dem.	*British History on Line* http://66.102.9.104/search?q=cache:Vs8yWKEA3IUJ:www.british-history.ac.uk/repo... (accessed 15.03.07).
1865	Chatham	proposed					-	Evans, *Steam Navy*, p.185.
<1874	Chatham						Dem.	Evans, *Steam Navy*, p.187.
1865	Leiden, Netherlands	De Prins van Oranje	10				Dem.	http://www.flickr.com/groups/fairbairncranes/ * (accessed 03 August 2012).
	Furness, Withy Hartlepool					steam	Dem.	D Martin, 'Swan Neck', at http://www.teeships.freeuk.com/03111dmswanneck.htm * (accessed 13.01.04).
c.1866	Gustemünde, Germany.(10).		2.5			hydraulic		J Weisbach and G Herrmann, *The Mechanics of Hoisting Machinery* ..., (ET of the second German ed. by K P Dahlstrom, 1907), pp.258-9.
1868	Dover	Fairbairn	50>20 (2009)			manual	Exists Grade II	A J Hazelfoot. *The Batsford Guide to the Industrial Archaeology of South-East England*, (1978), pp.37-8*; Listing Notice No. 685/0/10036 16-DEC-09, 'Wellington Dock and associated structures, including crane situated on Esplanade Quay'; http://www.dover.uk.com/news/2014/historic-crane-restoration-gives-heritage-trail-... *(accessed 2 September 2016).

	Venice Arsenal					manual	Exist.	'Asquith' *Old Machinery in Venice* at http://www.practicalmachinist.com/vb/showthread.php?t=114790 (accessed 02.06.08).* Image held by A King, Bristol Museums.* E-mail from E Bassett, 19 July 2016.*
1869		Fairbairn						J Bennison , *Essay,*(1869), Lancashire County Archive Ref. DDX/184 Essay No. 1; Hayward, 'Fairbairns', p.3.19;* R J M Sutherland, 'The birth of stress: a historical review' in W Addis (ed.), *Structural and Civil Engineering Design, (*1999) p.249.*
	Frankfurt am Oder	H Gruson Maschinen				manual	Exist.	http://www.flickr.com/groups/fairbairncranes/*(accessed 03 August 2012); L Simon, photograph 2012, at http://www.flickr.com/photos/lars_simon/7130814621/in/pool-fairbairncranes/*(accessed 17 July 2012).
	Pula, Croatia	J Korosi, Graz/ Andritz	15			manual	Exist.	Letter J Robinson to A King 28 October 2003 (Bristol Museums M Shed Archive). Illustration held by A King.*
<1871	Imperial Dockyard, Japan		30			manual	Dem.	K Tamagawa, 'The Historic Cranes in Japan, Part 1', *Cargo Handling JAPAN*, 4.5, 2004, 526-7*; T Hashimoto, 'Introducing a French Technological System: The Origin and Early History of the Yokosuka Dockyard', *East Asian Science, Technology and Medicine,* 16, 1999, p.53-72.
1870*	Kampen, Netherlands	De Prins van Oranje	9.5				Dem. 1969	Stichting Herbouw Kadekraan Campen, 'De kadekraan in vogelvlucht' at http://www.kadekraankampen.nl/index.php/fotos/category/3-animaties-en-modellen (accessed 17 July 2012)* - this website was revised on 23 July 2012 and no longer contains some material - it is replaced by http://www.kadekraankampen.nl/fotos/3d-renders/); (accessed 6 August 2012).
	Hull. (moved to present site 1901)					now electric but not in use	Exist. Grade II.	Hull City Council, Listing ref. No. 680-1/23/10041. Image in author's possession.*
	Dresden ? (doubtful)						Dem.	Dresden Ubigauer Uferkran' at http://www.dresden.city-map.de/oso12100 (accessed 08 August 2012); 'Schiffswerft Übigau' at http://www.dresdner-stadtteile.de/Nordwest/Ubigau/Schloss_Ubigau/Schiffswerft_Ub... (accessed 08 August 2012).
1875	Seville	Fairbairn	40			manual	Exist. Moved 2005	'Grúa Fairbairn' at http://es.wikipedia.org/wiki/Gr%C3%BAa_Fairbairn * (accessed 12.08.2010); http://www.flickr.com/photos/32957828@N00/303377826/sixes/m/in/pool-1461360... (accessed 17 July 2012).*
1875-6	Bristol	Stothert and Pitt	35	40	70	steam	Exist. Grade II*	A King, *Fairbairn Steam Crane 1876*, (City of Bristol Museum and Art Gallery – no date, but c.2000); Listing Notice Ref. 901-1/42/1317. Widely documented/ photographed.*
1876-8	Vlissingen, (Flushing) Netherlands	Cail, Halot & Cie, Brussels	50 or 75			manual. 1891 steam	Dem. 1940s	http://www.flickr.com/groups/fairbairncranes/ *(accessed 3 August 2012); *De Ingenieur* 19, 1904, 577.
	Calabar, Nigeria	Bedford Engineering Co.				manual	Existing in 2011	Graces Guide at http://www.gracesguide.co.uk/Bedford_Engineering_Co (accessed 23 January 2017)
c.1890	Scotia Engine Works, Sunderland.	Davis & Primrose, Leith	40			steam	Dem.	Advertisment. Bristol Museums M Shed Archive; *Grace's Guide: The Best of British Engineering 1750-1960s: Davis and Primrose* at http://www.gracesguide.co.uk/wiki/Davis_and_Primrose * (accessed 16 March 2010).

1891	Several locations for Admiralty	Cowans Sheldon	50					*Engineering*, 51, 1891, 664(illus.)*, 671(text).
1892	Maassluis, Netherlands	Begemann Machinefab					Dem. 1952	http://www.flickr.com/groups/fairbairncranes/ *(accessed 3 August 2012).
	Stockholm**					manual	Dem.	http://www.flickr.com/photos/mickeborg/6863034853/ (accessed 4 August 2012).*
	Valladolid,** Spain					manual	Exist.	http://www.flickr.com/groups/fairbairncranes/ *(accessed 3 August 2012); F Cortes, photograph 2010, at http://www.flickr.com/photos/harman_mad/4239534093/in/pool-1461360@N21/ *(accessed 17 July 2012).
	Langenargen, Bodensee					manual	Exist.	http://www.flickr.com/groups/fairbairncranes/ *(accessed 3 August 2012).
1900	Rorschach, Bodensee	Stahl	>4			electric	Exist. in use	http://www.flickr.com/groups/fairbairncranes/ *(accessed 3 August 2012); K Zwahlen, photograph 2010, at http://www.flickr.com/photos/gali_367/4930435539/in/pool-fairbairncranes/ (accessed 17 July 2012).
	Vienna						Exist.	http://www.flickr.com/groups/fairbairncranes/ *(accessed 3 August 2012).
c.1900	Visby, Gotland, Sweden		10>5			manual. 1921 electric	Exist. in use	L Lundquivist, photograph 2011, at http://www.flickr.com/photos/arkland_swe/6315398464/ * (accessed 4 August 2012).
1902	Flushing, Netherlands (?)	Cowans Sheldon	50					Cumbria County Council, Cowans Sheldon Archive, Ref 2616 5212; DB40/D/5215.
1905	Hartlepool	Cowans Sheldon	100	80	120	steam	Dem. 1975	SINE *Image Display* at http://www.sine.ncl.ac.uk/view_image.asp?digital_doc_id=2016 *(accessed 17 January 2004); SINE Project *Structure Details for Son of Swan Neck* at http://sine.ncl.ac.uk/view_structure_information.asp?struct_id=1227 *(accessed 10 August 2010); 'Port Cities – Hartlepool: A journey into the port half a century ago', at http://portcities,hartlepool.gov.uk/server.php?show=ConNarrative.138&chapterId=669 *(accessed 8 November 2005).
1908	Koog aan de Zann, NL						Exist.	http://www.flickr.com/groups/fairbairncranes/ * (accessed 3 August 2012).
1908	Chemnitz, Germany						Exist.	
1900-8	c.12 No. At various locations	Cowans Sheldon	20-75			steam/ electric		Cumbria County Council, 'List of Archive Documents from Cumbria CC related to Cowans Sheldon Civil Engineers of Carlisle', Ref. Nos. 6, 49, 64, 113, 133, 150, 174, 175, 176, 224, 259, 260; also 1647 2500, 2400 4797, 2443 4971, 2748 5586, 2883 5856/1-32
	Portsmouth							Evans, *Steam Navy*, Endpapers; Image held by A King, Bristol.
c.1905*	4 No. for Admiralty, Hong Kong.	Cowans Sheldon	20 and 50			steam/ electric		'Fairbairn Steam Crane' at http://en.wikipedia.org/wiki/Fairbairn_steam_crane (accessed 3 July 2013); Cumbria County Council, Cowans Sheldon Archive, Refs. 2809, 2904-5 5758 and 2906 5898. Image of one crane held by A King, Bristol.
1905	Gibraltar for Admiralty (2 or more)	Cowans Sheldon	30			electric		Cumbria County Council, Cowans Sheldon Archive, Ref. 2882 5857: DB40/D/5857.

1906	Devonport for Admiralty	Cowans Sheldon	75			electric		Cumbria County Council, Cowans Sheldon Archive, Ref. 2973-2974 6140/1-40; DB40/D/6140. A Earnshaw, *Carlisle's Crane Makers: The Cowans Sheldon Story*, (2004), p.62*.
	Devonport Coal Wharf							Burns, *Devonport Dockyard*, p.127*.
1908	Malta Naval Dockyard	Cowans Sheldon	50	65	60	steam	Dem.	*Engineering*, 19 February 1909, 240*. Another image in possession of A King, Bristol showing this crane or the following one.*
1908	Malta Naval Dockyard	Cowans Sheldon	20	60	53	steam	Dem.	*Engineering*, 19 February 1909, 240.
1908	Simons Bay, Cape Town, South Africa, for Admiralty	Cowans Sheldon	50 & 20			electric		Cumbria County Council, Cowans Sheldon Archive, Ref. 3049 6952 and 3050 6951; BB40/D/6951 and DB40/D/6952. Image held by A King, Bristol.
1908	London & NW Rly. Manchester	Cowans Sheldon	2.5 & 5			hydraulic	Dem.	Cumbria County Council, Cowans Sheldon Archive, Ref. 3013 6794.
c.1910	North Eastern Railway	Cowans Sheldon	2			hydraulic		Cumbria County Council, Cowans Sheldon Archive, Ref. 3095-3096 6794.
1913	Caledonian Rly. Goods Yard, Aberdeen	Babcock & Wilcox	20	30	36	manual	Now at NRM, York	'Northern Caledonian Relic Moves to National Museum', (Typescript, undated). Image from A King, Bristol.
1922	Leith Dock Com-missioners	Cowans Sheldon	8			electric		Cumbria County Council, Cowans Sheldon Archive, Ref. 4274 10692.

Bibliography

1. Works by William Fairbairn

1.1. Books by William Fairbairn

1831
Remarks on Canal Navigation illustrative of the Advantages of the use of Steam, as a Moving Power on Canals, (London: Longman, Rees, Orme, Brown & Green).

1836
Observations on Improvements of the Town of Manchester particularly as regards the importance of blending in those improvements, the Chaste and Beautiful with the Ornamental and Useful, (Manchester: Printed by R. Robinson).

1842
A List of Wheel Patterns, &c. belonging to William Fairbairn, Millwright, Engineer, Iron and Brass Founder, Canal Street, Manchester, (Manchester: printed by Simpson & Gillett). There were earlier and later editions.

1849
An Account of the Construction of the Britannia and Conway Tubular Bridges with a Complete History of their Progress, from the conception of the original idea, to the conclusion of the elaborate experiments which determined the exact form and mode of construction ultimately adopted, (London: John Weale, and Longman, Brown, Green & Longmans).

Illustrations of the Construction of the Britannia and Conway Tubular Bridges: Atlas to the Text in Large Octavo, Twenty Engravings, (London: John Weale).

1854
On the Application of Cast and Wrought Iron to Building Purposes, (London: John Weale).
Second Edition, *'Greatly Enlarged, with Corrections and Additions, to which is added A Short Treatise on Wrought Iron Bridges'*, 1857 (London: John Weale).
Third Edition, *with Additions, &c.*, 1864 (London: Longman, Green, Longman, Roberts & Green).
Fourth Edition, 1870 (London: Longman, Green & Co.).
American Edition, 1854 (New York: J. Wiley).
French Edition, 1857, translated by L. Perret-Porta: *De l'Application de la fonte, du fer et de la tôle dans les constructions* (Paris: Victor Dalmont).
Spanish Edition, 1857, translated by D.E. Saavedra: *Aplicación del hierro a las construcciones*, with *Suplemento,* 1859 (Madrid: D José C la Pena).

1856
Useful Information for Engineers; being a Series of Lectures delivered to the Working Engineers of Yorkshire and Lancashire; together with a Series of Appendices, containing the results of experimental inquiries into the Strength of Materials, the Causes of Boiler Explosions, etc. (London: Longman, Brown, Green & Longmans).

1859
Three Lectures on the Rise and Progress of Civil Engineering, and on Popular Education, (Derby: Printed by W. & W. Pike). These lectures are reprinted in *Useful Information for Engineers, Second Series.*

1860
Useful Information for Engineers, Second Series, (London: Longman, Green, Longman & Roberts). Second Edition, 1867.

1861
Iron – Its History, Properties, and Processes of Manufacture, (Edinburgh: Adam & Charles Black).
Second Edition, Revised and Enlarged, 1865 (Edinburgh: Adam & Charles Black).
Third Edition, Revised and Enlarged, 1869 (Edinburgh: Adam & Charles Black).
French Edition, 1864, translated by G Maurice, *Le Fer – Son Histoire, ses Propriétés et ses Différents Procédés de Fabrication,* (Paris: Eugène Lacroix).
This book was an expansion of Fairbairn's article, 'Iron' in the eighth edition of *Encyclopaedia Britannica* – see below.

1861
Treatise on Mills and Millwork. Part I. On the Principles of Mechanism and on Prime Movers comprising the accumulation and estimation of water power, the construction of water-wheels and turbines, the properties of steam, the varieties of steam-engines and boilers, and windmills. (London: Longman, Green, Longman & Roberts)
Second Edition, 1864 (London: Longman, Green, Longman, Roberts & Green).
Third Edition, 1871.
Fouth Edition, 1878, Parts I and II in one Volume (London: Longmans, Green & Co.).
American Edition, 1871, *Prime-Movers: comprising the accumulation of water-power; the construction of water-wheels and turbines; the Properties of Steam; the varieties of steam-engines and boilers and wind-mills,* (Philadelphia: Henry Carey Baird). This mainly comprises extracts from Part 1. Whilst there are adverts for this book as 'in press', evidence that it was actually printed is not currently to hand.

1863
Treatise on Mills and Millwork. Part II. On Machinery of Transmission and the Construction and Arrangement of Mills, comprising treatises on wheels, shafts, and couplings; engaging and disengaging gear; and mill architecture; and on corn, cotton, flax, silk, and woollen mills; to which is added a description of oil, paper, and powder mills, including a short account of the manufacture of iron, (London: Longman, Green, Longman & Roberts)
Second Edition, 1865 (London: Longmans, Green & Co.)
Third Edition, 1871.
Fouth Edition, 1878 – Parts I and II in one volume, (London: Longmans, Green & Co.).
American Edition, 1867, *The Principles of Mechanism and Machinery of Transmission, comprising the principles of mechanism, wheels and pulleys, strength and proportions of shafts, couplings for shafts, and engaging and disengaging gear* (Philadelphia: Henry Carey Baird). Reprinted 1870, 1871, 1872, 1876, 1903. This comprises extracts from Part I and more extensively from Part II.

1865
Treatise on Iron Ship Building: Its History and Progress as comprised in a series of experimental researches on the laws of strain; the strengths, forms, and other conditions of the material; and an inquiry into the present and prospective state of the navy, including the experimental results of the resisting powers of armour plates and shot at high velocities, (London: Longmans, Green & Co).

1866
Useful Information for Engineers. Third Series. As comprised in a series of lectures on the applied sciences; and in other kindred subjects; together with treatises on the comparative merits of the Paris and London International Exhibitions, on roofs, on the Atlantic Cable, and on the effect of impact on girders. (London: Longmans Green & Co.).

From 2010 generally poor quality facsimile reprints, often on a print-on-demand basis, have been published

1.2. Contributions by William Fairbairn to Other Books.

1857
Arago, F, *Biographies of Famous Scientific Men,* (English translation, London: Longman, Brown, Green, Longmans & Roberts). Fairbairn contributed a 'Note' following the biography of James Watt.

1858
Fairbairn, W., Messrs Forrester, Laird, J., Lay, O., and Messrs Seaward, *Steam Navigation: Vessels of Iron and Wood; the Steam Engine; and Screw Propulsion,* (London). Fairbairn's contribution is on the *Nevka.*

1860
Article 'Iron' in *Encyclopaedia Britannica,* Eighth Edition (Edinburgh: A. & C. Black)

1861
Anon. *The Engineer's, Architect's, and Contractor's Pocket-Book* (London: Lockwood & Co.)
This was an annual publication. The 1861 and some following editions included several extracts from *On the Application of Cast and Wrought Iron to Building Purposes,* and Fairbairn's paper 'On the Resistance of Tubes to Collapse'.

1864
Byrne, O, (ed.), *The Practical Metal-Worker's Assistant,* 2nd Edition, (Philadelphia: Henry Carey Baird). This included a section by Fairbairn on the 'Application of Iron to Ship-Building'. 'New revised and improved edition', 1874.

1867
Barlow, P., *A Treatise on the Strength of Materials,* 6th Edition, (London: Lockwood & Co).
This contains two reports and sundry other items by Fairbairn, The reports are 'On the Mechanical Properties of Specimens of the Iron and Steel Plates which have been subjected to experiment with Ordnance at Shoeburyness' and 'Strength of Wrought Iron Girders'.

1869
Scoffern, J., Truran, W., Clay, W., Oxland, R., Fairbairn, W., Aitkin, W.C., and Pickett, W.V., *The Useful Metals and their Alloys including mining ventilation, mining jurisprudence and metallurgic chemistry employed in the conversion of iron, copper, tin, zinc, antinomy, and lead ores with their application to the Industrial Arts,* (London: Houlston & Wright).
Fairbairn contributed chapters XIX to XXIII, on the applications of iron to ordnance, machinery, architecture, railway purposes and shipbuilding. This material was originally published in parts in *Orr's Circle of the Industrial Arts,* 1857.

1869
Fairbairn, W., 'The Rise and Progress of Manufactures and Commerce and of Civil and Mechanical Engineering in Lancashire and Cheshire' forming a substantial Supplement to 'Division IV' of Baines, T, *Lancashire and Cheshire, Past and Present: A History and a Description of the Palatine Counties of Lancaster and Chester, forming the North-Western Division of England, from the Earliest Ages to the Present Time (1867),* (London: William Mackenzie).

1869
Styffe, K., *The Elasticity, Extensibility, and Tensile Strength of Iron and Steel,* English Translation from the Swedish by C.P. Sandberg (London, John Murray).
Fairbairn contributed a Letter and some Notes.

1877
Pole, W. (edited and completed by) *The Life of Sir William Fairbairn, Bart. FRS, LLD, DCL. partly written by Himself,* (London: Longmans, Green & Co.). Fairbairn's autobiographical sections were written in 1851 and have been edited by Pole.
Abridged Edition, 1878 (London: Longmans, Green & Co.).
Facsimile reprint with 'Introduction' by A.E. Musson, (Newton Abbot, David & Charles, 1970).

1.3. Papers and Talks by William Fairbairn

Date	Organisation	Title	Main references
7 March 1837	Manchester Literary and Philosophical Society	'An Experimental Enquiry into the Strength and Other Properties of Cast iron from Various Parts of the United Kingdom'	*Memoirs of the Manchester Literary & Philosophical Society,* 1842, 171-273.
Sept.1837	BAAS, Liverpool	'On the Strength and Other Properties of Cast Iron obtained from the Hot and Cold Blast'	*BAAS1837*, pp.377-415; *MM*, 27, April - Sept 1837, pp.453-5; *Journal of the Franklin Institute*, Nov 1839, pp.334-45, 386-97.
1838	BAAS, Newcastle	'On the Application of Machinery to the Manufacture of Steam-Engine boilers, and other vessels of Wrought-Iron or Copper, subject to Pressure'	*BAAS1838*, p.160; *MM*, 29, April – Sept 1838, pp.473-4.
24 Jan.1839	Ancoats Lyceum	Presidential Address	*Address of William Fairbairn, Esq. President, to the Members of the Lyceum, Great Ancoats Street, 1839.*
1839	BAAS, Birmingham	'On the effect of weights acting for an indefinite time on bars of iron'	*BAAS1839*, p.s126.
1840	BAAS, Glasgow	'On the fan-blast as applied to furnaces'	*BAAS1840*, p.s199.
1840	BAAS, Glasgow	'On the Strength of Iron, and its application as a substitute for wood in ship-building'	*BAAS1840*, p.s201
1840	BAAS, Glasgow	'On raising water from low lands'	*BAAS1840*, p.s213; *CE&AJ*, 3, 1840, p.412.
29 Oct.1840	Manchester Geological Society	'On the Economy of raising Water from Coal Mines on the Cornish Principle'	*Transactions of the Manchester Geological Society*, 1, 1841, pp.179-194 and Plate 8.
17 Nov.1840	Manchester Literary and Philosophical Society.	'An Experimental Enquiry into the strength and other Properties of Anthracite Cast Iron'	*Memoirs of the Manchester Literary & Philosophical Society*, 1842, 524-60.
1841	Manchester Geological Society	'On the Salt mines of Austria'	*Manchester Geological Society – At the Third Annual Meeting*, 1841, 7 (no report –only title given).
24 June 1842	BAAS, Manchester	'On Combustion of Coal, with a view to obtaining the greatest Effect, and preventing the Generation of Smoke'	*BAAS1842*, pp.s107-8; *MM*, 37, July – Dec 1842, 31-2; *MG*, 29 June 1842.
21 March 1843	Institution of Civil Engineers, London	'Description of a Woollen Factory erected in Turkey'	*MPICE*, 1843, 125-6; *Life*, pp.173-4; Fairbairn, *M&MWII*, pp.183-4 and Plates 7 & 8.
30 April & 7 May 1844	Institution of Civil Engineers, London.	'Experimental researches into the properties of the Iron ores of Samakoff, in Turley, and of the Hæmatite Ores of Cumberland, with a view to determine the best means for reducing them into the cast and malleable states; and on the relative strength and other properties of cast-iron from the Turkish and other Hæmatite Ores'	*MPICE*, 3, 225-47; *Life*, p.173.
10 June 1844	Royal Scottish Society of Arts	Communication: 'On the Uses and Adaptation of Iron as a material for Building'	*CE&AJ*, 1844, p.249.
Sept.1844	BAAS, York.	'On the Economy of the Expansive Action of Steam in Steam-engines'	*BAAS1844*, p.s98; *MM*, 41, July – Dec 1844, pp.279-80.
Dec.1844	Manchester Literary and Philosophical society	'Consumption of Smoke'	*MG*, 28 December 1844.
1845	BAAS, Cambridge.	'On Railway Gradients'	*BAAS1845*, p.s93; *The Year-book of Facts in Science and Art*, (1846), pp.16-18.

1846	BAAS, Northampton.	'Experiments on the tubular bridge proposed by Mr. Stephenson for crossing the Menai Straits'	*BAAS1846*, pp.s107-8.
20 April 1847	Institution of Civil Engineers, London.	'On some Defects in the Principle and Construction of Fire-proof Buildings'	*MPICE*, 1847, 213-24 and Plate 22; *Life*, p.186; Fairbairn, *Application*. pp.279-84; *MG*, 17 March 1847.
9 Jan.1849	Institution of Civil Engineers, London.	'On Water-wheels with Ventilated Buckets'	*MPICE*, 1849, 45-66 + Plates 1-3; J Weale (publisher), *Papers on Engineering*, Vol.6, pp.3-20; *CE&AJ*, 1849, 60, 232-8; *The Practical Mechanic's Journal*, 2, 1849-50, 146-9, 178-81.
25 July 1849	Institution of Mechanical Engineers, Birmingham.	'On the Expansive Action of Steam, and a new construction of Expansion Valves for Condensing Steam Engines'	*MPIME*, 1849, 14-20; *MM*, 51, July – Dec 1849, 255-9; *CE&AJ*, 1849, 315-7; *The Practical Mechanic's Journal*, 2, 1849-50, 162-4; *Life*, p.319; *Dingler's Polytechnisches Journal*, 115, 1-6.
5 Feb.1850	Manchester Literary and Philosophical Society.	'An Experimental Enquiry into the Relative Power of the Locomotive Engine, and the Resistance of Railway gradients'	*Memoirs of the Manchester Literary & Philosophical Society*, 14, 1842, 149-78; *Life*, p.480.
12 March 1850	Institution of Civil Engineers, London.	'On Tubular Girder Bridges'	*MPICE*, 1850, 233-87 and Plate 11.
2 April 1850	Manchester Literary and Philosophical Society.	'On the Security and Limit of Strength of Tubular Bridges constructed of Wrought Iron'	*Memoirs of the Manchester Literary & Philosophical Society*, 1850, 179-95.
13 June 1850	Royal Society, London.	'An Experimental Enquiry into the Strength of Wrought Iron Plates and their Riveted Joints, as applied to Shipbuilding and vessels exposed to severe strain'	*Philosophical Transactions of the Royal Society*, 1850, pp.677-725 and Plates 55-8; Fairbairn, *UIfE*, pp.i-lix; *Life*, pp.402-5.
1850	BAAS, Edinburgh	'On a Wrought-iron Tubular crane', (communicated by Sir David Brewster)	*BAAS1850*, pp.s177-9.
1851	BAAS, Ipswich.	'On the Construction of Vessels exposed to severe strain'	*BAAS1851*, pp.s113-4.
23 & 24 April 1851	Leeds Mechanics' Institution.	'Two Lectures on the Construction of Boilers, and on Boiler explosions; with the Means of Prevention'	Fairbairn, *UIfE*, pp.1-47; W Fairbairn, *Two Lectures on the Construction of Boilers, and on Boiler explosions; with the Means of Prevention*, (1851); *The Engineer and Machinist*, .3, March-Dec 1851, pp.82-6, 117-22; *MM*, 54, Jan-June 1851, pp.436-7, 453-6, 507-9; *MM*, 55, July-Dec 1851, 12-5,28-3045-7; *Journal of the Franklin Institute*, 52.2, 1851, pp.128-38.
29 March 1852	Mechanics' Institution, Manchester	'The necessity of incorporating with the mechanical and industrial arts a knowledge of practical science'	Fairbairn, *UIfE*, pp.93-114; *MG* ,7 April 1852; *Manchester Times*, 3 April 1852.
2 April 1852	Mechanics' Institution, Manchester	'Metallic Constructions' (Iron Shipbuilding).	Fairbairn, *UIfE*, pp.115-138: *MG* ,10 April 1852; *Manchester Times*, 10 April 1852; *Mining Journal*, .22, 1852, p.182; *CE&AJ*,15, 1852, p.145 ; *Journal of the Franklin Institute*, 54.1,1852, pp.1-5.
1852	BAAS, Belfast	'On the Mechanical Properties of Metals, as derived from repeated meltings, exhibiting the maximum Point of Strength and the Causes of Deterioration'	*BAAS1852*, p.s125.
1852	BAAS, Belfast	'On the Tensile Strength of Unwrought Iron Plates at various Temperatures'	*BAAS1852*, p.s125.
1852	BAAS, Belfast	'On a New Tubular Boiler'	*BAAS1852*, p.s125; *CE&AJ*, 15, 1852, p.330; *MG*,18 September 1852.
1852	BAAS, Belfast	'Remarks on the Mini Rifle'	*BAAS1852*, pp.s125-6; *CE&AJ*, 15, 1852, 331; *The Practical Mechanic's Journal*, 5, 1852-53, pp.187-8.

1 March 1853	Institution of Civil Engineers, London	'Experiments on the Strength of Cast Iron smelted with Purified Coke' (associated with paper by F Crace Calvert)	*MPICE*, 12, 1853, pp.360-81; *MM*, 58, Jan-June 1853, pp.190-2.
1853	Institution of Mechanical Engineers.	'On a New Description of Winding Engine'	*MPIME*, 4, 1853, pp.137-142 and Plates 32-3; *Life*, p.482.
1853	Institution of Mechanical Engineers.	'On the Retardation and Stoppage of Railway Trains'	*MG*, 12 November 1853; *MPIME*, 4, 1853, pp.156-65.
1853	BAAS, Hull	'On the Progress of Mechanical Science' (Address as President of the Section)	*BAAS1853*, pp.s116-7; *Manchester Times*, 14 September 1853.
3 Oct.1853	Harpurhey, Manchester	'That this meeting desires to record with satisfaction, the attempt to establish, in this neighbourhood, a literary Institute'	*Manchester Times*, 5 October 1853.
9 May 1854	Institution of Civil Engineers, London	'Description of the Sliding Caisson at Her Majesty's Dockyard, Keyham, Devon'	C Manby (ed.) *Description of the Sliding Caisson at Keyham Dockyard by William Fairbairn*, (1857); *MPICE*,13, 1853-4 , pp.444ff; *MM*, 60, 1854, p.468; *MM*, 61, 1854, pp.8-10; *CE&AJ*, 17, 1854, p.237; *Life*, p.320.
July 1854	Institution of Mechanical Engineers, Birmingham.	'Description of an Improved Steam Travelling Crane'.	*MPIME*, 1854; *Railway and Commercial Gazette*, 24, p.823.
Sept.1854	BAAS, Liverpool	'On the Density of various bodies when subjected to enormous Compressing Forces' (associated with paper by Hopkins and Joule)	*BAAS1854*, p.s56; *Life*, pp.287-308.
Sept.1854	BAAS, Liverpool	'On the Solidification of Bodies under great Pressure' (Experiments with Hopkins and Joule).	*BAAS1854*, pp.s149-50; *Life*, 287-308.
Sept.1854	BAAS, Liverpool	'On the Consumption of Fuel and the Prevention of Smoke'	*BAAS1854*, p.s149; Fairbairn, *UIfE*, pp.47-92.
5 March 1855	Mechanics' Institution, Manchester	'Steam: its properties and application to the useful arts'.	*MG*, 7 March 1855; Fairbairn, *UIfE*, pp.139-61.
c.1855		'On Steam'.	Fairbairn, *UIfE*, pp.162-76.
c.1855		'On Steam and Steam Boilers'. (Two lectures).	Fairbairn, *UIfE*, pp.177-215.
6 March 1855	Oldham Lyceum, (at Oldham Town Hall).	'On the properties of steam and its application to the useful and industrial arts'.	*MG*, 10 March 1855.
7 March 1855	Mechanics' Institution, Manchester	'The strength and form of vessels calculated to ensure safety, and resist the elastic force of steam; the relative proportion of flue to furnace surface in boilers; and the relative value of high and low pressure steam'.	*MG*, 14 March 1855.
1855	BAAS, Glasgow	'On Boiler Explosions'.	Not reported in *BAAS1855*– only title given.
1855	BAAS, Glasgow	'On the Strength of Boiler Plates'.	Not reported in *BAAS1855* – only title given.
1855	BAAS, Glasgow	'On the Machinery of the Universal Exhibition of Paris'.	*BAAS1855*, p.s206; Board of Trade, *Machinery in the Paris Exhibition*, ('Blue Book'); *The Engineer*, 3, Jan-Jun 1857, p.372; *MM*, 63, Jul-Dec 1855, pp.482-4.
19 Dec.1855	Institution of Mechanical Engineers.	'Description of a New Construction of Pumping Engine'. (Fairbairn was unable to attend and the paper was read on his behalf.)	*MPIME*, 6, 1855, pp.177-182 and Plates 33-36; *The Engineer*, 1, Jan-Jun 1856, pp.183-4; *MG*, 28 December 1855.
1856	Manchester Literary and Philosophical Society.	'The Rise and Progress of Practical Science as exhibited at the Universal Exhibition in Paris'.	*The Engineer*, 1, Jan-Jun 1856, p.37.

1 April 1856	Manchester Literary and Philosophical Society.	'On the Comparative Value of various kinds of Stone, as Exhibited by the Powers of Resisting Compression'	*Memoirs of the Manchester Literary & Philosophical Society,* 19 / 2nd series 14, 1856, pp.31-47; Fairbairn, *UIfE2,* pp.129-144; Fairbairn, *Application,* (3rd ed. 1864), pp.196-7.
24 June 1857	Institution of Mechanical Engineers, Mechanics' Institution, Manchester.	'Description of Large Tubular Wrought-Iron Crane recently erected at Keyham Dockyard, Devonport'.	*MPIME,* 1857, pp.87-96 and Plates; *The Engineer,* 3, Jan-Jun 1857, p.505; *Newton's London Journal of the Arts,* 39(new series), 1858, pp.171-5; *Life,* p.321.
1857	BAAS Dublin	'On the Resistance of tubes to Collapse'	*BAAS1857,* pp.s215-9
9 March 1858	Manchester Literary and Philosophical Society.	'Experiments to determine the Properties of some mixtures of Cast iron and Nickel, similar in composition to meteoric iron'.	*Memoirs of the Manchester Literary & Philosophical Society,* 20/ 2nd series 15, 104-112; *The Engineer,* 5, Jan-Jun 1858, p.241; *MG,* 26 May 1858
6 April 1858	Manchester Literary and Philosophical Society.	'On the Comparative Temperature of the Climate of England and some parts of Italy'.	*Memoirs of the Manchester Literary & Philosophical Society,* 20/ 2nd series 15, pp.45-8; *Life,* p.380.
20 May 1858	Royal Society, London.	'On the Resistance of Tubes to Collapse'	*Philosophical Transactions of the Royal Society,* 1858, pp.349-413 and 2 Plates; *BAAS1857,* pp.215-9; *MM,* 3 July 1858, pp.5-7 and 31 December 1858, pp.8-9; *The Engineer's, Architect's and Contractor's Pocket Book,* (1863) pp.355-80; G H Love *Mémoire sur la Loi de Résistance des Conduits Intévieurs a Fumée dans les Chaudières a Vapeur Dédeite des Expériences de M. W. Fairbairn* (Neuilly 1859).
August 1858	Institution of Mechanical Engineers, Newcastle.	'On a Floating Corn Mill for the Navy'.	*MPIME,* 1858, 155-8; *The Engineer,* 6, Jul-Dec 1858, pp.171-2; *MM,* 69, 1858, p.243; *Life,* p.330; Fairbairn, *M&MWII,* pp.132-8.
August 1858	Institution of Mechanical Engineers, Newcastle	'On Improvements in Locomotive Axles and Couplings'.	*The Engineer,* 6, Jul-Dec 1858, pp.171-2;
25 Aug.1858	Institution of Mechanical Engineers, Newcastle	Response to Toast (at Dinner at Queen's Head Hotel, Newcastle).	*Mining Journal,* 28, 1858, p.571; *MM,* 11 September 1858, p.250.
Sept.1858	BAAS, Leeds.	'On the Progress of Mechanical Science'. (Presidential Address to the Mechanical Section).	*BAAS1858,* p.s201; *MM,* 69, Jul-Dec 1858, pp.317-9.
1858	probably London (possibly Royal Agricultural Society)	'On the Machinery Employed in Agriculture'.	Fairbairn, *UIfE2,* pp.162-185.
16 Nov.1858	Manchester Literary and Philosophical Society.	'Notes on some Experimental Apparatus for determining the Density of Steam at all Temperatures'. (Joule was in the Chair).	*Proceedings of the Manchester Literary and Philosophical Society,* 2, 1858, pp.70-3; *Life,* p.485.
Nov.1858	Blackburn Literary, Scientific and Mechanics' Institution, Blackburn	'The System of Education best adapted to Improve the Working Classes'.	W Fairbairn, *Three Lectures on the Rise and Progress of Civil and Mechanical Engineering and on Popular Education,* (1859), pp.43-50; Fairbairn, *UIfE2,* pp.ix-x, 145-61; *The Engineer,* 6, Jul-Dec 1858, p.407; *MG,* 3 Dec.1858.
1858	Derby Railway Institute et al	'On the Rise and progress of Civil and Mechanical Engineering and its Progress to the Present Century'.	Fairbairn, *Three Lectures,* pp.1-19; Fairbairn, *UIfE2,* pp.186-210.
1858	Derby Railway Institute et al	'On the Progress of Civil and Mechanical Engineering during the Present Century'.	Fairbairn, *Three Lectures,* pp.20-42; Fairbairn, *UIfE2,* pp.211-243.

3 Mar.1859	Manchester Literary and Philosophical Society.	'Experiments to determine the effects of different modes of treatment on cast iron for the manufacture of cannon'.	*Proceedings of the Manchester Literary & Philosophical Society*, 1859, pp.140ff; *MM*, 20 May 1859, pp.331.
22 March 1859	Manchester Literary and Philosophical Society.	'An Experimental Enquiry into the effect of Severe Pressure upon the Properties of Gunpowder'. (Joule was in the Chair).	*Proceedings of the Manchester Literary and Philosophical Society*,1859 , pp.117-9; *The Engineer*, 7, Jan-Jun 1859, p.248; *MM*, 8 April 1859, p.231; *Life*, p.486.
12 May 1859	Royal Society, London.	'On the Resistance of Glass Globes to Collapse from external pressure; and on the Tensile and Compressive Strength of various kinds of Glass'. (with Thomas Tate).	*Philosophical Transactions of the Royal Society*, 1859, pp.213-247; *Life*, p.273.
16 Sept.1859	BAAS Aberdeen.	'Experiments to determine the Efficiency of Continuous and Self-Acting Breaks [sic] for Railway Trains'.	*BAAS1859*, p.s76; *MG*, 22 September 1859; *Life*, p.411; *The Engineer*, 2 January 1860, pp.2-3.
19 Sept.1859	BAAS Aberdeen.	'Experimental researches to determine the Density of Steam at various temperatures'.	*BAAS1859*, pp.s233-5; *MG*, 22 September 1859; *MM*, 28 October 1859, pp.279-80; *The Engineer*, 28 October 1859, p.308.
13 Dec.1859	Manchester Literary and Philosophical Society.	'Some Experimental Researches on the Efficiency of Continuous and Self-Acting Railway Breaks [*sic*]'.	*Proceedings of the Manchester Literary and Philosophical Society*, 1859-60, pp.178-180.
7 Feb.1860	Manchester Literary and Philosophical Society.	'On the Strength of Iron Ships'.	*Proceedings of the Manchester Literary and Philosophical Society*, 1859-60, pp.200-3; *MG*, 18 February 1860.
14 Feb.1860	Liverpool Polytechnic Society.	'On the Strength of Iron Ships'.	*MG*, 18 February1860; *MM*, 17 February1860, p.103.
2 March 1860	Institution of Naval Architects.	'The Strength of Iron Ships'. (John Penn was in the Chair).	*Transactions of the Institution of Naval Architects*, 1, 1860, pp.11-81 and 2 Plates; *The Engineer*, 9, Jan-Jun 1860, pp.215-6; *MM*, 1860, pp.151, 158-61, 169-70, 210.
March 1860	Leeds Literary and Philosophical Society.	'The properties of Steam and its Application'.	*MG*, 12 March1860; *The Engineer*, 9, Jan-Jun 1860, pp.181-2.
3 April 1860	Manchester Literary and Philosophical Society.	'Memoir of the Late John Kennedy'.	W Fairbairn, *Memoir of the Late John Kennedy*, (1860).
17 April 1860	Institution of Civil Engineers, London.	'On the Efficiency of various kinds of Railway Breaks [*sic*], with Experimental Researches on their retarding Powers'.	*MPICE*, 20, 1860, pp.490-526; *Life*, p.411; *The Engineer*, 4 October 1861, p.207; Walker, *Unwin*, pp.38-41.
10 May 1860	Royal Society, London.	The Bakerian Lecture: 'Experimental Researches to determine the Density of Steam at different temperatures, and to determine the Law of Expansion of super-heated Steam'. (with Thomas Tate).	*Philosophical Transactions of the Royal Society*, 1860, pp.185-222 and 3 Plates; 'Ricerche sperimentali per determinare la densita' del vapour a tutte le temperature e la legge di dilatazione del vapour soprariscaldato', *Il Nuovo Cimento*, 12.1, 1860.
1860	BAAS Oxford.	'On the Density of Saturated Steam and on the Law of Expansion of Superheated Steam'.	*BAAS1860*, p.s210.
1860		'On the Construction of Iron Ships of Great Length'	*MM*, 9 November, 1860, pp.317-8.
2 April 1861	Manchester Literary and Philosophical Society.	'On the Temperature of the Earth's Crust as exhibited by Thermometrical Observations obtained during sinking of the deep mine at Dukinfield'.	*Proceedings of the Manchester Literary and Philosophical Society*, 2, 1861, pp.64ff; *MM*, 5, 19 April 1861, p.275.
4 Sept.1861	BAAS Manchester.	Presidential Address.	G Basalla, W Coleman and R H Kargon, *Victorian Science*, (1970), pp.129-57; *BAAS1861*, pp.li-lxvii; *MG*, 5 September 1861.

Sept.1861.	BAAS Manchester.	'Report on the Temperature of the Earth's Crust, as exhibited in the Dukinfield Deep Mine'.	*BAAS1861*, pp.s53-6; *Life*, p.488; *BAAS1870*, pp.31-2.
28 Oct.1861	Warrington Mechanics' Institute *soirée*.	'Science Self-reliance and Self-culture'	*MG*, 30 October 1861.
27March 1862	Opening of Patent File Machine and Patent File Manufacturing Company (Limited).	Toast to the new company.	*MG*, 28 March 1862.
3 April 1862	Royal Society, London.	'On the Law of Expansion of Superheated Steam'.	*Proceedings of the Royal Society of London*, 12, 1862, pp.53-7.
9 May 1862	Royal Institution, London	'On the Properties of Iron and its Resistance to Projectiles at High Velocities'.	*Proceedings of the Royal Institution*, 4, 1862, pp.491-502; *Life, pp.*345-60; *CE&AJ*, 25, 1862, pp.238-42; *MM*, 7, 1862, p.397.
June 1862	National Association for the Promotion of Social Science, London.	'On Public Health'.	G W Hastings (ed.), *Transactions of the National Association for the Promotion of Social Science, London Meeting 1862*, (1863), pp.59-66; *CE&AJ*, 25, 1861, pp.238-9.
2 Oct.1862	BAAS Cambridge.	Presidential Address to the Mechanical Science Section, 'On Guns and Iron-plated Ships'.	*BAAS1862*, pp.s178-82; *MM*, 8, 10 October 1862, 223-4; *Life*, p.394; *MG*, 4 October 1862.
Dec.1862	Liverpool School of Science. Prize-giving.	'Practical Science'.	*The Engineer*, 2 January 1863, p.3.
1862/3	*Popular Science Review.*	'The Great Exhibition Buildings'.	*Popular Science Review*, 1863, pp.317-26.
1862/3	*The Engineer.*	'The Machinery Department of the Exhibition'.	*The Engineer*, 7, 4 April 1862, pp.234-5; *Popular Science Review*, January1863.
24 Feb.1863	Institution of Civil Engineers, London.	'On the Reconstruction of the Dinting and Mottram Viaducts'.	C Manby and J Forrest (eds), *On the Reconstruction of the Dinting and Mottram Viaducts by William Fairbairn*, (1863); *MPICE*, 22, 1862-3, pp.327-35; *Life*, p.331.
Dec.1863	Literary and Philosophical Society, Newcastle-upon-Tyne.	'Iron and Its Application to the manufacture of Steam Engines, Millwork, and Machinery'.	Fairbairn, *UIfE3*, pp.98-124; W Fairbairn, *Two Lectures on Iron and Its Application to the manufacture of Steam Engines, Millwork, and Machinery, and on Natural Laws*, (1864), pp.1-16; *The Engineer*, 27 May 1864, pp.321-3.
Dec.1863	Literary and Philosophical Society, Newcastle-upon-Tyne.	'The Natural Laws, and the Thickness of the Earth's Crust Experimentally Considered'.	Fairbairn, *UIfE3*, pp.76-97; Fairbairn, *Two Lectures* , pp.17-28; *MM*, 1864, pp.347-9, 362-3; *The Engineer*, 3 June 1864, pp.334-5.
4 Feb.1864	Royal Society, London.	'Experiments to determine the effect of Impact, Vibratory Action, and long-term continued changes of load on Wrought-Iron Girders'.	*Philosophical Transactions of the Royal Society of London*, 1864, pp.311-325; *MM*, 4 November 1864, pp.313-4; *Life*, p.414.
Sept.1864	BAAS Bath.	'The Mechanical Properties of the Atlantic Telegraph Cable'.	*BAAS1864*, pp.408-15; *MM*, 23 September 1864, p.204; *MG*, 17 September 1864; *Life*, pp.381-3.
1864	*Quarterly Journal of Science*	'On the Construction and Mechanical properties of Submarine Telegraph Cables'	*Quarterly Journal of Science*, 1, 1864, pp.624-41; Fairbairn, *UIfE3*, pp.276-89.
23 Nov.1864	Society of Arts, London.	'Applications of Iron for the purpose of Naval Construction'.	*The Engineer*, December 1864, pp.337, 353-5; *MM*, 2 December, 1864, 387-8.

26 Jan.1865	Bolton Mechanics' Institute.	'On Labour: Its influence and Achievements'.	Fairbairn, *UIfE3*, pp.43-58; W Fairbairn, *A Lecture on Labour: Its Use, Influence and Achievements … to which is added a short memoir of the life of Samuel Crompton*, (1865).
Sept.1865	BAAS Birmingham.	'On the causes of the failure of deep-sea cables, and experimental researches on the permanency of their insulators'.	*BAAS1865*, pp.s178-184; Fairbairn, *UIfE3*, pp.317-325.
Sept.1865	BAAS Birmingham.	'On India rubber and Gutta Percha as insulators for submarine telegraphic cables'.	*BAAS1865*, p.s14. See also Fairbairn, *UIfE3*, pp.244-275.
	Inauguration of Southport Athenæum	'On Literary and Scientific Institutions'.	Fairbairn, *UIfE3*, pp.59-75.
17 April 1866	Aeronautical Society of Great Britain, London.	'Perseverance in meteorological experiments with a view to increasing our knowledge as to the law of storms and of elastic and magnetic phenomena, which enter so largely into the movements of elastic fluids when united to vapour and heat in the form of clouds.'	*MM*, Jan-Jun 1866, p.240.
April 1866	*Quarterly Journal of Science*	'Consideration of the Loss of the "London"'.	*Quarterly Journal of Science*, 3, 1866, pp.177-9.
1 Aug.1866	Institution of Mechanical Engineers	'Description of the removing and replacing of the iron columns in a cotton mill'.	*MPIME*, 1866, pp.181-5; *Life*, p.490.
1866	BAAS Nottingham.	'On the means employed for removing and replacing in a new position the iron columns of a fireproof cotton mill'.	*BAAS1866*, p.s141; *Life*, p.490.
6 Oct.1866	Rusholme Road Free Library, Manchester. Opening.	Brief Address.	*MG*, 8 October 1866.
Oct.1866	Social Sciences Congress, Manchester.	'Casualties from Boiler Explosions'.	*MG*, 7 October 1866.
Sept.1867	BAAS Dundee.	'Experimental Researches on the Mechanical properties of Steel'.	*BAAS1867*, pp.161ff. (details of 135 experiments).
16 Oct.1867	Hulme Working Men's Institute, Manchester. Prizegiving.	Address.	*MG*, 17 October 1867.
Jan.1868	*Quarterly Journal of Science*	'On the Mechanical Properties of Iron and Steel'	*Quarterly Journal of Science*, 5, 1868, pp.10-14.
29 July 1868	Institution of Mechanical Engineers, Leeds.	'Smoke Consumption'.	*MG*, 30 July 1868.
27 Oct.1868	Manchester Chamber of Commerce (Dinner at Queen's Hotel for Marquis of Salisbury).	Development in India.	*MG*, 30 July 1868.
Jan.1869	*Quarterly Journal of Science*	'Experimental Researches on the Mechanical Properties of Steel'	*Quarterly Journal of Science*, VI, January 1869, pp.21-8.
18 Mar.1869	Institution of Naval Architects, London.	'On the Law of Resistance of Armour Plates, composed of one or more thicknesses'. (Part 1).	S&A Tracts 334/6 at Manchester Central Reference Library (extract from *Transactions of the Institution of Naval Architects*); *Life*, pp.345-60.
19 Feb.1870	London Association of Foreman Engineers.	'State of engineering and role of the foreman'. (Whitworth also spoke).	*Observer*, 20 February 1870.
8 April 1870	Institution of Naval Architects, London.	'On the Law of Resistance of Armour Plates, composed of one or more thicknesses'. (Part 2).	*Transactions of the Institution of Naval Architects*, 11, 1870, pp.103ff.

1 June 1870	Owens College, Manchester.	Address at prize-giving.	*MG*, 2 June 1870.
Sept.1870	Presentation to Sir E W Watkin.	Address.	*The Railway News and Joint Stock Journal*, 14.351, 17 September 1870, p.322.
10 Jan.1871	Manchester Literary and Philosophical Society.	'On the Properties of Iron and Steel as applied to the Rolling Stock of Railways'.	*Proceedings of the Manchester Literary and Philosophical Society*, 10, 1871, p.86.
Feb.1871.	Manchester Association of Engineers	'Prosperity to the Association'.	*MG*, 17 February 1871; A C Dean, (ed.), *Some Episodes in the Manchester Association of Engineers*, (1938), pp.52-3.
31 Mar.1871	Institution of Naval Architects, London.	'Remarks on the Present and Past construction of the Navy'.	S&A Tracts 334/7 at Manchester Central Reference Library (extract from *Transactions of the Institution of Naval Architects*).
April 1871	*Quarterly Journal of Science.*	'On Steam Boiler Legislation'.	*Quarterly Journal of Science*, April 1871, pp.214-27; *Life*, p.278.
18 April 1871	Crewe Mechanics' Institute, *soirée.*	Address.	*MG*, 19 April 1871.
Pre-Aug.1871	*Nature.*	Railway Tires [*sic*].	*Nature*, reported in *MG*, 3 August 1871.
21 Dec.1872	Manchester Scientific and Mechanical Society	Presidential Address.	*Engineering*, 3 January 1873, p.18; *Transactions of the Scientific and Mechanical Society, Manchester*, 'Opening Meeting'.
24 April 1873	Royal Society, London.	'On the Durability and preservation of Iron Ships, and on Riveted Joints' (with Unwin).	*Proceedings of the Royal Society*, 21, 1873, pp.259-263; Walker, *Unwin*, p.64.
28 Oct.1873	Manchester Scientific and Mechanical Society	Presidential Address. (read by Prof. Osborne Reynolds)	*Life*, pp.419-20; *Transactions of the Scientific and Mechanical Society, Manchester*, 'Opening Meeting, October 1873'.

NOTE: The distinction between Papers/Talks in this Table and Reports in the following Table is, in some cases, arbitary: some items in the following Table could equally well have been included above.

1.4. Reports by William Fairbairn

Note: The distinction between Papers/Talks in the previous Table 1.3 and Reports in the following Table 1.4 is, in some cases, indistinct and arbitary: some items in the following Table could equally well have been included above .

Date	Report prepared for:	Title	Main references
March 1832		Drainage of Soham Mere, near Ely.	*Life*, p.157.
Nov.1835	Mill-owners on the River Don, Aberdeenshire	Reservoirs on the River Don.	*Life*, p.159.
1836	Mill-owners of County Down	'Reservoirs on the River Bann'.	W Fairbairn, *Reservoirs on the River Bann, in the County of Down, Ireland, for more effectually supplying the mills with Water*, (1836).
March 1837	Saddleworth Reservoirs	Report apportioning rates. (with John Raistrick)	*MG*, 29 March 1837.
1837	BAAS Liverpool.	'On the Strength and other properties of Cast Iron obtained from the Hot and Cold Blast'.	*BAAS1837*, p.377.
10 Jan.1840	Anti-Corn-Law League	Report on 'pavilion', Peter Street.	*Manchester Times & Gazette*, 14 January 1840.
1840		Drainage of Lake at Haarlem.	*Life*, p.180; *MM*, 33, 1840, 351-2; G D Dempsey, *Rudimentary Treatise on the Drainage of Districts and Lands*, (1849), pp.71-2; [Woodcroft] IX.
1841	Inspector of Factories	Factory safety	Parliamentary Papers, Report of Inspectors of Factories, p.24.
April 1841		Report on Perkins's Hot Water Apparatus for Warming Buildings.	
c.1841	Panama Canal	Feasibility study (with Gibbs)	*MPICE*, 1856, pp.407-8.
Nov.1841	Coroner's Court,Manchester	Report on boiler explosion at Jersey Mill. (with Richard Roberts).	*MG*, 13 November 1841; *CE&AJ*, 5, 1842, 51-2.
1842	BAAS Manchester.	'On the Strength and other properties of Cast Iron obtained from the Hot and Cold Blast'.	*BAAS1842*, p.3; *MM*, 37, Jan-Jul 1842, 110; *The Practical Mechanic & Engineer's Magazine*, 1, 1841-2, pp.146-9, 438; *CE&AJ*, 5, 1842, pp.314-5.
Jan.1843	Anti-Corn-Law League	Report on safety of 'pavilion' in Peter Street, Manchester	*MG*, 28 January 1843; *Life*, pp.466-7; T Wyke, *A Hall for All Seasons: A History of the Free Trade Hall*, (1996), p.10.
1843	BAAS Cork.	Report on the Consumption of Fuel and prevention of Smoke.	*BAAS1843*, p.294.
1843	BAAS Cork.	Report on the internal changes in the constitution of metals.	*BAAS1843*, p.294.
1844	BAAS York.	Report on the Consumption of Fuel and prevention of Smoke.	*BAAS1844*, pp.100-20; *Practical Mechanics' Journal*, 4, 1844-5, p.43; *Life*, pp.183-4.
1844	Samuel Holme.	'On the Construction of Fire-Proof Buildings'.	*Report of William Fairbairn, Esq., C.E. on the Construction of Fire-Proof Buildings with Introductory Remarks by Samuel Holme* (Liverpool 1844); *MG*, 27 November 1844 and 4 December 1844; *The Practical Mechanic and Engineer's Magazine*, 4, 1845, pp.105-8.

Oct 1844	Coroner's Court, Oldham.	Report on the Causes of the Fall of the Cotton Mill at Oldham, in October 1844.	*Application*, pp.274-9; *MG*, 6, 9 and 20 November 1844, and 9 July 1845; *MM*, 41, Jul-Dec 1844, pp.348-51 and 43, Jul-Dec 1845, pp.220-1; *CE&AJ*, 7, 1844, pp.429-32; *ILN*, 5, 9 November 1844, p.301; *The Sheffield & Rotherham Independent*, 9 November 1844.
Aug.1845	Liverpool Assizes.	Infringement of Newton's patent.	*Observer*, 31 August 1845.
Dec.1845	Coroner's Court, Bolton.	Report on Boiler Explosion at Rothwell & Kitt's Mill.	*The Times*, 27 December 1845; *MG*, 4 April 1846; *The Mining Journal*, 16, 1846,10; *ILN*, 7, 1845, p.412; A McEwen, *Historic Steam Boiler Explosions*, (2009), pp.1-4.
1846	Report to Robert Stephenson.	Summary of Results from Experiments relative to the proposed Bridge across the Menai Straits.	Fairbairn, *B&CTB*, pp.37-42; D Smith, 'Structural Model Testing and the Design of British Railway Bridges in the Nineteenth Century', *TNS*, 1977, pp.73ff.
Oct.1846	Coroner's Court, Ashton.	Report on Boiler Explosion.	*MG*, 31 October and 4 November 1846; *Manchester Times & Gazette*, 6 November 1846.
1847	Fall of 400ft chimney, Wigan.	Advice.	*The Times*, 27 January 1847; *MM*, 46, Jan-Jun 1847, p.117.
28 Aug.1847	Richard Turner, Dublin.	Report on proposed roof at Liverpool Lime Street station.	R Turner, 'Description of the Iron Roof over the Railway Station, Lime-street, Liverpool', *MPICE*, 9, 1850, p.210; Fairbairn, *UIfE3*, pp.216-20; R H G Thomas, *The Liverpool and Manchester Railway*, (1980), p.123.
1848-9	Application of Iron to Railway Structures.	Evidence and Letters from Fairbairn	*Report of the Commissioners appointed to enquire into the Application of Iron to Railway Structures*, (1849), pp.311-23, 403-11 and Plates.
1 April 1850		Valuation of Locomotives of the Blackburn, Darwen & Bolton Rairway.	J Marshall, *The Lancashire and Yorkshire Railway, Volume 3*, (1972), pp.37-8.
July 1850	Coroner's Court, Stockport.	Report on Collapse at Brinksway Mill	*MG*, 14 August 1850; *Manchester Times*, 10 August 1850; *The Engineer and Machinist*, 2, 1850-1, pp.212-3; *MM*, 53, Jul-Dec 1850, pp.130-5.
Dec.1850	Coroner's Court, Halifax.	Report on Boiler Explosion at Firth's Mill.	*The Times*, 16 December 1850; *The Engineer and Machinist* , 2, 1850-51, pp.343-4; *MG*,18 December 1850.
March 1851	Coroner's Court, Manchester.	Report on Boiler Explosion at Riga Street Saw-mill, Manchester.	*MG*, 29 March, 5 April, and 12 April 1851; *Observer*, 21 April 1851.
15 July 1851	To the Feoffees of the Manchester Free Grammar School	Report on proposed alterations to bridge over the Irk.	Manchester City Library Archive M516/30/3.
Aug. 1851	Assize Court, Manchester	Report on Boiler Explosion at Riga Street saw-mill, Manchester	*MG*, 20 August 1851.
12 Nov.1851	Arbitration, London	Alleged infringement of Gray's patent for Expansion Valve Gear.	R T Smith, 'John Gray and his Expansion Valve Gear', *TNS*, 50, 1978-9, pp.139-53.
March 1853	Coroner's Court	Report on Derailment at Clifton.	*MG*, 12 March 1853.
1853	Coroner's Court	Report on Locomotive Boiler Explosion at Longsight, Manchester.	*MG*, 12 March 1853; *Observer*, 14 March 1853.
April 1853	Dublin Exhibition	Report on gallery structure.	Fairbairn, *Application*, pp.119-24; *MG*, 30 April 1853.

1853	Smithfield Market, Manchester.	Adjudication of 'design & build' schemes.	*The Builder*, 11, 1853, pp.578, 773; *MG*, 17 December 1853; *Manchester Times*, 17 December 1853; *Manchester Examiner & Times*, 12 July 1854; J Schmiechen and Kenneth Carls, *The British Market Hall: A Social and Architectural History*, (1999), pp.117-8.
1853	BAAS Hull.	'Experimental Researchers to determine the Strength of Locomotive Boilers and the Causes which lead to Explosions'.	*BAAS1853*, pp.52-7; Fairbairn, *UIfE*, pp.lix- lxxiv; *Life*, p.264; *MM*, 59, 1853, pp.284-7 & 60, 1854, pp.293-5 & 393-5.
1853	BAAS Hull.	'On the Mechanical Properties of Metals as derived from repeated Meltings, exhibiting the maximum point of strength and the causes of deterioration'.	*BAAS1853*, pp.87-115; *MM*, 59, 1853, 286; *Life*, pp.405-6; *Manchester Times*, 9 October 1886.
Nov.1853	Coroner's Court, Blackburn	Report on Boiler Explosion.	*MG*, 16 November 1853
July 1854	Coroner's Court, Rochdale	Report on Boiler Explosion.	Fairbairn, *UIfE*, pp.lxxv-lxxxii; *MG*, 22, 26 and 29 July 1854; *Manchester Times*, 29 July 1854; McEwan, *Historic Explosions*, pp.15-19.
July 1854	Bolton Market	Report on failure of substructure.	*MG*, 5 July 1854, 24 March 1855, 14 April 1855, 20 December 1855; *The Builder*, 8 January 1853, 24; The Architectural History Practice Limited, 'Bolton Market Hall', 2004.
1 March 1855	Dean of Guild Court of Glasgow	Smoke Nuisance. (with J Leslie and R Johnston).	Fairbairn, *UIfE*, pp.lxxxv-xciv.
1855	Board of Trade	'On the Comparative Merits of the Machinery of the Paris Universal Exhibition, 1855'.	Fairbairn, *UIfE3*, pp.125-179.
1855	BAAS Glasgow.	'On Metals used for Artillery'. (Committee report presented by Fairbairn)	*BAAS1855*, p.100.
c1855	Woolwich Arsenal	'Experiments on the Iron Targets at the Arsenal, Woolwich'.	Fairbairn, *UIfE*, pp.xcv-xcviii.
1856	BAAS Cheltenham.	'On the Tensile Strength of Wrought Iron at various temperatures'.	Fairbairn, *UIfE2*, pp.96-128; *BAAS Report, 1856*, pp.405-22; *Life*, pp.483-4; W Fairbairn, 'Sulla resistenza allo strappamento del ferro malleabile a varie temperature', *Il Nuovo Cimento*, 7.1, Dec. 1858.
1857	Court, Merthyr Tydvil.	Boiler Gauges.	*MG*, 18 February 1857.
April 1857	Manchester Art Treasures Exhibition.	Report on Safety of the Building. (with E Walters and D Bellhouse).	*MG*, 28 April 1857.
1857	Coroner's Court, Darwen.	Report on Explosion of Locomotive Boiler.	*MG*, 2 February 1857.
Dec.1857	Coroner's Court, Huddersfield.	Report on Boiler Explosion.	*Observer*, 6 December 1857.
April 1858	Oxford Museum.	Report on stability of ironwork.	F O'Dwyer, *The Architecture of Deane and Woodward*, (1997), pp.264-7.
1858	Coroner's Court, Bacup.	Report on Boiler explosion at Dean, Rossendale.	McEwan, *Historic Explosions*, pp.36-9.
1858	Coroner's Court, Manchester.	Report on Boiler Explosion at Sharp Stewart & Co.	*The Engineer*, 16 July 1858, pp.37-8.
1858	BAAS Leeds.	'On Shipping Statistics'	*BAAS1858*, pp.239-60.

1858	BAAS Leeds.	The Patent Laws. (Committee report, presented by Fairbairn)	*BAAS1858*, pp.164-7.
1858	BAAS Leeds.	'On the Collapse of Glass Globes and Cylinders'.	*BAAS1858*, pp.174-6; *The Engineer*, 6, 1858, p.309.
1859	BAAS Aberdeen.	The Patent Laws. (Committee report, presented by Fairbairn)	*BAAS1859*, pp.191-3.
1860	BAAS Oxford.	'Experiments to determine the effect of Vibratory Action and long-continued changes of load upon Wrought Iron Girders'.	*BAAS1860*, pp.45-8; *Life*, p.413; Walker, *Unwin*, p.211; *CE&AJ*, 23, 1860, pp.257-60.
Nov.1860		Report on Bonelli's electric loom.	*The Engineer*, 2 November 1860, p.298.
1861	BAAS Manchester	'Experiments to determine the effect of Vibratory Action and long-continued changes of load upon Wrought Iron Girders', (continued from Oxford).	*BAAS1861*, pp.286-9; *Life*, p.413; *CE&AJ*, 24, 1861, pp.327-9.
1861	BAAS Manchester	'On the Resistance of Iron plates to Statical Pressure and the Force of Impact by Projectiles at High Velocities.'	*BAAS1861*, pp.280-6; *Life, pp.*345-60.
1861	Iron Armour Committee	'On the Mechanical Properties of Specimens of the Iron and Steel Plates subjected to Experiments with Ordnance, their Strength, Ductility, Resistance to Punching Force &c.'	P Barlow, *A Treatise on the Strength of Materials*, (New Edition 1867), pp.216-32; *Life*, p.359.
1862	Iron Armour Committee	'On the Resistance of different kinds of Shot to a Force tending to crush them, and the forms they assume under Pressure.'	*Life*, p.359..
1862	Iron Armour Committee	'Experiments in Punching'.	*Life*, p.359.
1862	1862 Exhibition	Report to the Commissioners of the 1862 Exhibition.	*Popular Science Review*, 1863, pp.317-26.
1862	1862 Exhibition	'On the Machinery department of the International Exhibition of 1862'.	Fairbairn, *UIfE3*, pp.180-203; *The Engineer*, 7, 4 April 1862, pp.234-5; *Popular Science Review*, January1863.
1863	Iron Armour Committee	'On the Mechanical Properties of Iron Plates'.	*Life*, p.359.
1863	Iron Armour Committee	'On the Manufacture of Armour Plates'.	*Life*, p.359.
1863	Iron Armour Committee	Experiments to test the value of annealing Armour Plates.	*Life*, p.359.
1863	Iron Armour Committee	'On the Tenacity of Cast-Iron Shot'.	*Life*, p.359.
1863	BAAS Newcastle	The Application of Gun Cotton to Warlike Purposes'. Committee report presented by Fairbairn.	*BAAS1863*, pp.1-33.
1864	Railway Department, Board of Trade	'Report of Mr. Fairbairn to the Board of Trade of his Experiments for ascertaining the Strength of Iron Structures, 1864'.	P Barlow, *A Treatise on the Strength of Materials*, (New Edition 1867), pp.232-50; *The Engineer*, 11 November 1864, pp.293-5.
1864	BAAS Bath.	The Application of Gun Cotton to Warlike Purposes'. (Second report) (committee included Fairbairn, Whitworth, Nasmyth, Russell, Anderson , Armstrong)	*BAAS1864*, p.75.

1864	Iron Armour Committee	'On the Mechanical Properties of Iron Plates'.	*Life*, p.359.
1864	Iron Armour Committee	'On the Manufacture of Armour Plates'.	*Life*, p.359.
1864	Iron Armour Committee	'General Summary of Results'.	*Life*, p.359.
1865	BAAS Birmingham	The Application of Gun Cotton to Warlike Purposes'. (Third committee report).	*BAAS1865*, p.264; *MM*, 22 September 1865, p.181.
1865	BAAS Birmingham	'The Strength of Materials considered in relation to the construction of Ships'. (with Thomas Tate)	*BAAS1865*, pp.243ff.
July 1866		Tests of Cross beams for the Lune Viaduct.	Barlow, *Treatise*, p.251.
Aug.1866	Wallasey Local Board	'Strengthening of New Brighton Landing Stage'.	*Liverpool Mercury*, 10 & 14 August 1866.
1867	Coroner's Court, Manchester	Report on Failure of Platform Hoist at Miles Platting.	*MG*, 30 January 1867, 2 March 1867.
Sept.1867	BAAS Dundee.	Experimental Researches on the Mechanical properties of Steel (details of 135 experiments).	*BAAS1867*, pp.161ff.; *The Engineer*, 13 September 1867, pp.233-7.
1869	BAAS Exeter	Experimental researches on the Mechanical Properties of Steel. (Steel manufactured by the Barrow Hæmatite Steel Co.).	*BAAS1869*, pp.96-150; *The Engineer*, 28 July-Dec.1869, pp.130-6; *Engineering*, August 1869, pp.152-3; W Fairbairn, *'An Experimental Enquiry into the Strength, Elasticity, Ductility, and other properties of Steel manufactured by the Barrow Hæmatite Steel Company'* (1869).
1869	BAAS Exeter	Investigation of steam boiler explosions. (committee included Fairbairn, Whitworth, Penn, Hick, Mason, Fletcher).	*BAAS1869*, pp.47-64; *The Engineer*, 28, July-Dec.1869, p.238; *MG*, 24 August 1869.
1870	BAAS Liverpool	Legislating on Steam Boiler Explosions.	*BAAS1870*, pp.1-9.
1871	BAAS Edinburgh	Legislating on Steam Boiler Explosions.	*BAAS1871*, p.166.
1872	BAAS Brighton	Legislating on Steam Boiler Explosions.	*BAAS1872*, p.57.

2. Main Archival Material

Bedfordshire & Luton Archives & Record Service, Bedford, Stoke Mills Sharnbrook Papers,X106/165a, 123-5, 204-8.

Birmingham Library: Boulton& Watt Collection, MS3147/3/446, 1-6, Portfolio 508.

British Library: Babbage Correspondence: Add MSS 37194 f.106.British Library: Western Manuscripts: Add MS 37461 f.378; Add Ms 37462 f.61.

Brotherton Library, Leeds: Gott Papers, MS193/149-153, 170-3.

Cambridge University Library, Stokes Papers, Add 7342, TR11-54, and 7656 various.

Cumbria County Council, Archive Documents related to Cowans Sheldon Civil Engineers of Carlisle. Ref. Nos. 6, 49, 64, 113, 133, 150, 174, 175, 176, 224,259, 260.

Mrs C Ferguson, Sion Mills, Letters (1835) and plan re water power at Herdman's Mill.

Imperial College, London: W C Unwin Papers, 1/79-114, 190.

Institution of Civil Engineers, London: Fairbairn-Stephenson Correspondence 1-376.

Institution of Civil Engineers, London: J.G. James Card Index.

John Rylands Library, Manchester: McConnel & Kennedy Papers, Cash Books 2 and 3; Ledgers 1 and 2

Lancashire County Archive, Preston: J Bennison, 'Essay', (1869), DDX/184.

London Metropolitan Archives: Records of Sun Fire Office, MS 11936/555/1252067-8 and /562/1273798.

Manchester Central Library: Executive Committee of the Manchester Art Treasures Exhibition, M6/2/1-47.

Merseyside Maritime Museum: Drawings of the hinged bridges serving the Liverpool Landing Stage.

Museum of Science and Industry, Manchester: Fairbairn Notebook.

National Archives, Kew: 'Returns and Plans of Iron Bridges, 1847', MT8/1.

Norsk Teknisk Museum, Oslo, Hjula Weavery Papers.

Scottish Record Office: Forth & Clyde Canal Navigation Papers, Ref. FCN 1/70-72.

Sheffield City Library: Bagshawe Collection, Ref. 654.

Wirral Archives, Birkenhead, Laird Papers, Nos.17-20.

Mr & Mrs Wood, Brereton Hall: Drawings of Brereton Corn Mill.

Fairbairn Locomotive Lists:

Davies, J., Typescript List, Woodridge, Queensland.

King, T., Typescript List, Stephenson Locomotive Society Library, Hersham, WL10308.

Page, M., Typescript List, Stephenson Locmotive Society Library, Hersham, L8733/18.

3. Main Newspapers, Periodicals and Transactions

Newspapers

Manchester Guardian
Manchester Times
The Times
The Observer

Periodicals

Civil Engineer and Architect's Journal
Engineer
Engineering
Illustrated London News
Industrial Archaeology Review.
Mechanics' Magazine
Mining Journal
Practical Mechanics' Journal

Transactions

Minutes of Proceedings of the Institution of Civil Engineers
Minutes of Proceedings of the Institution of Mechanical Engineers
Philosophical Transactions of the Royal Society
Proceedings of the Literary and Philosophical Society of Manchester
Reports of Meetings of the British Association for the Advancement of Science
Transactions of the Newcomen Society

4. Government Papers

1824 Report from the Select Committee on Artisans, Machinery and Combinations.

1841 Report of Inspectors of Factories.

1841 Report on the Practicality of Legislative Interference, to diminish the Frequency of Accidents to the Children and Young People engaged in Factories.

1841 Report of the Select Committee on the Operation of Laws affecting the Exportation of Machinery.

1842 Report on the Sanitary Condition of the Labouring Population of Great Britain by Edwin Chadwick.

1845 Select Committee on Railway Bills: Group 2: Chester and Holyhead Railway Bill.

1849 Report of the Commissioners appointed to inquire into the Application of Iron to Railway Structures.

1851 Report of Select Committee of House of Lords to consider Bills for Amendment of Law touching Letters Patent for Inventions.

1854 Report of the Committee on the Machinery of the United States 1855 and the Special Reports of George Wallis and Joseph Whitworth.

1856 Report from the Select Committee on Masters and Operatives (Equitable Councils of Conciliation).

1861 Report of the Joint Committee appointed by the Lords of the Committee of the Privy Council for Trade and the Atlantic Telegraph Company to inquire into the Construction of Submarine Telegraph Cables.

1864 Report of the Commissioners appointed to inquire into the Working of the Law relating to Letters Patent for Inventions.

1867–8 Report of Royal Commission on Scientific Instruction.

5. Theses and Dissertations

Burton, A.C. 'Lessons Learned in the Design and Erection of Box Girder Bridges from the West Gate Collapse', (MIT MEng 2007).

Bailey, M.R, 'Decision-making Processes in the Manufacturing Sector: The Independent Locomotive Industry in the 19th Century', (University of York DPhil 1999).

Collier, D.A, 'A Comparative History of the Development of the Leading Stationary Steam Engine Manufacturers of Lancashire, c.1800–1939', (University of Manchester PhD 1985).

Cookson, G, 'The West Yorkshire Textile Engineering Industry, 1780–1850', (University of York DPhil 1994).

Cruz, T.F dos S, 'Paranapiacaba: Aarquitetura e o urbanismo de uma Vila Ferroviária', (University of São Paulo MA 2007).

Dempster, B, 'Liverpool's Floating Landing Stages', (University of Liverpool, MSc 2003).

Dooley, S.C, 'The Development of Material-Adapted Structural Form', (École Polytechnique Fédérale de Lausanne, DSc 2004).

Glithero, J.P, 'The History of Engineering in Hyde to 1914', (UMIST PhD 2002).

Hayward, R.A, 'Fairbairns of Manchester', (UMIST MSc 1971).

Harrop, S.A, 'The Place of Education in the Genesis of the Industrial Revolution, with special reference to Stalybridge, Dukinfield and Hyde', (University of Manchester MEd 1976).

Lewis, J.H, 'The Development of the Royal Small Arms Factory (Enfield Lock) and its Influence upon Mass Production Technology and Product Design c.1820–c.1880', (University of Middlesex PhD 1996).

Moher, J.G, 'The London Millwrights and Engineers 1775–1825', (University of London PhD 1988).

Plampin, M.T.W, 'From Rio to Romola: Morality and Didacticism in the English Appreciation of Early Italian Art 1836–1863', (University of Plymouth PhD 2001).

Pilling, P.W, 'Hick Hargreaves & Co.: The History of an Engineering Firm c.1833–1839: A Study with Special Reference to Technological Change and Markets', (University of Liverpool PhD 1985).

Russell, P.E, 'John Frederick La-Trobe Bateman FRS, Water Engineer 18[illegible]0–1889', (UMIST MSc 1980).

Sibly, P, 'The Prediction of Structural Failures', (UCL DPhil 1977).

Smith, S, 'Thomas Cheek Hewes (1768–1832), An Ingenious Engineer and Mechanic of Manchester', (University of Manchester MSc 1969).

Stott, F.R, Town Hall Buildings in Tameside', (University of Manchester MA 1983).

Tuovinen, T, 'The Life and Work of Sir William Fairbairn: A Case Study in the History of the Engineering Profession in 19th-Century Britain', (Pro Gradu Thesis, University of Jyväskylä, Finland, 2000).

Warburton, B, 'Eaton Hodgkinson (1789–1861) and the Science of Strength of Materials (UMIST PhD 1971).

6. Main Internet Sites

flickr, 'Fairbairn Cranes'.

Grace's Guide to British Industrial History.

Oxford DNB on-line.

7. Primary Papers and Books – Pre-1914

(For works by William Fairbairn see 1 above).

Anderson, W, 'Description of a Six-ton Crane, with Curved, Diagonally Braced Jib, at the Pigeon-house Fort, Dublin', *Transactions of the Institution of Civil Engineers of Ireland*, 6, 1863, 188-92.

Anderson, W, 'The Interdependence of Abstract Science and Engineering', *MPICE*, 114, 1893, 255-282.

Anon., *A Sketch of the Foundation and of the Past Fifty Years' Activity of the Manchester Steam Users' Association for the Prevention of Steam Boiler Explosions and for the Attainment of Economy in the Application of Steam, together with an account of the Jubilee Celebration on the 14th February, 1905,* (Manchester, 1905).

Anon. 'Messrs. Stephenson and Fairbairn's Tubular Bridges', *North British Review*, 13, Feb.-Aug.1850, 399-446.

Anon. 'Obituary of John George Bodmer', *MPICE,* 28, 1868-9, 573-608.

Anon. 'Obituary of Alexander Wilson', *MPICE,* 30, 1870, 461-5.

Appleby, C J, 'On Mechanism for Utilising and Regulating Convict Labour', *Engineering*, 6, July-Dec.1868, 181 and 209.

Appleby, C J, *Appleby's Illustrated Handbook of Machinery. Section II – Hoisting Machinery,* (London, Spon, 1882).

Armstrong, R, *The Boilers of Steam Engines: Their Calculation, Construction, and Management with a view to the Saving of Fuel,* (J Weale, London, new ed. 1838).

Axon, W E A, *The Annals of Manchester: A Chronological Record from the earliest times to the end of 1885,* (Manchester, John Heywood, 1886).

Baker, B, *Long Span Railway Bridges,* (London, Spon, 1867).

Baker, T, *Memories of a Dissenting Chapel,* (London and Manchester, 1884).

Balgarnie, R, *Sir Titus Salt, Baronet: His Life and Its Lessons,* (London, Hodder & Stoughton, 1877).

Banfield, T.C, *Industry of the Rhine embracing a view of the social condition of the rural and manufacturing population of that district, 2, Manufactures,* (London, Charles Knight, 1848).

Banks, [I] (Mrs G.L), *The Manchester Man,* ([1876], Altrincham, Sherratt, 1970 ed.).

Barlow, P, *A Treatise on the Strength of Materials,* (London, Lockwood, 6th ed. 1867).

Barrow, J, *Tour of Austrian Lombardy, the North Tyrol and Bavaria in 1840.* (London, J Murray, 1841).

Barry, P, *Shoeburyness and the Guns: A Philosophical Discourse,* (London, Sampson, Low & Marston, 1865).

Barton, J, 'On the Calculations of Strains in Lattice Girders with practical deductions therefrom', *BAAS1852,*(1853), pp.123-4.

Barton, J, 'On the Economic Distribution of Material in the Sides or Vertical Portion of Wrought Iron Beams', *MPICE,* 14, 1855, 443-58.

Bateman, J F, 'Description of the Bann Reservoirs, County Down, Ireland', *MPICE,* 1, 1841, 168-70.

Bell, D, (ed.), *David Napier, Engineer, 1790-1869. An Autobiographical Sketch with Notes,* (Glasgow, James Maclehose, 1912).

Bennett, R, and Elton, J, *History of Corn Milling,* Vol.2, (London, Simpkin Marshall, 1899).

Bessemer, H, 'On the Manufacture of Malleable Iron and Steel without Fuel', *The Times,* 14 August 1856.

Bessemer, H, 'On the Manufacture of Cast Steel, its Progress, and Employment as a Substitute for Wrought Iron', *BAAS1865,* (1866), pp.165-173.

Bessemer, H, *Sir Henry Bessemer, F.R.S. An Autobiography,* (London, '*Engineering'*, 1905).

Bird, G.F, *The Locomotives of The Great Northern Railway, 1847-1902,* (London, Locomotive Publishing Co., 1903).

Blair, H, *Sermons,* 5 Volumes, (Edinburgh, 1820 ed.).

Bogardus, J, [and Thomson, J W], *Cast-Iron Buildings: Their Construction and Advantages,* (New York, 1856).

Böttcher, A, *Cranes: Their Construction, Mechanical Equipment and Working,* (Translated and Supplemented with English, American, and Continental Practice by A Tolhausen, London, Constable, 1908).

Bourne, J, *A Treatise on the Steam Engine in its Application to Mines, Mills, Steam Navigation, and Railways,* (London, Longmans, 1846).

Braithwaite, F, 'On the Fatigue and consequent Fracture of Metals', *MPICE,* 13, 1854, 463-74.

Brunel, I, *The Life of Isambard Kingdom Brunel, Civil Engineer,* (London, Longmans, 1870).

Buchanan, R, *Practical Essays on Mill Work and other Machinery,* ('revised into a 3rd edition with additions by George Rennie', London, J Weale, 1841).

Burgh, N.P, *A Practical Treatise on Boilers and Boiler-Making,* (London, Spon, 1873).

[Burnley, J, and Hogg, J, (eds)], *Fortunes Made in Business, A Series of Original Sketches Biographical and Anecdotic from the Recent History of Industry and Commerce, Volume 2,* (London, Sampson Low, Marston, 1884).

Burt, H.M, *Burt's illustrated guide to the Connecticut valley,* (Springfield MA, New England Publishing, 1867).

[Burtt, G.F,] *The Locomotives of the London, Brighton & South Coast Railway, 1839-1903,* (London, Locomotive Publishing Co, 1903).

Butterworth, E, *Historical Sketches of Oldham,* (Oldham, 1856 ed.).

Calvert, F.C, and Fairbairn, W, 'On the Increased Strength of Cast Iron produced by the use of Improved Coke; with a series of Experiments on the Strength of Cast Iron smelted with Purified Coke', *MPICE,* 12, 1853, 352-81.

Campbell, R, *A General Description of All Trades,* (London, 1747; facsimile reprint Newton Abbot, David & Charles, 1969).

Chadwick, J W, *Old and New Unitarian Belief,* (Boston MA, G H Ellis, 1894).

Clark, D.K, *Railway Machinery: A Treatise on the Mechanical Engineering of Railways: embracing the Principles and Construction of Rolling and Fixed Plant,* (Glasgow, Blackie & Son, 1855).

Clark, D K, *The Exhibited Machinery of 1862: A Cyclopaedia of the Machinery represented at the International Exhibition,* (London, Day & Son, 1864).

Clark, D K, *The Steam Engine: A Treatise on Steam Engines and Boilers,* (London, Blackie, 1891).

Clark, E, *The Britannia and Conway Tubular Bridges,* (London, Chapman & Hall, 1850).

Colburn, Z, 'American Iron Bridges', *MPICE,* 22, 1863, 540-556.

Colvin, S, and Ewing, J A, (eds.), *Papers Literary, Scientific, &c. by the Late Fleeming Jenkin, FRS, LLD,* (London, Longmans, 1887).

Conder, F.R, *Personal Recollections of English Engineers,* (London, Hodder & Stoughton, 1868).

Culmann, K, 'Der Bau der eisernen Brücken in England und Amerika', *AllgemeineBauzeitung,* 17, 1852, 163-222.

Culmann, K, *Die GraphischeStatik,* (Zurich, 1866).

Darbyshire, A, *An Architect's Experiences,* (Manchester, 1897).

Dempsey, G D, *Rudimentary Treatise on the Drainage of Districts and Lands,* (London, J Weale, 1849).

Dempsey, G D, *Tubular and other Iron Girder Bridges,* (London, J Weale, 1850).

Dircks, H, 'On the Smokeless Argang Furnace of Charles Wye Williams Esq', *Transactions of the Society for the Encouragement of Arts, Manufactures, and Commerce,* 55, 1845, 74-5.

Evans, O, *The Young Millwright and Miller's Guide,* (Philadelphia PA, 1795 and later editions).

Ewart, P, 'On the Measure of Moving Force', *Manchester Literary and Philosophical Society Memoirs,* 2nd series, 2, 1813, 105-258.

Fairbairn, A.H, *An Assize Sermon preached in Winchester Cathedral, on the Second Sunday in Advent, 1870, before the Hon. Sir Robert Lush, Knight, one of the Justices of the Court of Queen's Bench,* (1870).

Fairbairn Engineering Co Ltd and Thomas Beeley, *The Fairbairn Five-Tube Boiler,* (nd but c1873).

Fairbairn, T, *Truths and Tubes on Self-supporting Principles; A few words in reply to the Author of 'High-ways and Dry-ways',* (London, Longmans, 1849).

[Fairbairn, T], *Report of the Executive Committee of the Exhibition of Art Treasures of the United Kingdom held at Manchester in 1857.* (Manchester, 1859).

Faucher, L, *Manchester in 1844: Its present condition and future prospects,* (Manchester, Abel Heywood, 1844).

Fleming, S, *The Intercolonial. A Historical Sketch of the Inception, Location, Construction and Completion of the line of Railway uniting the Inland and Atlantic Provinces of the Dominion, with maps and numerous illustrations,* (Montreal, Dawson, 1876).

Fletcher, L.E, 'On the Lancashire Boiler, Its Construction, Equipment, and Setting', *MPICE,* 27, 1876, 59-126.

Gaskell, E, *North and South,* ([1855], Harmondsworth, Penguin, 1970 ed.).

Gaskell, P, *Artisans and Machinery: The Moral and Physical Condition of the Manufacturing Population considered with reference to Mechanical Substitutes for Human Toil,* (London, J.W Parker, 1836).

Gilbart-Smith, JW, 'Sir Arthur Fairbairn and his surroundings: the story of a deaf and dumb baronet', *Cassell's Magazine,* 1904, 606-12.

Glover, J, 'On the Decline of Shipbuilding on the Thames', *Journal of the Statistical Society of London,* 32.3, 1869, 288-292.

Glynn, J, *Rudimentary Treatise on the Power of Water,* (London, J Weale, 185[illegible]).

Goodrich, C R, (ed.), *Science and Mechanism: Illustrated by Examples in the New York Exhibition 1853-4,* (New York, G.P Putnam, 1854).

Grantham, J, *Iron, as a Material for Ship-Building; being a communication to the Polytechnic Society of Liverpool,* (Liverpool, 1842).

Grantham, J, *Iron Ship-building: with practical illustrations.* (Liverpool, Grantham, 1858).

Hair, T.H, *A Series of Views of the Collieries of Northumberland and Durham* (1844).

Hall, S.C, and Hall, A.M, *Ireland: Its Scenery, Character, &c.* (London, Jeremiah How, 1841-3).

Head, F.B, *High-ways and Dry-ways or, the Britannia and Conway Tubular Bridges,* (London, John Murray, 1849).

Heginbotham, H, *Stockport Ancient and Modern, Vol. 2,* (London, Sampson Low, Marston, Searle and Rivington, 1892).

[Heywood, I M, (ed.)], *Reminiscences, Letters, and Journals of Thomas Percival Heywood, Baronet, by his Eldest Daughter,* (Manchester, 1899).

Hill, J, *An Account of the Life and Writings of Hugh Blair DD,* (Philadelphia PA, JamesHumphreys, 1808).

Hodder, E, *The Life and Work of the Seventh Earl of Shaftesbury, K.G.,* Vol.2, (London, Cassell, 1886).

Hodgkinson, E, 'On the Transverse Strain, and Strength of Materials', *Memoirs of the Literary and Philosophical Society of Manchester,* 4, 1824, 225-84.

Hodgkinson, E, 'Theoretical and Experimental Researches to ascertain the Strength and Best Form for Iron Beams', *Transactions of the Manchester Literary and Philosophical Society,* 2nd Series 5, 1831, 407-544.

Hodgkinson, E, 'Experimental Researches on the Strength of Pillars of Cast Iron and Other Materials', *Philosophical Transactions of the Royal Society of London,* 130, 1840, 385-456.

Hodgkinson, E, 'Report on Experiments on the Resistance of Air', *BAAS1842,* p.211.

Hodgkinson, E, 'Some Account of the late Mr Ewart's paper on the Measure of Moving Forces and of the recent applications of principle of Living Forces to estimate the effectsof Machines and Movers', *Memoirs of the Manchester Literary and Philosophical Society,* Second Series 7, 1846, 137-56.

Hopkins, W, 'An Account of some Experiments on the Effect of Pressure on the Temperature of Fusion of Different Substances', *BAAS1854,* (1855), pp.s57-8

Hudson, JW, *The History of Adult Education in which is comprised a full and complete history of the Mechanics' and Literary Institutions, Athenaeums ... ,* (London, Longman, 1851).

Hughes, T, *Account of the Lock-out of Engineers, &c. 1851-2,* (Cambridge, Macmillan, 1860).

Hull, E, *The Coal-Fields of Great Britain,* (London, Edward Stanford, 1861).

Humber, W, *A Practical Treatise on Cast and Wrought Iron Bridges and Girders, as applied to Railway Structures, and to buildings generally,* (London, Spon, 1857).

Humber, W, *A Complete Treatise on Cast and Wrought Iron Bridge Construction including Iron Foundations,* (3rd ed. London, Lockwood, 1870).

Hunt, W.H, *Pre-Raphaelitism and the Pre-Raphaelite Brotherhood,* (London, Macmillan, 1905).

Husband, J, 'The Aesthetic Treatment of Bridge Structures', *MPICE,* 145, 1901, 145-216.

Jacob, G Le G, *The Raja of Saráwak: An account of Sir James Brooke, KCB, LLD, given chiefly through letters and journals,* (London, Macmillan, 1876).

James, J, *History of Worsted Manufacture in England,* (London, Longmans, 1857).

Jeaffreson, J.C, *The Life of Robert Stephenson, F.R.S.,* (London, Longmans, 1864).

[Jenkin, H.C.F,] 'Trade-Unions: How Far Legitimate', *North British Review,* 9(NS), 1868, 1-62.

Jessop, J & Son, *Catalogue,* (Leicester, 1892).

Johnson, J, *Clever Boys of our time who became Famous Men,* (London, Darton, c.1860).

Kay, J.P, *The Moral and Physical Condition of the Working Classes employed in the cotton manufacture in Manchester,* (London, James Ridgway, 2nd ed. 1832).

Kohl, J.G, *Journeys through England and Wales,* (English Translation, London, Chapman &Hall, 1844).

Laird, M, and Oldfield, R.A.K, *Narrative of an Expedition into the Interior of Africa by the River Niger in the Steam-Vessels Quorra and Alburkah, in 1832, 1833, and 1834.* (London, Richard Bentley, 1837).

Larmor, J, and Thomson, J, (eds.), *Collected Papers in Physics and Engineering by James Thomson,* (Cambridge University Press, 1912).

Lawson, W, *Steam in the Southern Pacific. The Story of Merchant Steam Navigation in the Australian Costal and Intercontinental Trades, and on the Ocean Liners of the Southern Pacific,* (Wellington, NZ, Gordon & Gotch, 1909).

Leigh, E, *The Science of Modern Cotton Spinning,* Vol.2, (Manchester, Palmer & Howe, 5th ed. 1882).

[Love, B], *Manchester As It Is,* (Manchester, Love & Barton, 1839).

Mackay, T, *The Life of Sir John Fowler, Engineer,* (London, John Murray, 1900).

Mann, T, *Buddenbrooks: The Decline of a Family,* (1901, English translations by H.T Lowe-Porter, 1924, and J.E Woods, 1993).

Marx, K, *Capital, A Critique of Political Economy. Vol. 1. The Process of Capitalist Production,* ([English Edition 1886] London, Lawrence & Wishart, 2003 ed.).

Meason, G, *The Official Illustrated Guide to the Great Northern Railway,* (London, Griffin Bohn, 1861).

von Meyer, G.H, 'Die Architectur der Spongiosa. (ZehnterBeitragzurMechanik des menschlichenKnochengerüstes)', *Archiv fur Anatomie, Physiologie und wissenschaftlicheMedicin,* 1867.

Mitchell, J, 'On the Construction and Works of the Highland Railway', *BAAS1867,* 151-61.

Mitchell, J, *Reminiscences of my Life in the Highlands,* (London, Unwin, 1884).

Moody, S.B, 'Description of a Water-wheel constructed by Mr. W. Fairbairn, MInstCE, and erected in Lombardy', *MPICE,* 3, 1844, 66-7.

Morin, A, *Résistance des Matériaux,* (3rd ed. Paris, Hatchette, 1862).

Morley, J, *The Life of Richard Cobden,* (London, T Fisher Unwin, One-volume Edition 1903).

Moseley, H, *The Mechanical Principles of Engineering and Architecture,* (London, Longman, 1843),

Murphy, W.S, *The Textile Industries. A Practical Guide to Fibres, Yarns & Fabrics in every branch of Textile Manufacture,* (London, Gresham, 8 volumes, 1910).

Murray, A, 'On the Construction and Proper Proportions of Boilers for the Generation of Steam', *MPICE,* 3, 1844, 331-42.

Murray, A, *The Theory and Practice of Shipbuilding,* (Edinburgh, A & C Black, 1861).

Nicholls, S, *In Memoriam. A Selection from the Letters of the Late John Ashton Nicholls, FRAS, &c.* (Manchester, 1862).

Official Catalogue of the Great Exhibition of the works of Industry of All Nations, 1851, (London, 1851).

Palgrave, FT, *Handbook to the Fine Art Collections,* (London, Macmillan, 1862).

Pike, W.T, *Manchester and Salford at the Close of the Nineteenth Century: Contemporary Biographies,* (Brighton, Pike, 1899).

Pole, W, 'Strength and Deflection of the Britannia Bridge considered as a Continuous Beam', in Clark, E, *The Britannia and Conway Tubular Bridges,* (London, Chapman & Hall, 1850).

Pole, W, (ed. and completor), *The Life of Sir William Fairbairn, Bart.... partly written by Himself,* (London, Longmans, Green, 1877).

Potter, T, *Concrete: Its Uses in Building from Foundations to Finish,* (London, Batsford, 3rd ed. 1908).

Powles, H.H.P, *Steam Boilers, Their History and Development,* (London, Constable, 1905).

Rankine, W.J.M, 'William Fairbairn' in *The Imperial Dictionary of Universal Biography,* Vol.2. (Glasgow, William Mackenzie, 1863).

Rankine, W.J.M, *A Manual of Machinery and Millwork,* (London, Charles Griffin, 6th ed. 1887).

Rathbone, E.F, *William Rathbone: A Memoir,* (London, Macmillan, 1905).

Rawlinson, R, *Report on Liverpool's Water Supply,* (Liverpool, 1846).

Rawson, R, 'Memoirs of the Late Eaton Hodgkinson', *Memoirs of the Manchester Literary and Philosophical Society,* 3rd Series, 2, 1865, 145-204.

Reach, A.B, (ed. by Aspin, C.), *Manchester and the Textile Districts in 1849,* ([1849] Helmshore Local History Society, 1972).

Reed, E.J, *Shipbuilding in Iron and Steel,* (London, John Murray, 1869).

Rennie, G, (ed.), *Practical Essays on Millwork and other Machinery. By Robertson Buchanan, Engineer,* (3rd ed. London, John Weale, 1841).

Roelker, C.R, 'Experimental Investigation of the Resistance of Flues to Collapse', *Van Nostrand's Magazine,* 34, 1881, 208-220.

Russell, J.S, *A Treatise on the Steam Engine from the Seventh Edition of the Encyclopaedia Britannica,* (Edinburgh, A & C Black, 1841).

Russell, W.H, *The Atlantic Telegraph,* (London, Day & Son, 1865).

St. John, S, *The life of Sir James Brooke, Rajah of Sarawak: from his personal papers and. correspondence,* (Edinburgh, Blackwood, 1879).

Saward, G, *The Trans-Atlantic Submarine Telegraph: A Brief Narrative of the Principal Incidents in the History of the Atlantic Telegraph Company,* (London, 1878).

[Scott, D], *The Engineer and Machinist's Assistant,* (London, Blackie & Son, 1850).

Shaw, W.A, *Manchester Old and New,* (London, Cassell, 1894),

Simon, H, 'Modern Flour-Milling in England', *MPICE,* 70, 1882, 191-233.

Slugg, J.T, *Reminiscences of Manchester Fifty years Ago,* (Manchester, J.E Cornish, 1881).

Smiles, R, *Brief Memoir of the Late William Muir,* (1888).

Smiles, S, *Lives of the Engineers, with an account of their Principal Works; comprising also a History of Inland Communication in Britain,* Vol.3, (London, John Murray, 1862).

Smiles, S, *Self-help; with illustrations of Conduct and Perseverance,* (London, John Murray, new ed. 1884).

Smiles, S, *Industrial Biography: Iron Workers and Toolmakers,* (London, John Murray, 1863).

Smiles, S, (ed.), *James Nasmyth, Engineer: An Autobiography,* (London, John Murray, 1895 ed.).

Smith, G.B, *Leaders of Modern Industry, Biographical Sketches,* (London, W H Allen, 1894).

Smith, R.A, *A Century of Science in Manchester,* (London, Taylor & Francis, 1883).

Sproule, J, (ed.), *The Irish Industrial Exhibition of 1863. A Detailed Catalogue of its Contents,* (Dublin, J McGlashan, 1853).

Stephenson, R, 'Iron Bridges', in *Encyclopaedia Britannica, 8th Edition,* (Edinburgh, A & C Black, 1856).

Stevenson, R.L, 'Memoir of Fleeming Jenkin', in S Colvin and J.A Ewing (eds), *Papers Literary, Scientific, &c. by the Late Fleeming Jenkin, FRS, LlD, with a Memoir by Robert Louis Stevenson,* (London, Longmans, Green, 1887).

Stoney, B.B, *The Theory of Strains in Girders and Similar Structures, with Observations on the Application of Theory to Practice and Tables of the Strength and other Properties of Materials,* (London, Longmans, Green, 1869).

Sutcliffe, J, *A Treatise on Canals and Reservoirs ...* (Rochdale, J Hartley, 1816).

Taylor, W.C, *Notes of a Tour in the Manufacturing Districts of Lancashire,* (London, Duncan & Malcolm, 2nd ed. 1842).

Thompson, J, *The Owens College: Its Foundation and Growth; and its connection with the Victoria University, Manchester,* (Manchester, J.E Cornish, 1886).

Thomson, H.W.B, *The choice of a profession: a comparative concise account and comparative review of the English professions,* (London, Chapman & Hall, 1857).

Thomson, J, *Collected Papers in Physics and Engineering,* (Cambridge University Press, 1912).

Todhunter, I and Pearson, K, *A History of the Theory of Elasticity and of the Strength of Materials from Galilei to the present time,* (Cambridge University Press, 1886).

Tomlinson, C, *Cyclopædia of useful arts, mechanical and chemical, manufactures, mining and engineering,* (London, G.Virtue,1852).

Tredgold, T, *The Steam Engine: its Invention, Progressive Improvement, an Investigation of its Principles, and its Application to Navigation, Manufactures, and Railways,* (London, J Weale, various editions).

Tuff, J, *Historical, Topographical and Statistical Notes of Enfield,* (Enfield, J H Meyers, 1858).

Turner, R, 'Description of the Roof over the Railway Station, Lime-street, Liverpool', *MPICE,* 9, 1850, 204-13.

Tyrrell, H.G. *History of Bridge Engineering,* (Chicago, Tyrrell, 1911).

Unwin, W.C, 'On the Resistance of Boiler Flues to Collapse', *MPICE,* 46, 1876, 225-241.

Unwin, W.C, *The Elements of Machine Design* (London, Longmans Green, 1877).

Unwin, W.C, (reporter), 'First Report to the Council of the Institution of Mechanical Engineers on the Form of Riveted Joints', *MPIME,* 1881, 301-68.

Unwin, W.C, *The Testing of Materials of Construction,* (London, Longmans, 1888; 2nded. 1899; 3rd ed. 1910).

Unwin, W.C, 'Strength of Alloys at Different Temperatures', *BAAS1889,* pp.746-7.

Unwin, W.C, 'Tensile Tests of Mild Steel; and the Relation of Elongation to the Size of the Test-Bar', *MPICE,* 155, 1904, 170-233.

Ure, A, *Philosophy of Manufactures or an Exposition of the Scientific, Moral, and Commercial Economy of the Factory System of Great Britain,* (London, Charles Knight, 1835); 3rd ed. edited by P.L Simmonds, (London, H.G Bohn, 1861).

Ure, A, *The Cotton Manufacture of Great Britain systematically investigated...,* (London, Charles Knight, 2 vols 1836).

Ure, A, *A Dictionary of Arts, Manufactures, and Mines; containing a clear exposition of their principles and practice,* (London, Longman,1853).

Valentine, J.S, 'Description of the Line and Works of the Railway from Lisbon to Santarem', *MPICE,* 18, 1859, 1-19).

Vicat, L.J, 'Note sur l'allongement progressif du fil de fersoumis à diverses tensions', *Annales des Ponts et Chausses,* 7.1, 1834, 40-4.

Vignoles, O.J, *Life of Charles Blacker Vignoles ... Soldier and Civil Engineer ... A Reminiscence of Early Railway History,* (London, Longmans, Green, 1889).

Vose, G.L, *Manual for Railroad Engineers and Engineering Students,* (Boston, Lea and Shepard, 1873),

Walmisley, A.T, *Iron Roofs. Examples of Design. Description, Illustrated with Working Drawings,* (London, Spon, 2nd ed. 1888).

Ward, A.W, 'Introduction' to E Gaskell, *Ruth and Other Tales,* (London, Smith Elder, Vol.III of *The Works of Mrs Gaskell,* 'The Knutsford Edition', 1906).

Ward, A.W, 'Introduction' to E Gaskell, *North and South,* (London, Smith Elder, Vol.IV of *The Works of Mrs Gaskell,* 'The Knutsford Edition', 1906).

White, G.S, *Memoir of Samuel Slater, the Father of American Manufactures, connected with a History of the Rise and Progress of the Cotton Manufacture in England and America,*(Philadelphia PA, 1836).

Williams, C.W, *The Combustion of Coals and the Prevention of Smoke Chemically Considered,* (London, J Weale, 1840).

Williams, C.W, 'Lecture on the Evaporative Power of Steam Boilers', in W Sturgeon (ed.), *The Annals of Electricity, Magnetism, and Chemistry; and Guardian of Experimental Science,* 8, 1842, 58-65, 161-7.

Wolff, J, 'Uber die innereArchitectur der Knochen und ihreBedeutungfür die FragevomKnochenwachsthum', *Virchows Arch Pathol.Anat.Physiol,* 50, 1870, 389-450; (ET and abridgement by Heller, M.O, Taylor, W.R, Aslanidis, N, and Duda, G.N, 'The Classic: On the Inner Architecture of Bones and its Importance for Bone Growth', *Clinical Orthopaedics and Related Research,* 468.4, 2010, 1056-65.

Wood, H.T, *A History of the Royal Society of Arts,* (London, Murray, 1913).

Woodcroft, B, *Brief Biographies of Inventors of Machines for the Manufacture of Textile Fabrics,* (London, Longmans, 1863).

[Woodcroft, B], 'Fairbairn and his Times', Ten articles in *The Engineer,* 44, July- Dec. 1877.

Young, C.F.T, *The Fouling and Corrosion of Iron Ships: Their Causes and Means ofPrevention, with Mode of Application to the Existing Iron-Clads,* (London, London Drawing Association, 1867).

8. Secondary Papers and Books – Post 1913

The secondary papers and books which have been referred to, and which have been written after 1913, number over seven hundred. They would take up twenty-five pages. For reasons of economy this part of the bibliography has been omitted. These high numbers are partly attributable to the fact that Fairbairn was active in so many areas of engineering – mills, waterwheels, boilers, steam engines, locomotives, shipbuilding, bridges, cranes, as well as experimental work. Inevitably many of these secondary papers and books are relevant to only one discrete area of Fairbairn's activities, or to a particular piece of his work. Almost all the papers and books are referred to in footnotes and thus readers with a specialist interest should be able to find references to them within the sections of the text relating to their specialist interest.

Index

The Index contains (1) most of the individuals, authors, organisations and subjects referred to in the text, and (2) many of the secondary authors referred to in the chapter-end notes. It does not include most of the primary references (which are accessible through chapter-end notes) nor generally the Appendices or Bibliography. The following appear in sub-sets: Boilers, Bridges, Ships, Steam Engines, Waterwheels (all including both the works of Fairbairn and others) and Patents, Exhibitions, Experiments, Explosions, Rivers, Fairbairn & Lillie, William Fairbairn and Thomas Fairbairn.